✓ Birds – General 54/4

Perspectives in ornithology

Perspectives in ornithology

**ESSAYS PRESENTED FOR
THE CENTENNIAL OF
THE AMERICAN ORNITHOLOGISTS' UNION**

Edited by Alan H. Brush and George A. Clark, Jr.
University of Connecticut

Sponsored by The American Ornithologists' Union

CAMBRIDGE UNIVERSITY PRESS

Cambridge
London New York New Rochelle
Melbourne Sydney

Published by the Press Syndicate of the University of Cambridge
The Pitt Building, Trumpington Street, Cambridge CB2 1RP
32 East 57th Street, New York, NY 10022, USA
296 Beaconsfield Parade, Middle Park, Melbourne 3206, Australia

First published 1983

Printed in the United States of America

Library of Congress Cataloging in Publication Data
Main entry under title:
Perspectives in ornithology.
   Includes index.
   1. Birds—Addresses, essays, lectures.   I. Brush,
Alan H. (Alan Howard), 1934–   .   II. Clark,
George A. (George Alfred), 1936–   .   III. American
Ornithologists' Union.
QL673.P47   1983   598   83–3931
ISBN 0 521 24857 4

Contents

v

vi   *Contents*

Contributors

Kenneth P. Able, Department of Biology, State University of New York, Albany, NY 12203, U.S.A.

John C. Avise, Department of Zoology, University of Georgia, Athens, GA 30602, U.S.A.

Luis F. Baptista, Department of Birds and Mammals, California Academy of Sciences, San Francisco, CA 94118, U.S.A.

Jon C. Barlow, Department of Ornithology, Royal Ontario Museum, Toronto, Ontario M5S 2C6, Canada

George F. Barrowclough, Department of Ornithology, American Museum of Natural History, New York, NY 10024, U.S.A.

William A. Calder III, Department of Ecology and Evolutionary Biology, University of Arizona, Tucson, AZ 85721, U.S.A.

William Conway, New York Zoological Society, Bronx Zoo, Bronx, NY 10460, U.S.A.

Joel Cracraft, Department of Anatomy, University of Illinois Medical Center, Chicago, IL 60680, U.S.A.

Stephen T. Emlen, Section of Neurobiology and Behavior, Cornell University, Ithaca, NY 14853, U.S.A.

Nancy J. Flood, Department of Zoology, University of Toronto, Toronto, Ontario, M5S 1A1, Canada

James R. Karr, Department of Ecology, Ethology, and Evolution, University of Illinois, Champaign, IL 61820, U.S.A.

John R. Krebs, Edward Grey Institute of Field Ornithology, Department of Zoology, Oxford, OX1 3PS, U.K.

Donald E. Kroodsma, Department of Zoology, Merrill Science Center, University of Massachusetts, Amherst, MA 01003, U.S.A.

Anthony J. Lednor, Laboratory of Ornithology, Cornell University, Ithaca, NY 14850, U.S.A.

Sarah Lenington, Institute of Animal Behavior, Rutgers University, Newark, NJ 07102, U.S.A.

J. David Ligon, Department of Biology, University of New Mexico, Albuquerque, NM 87131, U.S.A.

Larry D. Martin, Museum of Natural History, University of Kansas, Lawrence, KS 66045, U.S.A.

Ernst Mayr, Museum of Comparative Zoology, Harvard University, Cambridge, MA 02138, U.S.A.

Douglas W. Mock, Department of Zoology, University of Oklahoma, Norman, OK 73019, U.S.A.

J. P. Myers, Department of Ornithology, Academy of Natural Sciences, Philadelphia, PA 19103, U.S.A.

Dennis M. Power, Museum of Natural History, Santa Barbara, CA 93105, U.S.A.

Pat V. Rich, Department of Earth Sciences, Monash University, Clayton, Victoria 3168, Australia

Ian Rowley, CSIRO Division of Wildlife Research, Helena Valley, Western Australia 6056, Australia

Gerald F. Shields, Institute of Arctic Biology, University of Alaska, College, AK 99701, U.S.A.

Daniel Simberloff, Department of Biological Sciences, Florida State University, Tallahassee, FL 32306, U.S.A.

P. J. B. Slater, School of Biological Sciences, The University of Sussex, Brighton, BN1 9QG, U.K.

David W. Steadman, Division of Birds, Natural History Museum, Smithsonian Institution, Washington, DC 20560, U.S.A.

David W. Stephens, Chesapeake Bay Center for Environmental Studies, Smithsonian Institution, Edgewater, MD 21037, U.S.A.

William J. Sutherland, Department of Zoology, University of Liverpool, Liverpool, L69 3BX, U.K.

Sandra L. Vehrencamp, Department of Biology, University of California at San Diego, La Jolla, CA 92039, U.S.A.

Charles Walcott, Laboratory of Ornithology, Cornell University, Ithaca, NY 14850, U.S.A.

Glenn E. Walsberg, Department of Zoology, Arizona State University, Tempe, AZ 85287, U.S.A.

John A. Wiens, Department of Biology, University of New Mexico, Albuquerque, NM 87131, U.S.A.

Preface

The origin of this volume began in the deliberations of the Council and Centennial Committee of the American Ornithologists' Union. The committee, in its various versions, undertook the planning for the celebration of the one hundredth anniversary of the founding of the union. The charge to the original committee was broad in scope and reflected the thinking of many individuals. Our files on the activity of the committee go back to before 1975, but surely others have more complete memories and more extensive files. The history and activities of the committee may warrant their own documentation.

One of the suggestions of the committee, made early in its existence, was the production of a reviewlike volume to cover contemporary ornithological research. An additional group of authors was to attempt to indicate research directions for the future. Some even suggested trying to project to the union's bicentennial! As in the history of many committees, a compromise was reached. Most of these developments occurred during the 1979 annual meeting at College Station, Texas. In 1980, Walter Bock, representing the committee, asked the editors to begin preparation for the volume. General guidelines, which included a list of prospective topics and authors provided by a subcommittee, were our first source of direction. The diversity of topics suggested was staggering but nevertheless strongly influenced the present content and form of the volume. The intent of the project has been to provide a series of essays on research areas that are active and perhaps undergoing significant change. The initial plan was to follow each chapter with one or more commentaries. The latter were to present alternative viewpoints, provide specific criticisms of the chapter, or add to the primary presentation. As the project progressed, shorter chapters without commentaries were added in order to maintain the initially intended breadth of coverage.

From the start, it was obvious that not all the significant areas of ornithology could be included. The final selection of topics, authors, commentators, and all details of organization were left to the editors, and the committee was long ago absolved of any responsibility for the volume. It was possible to obtain a wide variety of contributions on aspects of contemporary ornithology. We have attempted to include different levels of biological organization and, in some cases, to identify interdisciplinary areas likely to grow in the near future. We tried also to include traditional subjects approached in nontraditional ways.

The contributions presented here are nontraditional in several ways. The authors were encouraged to discuss and speculate on their specialties encumbered by few editorial restrictions. They were asked to use the opportunity to reflect on their topics and to discuss the current situation so as to identify controversial or difficult questions. We encouraged unorthodox approaches that would reflect the excitement and the flux of contemporary ornithology. Consequently, certain chapters emphasize one particular conceptual view, which may differ from that in a commentary. This juxtaposition of contrasting interpretations should provide a basis from which readers can more critically evaluate the continuing discussions at the start of the second century of the union.

Numerous individuals have given generously of their time, opinions, and effort in the preparation of this volume. A complete enumeration of those who have aided would, unfortunately, be prohibitively lengthy. All authors have generously forgone any financial remuneration to reduce the final cost of the book. Our appreciation goes to all these individuals. In addition, we acknowledge specifically the substantial help given by the officers of the union, the council, and the centennial committee and its chairperson, Walter Bock. Preparation of the volume was supported by the National Science Foundation (Grant DEB-80-16544 to A.H.B.) and the staff of the Biological Sciences Group of the University of Connecticut. We thank Richard L. Ziemacki and his staff at Cambridge University Press, whose first consideration has been the quality of this volume. We have enjoyed the opportunity and adventure of editing this collection and hope that others will find the result to be both useful and appropriately provocative.

A. H. B.
G. A. C., Jr.

Storrs, Connecticut

Introduction

ERNST MAYR

A science is like a tree. It is forever growing and reaching out in all directions. New branches develop, but some earlier branches may stop growing, being shaded out by the new branches. However, there is one aspect in which science seems to differ from a tree: The older it gets the more vigorous its growth seems to be.

This is certainly true for ornithology. If one takes Frederick II's De Arte Vanandi as the birth of ornithology, our *Scientia amabilis* had already reached a venerable age when the American Ornithologists' Union (AOU) was founded in 1883. In the first 50 years after the birth of the AOU, ornithology was largely a self-contained science, concerned primarily in problems of interest only to ornithologists and with most of its publications being suitable only for ornithological journals. This is well reflected in an overview of ornithology presented on the occasion of the fiftieth anniversary of the founding of the AOU (Chapman and Palmer 1933), reflecting interests that do not depart drastically from the ornithology of 1883. By contrast, if one compares the subjects included in the present volume with the state of ornithology as recently as 1962 as reflected in the *Proceedings of the XIIIth International Ornithological Congress, Ithaca,* one notices a decided change. Entirely new branches of ornithology are now dominant that had hardly been heard of in 1962 or had not existed at all. The recency of much of this research is reflected in the bibliographies. For instance in the review on cooperative breeding strategies among birds by Emlen and Vehrencamp (Chapter 4) 42 among 46 citations were published in the 1970s and 1980s, four in the 1960s, and none earlier. In fact, the study of cooperative breeding strategies as a separate field (as compared to descriptive observations of naturalists) simply did not exist prior to the 1970s.

1

Recent changes in ornithology

How can one characterize the most important changes in ornithology in recent generations? These changes are profound to one who, like myself, published his first ornithological paper 60 years ago and has watched the changes through all these years and has perhaps, to a slight extent, contributed to these changes. I attempted once before (Mayr 1963a) to present a bird's eye view of the development of ornithology. Yet, even in the 20 years since then, there have been remarkable changes. If I had to pinpoint the most important of these changes in ornithological research in the last generation, I might single out the following developments without questioning in the least the existence of several others.

An accelerated shift from straight description to a concern with causal analysis. To be sure, some how and why questions were asked in ornithology from the very beginning. Yet it cannot be denied that the emphasis in earlier periods was on fact-finding and description. This is equally true for taxonomy, morphology, ecology, or behavior. However, as more and more facts became known, the available information was used increasingly to formulate theories concerning the causation of observed phenomena. This trend has accelerated noticeably in the last 25 years. Indeed, many papers are now published that are concerned entirely with theory or with model building. Discrepancies revealed during the testing of such theories or models lead to the posing of new questions, to new data collecting, and to new hypotheses. Most of the essays in this volume illustrate this modern approach in exemplary fashion.

A termination of the isolation of ornithology from other biological disciplines. In the earlier history of ornithology, authors were mostly concerned with problems of interest only to other ornithologists, such as details of the distribution of particular species, the validity or not of certain subspecies, the chronology of the migration of individual species, seasonal phenology, questions of nomenclature, and bird photography. Now most papers presented at an annual meeting of the AOU or at an international ornithological congress are of sufficiently broad biological interest that they could just as well be presented at meetings of zoologists, ethologists, ecologists, or other biological specialists.

A professionalization of ornithology. In the early years of the AOU, professional ornithologists hardly existed. Their number, however, has been growing steadily. Each of the contributions to this volume was prepared by a professional, most of the senior authors of the chapters being university professors. When one analyzes the program of the annual meetings of the AOU, one finds that about half or more of the lectures are presented by Ph.D. candidates or recent Ph.D's. As I pointed out recently (Mayr 1975), the most gratifying aspect is the number of colleges and universities at which ornithology is cultivated and young ornithologists are educated. This is in complete contrast to the early history of American ornithology – let us say the first 50 years of the AOU (1883–1933) – when virtually all leading ornithologists were museum curators or employees of government agencies. A distinguished group of amateurs were particularly active and prominent during the early history of the AOU. Ornithologists like Brewster or Bent were quite equivalent in standing to the best professionals, but most of the museum professionals (Ridgway and Chapman, for instance) had had no academic training whatsoever.

This increase in professionalism does not mean in the least that the days of the amateurs are ended. The finest work in recent years in bird speciation, for instance, was done by J. Haffer, an oil geologist. Modern population ecology often requires that a population be analyzed or monitored for many years (10–25 or more) and no Ph.D. candidate has that much time available. Such research can be undertaken only by a permanent research institution, like the Edward Grey Institute, or by an amateur who is able to monitor a local population year after year. For life history studies, faunistics, phenology, and migration research, as well as for much basic census work, the amateur is still indispensable. The recently completed great *Atlas of Australian Birds* is largely the work of amateurs. It is a pity that so few bird watchers realize how much of a lasting contribution to ornithological science they could make, if they would take up some project in one of these areas.

A bridging of the gap between biological disciplines. Perhaps the most striking development in recent years has been the breaking down of barriers among disciplines of biology. Behavior is studied increasingly often in the light of the ecological factors to which the respective species or individual is exposed. Ecological problems are now dealt with in terms of selection pressures, such as with an evolutionary viewpoint. Other such connections are evident in all the essays in this volume, such as the connection between physiology and ecology, life history and learning psychology (bird song), life history and sensory physiology (navigation), or population biology and biochemistry. What is particularly redeeming is that questions of evolutionary history or current selective advantage, are now asked almost routinely in connection with any ornithological problem.

As valuable and important as this approach is, it creates problems. In order to answer evolutionary questions, one often needs considerable technical knowledge in adjacent fields, like geology, genetics, or biochemistry. However, in view of the rapid advances made in these fields, it is usually inadvisable to rely on textbook information. Rather, one must consult the current periodical literature of these fields and such a search procedure is often difficult for a nonspecialist. Yet, such a bridging of fields has come to stay and has greatly enriched ornithology. As a philosopher of science might put it, such a broadened approach may permit us to advance from historical narrative to a truly causal analysis.

These new interdisciplinary connections are responsible for a considerable shift in emphasis in existing fields. In bird ecology, for instance, the emphasis has shifted from habitat selection to such problems as foraging strategies, resource partitioning, competition, and demographic factors. In the study of behavior, coverage has extended from courtship behavior to the behavior of communicating and to the effects of sexual selection. It is no longer possible to deal with a problem in bird taxonomy without considering the evolutionary ramifications.

When one looks at any volume of *The Auk* of former years, one can quickly determine the dominant research interests of the period. Faunistics, the phenology of migration, and new species or subspecies were dominant in the early issues, whereas these subjects occupy only a small fraction of space in a recent volume. At regular intervals, new interests, and sometimes entirely new fields, develop. This may be due either to the development of new techniques like the Kramer cage in bird navigation, telemetry, electronic sound recording, the radar observation of migrating birds, or the electrophoretic analysis of gene frequencies in populations. Other developments are due to conceptual shifts (like the renewed interest in sexual selection in the late 1960s). With individual reproductive success recognized as an important component of selective advantage, the study of mating systems (see Chapter 3) and cooperative breeding (see Chapter 4) have quickly blossomed into full-fledged research areas, strongly influenced by the ideas of Hamilton, Trivers, Alexander, and others usually referred to as sociobiologists.

Parental investment, sibling rivalry, female choice, inclusive fitness, the sensing of fitness, and reciprocal altruism are among the numerous concepts that have breathed new life into the field study of avian reproduction. There are enough healthy controversies on all these subjects to guarantee lively activity for many years to come. All these concepts permit numerous predictions that can be tested and for such tests birds are in many ways particularly suitable: Birds have signaling systems (color, postures, and vocalization) that are easily perceived by us, and they are mostly diurnal, easily marked individually, and usually have well-defined breeding seasons, which are all advantages for the observer. Against this are their longevity, slow sequence of generations, and small size of broods.

Comments on chapter topics

The chapters of this volume provide modern perspectives of some of the major scientific problems of modern ornithology. They focus not only on achievements but also – as is proper in a good scientific analysis – on open problems and controversies. Some of these concern my own research interests and, at the suggestion of the editors, I am offering a few comments on some additional topics.

Species and speciation

It is an historical fact that ornithologists such as Hartert, Stresemann, Rensch, and myself were among the leaders of the new systematics. The 1920s, 1930s, and 1940s were the most active period in population systematics. It was established that all species, except a few with exceedingly narrow ranges, consist of numerous local populations that, if contiguous, vary clinally. Geographically isolated populations often differed discontinuously from other populations of a species and have the potential of incipient species. Finally, when two previously isolated populations or subspecies reestablish contact secondarily and no reproductive isolation has been acquired during the preceding period of isolation, a hybrid belt or zone of secondary intergradation will be formed.

That geographic isolation can lead to the formation of full species had become evident ever since Gould's (1837) study of Darwin's bird collections from the Galapagos. Whether the same principles apply also to speciation on continents remained controversial until quite recently. It is, however, now firmly established that vegetational and physiogeographic barriers play the same isolating role on continents as water in the case of islands. Keast (1961) demonstrated this conclusively for the birds of Australia, Moreau and Hall (1970) for Africa, and Haffer (1974) for South America.

One of these conclusions, the explanation of the hybrid belts, has been challenged recently. Endler (1977) and others have advanced the thesis that such belts are the locale of a special process of speciation, designated as parapatric speciation. They postulate that these relatively narrow belts represent "ecological escarpments," that is rather abrupt transitions from one set of ecological conditions to another set, and that opposing selection forces are in the process of fracturing the continuity of populations along these lines. Eventually a total separation of the two halves of the ancestral species ensues and completes the parapatric speciation process.

This is not the place to refute this hypothesis. However, it may be mentioned that all the ornithological facts conflict with it. These hybrid belts virtually never follow any ecological escarpments, rather they occur where the expanding fronts of the populations would be expected to meet when they spread out from Pleistocene refuges. The methods of molecular biology now permit analysis of the distribution of genes in cross sections through such hybrid belts. Although work is still in its infancy there is now at least one instance, the contact zone of Myrtle and Audubon's Warblers (*Dendroica coronata*), where no increase in the variability of a limited number of enzymes was found in the hybrid zone (Barrowclough 1980).

Since the days of Darwin and Wallace, two theories of speciation have been competing with each other. According to A. R. Wallace, isolating mechanisms are built up through natural selection by a process called *reenforcement* when two incipient species meet. According to Darwin, natural selection cannot overcome the equalizing effect of hybridization in the contact zone and isolating mechanisms must, therefore, have been acquired as a by-product of the genetic restructuring of the isolated incipient species before contact was established. Most recent researches support Darwin's theory by showing that there is no evidence for a strengthening of isolating mechanism in hybrid zones. The only exception of which I am aware is the finding of Corbin et al. (1979) that at a point of recent contact between Baltimore and Bullock Orioles (*Icterus galbula*) in Nebraska there was frequent hybridization when the contact zone was first investigated but virtually none 15 years later.

To complete the discussion of systematics at the species level, it might be mentioned that the discovery of new species of birds continues unabated. Some 30 or 40 new species are still being discovered each decade. The area that is now most productive is the eastern (Amazonian) slope of the Andes, but an occasional spectacular discovery may be made almost anywhere in the world. Most of the recently discovered species seem to have an extremely restricted distribution like a new nuthatch known only from a single forest area in the Algerian part of the Atlas Mountains, a new grebe restricted to a few lakes in Patagonia, or a new swallow found only in one small district of Thailand.

The annual increment of 3 or 4 species is minimal compared to the total number of known species. The fact that the number of bird species is now given as more than 9,000, as compared to 8,600 or 8,700, is not due primarily to a large number of newly discovered species but rather to a slight change in the species concept. When the polytypic species concept was first consistently applied to nominal species of birds, there was a tendency to reduce "species" to subspecies rank, in order to denote relationship. Now that the superspecies concept is widely applied, it seemed legitimate to raise to the rank of allospecies all geographic isolates that have reached species level degree of phenotypic difference. The percentage of the 9,021 species of Recent birds (Bock and Farrand 1980) on a worldwide basis that are allospecies has not yet been determined.

Classification

For more than 200 years, one ornithologist after the other has proposed a new classification of birds. These classifications differed in three respects: (1) which characters were considered most important, (2) which groups were considered most primitive and which most advanced, and (3) which other family (or order) was considered as most closely related to a given family. This required subjective judgment, and, thus, in the absence of objective criteria, no consensus was ever reached (Mayr 1959).

Comparative anatomy, intelligently conducted, was the best guide and led to many important insights. Yet, the class of birds as a whole and even more so the individual orders are relatively so uniform in their basic anatomy that little morphological variation was available for taxonomic analysis. It was realized rather early that certain similarities, as between hawks and owls or between herons and storks, were apparently acquired independently as adaptations to the particular niche occupied by these taxa. At a later date, several authors expressed misgivings that warblers, flycatchers, thrushes, shrikes, titmice, timaliids, and other groups characterized by adaptations of bill and feet might be heterogeneous assemblages, based on convergent niche adaptations. But how should one determine this? It was obvious that this could be done only by discovering additional characters.

At one time, great expectations were held that behavior would yield important clues, and behavior was, indeed, helpful in determining the relationship of various genera of Anseriformes. On the whole, however, behavior as a clue to relationship was disappointing. This left two additional sources of information, the fossil record (see the following section) and biochemical characters.

The breakthrough finally was provided by molecular biology and Charles Sibley, more than anyone else, was responsible for it. Indefatigably he tried one method after the other. As long as he was using various methods of protein analysis (particularly egg white proteins), he was as often misled by his findings as he made lasting discoveries. However, when he turned to the DNA hybridization technique, he found a method that gave him precise, quantifiable results that could be reliably repeated (Sibley and Ahlquist, in press). Although all the work is not yet complete, important results are emerging. For instance, he has established what had been widely suspected: that the warblers, flycatchers, shrikes, robins, nuthatches, and tree creepers of Australia show no relationship to the corresponding adaptive types of the Eurasian region but are a convergent development. He proposes definite affinities for the lyrebirds, drongos, orioles, starlings, mockingbirds, swallows, white-eyes, larks, flowerpeckers, accentors, and pipits, all of them families, the nearest relatives of which had been previously uncertain.

No doubt Sibley's revolutionary proposals will be scrutinized thoroughly in coming years. The question will be raised whether a classification is most useful that is entirely based on DNA matching while not taking any components of the phenotype into consideration. Also, not all will accept Sibley's hypothesis that the molecular clock concept can be applied to DNA changes and thus assume that the rate of DNA turnover is the same in all phyletic lines and in all portions of the DNA. If this hypothesis should prove incorrect, the chronology of the splitting of lines may have to be modified considerably. The latest findings in DNA research indicate that there are many kinds of DNA, and it would seem almost certain that various phyletic lines will have their own special idiosyncrasies in genome evolution. Regardless what the answers to these questions will be, there is now every reason to believe that we are finally getting near the objective of every bird taxonomist: to have a relatively sound view of the phylogeny of birds. This, after all, is the most important step in the construction of a classification.

Fossil birds

The rarity of bird bones in fossil deposits has traditionally been deplored by paleontologists. Most birds, being adapted for flight, lack teeth (except for a few Mesozoic taxa) and have light bones that are easily crushed. Most species and genera of fossil birds are therefore based on just a few bones of the entire skeleton, often a single bone. Assignment of a new specimen to the wrong family or even order is by no means uncommon. As a consequence, quite a few Cretaceous fossils were assigned erroneously to modern orders, whereas more recent analyses have shown that all Cretaceous higher taxa became extinct and the modern orders arose in the Paleocene or later. Twenty years ago (Mayr 1963a), when I asserted that "not a single fossil bird has led to an improvement of the avian classification," no one attempted to refute my claim.

In the intervening years, our knowledge of fossil birds has increased almost by an order of magnitude. Nothing of the new discoveries has been as amazing as the discovery by Olson and others of large numbers of recently extinct types on islands, like the Hawaiian Islands, St. Helena, Ascension, Mauritius, and the Antilles. Although usually members of modern families, many of the extinct types were large and flightless. The evidence also indicates that they were exterminated by humans. One must conclude that the faunas of these islands had been far richer in species than one had assumed previously. This finding does not refute the thesis of island biogeography regarding species diversity on an island, but it does require a modification of certain numerical values in the equations of the equilibrium theory.

Archaeopteryx, the link between reptiles and birds, was described in 1861 and now no less than six specimens are known, most of them showing the greater part of the bird. Considering the excellence of the fossil material and the large number of workers that has studied it, one would expect general agreement on the question of the reptilian ancestor of birds, but this is not the case. According to Ostrom (and earlier authors), birds arose from coelurosaurian dinosaurs; according to Walker and Martin, from crocodilians; and according to Tarsitano and Hecht, from earlier archosaurians (pseudosuchians). It would seem that further comparative work will have to be done before it can be settled which of the opposing claims is valid. On the other hand, the arguments that the ancestors of birds had already feathers before they acquired flight are very convincing. Regal's (1975) hypothesis that feathers had not originated as a heat-conserving covering but rather as protection against heat and strong solar radiation is persuasive. Just when the ancestors of birds acquired endothermy is still an open question. At the present time, *Archaeopteryx* still remains the only known Jurassic bird.

The recent history of paleornithology has demonstrated the great importance of sound comparative anatomy. Unless each bony structure is identified correctly and its homology established clearly, it is quite impossible to draw sound conclusions on relationships. This is doubly important in birds where occupation of the same niche by different taxa so often has led to remarkable convergences.

The spectacular achievements of molecular analysis might suggest that henceforth one should work out the history of birds exclusively with the help of nucleic acid and protein analysis. This is, however, a mistaken notion. Since a large portion of fossil birds consists of extinct lineages, not available for biochemical analysis, and since the relationship of these extinct types to still existing birds is of the utmost importance for the understanding of the history of birds, the continued study of fossil birds is indispensable. There are, however, other reasons. Fossils supply interesting evidence on the adaptations and various radiations of birds in former geological periods, evidence not otherwise obtainable. Also, they are an indispensable source of information for the biogeographer. One can confidently predict that in the future the study of fossil birds will become a far more important branch of ornithology than it was in the past.

Biogeography

Simberloff (Chapter 11) entitles his survey of biogeography "The unification and maturation of a science." The biogeography of today is far more mature than it was 100 years ago. Contemporary biogeography asks deeper questions, and it has broadened by establishing intimate contact with ecology, geology, and other sciences. Yet, biogeography is perhaps less unified now than ever before, and there is hardly a single question in this field over which controversies are not raging. Most of the representatives of the various feuding schools seem little inclined to compromise. To be sure, some matters appear settled. The concept of the permanence of the oceans is thoroughly refuted, and no explanation of the zoogeographic regions can afford to disregard plate tectonics. Yet, each of the major subjects Simberloff deals with consists of unresolved controversies: vicariance versus dispersal, the equilibrium model of island biogeography, and the role of competition in faunal composition and in character divergence. In addition to basic conceptual differences among the opponents are major disagreements concerning methodology.

Part of the maturation of biogeography is that it is no longer considered sufficient simply to develop a reasonable scenario for an explanatory narrative. Instead, one should be able to test hypotheses and particularly to test the predictions that are engendered. To permit this, it is necessary to formulate conclusions as falsifiable hypotheses. One can hardly object to this demand even though there are many biogeographic arguments that cannot be settled in this manner. What falsifiable hypothesis, for instance, can settle the argument where to draw the border between the Oriental and the Australo–Papuan regions? Some authors, including most ornithologists, suggest that Weber's Line should be chosen, which separates the islands with more than 50% Oriental species from those with more than 50% Australo–Papuan species. Other authors prefer to recognize a separate island region, Wallacea, between the continental shelves of Asia and Australia, and accept the edge of the continental shelves as the borders of the Oriental and Australian regions. No falsifiable hypothesis can end this argument. Indeed, it is even possible that for groups (like mammals) that are poor in colonizing islands the Wallacea solution may be preferable, but for easy dispersers, like birds, there may be a line of faunal balance.

A far greater difficulty, and one that seems to be quite unresolved as Simberloff shows, is that the framing of testable null hypotheses is difficult, particularly when one is dealing with unique events of colonization and extinction and with species, each of which has uniquely different ecological niches and dispersal capacities. Two "random samples" from a pool of such species may be so different in composition that they cannot be compared. The assumption that "all species are the same" as far as dispersal and colonization are concerned inevitably leads to error. Let me illustrate this. The species per genus (S/G) ratios are often smaller in local biota than in whole countries (de Candolle, von Buch, and many later authors). Hence, it was argued that the ratio should also be smaller on islands than on mainlands, but often the opposite was found to be true. This overlooks completely the fact that only a limited number of genera supplies the island colonists and that it is not surprising, therefore, that islands might have larger S/G ratios than expected. The entirely unbiological assumption that island birds would be a random sample of mainland birds cannot but lead one to erroneous conclusions.

What lesson should we learn from this example? Certainly not that one should give up all attempts to falsify hypotheses. That would be throwing out the baby with the bathwater. Rather, what one should do is to give up treating species as if they were just so many pebbles on a beach, interchangeable for all practical purposes. It was a crucial weakness of David Lack's island biogeography when he assumed that all species had equal dispersal facilities and that their distribution on the islands of the Pacific was exclusively controlled by the availability of ecological niches.

To review the controversies presented by Simberloff would require a determination of the validity of the evidence on which certain inferences had been made, whether or not the uniqueness of a phenomenon or the heterogeneity of the samples preclude the use of statistical methods, what particular statistics are suitable for a given problem, and so forth. There is dim hope for a unification of biogeography until the problems are approached in a far more sophisticated manner than by many recent investigators or their critics. Also, it might be possible to achieve a better balance if workers in this area would not act as if zoogeography had been invented as recently as 1965. There are numerous papers in the classical literature that shed more light on certain biogeographic problems than the often very one-sided modern literature.

Biogeography has suffered from the inability of many authors to see that multiple answers are possible (Mayr 1976). No fauna is ever homogenous, indeed all faunas are composites, each element having a different history. Not only are the dispersal abilities of freshwater fishes very different from those of birds and tardigrades, but even among the birds there are good dispersers and those that never cross a water gap. Mean values are meaningless when each species, indeed each population, has its own history.

Discontinuous ranges

The explanation of discontinuous ranges has tempted naturalists since the mid-eighteenth century. One of the three solutions then proposed was made unacceptable by Darwin: that God had created the same species repeatedly at widely separated places. The two remaining explanations still form the basis of controversy today. According to one, an original continuity existed, subsequently broken by climatic or geological events, like continental drift. This was called *secondary discontinuity*. According to the other explanation, the discontinuity was established by a founder population established beyond the previous species border, these are *primary discontinuities*. Unfortunately, all too many participants in this controversy have favored a strict "either–or" choice among the two theories. Actually, it is becoming clearer that in poor dispersers many if not most discontinuities are secondary, in good dispersers they are primary. Second, the now observed pattern of distribution was in part established in the geological period in which a group had its major evolutionary "burst," hence earthworms or mayflies have a different pattern from mammals and birds. Finally, different patterns, or ratios of secondary to primary discontinuities, exist in different regions, continental versus insular, temperate versus tropical, and so forth. Single-track explanations are unable to cope with this pluralism and are, therefore, bound to be misleading.

Anyone who has noted the extraordinary number of Australian species of birds and other easy dispersers that turn up annually in New Zealand, or of American species that reach the British Isles, knows how important dispersal is. Curiously, some biogeographers have been blind to this factor. In order to explain patterns produced by dispersal across water gaps, such opponents of dispersal formerly built land bridges. In the 1930s and 1940s, G. G. Simpson and I, among others, fought the geologically unsupported building of land bridges. At that time the theory of the permanence of continents and oceans was held almost universally by geologists, and this theory permitted only shelf seas (like the Bering Straits) as locale for land bridges. Plate tectonics, however, completely revolutionized geology and much of biogeography. Since most movement of the plates took place in the Mesozoic and earliest Tertiary, these changes of the earth's crust had much less of an impact on the distribution of the higher taxa of birds (most of which evolved in the second half of the Tertiary) than on that of old groups of animals and plants. Hence, historical ornithogeography had to be much less rewritten than the zoogeography of fishes and certain lower invertebrates.

Dispersal

We still lack a good, I am tempted to call it *philosophical,* analysis of dispersal. First of all, it must be recognized that dispersal and colonization are two independent processes. Some organisms disperse easily but have great difficulties in getting established; with others, the opposite is the case. There is also the question as to what extent dispersal is a random phenomenon. Such calculations of probabilities, as were made by Simpson (1952) imply that given enough time any organism would eventually turn up anywhere in the world. Joseph Grinnell expressed this a long time ago by saying that every species of birds in the world would sooner or later turn up in California. David Lack also adhered to this faith and explained the absence of certain species of birds from certain islands as being caused by an absence of a suitable niche. No doubt this is the reason for certain absences, but actually there is a great difference among species in dispersal ability. The islands of the Pacific were colonized largely from New Guinea. Yet, among the more than 50 families of birds occurring in New Guinea, only 16 have provided the colonists that settled throughout Polynesia (Mayr 1965). Diamond has been able to classify the birds of Northern Melanesia into six dispersal classes, the species of each class having a characteristic pattern, caused by their dispersal ability. An analysis of faunal turnover on islands permits even the inference that island birds are selected for a much greater dispersal drive than mainland birds. Organisms are not like volcanic ashes or dust particles from the Sahara that are passively and indiscriminately dispersed. On the contrary, each species, up to a point, obeys its own specific rules of dispersal and colonization.

Wherever distribution is due to secondary discontinuity, the separated faunas and floras should all show a similar composition and degree of relatedness, regardless of the taxon considered. However, such a pattern is rarely encountered. Deviations from the expectancy of the vicariance theory can be readily explained by dispersal.

Equilibrium theory

I have held long-standing reservations concerning the use of mathematics in biology. Actually, I should not say *mathematics* but rather deterministic essentialistic mathematical models. When the demand is made for "statistically falsifiable dispersalist hypotheses," one wonders just exactly what is meant by this postulate. What assumptions must be made to permit the testing of certain postulates? Let us say an island of 100 km^2 in a certain climatic zone and distance from a mainland has an equilibrium number of 80 breeding species. Since this number is the product both of colonization and extinction, it is subject to a great deal of noise. If there was a series of climatically unfavorable years or the invasion of a dangerous predator (e.g., rats on Lord Howe Island) or pathogen (bird malaria on Oahu), more or less massive extinction would greatly depress the equilibrium figure. If the pool of appropriate potential colonists was exhausted in the source area, it might take a long time until the old equilibrium number was reestablished. Conversely, if the ecology of the island was improved or enriched for some reason or if there was an increase in faunal diversity in the source area resulting in increased colonization potential, an earlier equilibrium figure could be exceeded by a considerable margin.

What conclusions should one draw from these comments? Not at all that the equilibrium theory is faulty, but that one should not expect too great a mathematical precision in the figures of the equilibrium equations. In spite of its grandiose name, the theory of island biogeography (MacArthur and Wilson) was traditional knowledge among biogeographers. As soon as one realizes that faunas and floras are not static, but dynamic, that there is an input (colonization, immigration) and output (extinction, emigration) wherever there are biota, one has automatically arrived at the basic principle of island biogeography. The numerical values are mere icing on the cake.

The emphasis on prediction, falsification, and null hypotheses, ultimately deals with the precision of the accepted numerical value. Is there an opponent of the theory of island biogeography who would question that species diversity at a given locality is due to a balance between input and output? Surely not! Then what is the argument all about? It seems to concern mainly the causes of extinction. What is the relative contribution of density-independent factors, competition, the structure of the ecosystem, and other causes? Most of this cannot be approached by way of the equations of island biogeography. It requires the naturalist's knowledge of the ecological and dispersal potential of each individual species, a knowledge of its food niche, its competitors, its predators, its reproductive potential, and so forth. A purely statistical approach that acts as if all species were the same is bound to leave a large unexplained residue.

At this point, it is legitimate to question what is deterministic in a system with so much built-in noise, chance, and opportunism? How much deviation from a deterministically arrived at expectancy qualifies as a falsification? These are basic questions that should be carefully articulated and answered before sweeping claims are made. I must confess that I find the whole argument about falsification in biogeography badly flawed in this respect. At the same time I agree with those who demand that the hypotheses of biogeography must be tested. What I question are the methods and standards of testing (see also Mayr 1982).

Biochemical studies

It can hardly be questioned that no other recent advance in biology has had a deeper and broader impact than molecular biology. As far as ornithology is concerned, this has affected particulary two major areas, the study of the relationship of higher taxa and the analysis of intraspecific population structure. Broadly speaking, two kinds of molecules have proven most useful, nucleic acids and proteins. The former for the study of the relationship of higher taxa, the latter for the study of variation within populations, differences among populations, and degree of similarity of related species. Barrowclough (Chapter 7) concentrates on this latter area.

One of the interesting results of his analysis is that genetic differentiation among populations of a species seems to be smaller in birds than in other vertebrates and invertebrates. This almost certainly means that there is more gene flow in birds than in these other groups. A less acceptable hypothesis is that the genome of birds is more cohesive and less tolerant of new genetic variation than the genomes of other organisms.

There is regretful agreement among geneticists that we still do not yet understand the genetics of speciation. Barrowclough examines whether the study of enzyme genes, as revealed by electrophoretic analysis, might not shed light on this problem. The results, I think, are ambiguous. Various recent investigators have found in other organisms that the rate of change in enzyme genes does not differ materially in frequently and in rarely speciating phyletic lines. Enzyme genes seem minimally or not at all involved in speciation. After all, enzyme genes represent only a small fraction of the total DNA, even in groups with little repetitive DNA. It would seem desirable to identify one or several of the other classes of DNA as instrumental in speciation.

The mapping of the geographic ranges of literally thousands of species of birds has invariably confirmed that phenotypically incipient species are located in geographic isolates, either refuges or founder populations. Except for Hawaiian *Drosophila*, however, nothing is known about the genetics of such isolated populations. This ignorance is due to the fact that population geneticists, until very recently, had been remarkably uninterested in the problem of speciation. They are now developing some models, which not surprisingly confirm the overwhelming findings of the taxonomists.

The more or less independent evolutionary behavior ("mosaic evolution") of different morphological character sets has been confirmed in recent decades by almost all investigators. The same was found by Barrowclough for three sets of characters of Galapagos finches: plumage color, skeletal characters, and enzyme genes. Population geneticists and molecular biologists have tended in the past to act as if the behavior of enzyme genes could be considered as typical for all genes and molecules. I am quite convinced that this is an error. It can be predicted that mosaic evolution will also be found for different classes of DNA, just as much as for different classes of somatic characters. The tradition of the genetic literature to speak of genetic variation when only the variation of enzyme genes was studied is very misleading. This conclusion does not deprive enzyme genes of their value in certain types of investigations, such as the discovery of sibling species and the measurement of time of isolation. Among all molecules, changes in enzyme genes seem to support particularly well the concept of the molecular clock.

The avian genome

As compared with mammals, reptiles, amphibians, mollusks, and many groups of insects, the knowledge of the avian chromosomes has, until very recently, been exceedingly limited. The major reason is that in addition to a few (3–12) major chromosomes all species of birds have numerous (up to 60 pairs) microchromosomes. These are difficult to count and study. Furthermore, the macrochromosomes are relatively featureless when treated by traditional staining techniques. The situation has, however, been improving in the last 10 years owing to the introduction of new staining techniques and of various methods of molecular biology and genetic engineering. Among the discoveries already made possible by these methodological advances is that the total amount of DNA in birds is considerably smaller than in all other vertebrates, largely due to the relatively small amounts of repeated DNA. It is quite unknown, however, what in the life-style of birds should favor a reduction in the amount of repetitive DNA. Nor is there well-supported theory concerning the function of the repeated DNA and of the pattern of interspersion of unique and repeated DNA sequences. The detailed molecular study of avian genes, including the determination of base-pair sequences and the composition of genes sequences is only beginning. As Shields (Chapter 8) points out quite rightly, the abundant information that we have on morphology, behavior, and ecology of birds would seem of great value in any attempt to correlate genome variation with aspects of life history that are of selective significance.

Bird song

The learning of bird song, discussed by Slater (Chapter 12), is only a small part of the large field of avian vocal behavior, but a very important one. I do not think that I am exaggerating when I say that research in this area has made a greater contribution toward the invalidation of the rigid polarity of Galton's nature *or* nurture principle than almost any other research. It is a branch of behavioral biology particularly favorable for experimentation. Numerous approaches are possible, such as rearing of individuals in total acoustic isolation, crossfostering, reaction to playback of taped songs, and surgical or hormonal intervention.

The treatment of bird song has frequently suffered from a plethora of descriptive detail. Description, however, is merely a prelude to science. Real science is based on the formulation of how and why questions, as is well documented by the study of bird songs. The maturation of this field has been correlated closely with the development of challenging why questions. Why do males sing? To attract mates, to warn off rivals, or for some other reasons? Why do some species have numerous vocalizations, others very few? To what extent do specific vocalizations have an inferred selective significance, whereas others are, at least in part, merely a stochastic element? May the same song element have different functions in different contexts? Why do related species sometimes have very similar songs, when in other genera songs of closely related species are often remarkably different? What difference in life-style, habitat preference, and other ecological and phenotypic characters would seem responsible for these differences? These are just a sample of the questions raised by the student of bird song. Not all of these questions, perhaps only a few can be tested by experiment. The establishment of correlations between certain environmental conditions and certain types of songs is often the most productive approach. The predictions that can be derived from such correlations must, of course, be tested experimentally whenever possible.

Songs of birds raised in complete acoustic isolation are apparently never totally identical with the song of wild birds, but the differences are sometimes too slight to be appreciated by the human ear. Even in species in which the song is learned, there exists a genetic component for the species-specific song. If young birds of such species are simultaneously offered several songs during the learning period, including that of their own species, they usually unerringly learn that song. In some estrildine finches, however, the song of the foster father of a different species is learned, even when conspecific males sing in the same aviary.

The capacity for learning reaches its pinnacle in certain mimicking species, like parrots and mynahs, that imitate the human voice; starlings, mockingbirds, and marsh warblers that imitate primarily other kinds of birds; and that virtuoso, the lyrebird, who may imitate to perfection not only the most musical sounds, but also the raucous laugh of a Kookaburra or the sound of a chain saw or a squeaky bicycle. The lyrebird does not have a complex syrinx, and this confirms the conclusion of virtually all specialists that the variation of song and the constraints in learning it are not caused by the sound-producing apparatus but by a partially open genetic program that instructs the central nervous system what to select and what to reject.

Slater discusses many possible reasons why natural selection may have favored a role for learning in the development of bird vocalization. Degree of territoriality, fledgling dispersal, nature of the pair bond, sexual selection, and other life history attributes combine to favor one or another type of learning or a combination. It is now clear that the old question; Is song instinctive (=totally programmed genetically) or acquired (=totally learned without any genetic constraints) is misleading. In nearly all species, there is an open genetic program, which imposes variable constraints on what can be learned. Learning is, of course, only one of many aspects of bird vocalization. Among other facets perhaps none is as intriguing as to decipher the information content of bird songs and calls. Smith (1977) has recently reviewed this fascinating topic. The particular contribution he has made is the strict separation of the signal sent by the vocalizing individual and the message that this conveys to the recipient. Many components of the context may affect the message.

Bird navigation

How birds find their way on migration and are able to find their "home" area has been a puzzling problem for over 100 years. Some 30 years ago, a breakthrough seemed in sight. Gustav Kramer established the existence and precision of the sun compass, and Merkel and Wiltschko a dozen years later discovered magnetic orientation in birds. Now there is a veritable army of students of bird navigation, and new discoveries of remarkable abilities of birds are made all the time. What is so tantalizing about all this undoubted progress is that it all relates to the compass sense of birds: Birds can use several (how many is still controversial) clues that tell them, provided they know in what direction their home lies, how to maintain that direction. In other words, they can use a compass. A compass, however, is of no use without a map that gives location and the direction from your present position your home lies. This is what the students of bird navigation refer to as the *map sense*.

Walcott and Lednor (Chapter 13) present a condensed summary of many recent additions to our knowledge of the various components of the compass sense in birds, their interaction, and the affects of experimental changes in conditions. Yet, they also admit with resignation that one as yet has not a single clue on the nature of the map sense. In fact, no one has been able to suggest an experiment that would shed any light on the map sense. No one has so far been able to ask the right kind of questions. Since humans apparently have no trace of such a sense, or if we do, are entirely unconscious of it, it is particularly difficult to frame meaningful questions. This is not a unique situation in the history of science. For instance, biologists before Gregor Mendel had obtained most of Mendel's results in their crosses but had asked no relevant questions and had formulated no theories. As a result, they failed to develop a theory of inheritance. As far as the map sense of birds is concerned, we likewise seem at the present time without a single testable theory that has not already been thoroughly refuted.

In whatever direction we may look, we always find Vannevar Bush's statement confirmed that science is an endless frontier. Ornithology, however, is making a mighty contribution in pushing this frontier further and further back. The present volume is a splendid testimony for this.

Literature cited

Barrowclough, G. F. 1980. Genetic and phenotypic differentiation in a wood warbler (genus *Dendroica*) hybrid zone. *Auk 97*:655–668.

Bock, W. J., and J. Farrand. 1980. The number of species and genera of Recent birds. *Am. Mus. Novit. 2703*:1–29.

Chapman, F. M., and T. S. Palmer (eds.). 1933. *Fifty years' progress of American ornithology 1883–1933*. Lancaster, Pa.: American Ornithologists' Union.

Corbin, K. W., C. G. Sibley, and A. Ferguson. 1979. Genic changes associated with the establishment of sympatry in orioles of the genus *Icterus*. *Evolution 33*:624–633.

Endler, J. A. 1977. *Geographic variation, speciation, and clines*. Princeton, N. J.: Princeton University Press.

Gould, J. 1837. Description of new species of finches collected by Darwin in the Galapagos. *Proc. Zool. Soc. London 5*:4–7.

Haffer, J. 1974. *Avian speciation in tropical South America, with a systematic survey of the toucans (Ramphastidae) and jacamars (Galbulidae)*. Publ. No. 14, Cambridge, Mass.: Nuttall Ornithological Club.

Keast, A. 1961. Bird speciation on the Australian continent. *Bull. Mus. Comp. Zool. Harvard Univ. 123*:305–495.

MacArthur, R. H., and E. O. Wilson, 1967. *The Theory of Island Biogeography*. Princeton, N.J.: Princeton University Press.

Mayr, E. 1959. Trends in avian systematics. *Ibis 101*:293–302.

– 1963a. The role of ornithological research in biology. In *Proceedings, XIIIth International Ornithological Congress, Ithaca*, pp. 27–38.

– 1963b. *Animal species and evolution*. Cambridge, Mass.: Harvard University Press.

— 1965. The nature of colonization in birds. In H. G. Baker, and G. L. Stebbins (eds.). *The genetics of colonizing species.* pp. 30–47, New York: Academic Press.

— 1975. Materials for a history of American ornithology. In E. Stresemann (ed.). *Ornithology from Aristotle to the present*, pp. 365–396. Cambridge, Mass.: Harvard University Press.

— 1976. *Evolution and the diversity of life.* Cambridge, Mass.: Harvard University Press.

— 1982. Review of G. Nelson and D. E. Rosen (eds.), *Vicariance biogeography, Auk 99:*618–620.

Moreau, R. M., and B. P. Hall. 1970. *An atlas of speciation in African passerine birds.* London: British Museum (Nat. Hist.).

Regal, P. J. 1975. The evolutionary origin of feathers. *Quart. Rev. Biol. 50:*35–66.

Sibley, C. G., and J. E. Ahlquist. in press. The phylogeny and classification of the Passerine birds. In *Proceedings of the 18th International Ornithological Congress, Moscow.*

Simpson, G. G. 1952. Probabilities of dispersal in geologic time. *Bull. Am. Mus. Nat. Hist. 99:*163–176.

Smith, W. J. 1977. *The behavior of communicating.* Cambridge, Mass.: Harvard University Press.

1 Captive birds and conservation

WILLIAM CONWAY

Many unique and spectacular birds are now threatened with extinction or decimation. For this reason, captive bird collections and propagation are attracting new interest from conservationists, and the manipulative techniques of caring for captive birds are being examined for keys to the temporary salvation of birds faltering at the edge of extinction. Such technology includes the provision of polyvinyl chloride pipe nest boxes for Puerto Rican Parrots (*Amazona vittata*), the "hacking" of Bald Eagles (*Haliaeetus leucocephalus*) and Peregrines (*Falco peregrinus*), and even the crossfostering of Whooping Crane (*Grus americana*) eggs under Sandhill Crane (*Grus canadensis tabida*) foster parents. Fortunately, captive propagation of birds is becoming more predictable, more science than art. Within this evolution lies a potential for the survival of some species in unconventional preservation programs including ex situ preservation, with all the ecological and philosophical uncertainties that method can impose.

Most public collections of birds – zoological gardens – are located in population centers. For an increasing number of people, they comprise the only really significant experience of birdlife that they will ever have. During the past 15 years, new exhibit techniques have allowed zoo biologists to display birds in simulated habitat groups, to sustain them better, to breed them, and to do all these things in esthetically compelling ways. One result has been heightened public consciousness. Kellert (1979) reported that 46% of the American public had visited a zoo in the previous 2-year period and characterized the zoo as one of the most important and frequently used sources of contact with wildlife. However, he also noted that the interest of zoo goers was unlikely to be "scientific" and that their knowledge of wildlife was poor, although unusually supportive of conservation and humane measures.

Zoo educators used to complain that visits to dead animals in museums were thought to be educational fare for adults whereas visits to live ones in zoos were considered fit only for the amusement of chil-

23

dren. This is no longer the case. The majority of zoo visitors are adults. Despite enormous school group attendance, the ratio of adults to youngsters is about 2.5:1 in New York's Bronx Zoo. Nevertheless, few conservation organizations have taken advantage of the 125 million zoological garden audience to promote environmental education, a clear direction for the future. Nor have zoos themselves done as much as might be expected.

The usual role of captive bird collections has been educational and recreational (Conway 1969, 1971). In the next 25 years, the contributions of captive collections to avian conservation may become both more significant and more direct. Beyond the use of captives in ornithological research, collections can act as empirical laboratories for the development of such supportive conservation technologies as translocation, disease control, and physical care of traumatized rare birds. Well-known avicultural techniques including inducement of additional egg clutches, artificial insemination and incubation, foster parenting, handrearing, and supportive feeding already have given rise to important programs for wild birds of prey, cranes, pheasants, and waterfowl. At the same time, however, captive collections are sometimes themselves accused of being threats to rare bird populations, usually incorrectly.

The 172 zoos listed in the directory of the American Association of Zoological Parks and Aquariums (AAZPA) (Dillon 1980) house approximately 52,620 birds (N. A. Muckenhirn, unpub.). Of this number, about 70% are captive bred (based on projections from the International Species Inventory System (ISIS) data, N. Fleshness pers. comm.) In comparison, systematic collections in North America held approximately 4.3 million bird specimens in 1976 and were increasing those collections at the rate of about 100,000 each year, almost entirely with specimens from nature (King 1978). By way of example, the New York Zoological Society's live bird collection acquired 1,822 birds during 3 years, 1979–1981. Of these, 292 (16%) were imported from nature, 361 (20%) from other captive propagation programs, and 1,169 (64%) from the zoo's own breeding program. The number of bird specimens now in systematic collections of museums is probably greater than the combined acquisitions of all birds from nature by all North American zoos since 1864, when the first zoo opened in the United States. The same restraint is not true of the trade in pet birds.

For the 2 years 1980 and 1981, the pet trade averaged 693,675 birds imported into the United States (Anonymous 1982) of which almost 600,000 were caught in the wild. TRAFFIC (U.S.A.)'s David Mack (pers. comm) reports that approximately 350,000 of the imported birds are psittacines of which only about 80,000 are captive bred. Of the remainder, more than 200,000 are finches from Senegal, and there are also large numbers of finches imported from Europe and Asia. Zoos and other public collections of birds import less than 0.1% of the total of imported birds, perhaps 2,000 specimens. They cannot be said to be contributing to the diminution of bird populations.

At least 801 species of birds of 23 orders were bred in zoos according to my compilations from the 1979 *International Zoo Yearbook* census (Olney 1981). However, few birds designated "endangered," "rare," or "vulnerable" by International Union for Conservation and Nature–International Council for Bird Protection are housed in zoos. Only 37 listed species were bred, and, of these, only 17 successfully produced 25 young or more. Six large orders are particularly poorly represented in lists of birds bred in captivity; Procellariiformes, Charadriiformes (especially sandpipers), Cuculiformes (only Musophagidae are commonly bred), Caprimulgiformes (only the Tawny Frogmouth (*Podargus strigoides*) is usually maintained and bred), Apodiformes, and Piciformes. Several families in these orders are easily bred but unpopular in collections; others require special facilities (Conway 1967). Highly insectivorous birds, especially exceptionally aerial forms, such as swifts (Apodidae) and nightjars (Caprimulgidae), are difficult to care for, some migratory species present maintenance problems, and few lek breeders are commonly bred. Where a species is spectacular as well as rare, it may attract unusual efforts. At least 107 Rothschild's Mynahs (*Leucopsar rothschildi*) were reared in 1979, and the number in captivity is approaching the 900 mark.

Outside of zoos, few bird breeders have been willing to make their records available for study. Not even the various falcon-breeding programs have submitted their figures to the *International Zoo Yearbook* or ISIS although the records of the International Crane Foundation in Baraboo, Wisconsin, are included. The American Federation of Aviculturists now claims 30,000 members but attempts to census private collections have been largely fruitless (Shelton 1981). Although a few private aviculturists may contribute to serious propagation programs, and many are successful breeders of rare psittacines and pigeons, hobbyists as a group cannot yet be considered a reliable resource for conservation-oriented propagation. However, waterfowl and pheasant breeders, in particular, are notable exceptions to this generalization. Organizations such as the International Wild Waterfowl Association and the very active World Pheasant Association have developed serious and growing propagation programs. Unsurprisingly, the economic rewards of the pet trade recently have stimulated impressive commercial propagation efforts that, in a few instances, are beginning to approach the success of tropical fish breeders. The 80,000 or so captive-bred psittacines imported into the United States each year are mostly Cockatiels (*Nymphicus hollandicus*) and Budgerigars (*Melopsittacus undulatus*), but propagators in South Africa and elsewhere are even breeding commercial numbers of *Neophema* and *Platycercus* parakeets. Commercial propagation programs are also underway in the United States and one of the most ambitious of these is California's Behavioral Studies of Birds (BSB), Ltd. During 1981, BSB Ltd. produced 7,678 birds of 39 species including Golden Parakeets (*Aratinga guarouba*) and Salmon-crested Cockatoos (*Cacatua moluccensis*). It may be that such programs will reduce pressures upon wild bird populations.

Dividing the theoretical bird capacity of North American zoos, 52,600, by the number of threatened bird taxa, 433 (King 1981), suggests that 121 birds of each could be sustained and bred, but preservation is not so simple. Genetically, if not demographically, 121 would probably be an insufficient number to sustain over a long period of time for most bird species in a stable population except under unrealistically intensive management (Frankel and Soule 1981). Moreover, most present zoo facilities are ill designed for propagation, and the breeding of many endangered species is insufficently understood. Finally, zoos and other public collections receive a substantial part of their income from the "gate," whether through admission charges or in tax support. The devotion of all zoo space to endangered species, many comparatively unattractive, would require a major increase in public sophistication. In this light, it should be remembered that the animal exhibit spaces of all of the world's zoos would fit within the borough of Brooklyn very comfortably. Nevertheless, the outlines of the contributions captive breeding programs might make to the survival of vanishig species are beginning to appear. They derive almost as much from the development of captive propagation technologies as from gene banks of captive endangered birds.

With the gradual advance of aviculture from art and hobby to scientific technology, wild birds are being induced to breed in captivity after shorter periods of acclimatization, to expand their reproductive output, to live longer, and even, when opportunity permits, to adapt to nature more readily than might have been expected.

At the Fish and Wildlife Service's Patuxent Wildlife Research Center facilities, wild adult Andean Condors (*Vultur gryphus*) were successfully bred after only 4 years in captivity, during the first year separated from a community enclosure. Eventually, all six of the Condors retained from eight trapped in Argentina bred (J. Carpenter, pers. comm.). At the Bronx Zoo, a single pair of Andean Condors was induced to lay multiple clutches over a period of 4 years, thus far, producing eight viable young rather than a maximum of two that would have been expected from the every-other-year species breeding pattern in nature. Subsequently, Andean Condor chicks reared by hand at the zoo and parent-reared at Patuxent have been introduced to Condor habitat with encouraging, if preliminary, success in Peru. The technique of inducing replacement clutches is commonly utilized by aviculturists for many species. Bronx Zoo ornithologists have induced such unusual forms as Tawny Frogmouths and Green Woodhoopoes (*Phoeniculus purpureus*) to replace one clutch after another, artificially incubating the eggs, rearing the young by hand, and successfully breeding the progeny – and the progeny's progeny.

Countless past disappointments in the propagation of such monomorphic birds derive from errors no more complicated than keeping birds of only one sex together. The advent of fecal steroid, feather pulp, and blood chromosome sexing techniques has helped to overcome the problem, but now curators and veterinarians have also become more certain and experienced in laparoscopy. The application of artificial insemination methods worked out with domestic fowl has made it possible to overcome behavioral and even physical problems in some species, especially gallinaceous birds, birds of prey, and cranes.

Work aimed at developing techniques for initial evaluation of birds destined for breeding programs both cytogenetically and hormonally is now underway at several zoos, notably the San Diego Zoo. The development of optimal methods for sperm collection and its evaluation, for synchronization and recognition of the synchronous states of male and female birds in breeding programs, analysis of the effects of environment, including food, temperature, space, and light regimens, and, for females, better determination of the timing of ovulation and its relation to internal and external stimuli are clear directions for future research.

In several domestic mammals, notably cattle, embryo transfer techniques have reached a high degree of sophistication. In 1979, more than 17,000 bovine pregnancies resulted from embryo transfers whose objective, in most instances, was to increase the number of progeny of geneticaly desirable individuals (Seidel 1981). Desirable cows are superovulated with prostaglandins and fertilized naturally or artifically. A few days later, multiple embryos at the blastula stage are washed from the uterus. As many as six viable embryos may be obtained from a single cow and transferred to less desirable cows whose reproductive cycles have been synchronized so that they can be used as surrogates. Moreover, cattle embryos can be stored frozen for long periods and utilized in the future. In birds, inducement of replacement clutches combined with the use of foster parents, as in the *Grus americana* experiments in Idaho, or artificial incubation and hand rearing is analogous. Although some opportunities presented by the mammalian system are unique, the shipment of fertilized bird eggs at different stages of development is a common technique whose results can be improved, and so can those of artificial incubation. The bird egg may offer more husbandry possibilities than are presently apparent

Avian incubation is much more than keeping eggs warm, wiggling them about from time to time, and controlling humidity. Although some bird eggs hatch in incubators with little help beyond warming and wiggling, including such disparate forms as Chimney Swifts (*Chaetura pelagica*) and Andean Condors, the artificial incubation of others, crane eggs for example, is notably unreliable. Cade (pers. comm.) has found that hatchability of falcon eggs is affected by individual differences even between birds of the same species and that higher hatchability is associated with parental incubation. Trying to improve hatchability of crane eggs has led Archibald (pers. comm) and his co-workers to try intermittent doses of bright lights, chilling, and recorded crane calls; I confess to suspending an entire Bronx Zoo incubator on springs and subjecting it to "heart throb" vibrations. Despite the wealth of data collected by those monitoring bird nests, even with eggs containing sophisticated electronic data-reporting apparatus, the area begs for further study.

Captive animals constituting gene banks have been provided for several species of mammals now lost in nature. Of these, the Pere David's Deer (*Elaphurus davidianus*), European Bison (*Bison bonasus*), and Mongolian Wild Horse (*Equus przewalskii*) are best known. However, there are no breeding captive populations of birds lost in nature.

Most of the problems inherent in captive gene bank schemes are economic and sociopolitical. It is difficult to find ways to make money out of the long-term care of vanishing species, creatures that, because of their very rarity, have lost whatever ecological significance they may once have possessed. Keeping animals in their native habitats is the most parsimonious solution to the conservation problem. Where that is not possible, zoos and other public collections may be the only sources of support for their survival, a last resort. However, biological problems attend the long-term maintenance of necessarily small populations of wild animals in captivity. Inbreeding and domestication, in the sense of selection, inadvertent or not, for atypical forms more successful in captivity are serious concerns (Seal 1978). In the future, these dangers may be mitigated by the opportunity for renewal of original heterozygosity with stored sperm (and, perhaps, ova and embryos) from decades past. The first successful cryopreservation of wild bird sperm was recently achieved at Patuxent by George F. Gee. Extension of sperm preservation technology also holds forth the promise of facilitating the transport of birds by transporting sperm rather than delicate specimens. For the present, at least one coherent long-term propagation plan is already taking shape in the North American zoo community. It is called the *species survival plan* (SSP) of the AAZPA.

The SSP has begun its work on the basis of information from ISIS, subscribed to by most North American zoos, and housed at the Minnesota State Zoo. The program enjoys one almost full-time employee in the association's conservation coordinator and operates through AAZPA's Conservation of Wildlife Committee and its SSP Subcommittee. After analyzing the status of animals in nature and captivity, species are designated for SSP attention and preliminary management plans developed. Zoos holding the designated species are polled and a written commitment to dedicate their specimens to the plan is sought. Also, each zoo is asked for an appraisal of its maximum carrying capacity for the species. That done, the participating institutions and the Conservation Committee appoint a species coordinator and elect delegates to a species propagation group. It becomes the responsibility of this group to develop a preservation plan with the objective of sustaining the SSP-designated species over time with a minimum decline in genetic variability by maximizing its captive population and minimizing inbreeding. Transfers of stock between zoos are made upon the advice of the group with the object of gaining equal representation in the overall population of each founding line. It should be remembered that in a small intensively managed zoo population, where all family lines are caused to produce precisely the same number of offspring, the effective population size can be twice the number of breeding individuals: $N_e = 2(N - 1)$ (Frankel and Soule 1981). The propagation group promotes the exchange of husbandry information and, it is anticipated, will publish authoritative references on each species.

Criteria for selection of critical species are not yet adequately formalized. Beyond status in nature and a concern that unique forms not be lost, the realistic probability of success of a particular propagation program, the numbers presently held in captivity and the space and facilities available have to be considered. The SSP effort is new, but programs for mammals such as the Siberian Tiger (*Panthera tigris altaica*), Mongolian Wild Horse, Golden Lion Tamarin (*Leontopithecus rosalia*), Grevy Zebra (*Equus grevyi*), and Gaur (*Bos gaurus*) are underway. Among birds, the first species designated was the Rothschild's Mynah although the Humboldt's Penguin (*Spheniscus humboldti*), White-naped Crane (*Grus vipio*), and Andean Condor have also been selected, the latter because of the contribution the formalization of its propagation might make for better understanding the California Condor's (*Gymnogyps californianus*) prospects. Already, plans are afoot by ICBP to attempt a reintroduction of captive-bred Rothschild's Mynahs in their original homeland on Bali (S.D. Ripley, pers. comm.).

The saga of the Mauritius Kestrel (*Falco punctatus*) and Pink Pigeon (*Nesoensas* [= *Columba*] *mayeri*) have not been so promising as that of the mynah. In the aviary, the kestrel was soon bred but adults taken from nature and young hatched in captivity both were plagued by accidents and disease (Jones et al. 1981). About 15 birds continue to survive in nature, as aviculturist Carl G. Jones writes, "despite conservation efforts." In the 3,035 ha of remnant forest, about seven Pink Pigeons now survive but, in this instance, propagation efforts have been more promising. The birds are being bred in Mauritius in a government-supervised program and in the Jersey Wildlife Preservation Trust although eggs from both groups of birds suffer from low fertility and it may be that the founding groups were too small. Eight captive-bred birds have recently been sent to the Bronx Zoo in a further effort to avoid catastrophic loss in one installation.

As bird populations die more rapidly and obviously, there will be calls for rescue and salvage efforts. In the United States, five Dusky Seaside Sparrows (*Ammodramus maritimus nigrescens*), thought to be the total population, are now confined in Florida aviaries. They are all males. During 1981, Japanese ornithologists caught the last Japanese Crested Ibis (*Nipponia nippon*) in Japan for a captive propagation attempt; they had not bred successfully in nature for several years. Of course, joint efforts by the Fish and Wildlife Service and the National Audubon Society on behalf of the California Condor are in progress now. The work of Tom Cade at Cornell University in propagation and reintroduction of the Peregrine has attracted worldwide attention and so has that of George Archibald and Ron Sauey of the International Crane Foundation in Wisconsin. Here, such rare birds as Asiatic White Cranes (*Grus leucogeranus*) and Japanese Cranes (*Grus japonensis*) have reproduced. Both efforts and the distinguished propagation programs of Peter Scott's Wildfowl Trust in Great Britain and those of several zoos have prompted calls for the establishment of more captive populations of vanishing birds as insurance and rescue efforts for threatened species. Unfortunately, insufficient space and expertise is available to undertake most salvage efforts. Their development and refinement is a major responsibility for the next few years, while there is still time to obtain adequate foundling groups of unique species. Then it may be that captive collections can play a new role in relation to wildlife reserves.

Even the largest remaining blocks of wilderness are now being converted to other uses. At best, only isolated fragments are preserved; ecological islands with the same inherent potential for species impoverishment as those isolated in the sea (Terborgh 1974; Diamond 1975; Terborgh and Winter 1980). The largest of nature reserves are probably too small for the long-term care of large mammals (Soule et al. 1979) and perhaps birds. No matter how carefully protected against hunters, agriculturists, or lumberers, small populations of birds in small isolated reserves are likely to fall victim to genetic and demographic problems even if not to those of ecological succession (Morton 1977; Frankel and Soule 1981). For obvious genetic reasons, polygynous and lek species may be more vulnerable than monogamous breeders. There seems little chance that any wholly unmanaged wilderness will survive in 100 years so new ways of managing reserves to sustain at least some species that would otherwise be lost deserve examination.

Substitutes, avicultural in nature, are needed to make up for deficiencies in undersized reserves. Examples are artificial nesting sites, as has recently been accomplished for the Puerto Rican Parrot in the Loquillo Experimental Forest and as is traditional practice for Eastern Bluebirds (*Sialia sialis*) faced with competition from introduced House Sparrows (*Passer domesticus*), for Ospreys (*Pandion haliaetus*), and for many other species (Snyder 1977). In other instances, it may prove possible to sustain a particular species within an otherwise inadequate reserve with supportive feeding of nearly complete diets. Captive collections and technology could come to play a crucial part in the support of undersized reserves by acting to provide the substitute gene transfers of lost migratory corridors either through the introduction of captive birds, their eggs, or even through artificial insemination thus ensuring the retention of variability, or even enhancing it, in very small populations. Even demographic imbalances may be amenable to redress with the increasing sophistication of reintroduction techniques. Birds supported in this way would no longer be at liberty, no more so than the herds of wisent sustained in Poland's Bialoweza forest by supplementary feeding or the various populations of north-temperate birds sustained during periods of winter stress by backyard bird feeders, but they would still be with us. Indeed, the prospect of declining wild lands and the need for increasingly intense care and management to sustain many wild creatures in minimal spaces, often in the absence or diminution of critical ecological resources, suggests that some nature reserves will inevitably become megazoos or zooparks.

Today, many nature reserves are partially protected from human pressures by remote locations, whereas zoos, of course, are in the cities where the people they serve live. In the future, as human populations shift, grow, and further crowd nature reserve borders, zoological gardenlike park accessories may make sense. Such zooparks could act not only to sustain critically small animal populations in the adjoining reserves but also to enhance the visitor's experience and understanding. Their scientific staffs would provide medical care, recognizing and controlling outbreaks of disease, and ecological surveillance of those species most important to the welfare of the ecosystem. They would be well positioned to deal with needs for translocations, problems of cross-fostering, control of competing species, and maintenance of heterozygosity (where this was desirable and not damaging to the development of advantageous ecotypes). Their guard staffs would protect the zoopark, whereas their interpretive personnel would help to provide the philosophical and informational bases of continuing community support. This is nothing more than a projection of the directions in which some nature reserves and zoos are already headed, incorporating captive reservoirs and exhibits of critically endangered indigenous species. More controversial is the question of local responsibility for exotic species.

Although some exotic birds with no prospect of survival in nature may be sustained in captive collections, it may not be unreasonable to maintain some at liberty, away from their original range, the California Condor for instance. The California Condor's lost homeland near Los Angeles will not be returned. If what remains of it is protected and successful captive propagation programs are developed, the species might yet be supported. Moreover, there are other areas in the United States, apparently unattractive to dense settlement and in some instances where the Condor's Pleistocene relatives once soared, where it could be helped to live today. Two things, at least, are required for such an experiment to achieve success: vigorous captive reservoirs from which birds can be recruited and ongoing programs of protection and support, the latter primarily in the form of feeding stations and nest site development. All of these ideas, in more cautious form, are already a part of the thinking behind the Fish and Wildlife Service's recovery plan for the Condor.

The protection and focus of interest that such introduction and maintenance programs would bring could be helpful to the wildlife communities affected. This point of view contrasts with the concern that special programs for single species will take support away, that if only so much money was not being spent upon the Condor, it could be spent upon some more deserving community of less conspicuous species or, better, upon the preservation of some fast-diminishing habitat (Pitelka 1981). The same worry has been directed at the cost of captive breeding programs, but in most instances the competition for support is more imaginary than real. It is usually different money. The intricacies of attracting support to the cause of conservation are such that creatures of unusual interest must be made to serve the general wildlife community as flagships in support-raising efforts. However, highly managed programs are also opposed by those who find them inappropriate to treasured concepts of wilderness and wild things.

Although recent studies of humanity's progressive alteration of natural environments point ahead to massive extinctions of plants and animals and to the need for a fundamental and holistic change in perceptions of the future of biological diversity (Myers 1979; Barney 1980), biological conservation continues to evolve only piecemeal. No one is in charge. The beauty and rightness of wilderness scarcely permit the would-be conservationist to dwell upon its alteration, much less to be the agent of its manipulation. It is no wonder that the technology of captive care and intensive management are eschewed as a desecration by so many who love birds, that a "rather dead than bred" philosophy has attracted some very deeply caring adherents (Brower 1979; Stallcup 1981). Nor is it surprising that the threat of loss of wild animals forever has attracted a deeply dedicated cadre willing to attempt painstaking scientific management requiring the most time-consuming commitments to prevent that loss from happening, to preserve, as it were, options for the future.

An even more controversial conservation option is that of deliberate introduction of very rare exotic animals outside not only their natural habitats but even beyond their original continents. The history of successful introductions is mostly regrettable and almost invariably is of fast-breeding species whose potential effects upon new homes and native species have not been considered. No serious attempts ever seem to have been made to introduce large slow-breeding birds in the United States until the recent efforts of the Fish and Wildlife Service to establish a new population of Whooping Cranes in Idaho and Tom Cade's Peregrine restoration program in the eastern United States. These are exotic introductions in an instructive sense for they are attempts to establish these birds either where they have not been in recent time or in a habitat that is significantly different from that in which they once lived. Indeed, this latter point is frequently overlooked.

Humans are so dramatically altering the landscape that it is unrealistic in many situations to establish captive programs with the promise of restoring an endangered species to "the wild." Moreover, it is just as unreal to claim that every introduction will destroy "nature."

The largest and fastest growing of earthly habitats are those of agriculture, whether cultivated for crops or given over to pastoral livestock purposes. What great slow-breeding endangered and easily controllable birds could live in this inevitable foodscape in a symbiotic or minimally competitive fashion alongside people? An imaginative response to this question could spell the difference between life and extinction for a few species of compelling interest and beauty. It is the world's large slow-breeding forms that tend to lead endangered species lists. We need not think of all the agricultural land in terms of sterile cornfields. We should not forget rangeland, tropical forest crops and rice paddies, and managed timberlands often inhospitable to native species nor those modified areas being created on all sides by earth-moving and water-impounding activities.

During the next quarter century we should develop our ecological sophistication to a point where the results of the introduction of an exotic wild species can be anticipated and provided for. Certainly, the introduction of domestic animals to remaining wild lands will continue. If we must continually change environments and create new ones, we might also give thought to the ways such areas may be designed to accommodate needy wild creatures, native or not. Besides, we need to know better how plastic a bird may be in its behavioral ecology. Where food is a limiting factor upon population, bird populations might sometimes be increased to viable levels by provision of more food or more nesting sites. Could the suitability of uncharacteristic or second growth habitats be enhanced by the nature of nest site and food training offered to captive-bred birds designated for introduction or reintroduction schemes (Temple 1977)?

To evaluate the desirability of establishing and sustaining populations of birds in captivity, it is instructive to look back, for a moment, upon some of the birds already lost: Carolina Parakeet (*Conuropsis caroli-nensis*) and Passenger Pigeon (*Ectopistes migratorius*), Stephen's Island Wren (*Xenicus lyalli*) and Great Auk (*Pinguinus impennis*), Dodo (*Raphus cucullatus*) and elephant bird (*Aepyornis maximus*), Spectacled Cormorant (*Phalacrocorax perspicillatus*) and Heath Hen (*Tympanuchus c. cupido*), Pink-headed and Labrador Ducks (*Rhodonessa caryophyllacea* and *Camptorhynchus labradorius*). Would we be grateful to have any of these species now? Is there a place for them again in their original home or even elsewhere? Would they be worth seeing even if only in captivity? For some species the answers to these questions could be, yes. They lead to a fourth question: Must the nature of captivity remain the same as it is today?

Summary

Captive bird collections can have value to avian conservation beyond their traditional educational and recreational roles. Avicultural science already is in use in a variety of supportive bird preservation efforts, and advances hold forth the possibility of sustaining captive populations as insurance for some threatened species. North American zoos have begun coordinated breeding programs. As they identify ways of responding to the problems of sustaining small populations, their findings may aid the management of increasingly isolated, more zoo-like, nature reserves. Long-term preservation programs for rare birds at semiliberty are among the options that could develop from the application of avicultural conservation approaches.

Literature cited

Anonymous 1982. *TRAFFIC (USA)*, *4*(1):4

Barney, G. 1980. *The global 2000 report to the president*. Washington, D.C.: Government Printing Office.

Brower, K. 1979. Night of the condor. *Omni*, 14–15.

Conway, W. G. 1967. The opportunity for zoos to save vanishing species. *Oryx* *9*:154–160.

– 1969. Zoos: their changing roles. *Science 163*:48–52.

– 1971. The role of zoos and aquariums in environmental education. In J. A. Oliver (ed.). *Museums and the environment: a handbook for education*, pp. 171–176. New York: American Association of Museums, Arkville Press.

Diamond, J. M. 1975. The island dilemma: lessons of modern biogeographic studies for the design of natural reserves. *Biol. Conserv. 7*:129–146.

Dillon, C. 1980. *Zoological parks and aquariums of the Americas, 1980–81*. Oglebay Park, Wheeling, W. Va.: American Association of Zoological Parks and Aquariums.

Frankel, O. H., and M. E. Soule. 1981. *Conservation and evolution.* Cambridge: Cambridge University Press.

Jones, C. G., F. N. Steele, and A. W. Owadally. 1981. An account of the Mauritius Kestrel captive breeding project. *Avicult. Mag. 87*:191–207.

Kellert, S. R. 1979. Zoological parks in American society. In *American Association of Zoological Parks and Aquariums 1979 Annual Conference Proceedings*, pp. 88–126.

King, F. W. 1978. The wildlife trade. In H. P. Brokaw (ed.). *Wildlife and America*, pp. 253–271. Washington, D. C.: Council on Environmental Quality.

King, W. B. 1981. *Endangered birds of the world, the ICBP bird red data book.* Washington, D.C.: Smithsonian Institution Press.

Morton, E. S. 1977. Reintroducing recently extirpated birds into a tropical forest reserve. In S. A. Temple (ed.). *Endangered birds: management techniques for preserving threatened species*, pp. 379–384. Madison: University of Wisconsin Press.

Myers, N. 1979. *The sinking ark.* Oxford: Pergamon Press.

Olney, P. J. S. 1981. Birds bred in captivity and multi-generation births 1979. *Int. Zoo Yearb. 21:*271–304.

Pitelka, F. 1981. The condor case: an uphill struggle in a downhill crush. *Auk 98:*634–635.

Seal, U.S. 1978. The Noah's Ark problem: multigeneration management of wild species in captivity. In S. A. Temple (Ed.). *Endangered birds: management techniques for preserving threatened species*, pp. 303–314. Madison: University of Wisconsin Press.

Seidel, G. 1981. Superovulation and embryo transfer in cattle. *Science 211:* 351–358.

Shelton, L. C. 1981. American aviculturists, consumers or conservationists. In *American Association of Zoological Parks and Aquariums 1981 Annual Conference Proceedings*, pp. 112–114.

Snyder, N. F. R. 1977. Increasing reproductive effort and success by reducing nest site limitations: a review. In S. A. Temple (ed.) *Endangered birds: management techniques for preserving threatened species*, pp. 27–34. Madison: University of Wisconsin Press.

Soule, M. E., B. A. Wilcox, and C. Holtby. 1979. Benign neglect: a model of faunal collapse in the game reserves of East Africa. *Biol. Conserv. 15:*259–272.

Stallcup, R. 1981. Farewell, Skymaster. *Point Reyes Bird Observatory Newsletter. 53:*10.

Temple, S. A. 1977. Manipulating behavioral patterns of endangered birds, a potential management technique for preserving threatened species. In S. A. Temple (ed.). *Endangered birds: management techniques for preserving threatened species*, pp. 435–443. Madison: University of Wisconsin Press.

Terborgh, J. 1974. Preservation of natural diversity: the problem of extinction prone species. *BioScience 24:* 715–722.

Terborgh, J., and B. Winter. 1980. Some causes of extinction. In M. E. Soule and B. A. Wilcox (ed.). *Conservation biology: an evolutionary-ecological perspective*, pp. 119–134. Sunderland, Mass.: Sinauer Associates.

2 Research collections in ornithology – a reaffirmation

JON C. BARLOW AND NANCY J. FLOOD

Organized assemblages of natural curiosities presaging research centers first emerged in Europe in the sixteenth and seventeenth centuries. Small natural history museums and personal collections of this time were used primarily by a small group of scientifically pioneering amateurs. The first natural history museum open to the public was founded at Oxford, England, in 1683, followed 70 years later by the establishment of the British Museum in London. In North America, Charles Wilson Peale created the first natural history museum in the 1780s, in Philadelphia. Later additional Peale museums were established in Baltimore and New York. These museums failed in the 1820s after Peale's death, because of his emphasis on profit rather than perpetuity. Nonetheless, with zoological specimens arranged according to the Linnean classification, Peale's museums set the tone for the major research institutions founded later in the nineteenth century.

In the 1850s, Spencer Fullerton Baird, at the Smithsonian Institution, was faced with an immense influx of specimens from explorations in the American West. He realized that a U.S. national museum "would both increase public knowledge of fauna and flora and provide scholars with comparative material for biological research" (Alexander 1979:51). The dual purpose of museum collections has been echoed many times. At the dedication of the American Museum in 1877, for example, O. C. Marsh prophesied that although the extensive collections to be held there would enable the public to become better informed about the natural sciences, it would be the unobtrusive researchers studying the collections who would establish the museum's international reputation (Alexander 1979). Despite a long association between public displays and research, the importance of collections to ornithological research has recently come into question, both directly (Ricklefs 1980) and as evidenced by a trend away from museum-related studies among doctoral candidates within the last few years. The use of collections by researchers, however, still appears to be substantial. In replies to a 1977 questionnaire, 84.1% of professional ornithologists stated that they used collections either regularly or occasionally, and 71.5% of amateurs indicated the same (King and Bock 1978).

Although collection-based ornithological research has made significant contributions, it may now be necessary to reconsider and perhaps revise our approach to collecting and collection management. Serious financial constraints currently threaten the continued maintenance of many collections. Cutbacks or reductions in the hiring of staff make it increasingly difficult to maintain adequate service to the scientific and general communities. Museums are hard pressed to pay for the space, cabinets, and other equipment necessary for the storage of their current holdings, let alone prepare for future expansion. Inflation and decreased funding reduces the amount of ongoing collecting, processing, or research that can take place on already available material.

In addition to their financial difficulties, museums face pressure from another source. Ironically, the recent increase in public awareness of the environment and of wildlife – the same environment and wildlife that collections and curators are dedicated to preserving and understanding – has begun to pose significant problems for museums. First, there has been a marked decline in the number of young scientists entering those fields of study most closely associated with museums. Rather, there has been what some see as an overemphasis on ecology and ethology in the last decade (King and Bock 1978). This situation may soon lead to a shortage of qualified curators to make significant use of, and maintain adequately, the collections that come under their jurisdiction.

Secondly, the new public attitude has resulted in enactment of a profusion of laws designed to protect habitats and wildlife. This burgeoning of regulations, and of a bureaucracy to administer them, now frequently impedes scientific inquiry (especially that involving collecting) and is thus, paradoxically, sometimes detrimental to the overall conservation effort.

Despite these apparent setbacks, museums are, more than ever, necessary tools of science and society. The traditional function of collections, to be storehouses of reference material, is no less important today than it was 200 years ago. Important new methods of data collection and analysis have broadened the uses for specimens and have encouraged asking novel questions in a broad range of scientific disciplines. Modern museum-oriented research is far from restricted to the fields of taxonomy and systematics.

In addition, the current rate of habitat destruction, and the extinction or extirpation of increasing numbers of species all over the world, have made the role of collections as biological archives more important than ever before. This is especially true for tropical environments. Presently, closed tropical forests, which contain the greatest diversity of life on earth, are being cleared at a rate of $10-20 \times 10^6$ ha yearly, which amounts to an annual loss of 1–2%. Probably more than 3 million species of plants and animals – many currently unknown – reside in these forests, and thousands of migratory birds are seasonally dependent on them. The area of closed forest may shrink to only one-half of its present size by the year 2000 (U.S. Interagency Task Force on Tropical Forests 1980). As a result of this deforestation, one-third of the tropical organisms, over 1 million species of plants and animals, may become extinct by the end of this century and another 1 million may disappear during the course of the next (National Research Council 1980).

Reducing the rate of such destruction is obviously a primary goal. It must be recognized that much of the projected loss of both habitats and species is inevitable. Along with this, the opportunity to study them is lost as well. As a result, scientists and scientific bodies all over the world (e.g., the United Nations Environmental Program, L'Office de la Recherche Scientifique et Technique Outre-Mer, the Organization of Tropical Studies, the National Science Foundation, The National Academy of Sciences, the Association for Tropical Biology) have made tropical studies in all disciplines a matter of highest priority (National Research Council 1980). Even so, it will be impossible to discover, name, and study in detail in their original state, many of the unknown or little-known species before they disappear. A report to the Systematic Biology Program of the National Science Foundation recognized and described the role of collectors and collections in this situation, particularly well:

there is an urgency for increased emphasis on tropical biology before it is too late. This is not something that can await years of debate. We must act quickly for the forests and other tropical ecosystems are rapidly disappearing. . . . The field effort of collecting and documenting vanishing habitats must be undertaken with thorough coordination and planning befitting our last opportunity on earth to obtain pre-extinction representatives of unique species and data. (Stuessy and Thomson 1981:11–12)

There is thus currently more that collection-oriented biologists can and should do than ever before. It is also true, however, that there is now often relatively less funding available as well as more obstacles to overcome in the process of such work. It is, therefore, an appropriate time for museum personnel and supporters to consider their situation and resources, plan for future expansion of both their collections and their research potential, and discover cost-efficient methods of maintaining, and making available for use, their current holdings. Perhaps only by so doing can museums continue adequately to serve the scientific, as well as the general, public.

Uses of collections

Identification

Traditionally, the primary role of museum collections has been to provide a data base for identifying unknown specimens or species. The importance of collections in this regard is increasing. Collections are now used for this purpose extensively by scientists working in many fields as well as by students and nonscientific personnel. For example, official verification of the identity of illegally taken or held wildlife species often requires the use of museum specimens. Similar use of collections permits identification of the remains of birds that collide with aircraft, a process essential to eventual control of birds at airports and in flight paths.

Information derived from museum specimens has been used to develop the guides currently used in the field by scientists, wildlife managers, and bird watchers. Descriptions of distribution, structure, and sex or age class differences of birds rely heavily on the data stored in museum collections. The illustrations that accompany such manuals are often based almost wholly on the use of museum specimens. Identification of the avian remains found in stomach contents, pellets, feces, and in and around nests and dens has also frequently been facilitated by comparison with museum material (e.g., Platt 1956; Johnston 1975). As in past decades, studies of this nature continue to provide substantial insight into the food habits of many species. Similarly, identification of items from middens yields data on the former distribution and abundance of various species and on their use by native peoples and thus may be useful to both biologists and anthropologists (e.g., Howard 1929).

Finally, in almost all areas of scientific inquiry, whether theoretical or applied, proper identification of the organisms or species under study is essential. Undoubtedly, considerable geographic variation exists within taxa with respect to a variety of behavioral, anatomical, and physiological characteristics. Experimental results obtained for one species or subpopulation of a species may not, therefore, be equally valid for other species or subspecies. Proper identification of the taxa under investigation to the lowest possible taxonomic level is of great importance to the usefulness of any results. Much of the work in laboratories by ethologists, physiologists, endrocrinologists, and others employs material obtained from suppliers who may not know or care about the geographic origin of the organisms in their stock. Verification of the identity of the species or subspecies under study can be accomplished by comparison with museum material. In addition, samples of experimental animals can be deposited as voucher specimens in collections for use by later workers. If questions arise regarding significant variation within the taxa in question, examination of these samples will serve to indicate the particular race or geographic area to which the conclusions of the previous study apply (Parkes 1963).

Studies of communities, whether involving applied research into the impact of environmental perturbation or the testing of complex ecological theories, require knowledge of the components of the assemblage under investigation. Museum material, as well as published works based on such material, again provides the information necessary for such identifications. Several examples demonstrate the need for proper identification of the members of a community. After summarizing various studies conducted to determine the amount of environmental change in the San Francisco Bay area during the 60 years prior to 1973, Lee (1977) concluded that the unavailability of taxonomic information had made it impossible to identify accurately most of the invertebrate organisms sampled. As a result, most of the assessments of community structure, species diversity, and change in species composition that had been made were largely useless. In another example, substantial differences originally reported as signifying ecological changes in Chesapeake Bay over the course of only a few years were later discovered to have resulted from misidentification of the planktonic copepods involved (Bowman 1961). Although such insufficiencies of taxonomic reference material or similar misidentifications may be less common in birds, possible inaccuracies in some ecological studies of birds have been noticed (Austin et al. 1981), and the potential for errors is great in largely unknown regions. Without doubt, an inadequate taxonomic foundation can have deleterious effects on any work. "The increasing refinement and quantification of biological investigation requires comparative exactness in the identification of organisms used in both basic and applied research. Imprecision in identification neutralizes rigor. If any identifications are in error, published work becomes non replicative, and thus unscientific" (Carriker 1977).

Systematics

The problem of identification is actually twofold, particularly when the subject of investigation involves more than one species. At this point, questions of the phylogenetic relationships between the various taxa being studied often become important. Such matters are, of course, the purview of systematists and have been, or are being, resolved in large part through recourse to museum collections. Here, even more than for the simple identification of individuals or species assignment, workers in other scientific disciplines rely heavily on systematists.

Classifications are far from immutable, a factor that ecologists, ethologists, physiologists, and others frequently fail to consider, even though it can have serious implications for the results of their studies. This is especially true when such studies make critical assumptions regarding the degree of phylogenetic similarity or distance between the taxa being investigated.

Scientists outside the field of systematics often lack the training or knowledge to assess fairly the validity of taxonomic theory surrounding the organisms they study. Conclusions about competition in so-called natural groups can be sorely affected if the analyses are based on an erroneous taxonomic structure. Examples of systematic problems abound within the class Aves. The ordinal placement of the families Opisthocomidae and Phoenicopteridae, among others, have long been debated (e.g., Cracraft 1981). At a lower level, the genus *Aimophila* is thought by some to be actually a composite of two genera (e.g., McKitrick 1981), and the Olive Warbler (*Peucedramus taeniatus*), usually placed within the Parulinae (Emberizidae), has now been considered by some to be a member of the completely unrelated Sylviinae (Muscicapidae) (George 1962; Raikow 1978). These and other examples, the number of which is suggested by the extent of changes in the most recent revision of the American Ornithologists' Union (AOU) checklist (AOU Check-list Committee 1982), demonstrate an instability in avian systematics.

The possible effects of this instability are evident in many studies. An apparent lack of adaptive radiation within the West Indies, discussed by Lack (1976), may be more the result of a failure to identify correctly the degree of relatedness among several of the species resident there than of an actual absence of such radiation (Olson 1981). The numerous studies investigating competition in what are assumed to be natural groups rely heavily on the validity of the systematics of those groups. The attempts to relate morphological measurements of "related" sympatric species to possible ecological similarities or differences among those species are dependent upon the validity of studies that have determined the taxonomy of those species (Schoener 1965; Karr and James 1975; Ricklefs 1980). The work of avian systematists is far from complete, and the importance of the museum collections on which such studies are based is thus far from diminished.

General biology

As described, the provision of a basis for the identification of organisms and of the relationships among taxa is the most fundamental and pervasive use of collections. Museum material has contributed significantly to ornithological research in a number of other ways as well. As an example, the ratio of young to adult birds or of males to females in unbiased museum samples has been used to estimate rates of mortality and sex ratios in populations of various species (e.g., Pitelka 1951; Snow 1956; Rising 1970).

Studies of the sequence of plumages and molts of birds have traditionally employed museum study skins. Data relevant to such studies can also be applied to a variety of other questions, such as the timing and duration of molts and the length and degree of synchrony of breeding seasons. These are important aspects of the reproductive biology of any species.

Museum collections are also important in the study of migration. Specimen labels have often been used to pinpoint the presence of a species at a specific site at a given time. Because of the relative paucity of observers in some areas, the migratory patterns and wintering areas of birds south of the United States are still known chiefly from museum specimens (Eisenmann 1955; Parkes 1963). Relating information on the sex or state of molt exhibited by specimens to the date and location of collection can provide evidence of differences in migration rates, traveling routes, or wintering areas among age and/or sex classes. Skeletal or alcoholic material have traditionally been useful for a variety of nontaxonomic purposes including analyses of functional anatomy.

Wildlife management

The Migratory Bird Treaty of 1918 is one of the most widely used series of laws affording protection to wildlife in North America. The designation of taxa to which the act pertains relies heavily on the official AOU checklist of North American bird species, as well as on other current taxonomic literature. Studies of museum specimens provided the basis for most of these names. In order to be afforded federal U.S. protection, a species listed in the act must be found in the United States. Museum specimens often have provided the evidence necessary for the addition of a species to the protected list (e.g., White and Baird 1977).

The Endangered Species Act of 1973 is another of the important laws protecting wildlife in the United States. Among other things, the act provides for the safeguarding of threatened populations of a species, that is, for the protection of a subspecies or geographic race. Aldrich's (1972) description of the Mississippi Sandhill Crane (*Grus canadensis mississippiensis*) as a distinct and threatened taxon, a study based in part on the use of museum specimens, paved the way for the addition of this crane population to the endangered species list.

The act also includes the provision that, in the case of species that are difficult to distinguish, common species may receive the same protection as the rare ones. The opportunity for valuable input by taxonomists and others using museum specimens in this situation is obvious. Accurate taxonomic discrimination and characterization of taxa are necessary in such circumstances to ensure the protection of endangered forms as well as to prevent wastage of conservation funds and time on healthy taxa.

Many foreign species have been introduced, either accidentally or on purpose, into North American ecosystems (Long 1981). Although most have failed to become established or widespread, some have flourished and spread across the continent, having various effects on the indigenous fauna. Particularly in the 1950s and 1960s considerable effort was invested in the introduction of exotic game species into the United States. The more carefully planned among these introductions employed those species judged most likely to become adapted to, and flourish in, new areas. The rationale for such judgments was to a great extent based on an early paper on galliforms (Aldrich 1946), which pointed out the relationship between the morphological characters on which much avian taxonomy is based and many of the physiological or ecological characteristics of a species. These characteristics may determine whether or not a population will survive in a particular environment. Museum-based studies can play a valuable part in the selection of stock from climatic regions similar to those where introductions are planned, as well as in predicting the outcome of introduction experiments.

Introduced populations of unknown origin pose particular problems for wildlife managers. Identification of such taxa to the lowest possible taxonomic level, a process facilitated by comparison with museum specimens, provides at least a general indication of the area of origin and thus of the climatic and habitat requirements of the population (e.g., Banks and Laybourne 1968). Continued monitoring of such introduced species (involving the collection of specimens) and comparison with populations of the same taxa remaining in their place of origin make it possible to evaluate any evolutionary changes that occur in the species in its new location (e.g., Johnston and Selander 1971; Barlow 1973, 1980).

Finally, because date and place of collection are data ordinarily associated with any museum specimen, large collections or combinations of collections can provide information on faunal change. Knowledge of the requirements of the various species present at a site can then provide insight into the environmental changes that may have produced any faunal alterations (e.g., Hayward et al. 1963). Such an approach is, of course, valuable to theoretical researchers as well as to those interested in the impact of man-caused alterations (e.g., Rising 1970). Collections have contributed to work involved in establishing the effect of environmental change or contamination on the structure of individual organisms, for example in demonstrating the influence of pesticides on eggshell thickness (e.g., Anderson and Hickey 1972; Klaas et al. 1974). The proven value of museum material implies that if we intend to analyze the effects of some current environmental change in the future, it would be prudent to amass representative samples of the present fauna. In these days of increasing loss of, or damage to, wildlife habitats, acquisition of museum material becomes of augmented importance.

Teaching

Finally, museum collections play a recognized and important role in ornithological education. In addition to the many graduate students who use collections in their research, the training of undergraduates in avian biology relies to a large extent on the resources housed in museums. Sixty percent of institutions sampled in one study reported that they conducted some type of formal student training within their facilities (King and Bock 1978). The majority (75%) of these facilities maintain separate teaching collections for use by their students and other institutions or to loan to outside teachers. Another survey conducted as part of the workshop revealed that the preparation of study skins and the identification of birds – activities predominantly conducted in museums – constituted approximately 30% of the "typical" ornithology course. The recognized value of collections to these teaching functions is demonstrated by the fact that almost 75% of student respondents felt that their access to study skins was inadequate.

Priorities and needs

Myriad problems beset collections and collection-based research. Although most have their roots in the financial constraints currently facing educational and cultural institutions, others are the result of past museum practice or of increased bureaucratic interference and injudicious conservation policies. These problems can, in general, be divided into two categories: those dealing with the physical support of collections and those concerned with the human resources responsible for their maintenance.

Physical resources

One of the most important physical resources necessary to sustain active collection-based research in ornithology is simply space. Room is necessary for the storage of specimens, the equipment and staff needed to prepare and curate them, and the research carried on within them. Such space is, unfortunately, usually in short supply (King and Bock 1978). Lack of space is critical for anatomical specimens in particular. Because of an emphasis on the use of skins for research during the first half of the twentieth century, the preservation and careful storage of anatomical materials was largely neglected. Now, however, with the advent of various statistical procedures, sophisticated biochemical techniques, and new methods for the analysis of functional morphology, the demand for fresh skeletal and alcoholic material has increased. Unfortunately, the relative suddenness of this increased need for the preparation and storage of such material has left most museums without sufficient space or staff to allow for immediate expansion in this area. In addition to facing space limitations, many collections are housed in older buildings with insufficient environmental controls and only limited protection from damage by water, fire, dust, insects, or temperature extremes (King and Bock 1978). A complete solution to such problems, although of immediate and pressing concern, is unlikely in the light of present financial constraints.

As discussed in the introduction to this chapter, the past two decades have witnessed the rise of a number of new methods of data collection and analysis. However, the majority of institutions housing ornithological collections still lack most of the equipment necessary to take advantage of these techniques. Such facilities include reflectance spectrophotometry, sound spectrographs, X-ray machines, high quality light and/ or electron microscopes, and electrophoretic laboratories. Whereas access to computers is also necessary to take advantage of the revolution in methods of data analysis that has occurred during the last two decades, some of the institutions responding to the 1977 questionnaire still lacked even desk calculators (King and Bock 1978). The acquisition of such equipment, or of access to it, is a matter of highest priority. If museum collections are to fulfill their purpose adequately, investigators must be able to conduct research with the same degree of quantification and precision currently employed in other fields. The expense involved is obviously large. Sharing of particular facilities by several institutions might reduce the purchase and maintenance costs for any single museum to manageable levels.

Human resources

The most pressing problem faced by ornithological collections today involves the associated human resources (King and Bock 1978). At all levels, from curators to technicians and secretarial staff, there is a great need for increased numbers of qualified personnel. Augmented assistance with the secretarial, clerical, and library work associated with the maintenance of research collections is considered a major need by many institutions. Personnel shortages are particularly critical with respect to paraprofessional staff. Serious backlogs in specimen preparation, cataloging, and loan processing occur as a result of insufficient technical staff. The need for additional technical staff with advanced qualifications has escalated in the past decade. The advent of new methods for the collection and analysis of data has necessitated training people to preserve specimens suitable for biochemical research, electron microscopy, and detailed anatomical studies, as well as for traditional studies based on skins. At least some technical personnel must now be capable of operating the sophisticated instrumentation and machinery pertinent to these new techniques.

An issue of almost equal magnitude, however, involves the shortage of curatorial personnel. The number of professional ornithologists with a primary interest in collection-oriented fields is declining rapidly. Of the 48 Ph.D.-level scientists employed by institutions surveyed (King and Bock 1978), only 29% received their doctorates during the 1970s, whereas the remainder graduated prior to that. From among this remaining 71%, all but one listed their major research interests or speciality as one of the collection-oriented disciplines: systematics, evolution, zoogeography, anatomy, or paleontology. Conversely, of those who received their Ph.D.s during the 1970s, 43% considered their training and primary interest to be in ecology, ethology, or other fields that may not make primary use of museum resources. Among the ornithological dissertations completed in the last 5 years, only 7.2% were in the traditionally museum-related fields of systematics, anatomy, and paleontology, whereas 79% dealt with ecological or ethological problems. The trend toward a decline in professionals whose research involves museum specimens and who would, therefore, be qualified and interested in the proper care and expansion of collections poses a serious threat to the future of such collections (King and Bock 1978; Steussy and Thomson 1981).

There appears to be no basic lack of interest in systematics and other museum-oriented fields; 25% of undergraduate respondents to the Ornithological Education Panel of the workshop listed such topics as their primary area of interest. The importance and use of collections should escalate in the future in view of proposals for major new studies in the tropics, as well as an increased demand for environmental impact assessments (Steussy and Thomson 1981).

The nature and scope of this problem is summed up in the report of the Panel on Systematics Collections:

If collections continue to suffer from underfunding as well as from a limited supply of, and employment opportunities for, trained systematists and collection-oriented ornithologists, the very existence of these collections is in jeopardy. . . . Museums, particularly university museums must train an adequate number of Ph.D. and paraprofessional students to keep collections supplied with appropriate staffing. . . . The greatest immediate need of systematic collections in ornithology is the augmentation of curatorial staffs. This is particularly true of paraprofessional positions but also includes secretarial, clerical and librarian assistance and professional staff. Without additional personnel, it will not be possible to maintain collections properly or to serve the user community. (King and Bock 1978)

An additional drain on museum resources and a matter of concern to many professional ornithologists is the increasing profusion of state, provincial, and federal laws regarding wildlife. Although efforts to preserve natural resources are, in general, commendable, many recently promulgated regulations seriously inhibit collection for the purposes of scientific inquiry. In fact, such laws may be detrimental to conservation efforts in the long run. It is an incontrovertible fact that specimens of birds are as essential to research in some fields as are wild, free-living populations. The removal of the small numbers (compared to that removed by hunting or as a result of natural mortality) of birds needed for scientific research has no effect on the overall present or future size of avian populations (Aldrich et al. 1975). Only about 0.00021% of the annual avian mortality in the United States, for example, is the result of taking birds for scientific and educational purposes (King and Bock 1978).

Early legislative acts to protect birds realized the importance of scientific studies and included clauses to encourage their continuance. Recently, however, the increase in public awareness of environmental problems has led to promulgation of a large number of new regulations and has fostered the development of a restrictive attitude toward resource protection. The original intent of the early conservation laws has thus often been lost. The rapid expansion of political structures to interpret and enforce the wildlife preservation acts has occurred largely without reference to, or consultation with, the scientific community. This explosion of bureaucracy has widened the gap between those who administer the laws and the field workers who must cope with them, to the point where scientific inquiry is hampered significantly.

The once and future science

In the past, groups of systematists from many disciplines trooped into the field together, surveying regional biotas, collecting specimens, recording natural history information, and generally sharing equipment, food, sites, and expenses, as well as experiences. The field investigations of the California Academy of Science in the late nineteenth century and the first decades of the twentieth century were classics of this sort. Later, university museums sponsored short-term data-gathering trips, staffed mainly by graduate students, that visited areas where sampling biota afforded interesting insights into the geographic variation, distribution, and interactions of animals. The museums at the University of California (Berkeley), the University of Kansas, and the University of Michigan were among the leaders in this practice.

These activities fostered in microcosm a cooperation among biologists that must be embraced today and in the future with ever greater vigor. Burgeoning bureaucracy at every level, dwindling funds for research, and a new protective attitude toward antiquities and biota have in large measure dampened the cooperative spirit between systematic ornithologists and other scientists. The difficulties faced by

teams of biologists in obtaining the requisite permits within a reasonable period so that a coordinated field effort can be mounted, now
tend to frustrate even the most tenacious field workers. In this regard, the present and future are as one: Once we accept that the
economic and bureaucratic climate are not going to ameliorate in the
next few decades, it becomes possible to reorganize priorities and
develop a new approach to maximizing the opportunities to acquire
specimens and data whenever an opportunity presents itself. To this
end, institutionally based ornithologists need to share their good for-
tunes when they obtain access to contacts and field opportunities with
scientists in other disciplines. In the spirit of past decades, ornithologists must plan cooperative ventures that reduce finances and bureaucracy to a minimum and yield the largest return in data for the
smallest expenditure of resources. Cooperation among institutions
perforce must extend beyond the sharing of personnel and field opportunities to include common use of major equipment, for example,
electron microscopes, biochemical laboratory facilities, computer hard-
and software, and spectrographic equipment. Such sharing is predi-
cated upon the idea of reducing costs. The proximity of institutions,
common interests of research, and expertise of investigators are thus
some of the prominent factors that should facilitate professional altruism of this nature.

The need to maximize returns also has implications for the number
of specimens collected and the way in which they are preserved. It is
clear that random specimen acquisition is no longer practical and of
uncertain reward. It follows that collecting has become goal oriented,
centered on procurement of specimens for specific research projects.
Sample size can now be predetermined – computed to anticipate requirements of specific statistical procedures. Unless the intent of an
investigation dictates otherwise (e.g., anatomical studies based on material preserved whole in spirits), individual birds should be preserved as
pelts (one wing, one leg, and the tail retained as part of the pelt, and
the beak carefully measured at the time of preparation) and partial
skeletons (lacking only the distal elements of the wing and leg left in
the pelt). Viscera, including stomachs from which the contents should
be removed, identified, and stored, and tissues for biochemical assessment should be retained as well. Studies that have employed this preparation format allow measurement of the covariation of genomic and
morphometric characters. Again, in the face of the fiscal and bureaucratic limits of the forseeable future, there seems no other way to

proceed than one that maximizes the informational content of individual specimens.

A boon to systematic studies has come from the recent computerization of specimen data. Across-institution summaries of anatomical material indicate the location and numbers of various taxa, as well as sexes, and their distribution, thus indicating the strengths and weaknesses of existing collections. Similar summaries of study skin holdings facilitate generation of lists by geographic unit, taxa, or age both among and within collections. National and international computer linkups allow or will allow rapid searches of the holdings of many museums simultaneously. For poorly represented species, condition reports of existing material represent a refinement that will eliminate borrowing of material with damaged or missing characters while at the same time indicating the need for fresh collecting or allowing investigators to use alternate taxa where possible.

Many authors have highlighted the importance of studies in the American tropics (see Keast and Morton 1980). Here museums will continue to play a vital role to chronicle distribution, abundance, and variation, and, in conjunction with ecologists and ethologists, measuring parameters of ecology and behavior vital to the preservation of habitats critical to the survival of both resident and migratory species.

Human land and resource use patterns have influenced the ecology of the Americas since the end of the Pleistocene. Certain areas have had an altered ecology for the past 10,000 years (Streuver and Holton 1979). Humans have altered riparian and adjacent forest patterns at specific sites for intervals of 200–400 years until the 1600s. By the mid-nineteenth century, clearing of land for agricultural purposes affected distributional patterns of native birds in North America in general. Today habitat attrition in conjunction with a general environmental deterioration has effected many changes in population size and species distributions. Accumulation of collections in North America has in some measure marked these changes. Faunal changes over time can be useful predictors of the future. Now largely untouched regions of Latin America are under assault (Terborgh 1980). The responsibility of systematists to document and quantify the distributions, migrations, and ecology of birds that are in a constant state of flux as avian habitats are altered must extend to the tropical Americas (Tramer 1981). Implementing measures to fulfill this responsibility is only one of the ways to show our resolve to realize the dynamic potential inherent in collection-based research.

Literature cited

Aldrich, J. W. 1946. Significance of racial variation in birds to wildlife management. *J. Wildl. Manage. 10:*86–93.

– 1972. A new subspecies of Sandhill Crane from Mississippi. *Proc. Biol. Soc. Wash. 85:*63–70.

Aldrich, J. W., et al. 1975. Report of the American Ornithologists' Union *ad hoc* committee on scientific and educational use of wild birds. *Auk Suppl.* 92(3), 1A–27A.

Alexander, E. P. 1979. *Museums in motion.* Nashville, Tenn.: American Association for State and Local History.

Anderson, D. W., and J. J. Hickey. 1972. Eggshell changes in certain North American birds. In *Proceedings of the XV International Ornithological Congress,* pp. 514–540.

AOU Check-list Committee. 1982. Thirty-fourth supplement to American Ornithologists' Union Check-list of North American birds. *Auk Suppl. 99*(3), 1CC–16CC.

Austin, G. T., et al. 1981. Ornithology as science. *Auk 98:*636–637.

Banks, R. C., and R. C. Laybourne. 1968. The Red-whiskered Bulbul in Florida. *Auk 85:*141.

Barlow, J. C. 1973. Status of the North American population of the European Tree Sparrow. *Ornithol. Monogr. 14:*10–23.

– 1980. Adaptive responses in skeletal characters of the New World population of *Passer montanus.* In R. Nöhring (ed.). *Acta XVII Congressus Internationalis Ornithologici,* pp. 1143–1149. Berlin: Verlag der Deutschen Ornithologen-Gesellschaft.

Bowman, T. E. 1961. The copepod genus *Acartia* in Chesapeake Bay. *Chesapeake Sci. 2:*206–207.

Carriker, M. R. 1977. The crucial role of systematics in the identification and control of pollution. In *National estuarine pollution and control assessment, proceedings of a conference.* Washington, D.C.: Government Printing Office.

Cracraft, J. 1981. Toward a phylogenetic classification of the Recent birds of the world (Class Aves). *Auk 98:*681–714.

Eisenmann, E. 1955. The species of Middle American birds. *Trans. Linn. Soc. N.Y. 7:*1–128.

George, W. 1962. The classification of the Olive Warbler, *Peucedramus taeniatus. Am. Mus. Novit. 2103:*1–41.

Hayward, C. L., M. L. Killpack, and G. L. Richards. 1963. Birds of the Nevada test site. *Brigham Young Univ. Sci. Bull. Biol. Ser. 3*(1):1–27.

Howard, H. 1929. The avifauna of Emeryville Shellmound. *Univ. Calif. Publ. Zool. 32:*301–394.

Johnston, D. W. 1975. Ecological analysis of the Cayman Island avifauna. *Bull. Fla. State Mus. Biol. Sci. 19:*235–300.

Johnston, R. F., and R. K. Selander. 1971. Evolution in the House Sparrow. II. Adaptive differentiation in North American populations. *Evolution 25:*1–28.

Karr, J. R., and F. C. James. 1975. Eco-morphological configurations and convergent evolution in species and communities. In M. L. Cody and J. M.

Diamond (eds.). *Ecology and evolution of communities*, pp. 258–291. Cambridge, Mass.: Belknap Press.

Keast, A., and E. S. Morton (eds.). 1980. *Migrant birds in the neotropics: ecology, behavior, distribution and conservation. Symposium of the National Zoological Park.* Washington, D.C.: Smithsonian Institute Press.

King, J. R., and W. J. Bock. 1978. *Workshop on a national plan for ornithology, final report.* Washington, D.C.: National Science Foundation/The Council of the American Ornithologists' Union.

Klaas, E. E., H. M. Ohlendorf, and R. G. Heath. 1974. Avian eggshell thickness: variability and sampling. *Wilson Bull. 86:*156–164.

Lack, D. 1976. *Island biology illustrated by the land birds of Jamaica.* Oxford: Blackwell Scientific Publications.

Lee, W. L. 1977. The San Francisco Bay Project: a new approach to using systematics. *Assoc. Systemat. Collect. Newslett. 5:*15–17.

Long, J. L. 1981. *Introduced birds of the world.* New York: Universe Books.

McKitrick, M. C. 1981. Old specimens and new directions: a comment. *Auk 98:*196.

National Research Council, Committee on Research Priorities in Tropical Biology. 1980. *Research priorities in tropical biology*, pp. xii, 1–116. Washington, D.C.: National Academy of Sciences.

Olson, S. L. 1981. The museum tradition in ornithology – a response to Ricklefs. *Auk 98:*193–195.

Parkes, K. C. 1963. The contribution of museum collections to knowledge of the living bird. *Living Bird 2:*121–130.

Pitelka, F. A. 1951. Speciation and ecologic distribution in American Jays of the genus *Aphelocoma. Univ. Calif. Berkeley Publ. Zool. 50:*195–464.

Platt, D. 1956. Food of the Crow, *Corvus brachyrhynchos* Brehm, in south-central Kansas. *Univ. Kansas Publ. Mus. Nat. Hist. 8*(8):477–498.

Raikow, R. J. 1978. Appendicular myology and relationships of the New World nine-primaried oscines (Aves:Passeriformes). *Bull. Carnegie Mus. Nat. Hist. 7:*1–43.

Ricklefs, R. E. 1980. Old specimens and new directions: the museum tradition in contemporary ornithology. *Auk 97:*206–207.

Rising, J. D. 1970. Morphological variation and evolution in some North American orioles. *Syst. Zool. 19:*315–351.

Schoener, T. W. 1965. The evolution of bill size differences among sympatric congeneric species of birds. *Evolution 19:*189–213.

Snow, D. W. 1956. The annual mortality of the Blue Tit in different parts of its range. *Br. Bird 49:*174–177.

Streuver, S., and F. A. Holton. 1979. *Koster: Americans in search of their past.* Garden City, N.Y.: Doubleday, Anchor Press.

Stuessy, T. F., and K. S. Thomson (eds.). 1981. *Trends, priorities and needs in systematic biology; A report to the Systematic Biology Program of the National Science Foundation.* Lawrence, Kans.: Association of Systematics Collections.

Terborgh, J. W. 1980. The conservation status of neotropical migrants: present and future. In A. Keast and E. S. Morton (eds.). *Migrant birds in the neotropics. Symposium of the National Zoological Park*, pp. 21–30. Washington, D.C.: Smithsonian Institution Press.

Tramer, E. 1981. Response: in the eye of the beholder. *Auk 98*:638.

U.S. Interagency Task Force on Tropical Forests. 1980. *The world's tropical forests: a policy, strategy, and program for the United States*, Department of State Publication 9117, pp. 1–53. Washington, D.C.: Government Printing Office.

White, C. M., and W. M. Baird. 1977. First North American record of the Asian Needle-tailed Swift, *Hirundapus caudacutus. Auk 94*:389.

3   On the study of avian mating systems

DOUGLAS W. MOCK

At the founding of the American Ornithologists' Union, a list of scientific priorities would not have included the study of bird mating systems. However, natural history anecdotes on avian breeding habits were accumulating as incidental by-products of more pressing field activities. For example, in the first issue of the new journal, *The Auk*, J. W. Banks (1884) documented male parental investment in the Broad-winged Hawk, *Buteo "pennsylvanicus"(=platypterus)* as follows: "The male parent was sitting on the nest at the time I approached it, and when I began to climb the tree, he flew to a bough some seventy yards off, where he was shot. His stomach contained the partially digested remains of three unfledged thrushes."

In the century since, numerous developments have shifted ornithological priorities away from basic collecting and toward increasingly theoretical issues. The maturing of Darwinism and its marriage to Mendelian genetics produced "neo-Darwinism," which begot a family of new disciplines, such as ethology, population biology, and sociobiology (Wilson 1975; Lloyd 1979). The male Broad-winged, immortalized above (and presumably as a specimen), might now be remembered primarily for his paternal contributions of nest guarding, brooding, and thrush hunting.

On a modern theoretical level, sexual reproduction itself is seen as a paradoxically inefficient means of replicating genes; yet it is a system that a great many organisms use (or are stuck with; Williams 1975). The genetic cost of reduction division (meiosis) in gametogenesis is unavoidable and the return to diploidy generally requires the breeding individual to cooperate with a member of the opposite sex, at least enough to produce zygotes. Moreover, one sex (females; defined as the producers of large zygotes) is a coveted resource for the other (males, producers of tiny zygotes), which generates an endless variety of sexual competition, attraction, and reciprocity (Alexander and Borgia 1979). Just how this is arranged, and the consequences of the sexual struggles,

55

fall under the rubric of mating system research, which is itself a subset of Darwin's theory of sexual selection.

Sexual selection can be defined as the within-sex variance in mating success that results from intrasexual competition and/or opposite-sex choosiness (cf. Trivers 1972; Wade and Arnold 1980). More simply, it can be characterized as "when one sex becomes a limiting factor for the other" (Emlen and Oring 1977). Typically, females are the limiting sex and males compete among themselves to obtain as many females as possible (or as many as they can afford), such that nearly all females reproduce roughly equally, whereas some males enjoy great success and others fail (Bateman 1948). Clearly, the major preoccupation of each male is to be among the relatively successful.

The social manifestations of sexual selection are so incredibly diverse that a distinction is frequently made between its two processes. *Epigamic* (or *intersexual*) *selection* refers to the ways in which either sex, but predominantly the limiting sex, exerts its powers of preference: This is held primarily responsible for extremes in male adornment (e.g., the colossal antlers of the Irish Elk and the flamboyance of bird-of-paradise plumage). By contrast, *intrasexual selection* refers to the competition among males (less commonly among females) and leads to such anomalies as extreme sexual dimorphism in size (e.g., elephant seals; LeBoeuf 1974), modification of male genitalia to remove the gametes of other males from the female tract (damselflies; Waage 1979), and male worms that forcibly seal over the sexual orifices of other males with glue (Abele and Gilchrist 1977). Mating system research recombines those two selection processes in a methodical consideration of the behavioral strategies used to obtain mates (Emlen and Oring 1977). Thus, mating system study includes the behavior of mate acquisition, the reproductive success achieved, the nature of social bonds between mates (if any), and the division of any parental care between the mates.

Ornithological research has played a uniquely important role in the development of mating systems ideas (see Emlen 1981a), beginning with the pioneering comparative work of Crook (1964) and Lack (1966, 1968). The enormous preexisting literature on avian natural history was a rich source of material for preliminary testing of ideas (e.g., see Verner and Willson 1969) and quickly allowed synthesis and formal hypothesizing about how ecology and mating systems are related (Verner 1964; Verner and Willson 1966; Orians 1969a). The reader is referred to a large number of recent reviews on the facts and theories of avian mating systems, which will be mentioned here only in passing (see Crook 1964; Lack 1968; Orians 1969a; Wilson 1975; Emlen and

Oring 1977; Wittenberger 1979, 1981; Wittenberger and Tilson 1980; Krebs and Davies 1981; Oring 1982; Handford and Mares, in press).

In this chapter, I wish to explore why ornithology was so pivotal in the development of this exciting new field. Specifically, I want to scrutinize the biological attributes of birds (relative to other animals) and their convenience for various research approaches to the topic of mating systems. In addition, I will address the issue of genetic paternity, which is the most crucial hurdle lying in the path of future progress in avian mating system research.

Background and terminology

Four major categories of mating systems are widely recognized, each of which can be divided further according to ecological (Emlen and Oring 1977) and/or temporal classifications (Wittenberger 1979, 1981). Attempts to rename these standard categories (e.g., Selander 1972) have failed to gain acceptance, indicating that the current usages are stable for the time being. Here I present the systems in order of increasing male investment, which reflects both the historical and biologically relevant fascination with that sex's relative economy of contribution: promiscuity, polygyny, monogamy, and polyandry.

Promiscuity

Though this term is something of a misnomer insofar as it connotes a lack of choosiness in sexual partners, it is commonly used for polygynous systems in which no pairbonds form, and males contribute only gametes to the production of offspring. Such species are usually precocial (exceptions include birds of paradise, manakins, cocks of the rock, and some brood parasites). In many species, males congregate in traditional display arenas (leks), where they are visited by females seeking inseminations. There are many hypotheses for how lek behavior evolved (e.g., Wiley 1974; Wittenberger 1979), mostly centering on the issue of whether or not females are "imposing" that pattern on males so as to gain opportunities for mate choice (see review by Bradbury 1981).

Polygyny

Polygyny occurs when one male mates with at least two females, more or less concurrently. The essential distinction from promiscuity is that the polygynous male has some kind of social relationship with the females and contributes postzygotically to the offspring, whereas the promiscuous male does not. Admittedly, the male's continuing support may be highly indirect (such as territorial defense that vouchsafes sufficient food) and may be distributed unevenly among his different broods.

Inasmuch as paternal investment is limited, the addition of more females and broods is assumed generally to increase competition for male care, and perhaps other limited resources as well (e.g., food) among the females. Most discussions, therefore, assume that resident females suffer significant reductions in their reproductive success whenever a new female joins the harem (Orians 1969a). Polygyny is seen as a system profiting the resource-controlling males, but to some extent imposed on females. For the prospecting female, joining a harem is presumed advantageous only when some facet of that particular male more than compensates for her having to share his help. The *polygyny threshold model* introduced this point, proposing that differences in male-controlled resources (territory quality) could be the currency of that compensation (Verner 1964; Verner and Willson 1966; Orians 1969a). Alternatively, Weatherhead and Robertson (1979) argued that genetic differences between males could provide the compensation by improving the reproductive value of male offspring (these "sexy sons" are assumed to inherit some of their father's successful attributes).

The logic of the polygyny threshold model is based on game theory and individual selection. The joining female must choose the lesser of two evils (e.g., monogamy with a pauper vs. polygyny with a tycoon) whenever males can monopolize resources critical to her. The system need be beneficial to the female only in relation to her true alternatives and her ability to monitor those alternatives accurately (Lenington 1980).

Recently, several authors have challenged the assumption that female reproductive success is depressed under polygyny, arguing instead that it may be unaffected (Nolan 1978; Carey and Nolan 1979) or actually increased (Altmann et al. 1977; Lenington 1980; Searcy and Yasukawa 1981). Pleszczynska and Hansel (1980), however, show how single-season measurements can be misleading indicators of the female's lifetime reproductive success. The issue is clouded further by documentation that the male's contribution to secondary broods may vary with the relative harshness of ecological conditions. For example, a male Bobolink (*Dolichonyx oryzivorus*) may assist only his first mate when ecological conditions are severe but contribute to secondary broods when conditions are more benign (Wittenberger 1980). Similarly, male Red-winged Blackbirds (*Agelaius phoeniceus*) in the Midwest commonly provide parental care, but males in the far western parts of the species' range seldom do so (Payne 1979).

These refinements notwithstanding, the influence of the polygyny threshold model has been extraordinarily important in shaping our view of mating systems, both in birds and in other taxa (e.g., Downhower and Armitage 1971; Downhower and Brown 1981). Within polygynous birds, various predictions from the model have been subjected to field tests (e.g., Verner 1964; Zimmerman 1966; Verner and Engelsen 1970; Martin 1974; Wittenberger 1976; Pleszczynska 1978; Carey and Nolan 1979; Garson 1980; Patterson et al. 1980; Lenington 1980), and its logic has been extended to nonpolygynous mating systems as well (e.g., Emlen and Oring 1977; Graul et al. 1977; Wittenberger and Tilson 1980; Gowaty 1981a).

Monogamy

Monogamy may be defined as "a prolonged association and essentially exclusive mating relationship between one male and one female" (Wittenberger and Tilson 1980). This definition embraces considerable social diversity under the general heading of monogamy. Yet, despite the system's widespread occurrence and its importance for *Homo sapiens*, there has been remarkably little research devoted to the exploration of that diversity.

The neglect of monogamy – presumably because the manifestations of sexual selection are less extreme under such reproductive balance – is particularly unfortunate for ornithology. An estimated 91% of all avian taxa are primarily monogamous (Lack 1968), making birds by far the most monogamous of organisms. Understanding monogamy is critical to understanding avian sociality, and, conversely, birds are the taxon of choice for the study of this mating system.

Avian monogamy is also a potential showcase for examining the balance of cooperation versus selfishness between two (usually unrelated) individuals that must work in concert. Contributing to the confusion is a held-over term, pairbond, that has focused inordinate attention on the mutualistic side of male–female coevolution. Dawkins and Krebs (1978) present a similar discussion of how classical "mutualistic" definitions of communication have biased the study of that phenomenon toward group-selectionist interpretations.

The descriptive term pairbond has endured tenaciously, without even an accepted operational definition, and apparently has misled many. For example, in his discussion of pairbonds, Lack (1968:29) contended that "almost certainly, the main advantage of monogamy is that two parents can feed the young and hence can raise more off-spring than if the female does so alone." The question follows, advantage to whom? It is apparent that Lack meant to the pair as a unit. Many contemporary biologists would draw a similar conclusion, but from a different theoretical route – one that focuses on the male's "decision." Thus, Lack's statement would be modified to emphasize that biparental care must be so essential that the male gains more in the long run by investing heavily in his first brood than if he channeled the same time and effort to siring additional broods. Alternatively, his decision might be shaped by reduced availability of fertilizable females or by various ecological and phylogenetic factors (Emlen and Oring 1977) that severely limit the potential value of his parental care (e.g., inability to lactate). In short, his decision hinges on which of his true reproductive options offers the best payoff (Maynard Smith 1977). Though his aid might enable "the pair" to double their season's output, he is expected to choose a nonmonogamous behavior if doing so triples his season's output.

If even monogamous mates are best viewed as selfish entities, the potential array of how they can practice mutual reproductive exploitation is huge. Of course, this cynical view of monogamy does not preclude cooperation and generosity between mates in any way. It merely qualifies that such tendencies are expected only when they fit the genetic interests of the individuals involved. In this light, the pairbond is seen more as a "grudging truce," in which acts of generosity are regarded with mild skepticism.

A great variety of questions that have not been covered well in the extant literature are raised by this view. There are virtually no data, for example, about how an individual's *share* of the total parental care varies with age or mate changes. We know little about how different pairs in a population vary in the division of parental labor, much less why (but see Pierotti 1981). Also, the evolutionary significance of permanent (or regularly renewed) pairbonds has received relatively little research attention. In two gull studies (Coulson 1966; Mills 1973), reproductive success was higher in individuals that retained the same partners than for those mating with new partners but in a similar study with geese, no strong correlation between mate retention and reproductive success was found (Cooke et al. 1981). In addition, long-term bonds may have numerous advantages outside the breeding season (Scott 1980), but this has been little studied.

Despite a paucity of discussion, there is ample reason for expecting sexual selection to operate routinely, if subtly, in monogamous birds. Many factors that contribute to within-sex variance in reproductive success are bona fide sexual selection pressures. For example, if territory ownership is required for breeding and there is a dearth of territories (e.g., Nolan, 1978), then selection will favor phenotypic traits conferring ownership. Removal experiments demonstrating the existence of a socially disenfranchised ("floater") subset of the population – that will breed if given the chance (reviewed in Brown 1969) – attest to the pervasiveness of sexual selection in monogamous birds. Any time there are reproductive have-nots competing for breeding opportunities, intrasexual selection must be present.

Beyond simple social exclusion of floaters, other mechanisms can increase the intensity of sexual selection in monogamy. If certain territories allow larger clutches, greater success in renesting or higher fledging rates than others, competition can focus on these differences in territory quality.

Similarly, epigamic selection can operate on differences in mate quality (Darwin 1871). O'Donald (1980a,b) has shown that early-breeding female Arctic Skuas (*Catharacta* [=*Stercoraria*] *skua*) are more successful than later-breeding females and that resulting patterns of assortative mating can produce significant effects of sexual selection on the population. Assortative mate choice within monogamy has also been demonstrated with captive pigeons (Burley 1977). Finally, extra-pair copulations and consequent sperm competition may allow sexual selection to operate in apparently monogamous birds. The fact that these effects are less exaggerated than in nonmonogamous mating systems should not lead to their being overlooked entirely.

Polyandry

In polyandry, a female mates with two or more males, which then perform most of the parental care (Jenni 1974). Thus, a complete reversal of the sex roles established by anisogamy is found. Not surprisingly, polyandry is extremely rare in birds and other vertebrates. As in the riddle of monogamy, the generous contribution of male parental investment requires ecological explanation. Though we may never know for certain what historical conditions led to sex-role reversal and on to polyandry, two plausible hypotheses have been advanced. According to the fluctuating food hypothesis (Graul et al. 1977), periods of relatively low food availability could promote parental role reversal if female energy reserves were depleted drastically by egg production. Faced with an exhausted mate, the monogamous male's best reproductive option might easily have been to increase postzygotic care substantially while the female recuperated (a response that sometimes occurs in monogamous gulls; Pierotti 1981). Over many low-food generations, males could become adapted to a binding high-investment strategy (monogamy with "reverse" dimorphism and roles; e.g., Howe 1975). From that stable equilibrium, a second shift to relatively high-food conditions would give the female ecological options, including the production of more eggs in a second nest, which she might tend ("double-clutching") or which she might leave to a second male (biandry). The point is that once the tables are turned and males are specialized as heavy contributors of parental investment, "phylogenetic inertia" (Wilson 1975) can lift the constraints of anisogamy from females and place them in a rare position of holding multiple mates simultaneously.

Alternatively, Maxson and Oring (1980) have proposed that the essential role reversal could have arisen from high predation pressure on nests that forced the male into parental duties so his mate could double-clutch, thus doubling both mates' probability of fledging young.

Simultaneous polyandry is highly developed in American Jacanas (*Jacana spinosa*), where even the important issue of paternity may be flagrantly clouded by the female's behavior. Jenni and Collier (1972) reported one female copulating with three of her mates within a 20-minute period, suggesting that those males are, in effect, forced to care for genetically unrelated progeny (but presumably lacking a better option).

Getting back to the first step of role reversal, it is striking that all polyandrous birds have highly precocial young (Lack 1968). This is consistent with the view that role reversal can occur only when uniparental care is not merely possible but relatively easy.

Pigeon holing and mating system classifications

As in any system of categorization, information is lost when mating systems are assigned to type. There are many birds, for example, that breed in ways that defy the definitional boundaries presented here (e.g., Nolan 1978). Some ratites practice what Wittenberger (1981) calls polyandry–polygyny: The male mates with all members of a harem, which lay communally in a single nest. Those females then move on to another male while the first provides all postzygotic care of the precocial young (Lancaster 1964; Bruning 1974; Handford and Mares, in press). In another variation, the Tasmanian Native Hen (*Tribonyx mortierii*) exhibits a system of polyandry in which the harem males are full siblings (Maynard Smith and Ridpath 1972). Cooperation among brothers has also been postulated in the Rio Grande Turkey (*Meleagris gallopavo*), in which harems of females are successfully courted by pairs or trios of males (Watt and Stokes 1970). There is, however, no direct evidence showing that these male turkeys are actually kin (Balph et al. 1981). Finally, the classifier encounters difficulty when confronted by the breeding "trios" or female–female pairs reported for several gull species (Hunt and Hunt 1977; Ryder 1978; Ryder and Somppi 1979).

Exceptional patterns aside, the actual assignment of a species to a mating system is seldom straightforward anyway. In fact, there seems to be a bias toward the exotic, such that a species showing any departure from monogamy is likely to be listed quickly in the nonmonogamous category. For example, in some populations an individual "polyandrous" Spotted Sandpiper (*Actitis macularia*) is more likely to breed monogamously than as part of a simultaneous polyandry set (Oring and Knudson 1972; Hays 1972; Oring and Maxson 1978). There is no accepted standard for how behaviorally polygamous a species (or population) must be to escape assignment to monogamy. The point is that birds may well be more monogamous than the 91% figure that a taxon-by-taxon analysis would suggest (Lack 1968). Even here I must hedge because genetic monogamy need not accord with social monogamy (Gowaty 1981a), so all those birds I just pushed over into monogamy may or may not reproduce that way. The point is that this whole discipline suffers greatly from vague, nonoperational terminology, as illustrated by pairbond and by the mating system categories themselves.

The paternity issue

The inevitable hiatus between theory and measurement is particularly exaggerated in the study of avian mating systems. Historically, this stems from the marriage of theoretical population genetics logic (e.g., Fisher 1958; Williams 1966) to descriptive "data" emanating from the field naturalists' tradition in classical ornithology. In addition, and potentially much more frustrating, the theoretical requirements are virtually impossible to meet in the field. We simply have no way of directly measuring genetic fitness.*

Generally, field researchers have worked around this problem by

---

*Gary Schnell once quoted a remark that I've come to think of as *Selander's law:* "You can't measure fitness in the field, you have to take it into the lab." This must be linked to *Selander's corollary:* "You can't measure it in the lab, either."

assuming that "fitness," which is defined variously and abstractly in terms of contributions to the gene pool, can be estimated reasonably by measuring reproductive success. Unfortunately, there is little agreement on how (and, specifically, when) that should be measured. Howard (1979) clarified the situation considerably by creating a chronological hierarchy of ERS (estimated reproductive success). He recognized six distinct and nonoverlapping periods at which ERS might be measured in a single breeding effort. Primary ERS includes the presence or absence of suitable sexual partners, basically asking how many opportunities the individual has to mate. Secondary ERS refers to the actual number of copulations he or she obtains. Tertiary ERS is defined as the number of zygotes produced per individual, and so on. The full scheme is summarized in Figure 3.1, which shows that even 6° ERS (number of offspring from this breeding effort that survive to sexual maturity) is but an arbitary measurement point leading to consideration of how the next generation fares. Obviously, each step in this progression increases the validity of conclusions drawn from the data, but each is harder to measure—and, therefore, less accurate—in the field. One of the great virtues of birds for the study of mating systems is that ornithologists routinely measure to the 5° ERS level and sometimes well beyond (see the next section).

Lest we grow complacent, though, it must be pointed out that although 5° ERS measurement is satisfyingly commonplace for female birds, males usually are measured only to the 2° ERS. That is, we record female fitness in terms of how many young survive to fledging or independence but are quite lucky even to see how many copulations each male obtains. Instead of measuring paternity, we assume it. The data, however, are frequently treated as essentially equivalent for subsequent discussions of how sexual selection affects behavior. Obviously, our extrapolations of the field data lack sufficient resolution for confident analyses of within-sex variance in reproductive success.

Inasmuch as mating system research is based on sexual selection theory, the accurate assignment of paternity stands out as the single most important hurdle to be crossed. Progress will be quite limited until this profoundly technical problem is solved routinely and accurately in the field. Otherwise, the work will be confounded by uncertainty. Even the concept of "monopolizability of females" loses much utility in the light of the discovery by Bray et al. (1975) that 69% of the eggs laid by harem females of vasectomized male Redwings were fertile.

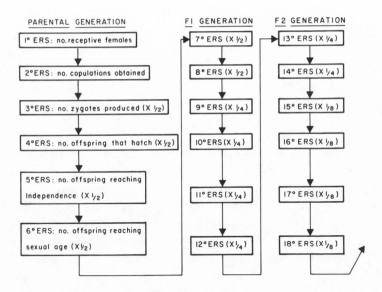

Figure 3.1. Summary of Howard's (1979) scheme for calculating estimated reproductive success (ERS) in natural populations of sexual, diploid vertebrates.

Bases and consequences of sperm competition

Though seemingly impoverished relative to insects in their special adaptations for sperm competition (Parker 1970), birds must cope with that pressure as a consequence of internal fertilization. The simple fact that viable sperm from different males can be introduced in close succession into the female reproductive tract makes sperm competition inevitable. Sperm need only be in the oviduct during the brief period following ovulation and preceding albumen deposition – a "time window" of about 75 minutes (Sturkie 1976). The additional abilities of sperm to remain viable there for an extended period accentuates the competition. Elder and Weller (1954) showed that domestic Mallard (*Anas playtyrhynchos*) hens could produce fertilized eggs 17 days after being isolated from all males, though the sperm's viability decreased sharply after 6 to 8 days. Zenone et al. (1979) reported an effective period of nearly 6 days for sperm in Ring Doves (*Streptopelia risoria*). Relatively little is known about the function of uterovaginal "sperm-host" glands in birds (see Compton et al. 1978).

The social consequences of sperm competition in birds are intensified by the high male postzygotic investment found in most species (Trivers 1972; Gladstone 1979; Burger 1981). A male that can "cuckold" another is, in effect, parasitizing the parental investment of that second male. Cuckoldry is thus analogous (in terms of intrasexual competition) to the theft of parental care by brood parasites like cowbirds (Payne 1977; Gowaty 1981b; Power et al. 1981). Theoretically, at least, a male knowing that his probability of fathering the offspring is low should be relatively tempted to desert (Trivers 1972) or to adjust his subsequent investment downward. This choice should depend on his reproductive alternatives.

Behavioral manifestations of sperm competition in birds

There are abundant indications that paternity is of great concern to male birds. In contrast to insects (Parker 1970) and mammals (Dewsbury and Baumgardner 1981; Baumgardner et al. 1982; Dewsbury, in press), there is no evidence of mating plugs, prolonged copulations ("tandems" that exclude other males), copulatory "ties," or sperm displacement mechanisms in birds (McKinney et al., in press-a). Instead of the morphological and physiological assurances evolved by other taxa, male birds rely more on their great avian mobility to maximize confidence of paternity through behavioral means. This usually makes the field researcher's job more difficult, again underscoring the inadequacy of using 2° ERS as the measurement of male fitness: One can seldom be sure of witnessing all copulations in which a bird participates.

In the following discussion, I will use the proposed descriptive terminology of McKinney et al. (in press-a) instead of potentially misleading terms like "rape." EPC refers to all extra-pairbond copulations and is divided into two categories: (1) FC (forced copulations) includes all EPCs in which the female appears to resist coition, and (2) PC (promiscuous copulations) for times when she does not. The use of female resistance as an operational criterion is not without problems, of course, but at least it draws attention to the female's pivotal role and potential willingness to participate in EPCs. It also dichotomizes the primary strategies that males can use, namely force (FCs) or more positive inducements such as genetic (Williams 1975) or material rewards (Thornhill 1980). Finally, McKinney et al. (in press-a) define FPC (forced pair copulation) as forceful insistence on copulation by the female's own mate, which has been reported in the literature as an anticuckoldry tactic (Barash 1977).

Forced copulations have been reported in nearly 100' avian species, including 37 species of waterfowl, and may prove to be very widespread once field observers are sensitized to watch for them (McKinney et al., in press-b). The widespread occurrence of FCs and the apparent energy that males must expend to obtain them suggests strongly that the primary function is to increase the male's fitness via extra fertilizations. Only recently, however, has there been any evidence that FCs can fertilize females. Using genetically marked captive Mallards, Burns et al. (1980) demonstrated that FCs produced broods of mixed paternity, even in the face of sperm competition.

On the purely behavioral level, though, there is considerable evidence suggesting that many male birds practice a *mixed reproductive strategy* (MRS) (Trivers 1972) – that is, mating monogamously with one primary female in whose offspring they invest postzygotically while not passing up the opportunity for additional copulations. The following five behavioral phenomena are appropriately viewed as being "consistent with" the MRS hypothesis, if not direct evidence for it. Taken together with the theoretical arguments, they seem to indicate that the issue of avian paternity should become almost as important to students of mating systems as it is to male birds.

Males are selective in FC choices. In general, males attempt FCs when the target female is fertilizable, suggesting both that the ultimate cause is male fitness and that they have some way of monitoring female laying schedules. The selectivity of male timing has been reported in wild Pintails (*Anas acuta*) (Derrickson 1977, 1978), captive Green-winged Teal (*Anas crecca*) (McKinney and Stolen 1982), and captive Mallards (*Anas platyrhynchos*) (Cheng et al. 1982, in press). Some workers have suggested that males may cue in on undisguisable details of female behavior (e.g., recognizably different flight when carrying an egg) to identify fertilizability in even unfamiliar females (Beecher and Beecher 1979; McKinney and Stolen 1982). In colonial birds, the female's laying phenology is presumed harder to conceal from neighboring males (Gladstone 1979; Fujioka and Yamagishi 1981).

Males are cautious and/or slow in mate selection. It has been suggested that males may reduce their risk of cuckoldry by delaying pair formation with females whose recent sexual history is suspect. Female Ring Doves that have been in the company of other males elicit weaker courtship responses from new males than do "untainted" females (Erickson and Zenone 1976; Zenone et al. 1979). These authors also suggest that "aggressive courtship" by the male may be a stalling tactic that reduces his risk of being cuckolded.

Males "mate guard" females during fertilization period. Field data from a growing number of monogamous bird species show that the male's attentiveness to his mate increases sharply when ova can be fertilized and drops as soon as the clutch is complete. This apparent monitoring of female activities has been interpreted as necessary for additional anticuckoldry measures, such as aggressive interference with FCs (see next paragraph). It, too, will probably turn out to be extremely widespread now that field workers are watching for it (e.g., Wolf and Wolf 1976; Hoogland and Sherman 1976; Derrickson 1977; Mineau and Cooke 1978, 1979; Birkhead 1978, 1979; Beecher and Beecher 1979; Power 1980; Power et al. 1981; Fujioka and Yamagishi 1981; Lumpkin et al., 1982; McKinney et al., in press-b).

Males aggressively oppose FCs to their own mates. In several species of waterfowl, mated males have been reported to interfere actively with FC attempts on their mates by other males. Either through vigorous assault or simply by interposing his body between the female and her attacker, the paired male can sometimes prevent FCs unless too many other males are involved (Derrickson 1977; Barash 1977, Mineau and Cooke 1978; Wishart and Knapton 1978). In Cattle Egrets (*Bubulcus ibis*), Fujioka and Yamagishi (1981) described a linear dominance hierarchy among neighboring males in the nesting colony: Only high-ranking males were able to oppose EPCs successfully.

Males reinseminate after mate's FC. Mated males witnessing FCs with their own mates have been reported to insist on an immediate copulation (and presumably a chance to compete with any sperm introduced) (see also Cheng et al., in press). This behavior has been described in both wild (Derrickson 1977) and semiwild populations of ducks (Barash, 1977, reported on Mallards in a campus fountain). Given some serious doubts about some of the behavioral criteria used and unanswered questions about the potential value of FPCs (see McKinney et al. 1978b; Hailman 1978), the first need is for additional field documentation in species that practice FPCs frequently.

Several related points deserve mention in any discussion of paternity, MRSs, and FC. One is the tricky question of female resistance versus solicitation. It is fairly simply to suggest reasons why females "should" resist and/or why they "should" invite EPCs. Quite probably, both tactics occur in some female reproductive strategies. Proposed benefits to the female for EPCs include genetic diversity within the brood (Williams 1975), genetic advantages for some or all progeny (e.g., Weatherhead and Robertson 1979), and/or material gain in the form of presented (Thornhill 1980) or available resources (cf. Wolf

1975). Conversely, EPC costs to the female range from injury to self and clutch to reduction or loss of male investment (Trivers 1972).

Documentation of actual female benefits and costs for EPCs is sparse. For the benefits listed here, only Wolf (1975) presents avian field data supporting the interpretation. Costs, however, have been documented in physical injury and death to assaulted females (Huxley 1912) and in potential damage to eggs being incubated, as when FCs occur on the nest during early laying (Mineau and Cooke 1978; Fujioka and Yamagishi 1981). Significant losses of reproductive success through FC attempts are very hard to document quantitatively in the field, simply because the events are so rare. Despite their infrequency, however, such catastrophes (death of female or loss of eggs) can constitute a potent selection pressure. There are a variety of social sanctions predicted for female infidelity, but they are either purely conjectural (e.g., male deserts unfaithful mate; Trivers 1972) or insufficiently documented (e.g., male attacks unfaithful mate; Barash 1976; see critiques by Morton et al. 1978; Gould 1978; Power and Doner 1980; Gowaty 1981b).

On balance, the descriptive literature seems to suggest that female birds are more overtly resistant than encouraging to EPCs. For example, female waterfowl flee and hide from male pursuits (McKinney 1965; Mckinney and Stolen 1982; McKinney et al., in press a,b). Bingham (1980) even describes a female Mallard choosing to perch in a tree, where males could not complete an FC attempt. These observations do not, however, refute the possibility that females may also invite EPCs, a process that might be considerably less spectacular and thus more difficult to document convincingly. Obviously, the subject deserves closer attention in the field.

In addition, there may be many situations in which the female has little or nothing at stake when approached by an ardent male. If, for example, she has already completed laying, the genetic and social consequences of infidelity may be totally relaxed, and she may do better energetically by submitting than by launching an expensive avoidance response. Alternatively, she may have good reason to believe that her mate is away, which may relax the social consequences for being caught in the act. In other words, even detailed quantitative evidence showing a lack of female resistance cannot, by itself, be interpreted safely as evidence for female willingness.

Resolution of the avian paternity problem

Presently, there are two diagnostic techniques available for assessing paternity in birds. Where genetic markers are available, specific tests of sperm competition can be designed. Probably, most of these will be laboratory or captive studies to validate specific points (e.g., Burns et al. 1980). The techniques of protein electrophoresis, however, are also applicable to field problems of paternity and are, therefore, the more generally promising route (Sherman 1981). Protein polymorphisms have been used to resolve human paternity cases for many years and have been used as the basis for recent documentation of multiple-male paternity in litters of mice (Birdsall and Nash 1973), squirrels (Hanken and Sherman 1981), and bats (McCracken and Bradbury 1977). Though long regarded as having insufficient heterozygosity for such paternity exclusion assays, many birds have been shown to have sufficient heterozygosity (Barrowclough and Corbin 1978; Chapter 7). Other genetic analyses may also be adaptable for paternity assays.

A final irony of the burgeoning cuckoldry literature is that most of it has appeared prior to the concrete demonstration of that phenomenon in any bird! Perhaps the closest case was the discovery that harem females of vasectomized male Redwings still laid fertile eggs, obviously having obtained some sperm from males other than their mates (Bray et al. 1975). However, in that study the mated males could not introduce competing sperm, which normally might constitute effective defense against cuckoldry. The rest of the published literature has consisted of behavioral descriptions interpreted around cuckoldry – usually as the most parsimonious hypothesis, but too often as the only one presented. Recently, Gowaty and Karlin (pers. comm.) detected one case of mixed paternity in Eastern Bluebirds (*Sialia sialis*) (1 out of 20 males cared for offspring not his own) for the first conclusive demonstration of what may prove to be a pervasive phenomenon in birds.

As protein analyses of paternal relatedness become increasingly common, a fascinating new topic may open as a bonus: female MRSs. As suggested earlier, cryptic intraspecific brood parasitism (Yom-Tov et al. 1974, Anderson and Eriksson 1982) constitutes a female reproductive option analogous to cuckoldry in that it involves the theft of parental care. Thus, some females may be able to adopt the MRS of laying one primary clutch in which they will invest further while also seeking opportunities to place additional eggs in other nests. Such parasitism might be done surreptitiously – in which case protein analysis could reveal that neither of the care-providing adults is a genetic parent – or with the cooperation of the (rewarded male?) nest owner. Significantly, in the analysis of bluebird relatedness mentioned earlier, Gowaty and Karlin (pers. comm.) found mixed maternity but not mixed paternity in four nests. Provocative patterns have been suspected for "dump nesting" species of waterfowl, in which many females contribute to a single nest (see Andersson and Eriksson 1982; Chronister 1982), but their frequency and importance will have to be documented electrophoretically.

The "ideal" mating systems animal

Because ornithologists have played such an important role in the development of mating systems ideas, it is fair to ask, why birds more than other groups? Is it merely an historical artifact, stemming from the relatively early development of ornithology? That is, might the conspicuousness of birds and their intrinsic allure to decades of amateurs have produced a snowballing effect in which theory came to overlie tomes of descriptive natural history? Or are there fundamental biological features of birds that make them particularly well suited to the passing of mating system questions? This latter question is crucial for present-day scientists, and especially for students, because in the long run it will dictate whether major developments continue to emanate from ornithologists or whether this lively topic will become the principal domain of mammalogists, herpetologists, entomologists, and so forth. To put it another way, are we still on the crest of the wave or have we had most of our ride already? How likely is it that the *bi*centennial celebration of the American Ornithologists' Union will include a contribution or two from this area (or its descendants)?

The approach I have chosen is based on the kind of daydreaming that sometimes intrudes on a scientist's thoughts, particularly after a sticky problem has resisted solution. If you had a magic wand and could specify the biological features of a study animal that would make your mating system research as productive as possible, what would you want? Finally, how do birds compare with this ideal animal and thus, indirectly, with other taxa?

Here I list 17 considerations as a starting point for discussion.

1. Mating system diversity within taxon. Overall, it is useful to have a great variety of mating systems exhibited by the animals under study so comparisons can be used analytically for teasing apart causal factors. As a class, birds seem overwhelmingly monogamous; one might prefer, for example, closer balance among the four main mating system categories. Compared to other animal classes, however, birds are among the most balanced of groups. Also, by having monogamy as the predominant pattern, the class is "centered" with male-biased and female-biased systems available peripherally for study. For example, though rare in birds, polyandry is virtually nonexistent in mammals and insects. Teleost fish may be the only other group with the balance and diversity of birds.

Similarly, it is convenient if subtaxa contain high diversity in mating systems because then the comparisons can be made between closely related species (holding much phylogenetic inertia constant). In the avian mating system literature, shorebirds, jays, blackbirds, ducks, and grouse in particular have been worked successfully by this comparative means.

2. Quantifiable life history parameters and accurate ERS. A major advantage of birds for the study of breeding behavior and ecology is that the season's reproductive output, the eggs and later the brood, can be counted and recounted for various measures of success. The use of a discrete and easily located oviposition site has made the collection of breeding success data routine for ornithologists (e.g., Perrins 1979).

3. Short generations. Long-lived animals are inconvenient for quantification (or estimate) of inclusive fitness, simply because field projects seldom persist for a whole lifetime. To borrow from Henry Horn's (1971) facetious operational definition of forest "climax" as "that stage of plant succession where no significant change occurs during the lifetimes of several research grants," our ideal mating system animal should not live long relative to a reasonable project length. Within limits, the shorter, the better.

One highly provocative recent study has shown how avian longevity can be accommodated if the population is highly philopatric and the scientists are patient. The long-term study of Great Tits (*Parus major*) near Oxford, with meticulous banding records over three decades, continues to pay remarkable dividends with information on how age, status, and mobility affect mate choice (e.g., Greenwood et al. 1979). Similarly, a recent analysis of how Great Tit male song repertoires influence fitness followed the effects through four generations of descendants (McGregor et al. 1981), to the 24° ERS level! Of course, had paternity been established along the way, this feat would have been even more impressive.

4. Unconstrained parental investment (PI) patterns. One of the inherent limitations of mammalian mating system diversity is that females are phylogenetically strapped by internal gestation and subsequent location into a high PI pattern. By contrast, male fish, anurans, and birds often can and do provide equally valuable parental care. In insects, parental investments by the two sexes are sometimes reversed prezygotically by the evolution of expensive and nutritious spermatophores (see Gwynne 1981), which adds considerable leverage to comparative studies.

5. Sensory perceptions very similar to those of humans. The ideal animal for study would probably not be capable of perceiving information from its surroundings undetectable to a human observer. It certainly would not rely heavily on olfaction as many mammals, fishes and insects do. In short, it should have the same "umwelt" (perceived world) as its "kumpan" (associate).

6. Observable. The observability of an animal contributes to its ease of behavioral study. My ideal animal would be terrestrial (vs. aquatic), diurnal, relatively large (e.g., heron sized), tame, colonial, and particularly fond of open habitats. It also would be relatively immobile so that daily movements could be monitored easily.

7. Discrete demes. If the species were divided up into a series of local populations with little movement or breeding between groups, a variety of sex ratio questions could be addressed by using naturally skewed or experimentally manipulated demes for comparisons.

8. Fundable. My idealism in this exercise does not extend to generous research support, so the target animal should be either extremely common (i.e., an economic pest) or extremely rare (i.e., endangered). The former condition is certainly preferable so the study design can include manipulating individuals (e.g., capture and color marking) and adjusting population compositions.

9. Recognizable individuals. Ideally, the members of the population should exhibit sufficient phenotypic variability for individual recognition without marking (e.g., Bertram 1976). The sexes should be distinguishable (cf. Burley 1981a) as should various categories of age (e.g., Orians 1969b; Recher and Recher 1969) or social role (e.g., Hogan-Warburg 1966). Furthermore, individuals should be easily captured for color marking. It is important also that the color marking itself have no effect on mating choice (cf. Burley 1981b) or other social functions (Watt 1982).

10. Polymorphic blood proteins. If genetic relatedness can be ascertained with a minimum of trauma to the animals, many conclusions about mating systems can be strengthened enormously (Birdsall and Nash 1973; Foltz 1981; Sherman 1981; this chapter). Though there are ways of taking small muscle biopsies for electrophoretic assay (Baker 1981), blood is easier. Of course, a truly ideal animal should probably have sufficient visible genetic "markers" to make protein analysis unnecessary (Bateman 1948).

11. Extensive background literature. Although our primary concern here is future mating system research, there are many times when seemingly unrelated and incidental information on the natural history of a species, or its relatives, competitors, predators, and food is extremely valuable. Ornithologists are particularly blessed with a huge literature and active amateur support that has existed for a century or more.

12. Optimal breeding synchrony. The populations under study should breed in enough synchrony to allow a discrete field season, but not so explosively as to be unrecordable (e.g., if everyone mates in the same hour). Again, it would be ideal to have significant variation in breeding synchrony between demes for study of how sex ratio affects monopolizability of mates.

13. Parental care. For the study of mating systems other than promiscuity, the ideal mating system animal should exhibit postzygotic parental care, including variable degrees of biparental and even multiple-adult care (Emlen 1978, 1981b; Brown 1978; Brown and Brown 1981; Chapter 4). In general, within-taxon male parental investment patterns should be variable enough to foster a diversity of reproductive strategies for each sex.

14. Quantifiable parental investment. Theories of optimal foraging and prudent resource defense have received substantial verification and refinement from field studies of animals whose resource requirements are relatively simple. By virtue of being easily converted to calories, nectar has proven a particularly instructive resource base (e.g., Stiles 1971; Gill and Wolf 1975; Carpenter and MacMillen 1976; Pyke

1979). Similarly, application of parental investment theory faces measurement problems that can be partially ameliorated if the units of PI are simple. An ideal study species might invest primarily in quantifiable commodities such as food, which can be analyzed for nutritive value and difficulty in collection, or time invested. Specifically, the ideal subject would not incur significant parental risks (e.g., physically defending the young against formidable predators) because that important kind of investment is almost impossible to compare with more mundane "costs."

15. Quantifiable resources. In a similar vein, it is often desirable to measure the supposedly limiting resources for which reproductives are competing. Nectar is probably as measurable as any primary food known, so my ideal animal would probably be nectarivorous. Alternatively, it might be nice to concentrate on an animal that is not food limited but which faces a shortage of some manipulable commodity such as cavity nest sites (e.g., Gowaty 1981b).

16. Reproduction in captivity. There are many kinds of questions that can be answered only in controlled laboratory conditions, so ability to flourish and breed in captivity is obviously desirable.

17. Miscellaneous shortcuts. Finally, there are certain details of a species' reproductive biology that could contribute to research efficiency and yield. For example, reliable behavioral clues that allow discrimination of successful versus unsuccessful copulations in the field (Barash 1977; McKinney et al. 1978b) or existing information on sperm competition (Parker 1970) would be extremely valuable. Similarly, it would be useful to know many aspects of population structure (e.g, age-specific survivorship probabilities) or preexisting correlations between easily measured factors and hard-to-measure ones (e.g., that fledgling weight predicts longevity).

Birds, the "ideal," and other groups

Though this less than balanced treatment may draw accusations of taxon chauvinism, I am not blindly optimistic about the future of ornithological contributions to mating system research. All groups of animals have attributes and handicaps; my intent here is merely to speculate as to where the birds may lead us and perhaps where they cannot. On the plus side, birds are "centered" in mating system diversity (by being mostly monogamous), flexible in the division of parental investment between sexes, relatively similar to human "umwelt," highly observable, individually recognizable (with bands, etc.), sufficiently polymorphic for paternity assays, rich in the extent and variety of parental care patterns, and anchored by a huge natural history literature. These are major research virtues. On the negative side, they are too long-lived and too mobile for convenience.

Although reaping an evolutionary bonanza, aerial mobility created twin headaches for mating system researchers. First, it greatly complicates the identification and measurement of key ecological resources, such as food, that affect so many of the hypotheses at issue. Second, it frustrates our ability to keep track of the participants in the reproductive race. For example, the potential value of the *operational sex ratio* (OSR) concept will be particularly hard to realize with birds. Emlen and Oring (1977) defined OSR as "the average ratio of fertilizable females to sexually active males *at any given time.*" It is, thus, an index of the intensity of sexual competition and it requires the continuous monitoring of all local reproductives during the prezygotic period. This is much more feasible in nonvolant organisms like frogs (Emlen 1976) and crustacea (Birkhead and Clarkson 1980). Quite probably, applications of OSR will be possible only for bird species that are sexually dimorphic and live in discrete and conspicuous breeding populations. The fact that most colonial birds are sexually monomorphic (Lack 1968) severely limits the list of suitable subjects, but certain dense populations of sexually dimorphic icterids should be considered for such an approach.

Other taxa, of course, have their own advantages. Insects, in particular, have emerged as extremely useful for studying prezygotic aspects of sexual selection (e.g., Blum and Blum 1979) and offer a fantastic variety of short-lived and relatively sedentary subjects for consideration. Anurans harbor a fascinating suite of species in which males provide all postzygotic care (Wells 1981). Teleost fishes, with their array of internal versus external fertilization, protogyny, and variety of male investment patterns, would be close to the ideal but are inconveniently aquatic!

Birds are suited to those topics for which their attributes are most useful and their liabilities surmountable. They are particularly good for field studies of biparental care and monogamy. They display exciting social diversity, culminating in cooperative/communal breeding systems (Chapter 4) and may possess the necessary preadaptations for reciprocal altruism (Trivers 1972; Alexander 1974; Ligon and Ligon 1978; Emlen 1981b).

In any case, it is to be hoped that participants in the bicentennial celebration of the American Ornithologists' Union will find all the concerns outlined in this chapter quaint and amusing. Not only are they likely to regard paternity as a long-solved technical detail, they will probably find our present level of theoretical sophistication very crude. Quite possibly, some basic premise of our current thinking will be considered totally fallacious so that many of the questions we ask now will have been scrapped as inappropriate. The "cutting edge" of science moves so rapidly, especially in recent decades since it became a professional enterprise – embracing those of us who would have starved as ornithologists a century ago – that we can only do our best. After all, Darwin presumably did not join the *Beagle* with aspirations for overhauling the biological perspectives of Western civilization. Serendipity played a central role. Yet, after well over a century of debate and scrutiny, his general principles remain essentially intact and even his geospizine finches continue to supply exciting answers to our questions.

Maybe the bicentennial celebrants won't laugh so hard after all.

Acknowledgments

I thank P. L. Schwagmeyer for assistance in various stages of the preparation of this article. The National Science Foundation (Grant BNS 7906059) and the University of Oklahoma provided support for my research on avian monogamy. An earlier draft profited from criticisms by P. A. Gowaty, W. D. Graul, F. McKinney, L. W. Oring, and J. S. Quinn. Finally, I thank the members of the 1981 Social Behavior Seminar at OU for discussions and reviews of many topics presented here.

Literature cited

Abele, L., and S. Gilchrist. 1977. Homosexual rape and sexual selection in acanthocephalan worms. *Science 197*:81–83.

Alexander, R. D. 1974. The evolution of social behavior. *Annu. Rev. Ecol. Syst. 5*:324–383.

Alexander, R. D., and G. Borgia. 1979. On the origin and basis of the male-female phenomenon. In M. S. Blum and N. A. Blum (eds.). *Sexual selection and reproductive competition in insects,* pp. 417–440. New York: Academic Press.

Altmann, S. A., S. S. Wagner, and S. Lenington. 1977. Two models for the evolution of polygyny. *Behav. Ecol. Sociobiol. 2*:397–410.

Andersson, M., and M. G. Eriksson. 1982. Nest parasitism in goldeneyes *Bucephala clangula:* some evolutionary aspects. *Am. Nat. 120*:1–16.

Baker, M. C. 1981. A muscle biopsy procedure for use in electrophoretic studies of birds. *Auk 98:* 392–393.

Balph, D. F., G. S. Innis, and M. H. Balph. 1981. Kin selection in Rio Grande turkeys: a critical assessment. *Auk 97:*854–860.

Banks, J. W. 1884. Nesting of the Broad-winged Hawk (*Buteo pennsylvanicus*). *Auk 1:*95–96.

Barash, D. P. 1976. The male response to apparent female adultery in the Mountain Bluebird, *Sialia curricoides:* an evolutionary interpretation. *Am. Nat. 110:*1097–1101.

– 1977. Sociobiology of rape in Mallards (*Anas platyrhynchos*): responses of the mated male. *Science 197:*788–789.

Barrowclough, G. F., and K. W. Corbin. 1978. Genetic variation and differentiation in the Parulidae. *Auk 95:*691–702.

Bateman, A. J. 1948. Intrasexual selection in *Drosophila*. *Heredity 2:*349–368.

Baumgardner, D. J., T. G. Hartung, D. K. Sawrey, D. G. Webster, and D. A. Dewsbury. 1982. Muroid copulatory plugs and female reproductive tracts: a comparative investigation. *J. Mamm. 63:* 110–117.

Beecher, M. D., and I. M. Beecher. 1979. Sociobiology of Bank Swallows: reproductive strategy of the male. *Science 205:*1282–1285.

Bertram, B. C. R. 1976. Kin selection in lions and evolution. In P. Bateson and R. Hinde (eds.). *Growing points in ethology*, pp. 281–301. Cambridge: Cambridge University Press.

– 1979. Ostriches recognize their own eggs and discard others. *Nature (London) 279:*233–234.

Bingham, V. P. 1980. Novel rape avoidance in the Mallard. *Wilson Bull. 92:*409.

Birdsall, D. A., and D. Nash. 1973. Occurrence of successful multiple insemination of females in natural populations of deer mice (*Peromyscus maniculatus*). *Evolution 27:*106–110.

Birkhead, T. R. 1978. Behavioural adaptations to high nesting density in the Common Guillemot (*Uria aalge*). *Anim. Behav. 26:*321–331.

– 1979. Mate guarding in the Magpie (*Pica pica*). *Anim. Behav. 27:*866–874.

Birkhead, T. R., and K. Clarkson. 1980. Mate selection and precopulatory guarding in *Gammarus pulex*. *Z. Tierpsychol. 52:*365–380.

Blum, M. S., and N. A. Blum (eds.). 1979. *Sexual selection and reproductive competition in insects*. New York: Academic Press.

Bradbury, J. W. 1981. The evolution of leks. In R. D. Alexander and D. W. Tinkle (eds.). *Natural selection and social behavior*, pp. 138–169. New York: Chiron Press.

Bray, O. E., J. Kennelly, and J. L. Guarino. 1975. Fertility of eggs produced on territories of vasectomized Red-winged Blackbirds. *Wilson Bull. 87:*187–195.

Brown, J. L. 1969. Territorial behavior and population regulation in birds: a review and reevaluation. *Wilson Bull. 81:*293–329.

– 1978. Avian communal breeding systems. *Annu. Rev. Ecol. Syst. 9:*123–155.

Brown, J. L., and E. R. Brown. 1981. Kin selection and individual selection in babblers. In R. D. Alexander and D. W. Tinkle (eds.). *Natural selection and social behavior*, pp. 244–256. New York: Chiron Press.

Bruning, D. F. 1974. Social structure and reproductive behavior of the Greater Rhea. *Living Bird 13:*251–294.

Burger, J. 1981. Sexual differences in parental activities of breeding Black Skimmers. *Am. Nat. 117:*975–984.

Burley, N. 1977. Parental investment, mate choice, and mate quality. *Proc. Natl. Acad. Sci. USA 74:*3476–3479.

– 1981a. The evolution of sexual indistinguishability. In R. D. Alexander and D. W. Tinkle (eds.). *Natural selection and social behavior,* pp. 121–137. New York: Chiron Press.

– 1981b. Sex ratio manipulation and selection for attractiveness. *Science 211:*721–722.

Burns, J. T., K. Cheng, and F. McKinney. 1980. Forced copulation in captive Mallards. I. Fertilization of eggs. *Auk 97:*875–879.

Carey, M., and V. Nolan, Jr. 1979. Population dynamics of Indigo Buntings and the evolution of avian polygyny. *Evolution 33:* 1180–1192.

Carpenter, F. L., and F. E. MacMillen. 1976. Threshold model of feeding territoriality and test with a Hawaiian Honeycreeper. *Science 194:*639–642.

Cheng, K., J. T. Burns, and F. McKinney. 1982. Forced copulation in captive Mallards (*Anas platyrhynchos*). II. Temporal factors. *Anim. Behav. 30:*695–699.

– in press. Forced copulation in captive Mallards. III. Sperm competition. *Auk.*

Chronister, C. D. 1982. Egg laying and incubation behavior of Black-bellied Whistling Ducks. M. S. thesis, University of Minnesota, Minneapolis.

Compton, M. M., H. P. Van Krey, and P. B. Siegel. 1978. The filling and emptying of the uterovaginal sperm-host glands in the domestic hens. *Poul. Sci. 57:*1696–1700.

Cooke, F., M. A. Bousfield, and A. Sadura. 1981. Mate change and reproductive success in the Lesser Snow Goose. *Condor 83:*322–327.

Coulson, J. C. 1966. The influence of the pair-bond and age on the breeding biology of the Kittiwake gull *Rissa tridactyla. J. Anim. Ecol. 35:*269–279.

Crook, J. H. 1964. The evolution of social organization and visual communication in the weaver birds (Ploceinae). *Behaviour Suppl. 10:*1–178.

Darwin, C. 1871. *The descent of man and selection in relation to sex.* London: Murray.

Dawkins, R., and J. R. Krebs. 1978. Animal signals: communication or manipulation? In J. R. Krebs and N. B. Davies (eds.). *Behavioural ecology: an evolutionary approach,* pp. 282–309. Sunderland, Mass.: Sinauer Associates.

Derrickson, S. R. 1977. Aspects of breeding behavior in the Pintail (*Anas acuta*). Ph.D. thesis, University of Minnesota, Minneapolis.

– 1978. Mobility of breeding Pintails. *Auk 95:*104–114.

Dewsbury, D. A. in press. Sperm competition in muroid rodents. In R. L. Smith (ed.). *Sperm competition and the evolution of animal mating systems.* New York: Academic Press.

Dewsbury, D. A., and D. J. Baumgardner. 1981. Studies of sperm competition in two species of muroid rodents, *Behav. Ecol. Sociobiol. 9:*121–134.

Downhower, J. R., and K. B. Armitage. 1971. The Yellow-bellied Marmot and the evolution of polygamy. *Am. Nat. 105:*355–370.

Downhower, J. F., and L. Brown. 1981. The timing of reproduction and its behavioral consequences for Mottled Sculpins. In R. D. Alexander and

D. W. Tinkle (eds.). *Natural selection and social behavior,* pp. 78–95. New York: Chiron Press.

Elder, W. H., and M. W. Weller. 1954. Duration of fertility in the domestic Mallard hen after isolation from the drake. *J. Wildl. Manage. 18:*495–502.

Emlen, S. T. 1976. Lek organization and mating strategies in the Bullfrog. *Behav. Ecol. Sociobiol. 1:*283–313.

– 1978. The evolution of cooperative breeding in birds. In J. R. Krebs and N. B. Davies (eds.). *Behavioural ecology: an evolutionary approach,* pp. 245–281. Sunderland, Mass.: Sinauer Associates.

– 1981a. The ornithological roots of sociobiology. *Auk 98:*400–403.

– 1981b. Altruism, kinship, and reciprocity in the White-fronted Bee-eater. In R. D. Alexander and D. W. Tinkle (eds.). *Natural selection and social behavior,* pp. 217–230. New York: Chiron Press.

Emlen, S. T., and L. W. Oring. 1977. Ecology, sexual selection, and evolution of mating systems. *Science 197:*215–223.

Erickson, C. J. and P. G. Zenone. 1976. Courtship differences in male Ring Doves: avoidance of cuckoldry? *Science 192:*1353–1354.

Fisher, R. A. 1958. *The genetical theory of natural selection,* 2nd ed. New York: Dover.

Foltz, D. W. 1981. Genetic evidence for long-term monogamy in a small rodent, *Peromyscus polionotus. Am. Nat. 117:*665–675.

Fujioka, M., and S. Yamagishi. 1981. Extramarital and pair copulations in the Cattle Egret. *Auk 98:*134–144.

Garson, P. J. 1980. Male behaviour and female choice: mate selection in the wren? *Anim. Behav. 28:*291–502.

Gill, F. B., and L. L. Wolf. 1975. Economics of feeding territoriality in the Golden-winged Sunbird. *Ecology 56:*333–345.

Gladstone, D. E. 1979. Promiscuity in monogamous colonial birds. *Am. Nat. 114:*545–557.

Gould, S. J. 1978. Sociobiology and the theory of natural selection. In G. W. Barlow and J. Silverberg (eds.). *Sociobiology: beyond nature/nurture? American Association for the Advancement of Science Symposium,* Vol. 35; pp. 257–269.

Gowaty, P. A. 1981a. An extension of the Orians-Verner-Willson model to account for mating systems besides polygyny. *Am. Nat. 118:*851–859.

– 1981b. Aggression of breeding Eastern Bluebirds (*Sialia sialis*) toward their mates and models of intra- and interspecific intruders. *Anim. Behav. 29:* 1013–1027.

– in press. Male parental care and apparent monogamy among Eastern Bluebirds. *Am. Nat.*

Graul, W. D., S. R. Derrickson, and D. W. Mock. 1977. The evolution of avian polyandry. *Am. Nat. 111:*812–816.

Greenwood, P. J., P. H. Harvey, and C. M. Perrins. 1979. Mate selection in the Great Tit (*Parus major*) in relation to age, status and natal dispersal. *Ornis Fenn. 56:*75–86.

Gwynne, D. L. 1981. Sexual difference theory: Mormon Crickets show role reversal in mate choice. *Science 213:*779–780.

Hailman, J. P. 1978. Rape among Mallards. *Science 201:*280–281.

Handford, P., and M. A. Mares. in press. The mating systems of ratites and tinamous: an evolutionary perspective. In C. Nelson and C. Hubbs (eds.). *W. F. Blair commemorative volume.* Austin, Tx.: University of Texas Press.

Hanken, J., and P. W. Sherman. 1981. Multiple paternity in Belding's Ground Squirrel litters. *Science 212:*351–353.

Hays, H. M. 1972. Polyandry in the Spotted Sandpiper. *Living Bird 11:*43–58.

Hogan-Warburg, A. J. 1966. Social behaviour of the Ruff *Philomachus pugnax* (L.). *Ardea 54:*109–229.

Hoogland, J. L., and P. W. Sherman. 1976. Advantages and disadvantages of Bank Swallow coloniality. *Ecol. Monogr. 46:*33–58.

Horn, H. S. 1971. *The adaptive geometry of trees,* Monographs in Population Biology No. 3. Princeton, N.J.: Princeton University Press.

Howard, R. D. 1979. Estimating reproductive success in natural populations. *Am. Nat. 114:*221–231.

Howe, M. A. 1975. Behavioral aspects of the pair bond in Wilson's Phalarope. *Wilson Bull. 87:*248–270.

Hunt, G. L., Jr., and M. W. Hunt. 1977. Female-female pairing in Western Gulls (*Larus occidentalis*) in southern California. *Science 196:*1466–1467.

Huxley, J. S. 1912. A "disharmony" in the reproductive habits of the wild duck (*Anas boschas* L.). *Biol. Zentralbl. 32:*621–623.

Jenni, D. A. 1974. The evolution of polyandry in birds. *Am. Zool. 14:*129–144.

Jenni, D. A., and G. Collier. 1972. Polyandry in the American Jacana (*Jacana spinosa*). *Auk 89:*743–765.

Krebs, J. R., and N. B. Davies. 1981. *An introduction to behavioural ecology.* Sunderland, Mass.: Sinauer Associates.

Lack, D. 1966. *Population studies of birds.* Oxford University Press (Clarendon Press).

– 1968. *Ecological adaptations for breeding in birds.* London: Methuen & Co.

Lancaster, D. A. 1964. Biology of the Brushland Tinamou, *Nothoprocta cinerascens. Bull. Am. Mus. Nat. Hist. 127:*269–314.

LeBoeuf, B. J. 1974. Male-male competition and reproductive success in elephant seals. *Am. Zool. 14:*163–176.

Lenington, S. 1980. Female choice and polygyny in Redwinged Blackbirds. *Anim. Behav. 28:*347–361.

Ligon, J. D., and S. H. Ligon. 1978. Communal breeding in Green Woodhoopoes as a case for reciprocity. *Nature (London) 276:*496–498.

Lloyd, J. E. 1979. Insect behavioral ecology: coming of age in bionomics *or* Compleat biologists have revolutions too. *Fla. Entomol. 63:*1–4.

Lumpkin, S., K. Kessel, P. G. Zenone, and C. J. Erickson. 1982. Proximity between the sexes in Ring Doves: social bonds or surveillance? *Anim. Behav. 30:*506–513.

Martin, S. G. 1974. Adaptations for polygynous breeding in the Bobolink, *Dolichonyx oryzivorus. Am. Zool. 14:*109–119.

Maxson, S. J., and L. W. Oring. 1980. Breeding season time and energy budgets of the polyandrous Spotted Sandpiper. *Behaviour 74:*200–263.

Maynard Smith, J. 1977. Parental investment – A prospective analysis. *Anim. Behav. 25:*1–9.

Maynard Smith, J., and M.G. Ridpath. 1972. Wife-sharing in the Tasmanian Native Hen, *Tribonyx mortierii:* a case of kin selection? *Am. Nat. 106:*447–452.

McCracken, G. F., and J. W. Bradbury. 1977. Paternity and genetic heterogeneity in the polygynous bat, *Phyllostomus hastatus*. *Science 198:*303–306.

McGregor, P. K., J. R. Krebs, and C. M. Perrins. 1981. Song repertoires and lifetime reproductive success in the Great Tit (*Parus major*). *Am. Nat. 118:* 149–159.

McKinney, F. 1965. Spacing and chasing in breeding ducks. *Wildfowl Trust Annu. Rept. 16:*92–106.

McKinney, F., W. R. Siegfried, I. J. Ball, and P. G. H. Frost. 1978a. Behavioral specializations for river life in the African Black Duck (*Anas sparsa* Eyton). *Z. Tierpsychol. 48:*349–400.

McKinney, F., J. Barrett, and S. R. Derrickson. 1978b. Rape among Mallards. *Science 201:*281–282.

McKinney, F., K. M. Cheng, and D. J. Bruggers. in press-a. Sperm competition in apparently monogamous birds. In R. L. Smith (ed.). *Sperm competition and the evolution of animal mating systems.* New York: Academic Press.

McKinney, F., S. R. Derrickson, and P. Mineau. in press-b. Forced copulation in waterfowl. *Behaviour.*

McKinney, F., and N. P. Stolen. 1982. Extra-pair-bond courtship and forced copulation among captive Green-Winged Teal (*Anas crecca carolinensis*). *Anim. Behav. 30:*461–474.

Mills, J. A. 1973. The influence of age and pair-bond on the breeding biology of the Red-billed Gull *Larus novahollandiae scolipinus. J. Anim. Ecol. 42:*147–169.

Mineau, P., and F. Cooke. 1978. Territoriality in Snow Geese or the protection of parenthood – Ryder's and Inglis's hypotheses re-assessed. *Wildfowl 30:*16–19.

– 1979. Rape in the Lesser Snow Goose. *Behaviour 70:*280–291.

Morton, E. S., M. S. Geitgey, and S. McGrath. 1978. On bluebird "responses to apparent female adultery." *Am. Nat. 112:*968–971.

Nolan, V., Jr. 1978. *The ecology and behavior of the Prairie Warbler* Dendroica discolor, Ornitholog. Monograph No. 26, Washington, D. C.: American Ornithologists' Union.

O'Donald, P. 1980a. Sexual selection by female choice in a monogamous bird: Darwin's theory corroborated. *Heredity 45:*201–217.

– 1980b. Genetic models of sexual and natural selection in monogamous organisms. *Heredity 44:*391–415.

Orians, G. H. 1969a. On the evolution of mating systems in birds and mammals. *Am. Nat. 103:*589–603.

– 1969b. Age and hunting success in the Brown Pelican. *Anim. Behav. 17:* 316–319.

Oring, L. W. 1982. Avian mating systems. In D. S. Farner, J. King, and K. C. Parkes (eds.) *Avian Biology,* Vol. 6, pp. 1–92. New York: Academic Press.

Oring, L. W., and M. L. Knudson. 1972. Monogamy and polyandry in the Spotted Sandpiper. *Living Bird 11:*59–74.

Oring, L. W., and S. J. Maxson. 1978. Instances of simultaneous polyandry by a Spotted Sandpiper. *Ibis 120:*349–353.

Parker, G. A. 1970. Sperm competition and its evolutionary consequences in the insects. *Biol. Rev. 45:*525–567.

Patterson, C. B., W. J. Erckman, and G. H. Orians. 1980. An experimental study of parental investment and polygyny in male blackbirds. *Am. Nat.* *116:*757–759.

Payne, R. B. 1977. The ecology of brood parasitism in birds. *Annu. Rev. Ecol. Syst. 8:*1–28.

– 1979. Sexual selection and intersexual differences in variance of breeding success. *Am. Nat. 114:*447–452.

Perrins, C. M. 1979. *British tits.* London: Collins.

Pierotti, R. 1981. Male and female parental roles in the Western Gull under different environmental conditions. *Auk 98:*532–549.

Pleszczynska, W. K. 1978. Microgeographic prediction of polygyny in the Lark Bunting. *Science 201:*935–937.

Pleszczynska, W. K., and R. J. C. Hansel. 1980. Polygyny and decision theory: testing of a model in Lark Buntings (*Calamospiza melanocorys*). *Am. Nat. 116:*821–830.

Power, H. W. 1980. Male escorting and protecting females at the nest cavity in Mountain Bluebirds. *Wilson Bull. 92:*509–511.

Power, H. W., and C. G. P. Doner. 1980. Experiments on cuckoldry in the Mountain Bluebird. *Am. Nat. 116:*689–704.

Power, H. W., E. Litkovich, and M. P. Lombardo. 1981. Male Starlings delay incubation to avoid being cuckolded. *Auk 98:*386–387.

Pyke, G. H. 1979. Optimal foraging in bumblebees: rule of movement between flowers within inflorescences. *Anim. Behav. 27:*1167–1181.

Recher, H. F., and J. A. Recher. 1969. Comparative foraging efficiency of adult and immature Little Blue Herons. *Anim. Behav. 17:*319–321.

Ryder, J. P. 1978. Possible origins and adaptive value of female-female pairing in gulls. In *Proceedings of the Colonial Waterbird Group,* Vol. 2, pp. 138–145.

Ryder, J. P., and P. L. Somppi. 1979. Female-female pairing in Ring-billed Gulls. *Auk 96:*1–5.

Scott, D. K. 1980. Functional aspects of the pair bond in winter in Bewick's Swans (*Cygnus columbianus bewickii*). *Behav. Ecol. Sociobiol. 7:*323–327.

Searcy, W. A., and K. Yasukawa. 1981. Does the "sexy son" hypothesis apply to mate choice in Red-winged Blackbirds? *Am. Nat. 117:*343–348.

Selander, R. K. 1972. Sexual selection and dimorphism in birds. In B. G. Campbell (ed.). *Sexual selection and the descent of man, 1871–1971,* pp. 180–230. Chicago: Aldine.

Sherman, P. W. 1981. Electrophoresis and avian genealogical analyses. *Auk 98:*419–421.

Stiles, F. G. 1971. Time, energy, and territoriality of the Anna Hummingbird (*Calypte anna*). *Science 173:*818–821.

Sturkie, P. D. 1976. *Avian physiology,* 3rd. ed. New York: Springer-Verlag.

Thornhill, R. 1980. Rape in *Panorpa* scorpionflies and a general rape hypothesis. *Anim. Behav. 28:*52–59.

Trivers, R. L. 1972. Parental investment and sexual selection. In B. Campbell (ed.). *Sexual selection and the descent of man, 1871–1971,* pp. 136–179. Chicago: Aldine.

Verner, J. 1964. The evolution of polygyny in the Long-billed Marsh Wren. *Evolution 18:*252–261.

Verner, J., and G. H. Engelsen. 1970. Territories, multiple nest building and polygyny in the Long-billed Marsh Wren. *Auk 87*:557–567.

Verner, J., and M. F. Willson. 1966. The influence of habitats on mating systems of North American passerine birds. *Ecology 47*:143–147.

— 1969. Mating systems, sexual dimorphism and the role of male North American passerine birds in the nesting cycle. *Ornithol. Monogr. 9*:1–76.

Waage, J. K. 1979. Dual function of the damselfly penis: sperm removal and transfer. *Science 203*:916–918.

Wade, M. J., and S. J. Arnold. 1980. The intensity of sexual selection in relation to male sexual behaviour, female choice, and sperm precedence. *Anim. Behav. 28*:446–461.

Watt, D. J. 1982. Do birds use color bands in recognition of individuals? *J. Field Ornithol. 53*:177–179.

Watt, C. R., and A. W. Stokes. 1971. The social order of turkeys. *Sci. Am. 224*:111–118.

Weatherhead, P. J., and R. J. Robertson. 1979. Offspring quality and the polygyny threshold: the "sexy son hypothesis." *Am. Nat. 113*:201–208.

Wells, K. D. 1981. Parental behavior of male and female frogs. In R. D. Alexander and D. W. Tinkle (eds.). *Natural selection and social behavior*, pp. 184–197. New York: Chiron Press.

Wiley, R. H. 1974. Evolution of social organization and life-history patterns among grouse. *Q. Rev. Biol. 49*:201–227.

Williams, G. C. 1966. *Adaptation and natural selection: a critique of some current evolutionary thought*. Princeton, N.J.: Princeton University Press.

— 1975. *Sex and evolution*, Monographs in Population Biology 8. Princeton, N.J.: Princeton University Press.

Wilson, E. O. 1975. *Sociobiology*. Cambridge, Mass.: Harvard University Press.

Wishart, R. A., and R. W. Knapton. 1978. Male Pintails defending females from rape. *Auk 95*:186–187.

Wittenberger, J. F. 1976. The ecological factors selecting for polygyny in altricial birds. *Am. Nat. 110*:779–799.

— 1979. The evolution of mating systems in birds and mammals. In P. Marler and J. G. Vandenberg (eds.). *Handbook of behavioral neurobiology*, Vol. 3; *Social behavior and communication, pp. 271–349*. New York: Plenum.

— 1980. Feeding of secondary nestlings by polygynous male Bobolinks. *Wilson Bull. 92*:330–340.

— 1981. *Animal social behavior*. Boston: Duxbury Press.

Wittenberger, J. F., and R. L. Tilson. 1980. The evolution of monogamy. *Annu. Rev. Ecol. Syst. 11*:197–232.

Wolf, L. L. 1975. "Prostitution" behavior in a tropical hummingbird. *Condor 77*:140–144.

Wolf, L. L., and J. S. Wolf. 1976. Mating system and reproductive biology of Malachite Sunbirds. *Condor 78*:27–39.

Yom-Tov, Y., G. M. Dunnet, and A. Anderson. 1974. Intraspecific nest parasitism in the Starling *Sturnus vulgaris. Ibis 116*:87–90.

Zenone, P. G., M. E. Sims, and C. J. Erickson. 1979. Male Ring Dove behavior and the defense of genetic paternity. *Am. Nat. 114*:615–626.

Zimmerman, J. L. 1966. Polygyny in the Dickcissel. *Auk 83*:534–546.

Commentary

SARAH LENINGTON

Dr. Mock has beautifully summarized the advances made by ornithology in our understanding of mating systems. Rather than cover the same material in this commentary, I propose to focus on some of the problems that have arisen in studies of avian mating systems.

As indicated in the preceding chapter, birds are close to an ideal organism for the study of questions regarding the evolution of social organization. Despite their considerable virtues in this regard, however, much of the research on mating systems remains inconclusive and has produced contradictory conclusions that are difficult to resolve. This is particularly apparent in those cases where a single species such as the Red-winged Blackbird (*Agelaius phoeniceus*) has been studied by a variety of researchers in different locations. Since the pioneering work of Orians (1961) on social organization of Redwings, they have virtually become the *Drosophila* of mating systems research. Despite the extent of both theoretical and empirical effort directed toward elucidating the basis for polygynous matings in this species (Orians 1972, 1980; Holm 1973; Blakley 1976; Altmann et al. 1977; Caccamise 1977; Weatherhead and Robertson 1977, 1979; Searcy 1979; Yasukawa 1979; Lenington 1980; Heissler 1981; Wittenberger 1981; Yasukawa and Searcy 1981), the studies often lead to opposing conclusions. For example, several authors have concluded that harem size in Redwings is correlated with one or more aspects of territory quality (Holm 1973; Searcy 1979; Lenington 1980; Orians 1980). Weatherhead and Robertson (1977), however, found no relationship between these two variables. Some authors assume that harem size is a good indicator of female preference and assert that females prefer males with high-quality territories (Holm 1973; Searcy 1979; Orians 1980). Others dispute this assumption and report little or no association between female preference and territory quality (Lenington 1980, unpub.). Some studies report an increase in female reproductive success as a function of harem size (Orians 1972, 1980; Holm; 1973). Others find no relationship or a negative relationship between harem size and female reproductive success (Caccamise 1977; Weatherhead and Robertson 1977; Lenington 1980).

To some extent, these contradictory findings may reflect genuine differences among populations of the same species. To a large degree, however, the contradictions also arise from problems in definition and logic. Although these problems are by no means confined to the research on Redwings, in the discussion that follows I give particular emphasis to Redwing data, because for reasons mentioned earlier, the problems are particularly evident in studies of this species. The most serious difficulties in studies of avian mating systems tend to fall into three categories: the definition of territory quality, the assessment of female preference, and the measurement of reproductive success.

Definition of territory quality

Since the publication in 1969 of Orians's classic paper, *On the Evolution of Mating Systems in Birds and Mammals,* the *polygyny threshold model* (PTM) has held center stage in discussions about the basis of avian mating systems. Most studies of polygynous birds have attempted to test this model. The interpretation of "territory quality," however, has posed particular difficulty in these tests. The first problem with interpreting territory quality lies in the definition of territory.

Holm (1973) interpreted territory quality in terms of differences between marshes. She compared marshes where the vegetation was primarily cattails with those where the vegetation was primarily bullrushes. She did not report on variability among territories within a marsh. Other workers (Searcy 1979; Lenington 1980) defined territory as the entire area defended by an individual male or an arbitrarily chosen section within a marsh. Still others (Weatherhead and Robertson 1977) ignored characteristics of territories as a whole and use "territory" to refer to nest sites. The use of the term "territory" to refer to three very different ecological levels tends to obscure the possibility that different processes operate on each of these levels. It is possible, for example, that females engage in a hierarchical decision-making process, first choosing a general locale in which to nest (e.g., a particular marsh or field), then choosing a particular territory within a general locale, and, finally, choosing a particular nest site within a territory. If this were the case, there is no reason why the criteria for assessing quality at one level will necessarily correspond with the criteria at another level. Available data on Redwings indicate that predictors of harem size between marshes are not the same as predictors of harem size within marshes (Lenington 1980). Alternatively, it is possible that females ignore one or more of these levels when choosing mates. In either case, the question is an empirical one that could lead to modification or rejection of the PTM for a particular population or species. Little is gained by referring to all three ecological levels as "territory" and ignoring the different implications for the process of mate choice entailed by choice at different levels.

Attention to the implications of differences in ecological level can potentially resolve some of the contradictions in Redwing data. Thus, the contradiction between Weatherhead and Robertson (1977), who claimed that harem size was not associated with territory quality, and other authors who found the opposite result is more apparent than real. Territory in each of these cases is used to refer to a different ecological level. The contradiction is resolved if Weatherhead and Robertson's conclusion is restated to claim that harem size is uncorrelated with nest-site quality, a result supported by Lenington (1980), who found that harem size was correlated with territory quality but not nest-site quality.

A second problem in the assessment of territory quality is in the methods used to measure quality. Most studies base their assessment of territory quality on a prediction of the PTM that the best territories will contain the largest harems. Thus, territories with large numbers of females are referred to as "high-quality territories" and those with few females as "poor-quality territories." This prediction of the PTM, however, is based on the assumption that females can settle in whichever territory they choose. This assumption is rarely tested yet there are compelling reasons to believe it may commonly not be valid. If, as originally emphasized by Orians (1969), a female's reproductive success is decreased by harem membership, one would expect resident females to attempt to exclude newly arriving females from their territory. Nero (1956) and Lenington (1980) witnessed resident females driving incoming females from territories. Holm (1973) suggested, in fact, that competition among females was a major factor in limiting harem size on territories. If competition among females is prevalent, the best territories will not necessarily obtain the most females. Indeed, it is possible to envision the opposite situation. If older experienced females return to the breeding ground early (Blakley 1976; Crawford 1977), they may obtain space on the most desirable territories. Later when younger or less competitively experienced females arrive, they may be prevented from settling on good territories and may be forced to settle on undesirable territories or in inferior habitats. In such a situation, using harem size to assess either territory quality or female preference could produce highly misleading results. Unless one can be certain that competition among females does not occur, it is preferable to test whether or not territory quality is associated with harem size rather than assuming such a relationship. In order to test this relationship, one must assess territory quality independently of harem size. Since quality in evolutionary models is synonymous with reproductive success, the most meaningful way to do this is to examine the correlation between territory characteristics and female reproductive success while controlling for harem size.

Conversely, when examining the relationship between harem size and reproductive success, it is necessary to control for territory quality. Otherwise, positive correlations between harem size and reproductive success might simply be due to a correlation between harem size and territory quality, whereas harem membership per se might have no effect or a negative effect on female fitness (Altmann et al. 1977).

Assessment of female preference

Most models for the evolution of mating systems, including the PTM, assume that females will preferentially select the highest quality territories. Two methods are commonly used to assess female preference.

It is sometimes assumed that females prefer territories that ultimately obtain the largest number of females (Holm 1973; Orians 1972, 1980; Searcy 1979). This method leads to the same problems noted earlier in the use of harem size to infer territory quality. Competition among females may make harem size useless as a measure of either female preference or territory quality.

The second method for assessing female preference is to use the order in which females settle on territories. Quite legitimately, it is assumed territories that obtain the first females are territories preferred by females. Some authors (i.e., Orians 1980) do not record female settlement patterns directly but instead infer order of choice from order of nesting. This practice relies on the assumption that the order in which females arrive on territories is highly correlated with the order in which they build nests. However, in the one study to test this assumption, the order of arrival between territories correlated poorly with the order of nest initiation ($\tau = +0.231$) (Lenington 1980). Unless it can be shown that for a particular population, order of arrival between territories is strongly correlated with order of nesting, the only way to assess female preference is to directly observe the order in which females settle on territories. When this was done, the results were surprising and not those predicted by most of the commonly discussed models for mate choice. Territories preferred by females were neither those to obtain the largest harems nor the highest quality territories (Lenington 1980, unpub.).

Measurement of reproductive success

Reproductive success is the critical variable for discriminating among models of mate choice. Depending upon the model being tested, it is necessary to know the relationship between reproductive success and territory characteristics, harem size, male characteristics, or female characteristics (Orians 1969; Elliott 1975; Altmann et al. 1977; Searcy 1979; Weatherhead and Robertson 1979; Lenington 1980; Heissler 1981). Many problems in the measurement of reproductive success may be difficult or impossible to solve. These include such issues as the existence of confounding variables not recorded in a study or the fact that fitness refers to lifetime reproductive output, which may be impractical to measure. Other problems in the measurement of reproductive success may be more tractable, and it is these I propose to address.

Commonly, reproductive success is defined as the total number of young fledged from a given nest. On the surface, this seems a reasonable practice because the number of young fledged may be for many birds the closest we can come to measuring fitness. However, the number of young fledged is the outcome of several independent sources of mortality, primarily, predation, starvation, disease, and weather. There is no reason why factors that affect one of these processes will necessarily affect others. For example, cattail density of territories may influence the probability of predation on Redwing nests (Lenington 1980, unpub.; R. Eunash, pers. comm.), but it has no effect on starvation. If there is considerable variability among nests in mortality due to starvation, the "noise" thus generated may mask the effect of cattail density if one simply correlates cattail density with total number of young fledged. In one marsh on which I worked, the correlation between cattail density and nesting success (a measure of predation) was $+0.518$ ($p < 0.05$), whereas the correlation between cattail density and total number of young fledged was $+0.203$ ($p > 0.05$). Thus, when authors report no correlation between number of young fledged and various ecological characteristics (Holm 1973; Weatherhead and Robertson 1977), it is difficult to know if these variables really have no effect on fitness or if an effect is being obscured by combining sources of mortality.

Although it is often difficult to assess sources of mortality, for many purposes success or failure of nests is an adequate measure of mortality due to predation (Lack 1968; Ricklefs 1969) and the number of young fledged from successful nests an adequate measure of starvation (Emlen and Demong 1975). In some Redwing studies, it has been possible to go farther and identify mortality due to other factors (Haigh 1968; Robertson 1972; Caccamise 1976; Cronmiller and Thompson 1980). Thus, the task of examining the effect of independent variables on each source of mortality separately should be feasible.

In conclusion, I want to emphasize again that this discussion does not just apply to the research on Redwings. The same comments could be made about many studies of avian mating systems. Attention to such issues in the future, however, may make it possible to either resolve contradictory findings or at least to ascribe the contradictory findings to genuine variability in the animal rather than to variability in the analysis of data.

Literature cited

Altmann, S., S. W. Wagner, and S. Lenington. 1977. Two models for the evolution of polygyny. *Behav. Ecol. Sociobiol. 2:*397–410.

Blakley, N. R. 1976. Successive polygyny in upland nesting Redwinged Blackbirds. *Condor 78:*129–133.

Caccamise, D. F. 1976. Nesting mortality in the Red-winged Blackbird. *Auk 93:*517–534.

– 1977. Breeding success and nest site characteristics of the Red-winged Blackbird. *Wilson Bull. 89:*396–403.

Crawford, R. D. 1977. Breeding biology of year-old and older female Redwinged and Yellow-headed Blackbirds. *Wilson Bull. 89:*73–80.

Cronmiller, J. R., and C. F. Thompson. 1980. Experimental manipulation of brood size in Red-winged Blackbirds. *Auk 97:*559–565.

Elliott, P. F. 1975. Longevity and the evolution of polygamy. *Am. Nat. 109:* 281–287.

Emlen, S. T., and N. J. Demong. 1975. Adaptive significance of synchronized breeding in a colonial bird: a new hypothesis. *Science 188:*1029–1031.

Haigh, C. 1968. Sexual dimorphism, sex ratios, and polygyny in the Redwinged Blackbird. Ph.D. dissertation, University of Washington.

Heissler, I. L. 1981. Offspring quality and the polygyny threshold: a new model for the "sexy son" hypothesis. *Am. Nat. 117:*316–328.

Holm, C. H. 1973. Breeding sex ratios, territoriality, and reproductive success in the Red-winged Blackbird (*Agelaius phoeniceus*). *Ecology 54:*356–365.

Lack, D. 1968. *Ecological adaptations for breeding in birds.* London: Methuen & Co.

Lenington, S. 1980. Female choice and polygyny in the Redwinged Blackbird. *Anim. Behav. 28:*347–361.

Nero, R. W. 1956. A behavior study of the Red-winged Blackbird. II. Territoriality. *Wilson Bull. 68:*129–150.

Orians, G. H. 1961. The ecology of blackbird (*Agelaius*) social systems. *Ecol. Monogr. 31:*285–312.

— 1969. On the evolution of mating systems in birds and mammals. *Am. Nat. 96:* 257–263.

— 1972. The adaptive significance of mating systems in the Icteridae. In *Proceedings of the XV International Ornithological Congress,* pp. 389–398.

— 1980. *Some adaptations of marsh-nesting Blackbirds.* Princeton, N.J.: Princeton University Press.

Ricklefs, R. E. 1969. An analysis of nesting mortality in birds. *Smithson. Contrib. Zool. 9:*1–48.

Robertson, R. J. 1972. Optimal niche space of the Redwinged Blackbird (*Agelaius phoeniceus*). I. Nesting success in marsh and upland habitat. *Can. J. Zool. 50:*247–263.

Searcy, W. A. 1979. Female choice of mates: a general model for birds and its application to Red-winged Blackbirds (*Agelaius phoeniceus*). *Am. Nat. 114:*77–100.

Weatherhead, P. J., and R. J. Robertson. 1977. Harem size, territory quality and reproductive success in the Redwinged Blackbird (*Agelaius phoeniceus*). *Can. J. Zool. 55:*1261–1267.

— 1979. Offspring quality and the polygyny threshold: the "sexy son hypothesis." *Am. Nat. 113:*201–208.

Wittenberger, J. F. 1981. Male quality and polygyny: the "sexy son hypothesis" revisited. *Am. Nat. 117:*329–342.

Yasukawa, K. 1979. Territory establishment in Red-winged Blackbirds: importance of aggressive behavior and experience. *Condor 81:*258–264.

Yasukawa, K., and W. A. Searcy. 1981. Nesting synchrony and dispersion in Red-winged Blackbirds: is the harem competitive or cooperative? *Auk 98:* 659–668.

4 Cooperative breeding strategies among birds

STEPHEN T. EMLEN AND SANDRA L. VEHRENCAMP

The cooperative rearing of young is a topic of considerable interest to both biological and social scientists. Such behavior reaches its extreme development in many eusocial insect societies, where vast numbers of individuals live their entire lives as sterile workers, rearing young but never themselves becoming reproductives (Wilson 1971). Such sterile castes have not yet been reported among vertebrates. However, there are numerous instances in which individuals of vertebrate species (most of them avian) forego breeding for a significant portion of their adult lives and spend such time helping to rear offspring that are not genetically their own.

There are three fundamental questions surrounding the topic of cooperative breeding in animals. First, what role have ecological factors played in promoting the development of such aid-giving societies? Second, how can such seemingly altruistic behavior be explained in terms of natural selection theory? Third, what behavioral tactics will members of such societies adopt to maximize their own fitness when interacting with others?

In this chapter, we will attempt to address each of these three topics. Before doing so, however, it is necessary for us to define our terms. We use "group" to describe any long-lasting association of more than two individuals (Rowley et al. 1979). An "auxiliary" is any mature, non-breeding member of a reproducing group. It may or may not provide aid in the rearing of young. "Cooperative breeding" refers to any case where more than two birds provide care in the rearing of young. These can be subdivided into two types: helper-at-the-nest systems and communal breeding systems. In helper-at-the-nest systems, auxiliaries contribute physically, but not genetically, to the young being reared. Auxiliaries serve as helpers, but they do not engage in sexual activity with the breeding pair. Communal breeding systems are those in which parentage of the offspring is shared. In the case of shared paternity,

93

more than one male has a significant probability of fathering some of the offspring. With shared maternity, multiple females have a significant probability of contributing eggs to a communal clutch. Although this subdivision is by no means absolute, we feel that it is useful in modeling questions about ecological determinants, as well as behavioral tactics, that are important for the evolution and maintenance of cooperative breeding systems.

A potpourri of cooperative breeders

It is useful to begin our review with a brief introduction of the types and diversity of cooperative breeding systems that have been described for birds. Abbreviated synopses of the nesting biology of six selected species are provided.

Florida Scrub Jay *(Aphelocoma coerulescens)*

The Scrub Jay is a member of the family Corvidae with a range extending widely across the American Southwest and with a small population inhabiting the relict oak–scrub habitat of peninsular Florida. Suitable habitat is scarce and patchy in Florida, leading to disjunct and isolated small populations. It is in these populations that helping behavior has been described (Woolfenden 1973, 1975, 1981).

Scrub Jays in Florida live in stable groups comprised of a monogamous pair and their young of the previous 1 or more years. These groups are sedentary and inhabit year-round, group-defended territories. Group size averages three birds. Roughly half of the groups consist of a pair alone; the other half may have from one to six helpers (average = 1.8).

Three-quarters of these helpers assist at the nests of one or both of their genetic parents, confirming the hypothesis that most helpers come from the ranks of young that remain on their natal territories. These auxiliaries do not engage in sexual activities with the breeding (parental) pair. Thus, incest does not occur, and the helper does not share in the genetic parentage of the nestlings that it helps to rear. It does, however, make important contributions to the feeding and predator protection of these nestlings. Woolfenden's (1981) calculations demonstrate that the presence of helpers increases both the group's reproductive success (highly significantly) and the per capita reproductive success (not significantly).

Tasmanian Native Hen *(Tribonyx mortierii)*

The Tasmanian Native Hen is an Australian member of the rail family studied by Ridpath (1972). He describes a social system in which two males frequently pair with the same female. The trio then defends a permanent territory and cooperates in the incubation of eggs and care of the young. An interesting twist in this system is that the typical trio consists of two genetic brothers and an unrelated female (Maynard Smith and Ridpath 1972). Young birds normally disperse or "rove" prior to their first breeding season. Whether an individual becomes a breeder in a pair, or a cobreeder in a trio, depends in large part upon whether it disperses solitarily or together with its sibling. There is some evidence that trios are better able to obtain and defend suitable territories, which are in short supply.

Ridpath's data demonstrate that both brothers have sexual access to the female mate; the alpha male performs approximately two-thirds of all copulations, the beta male, one-third. Thus, there is shared paternity of the clutch, and *T. moteierii* would be classified as a communally breeding species.

Reproductive success in Ridpath's study was influenced both by group size and age of the breeding birds. For both first-year and older birds, trios significantly outproduced pairs. The per capita reproductive success of trios, however, was greater than for pairs only for first-year birds.

Mexican Jay *(Aphelocoma ultramarina)*

The Mexican Jay has been studied for many years in southeastern Arizona by Brown and his co-workers (Brown 1963, 1970; Brown and Brown 1980, 1981a; Trail et al. 1981). The species typically lives in groups of between 8 and 18 individuals and, like the examples cited previously, resides on permanent, year-round territories that are defended by all. The geneological composition of these groups is more complex than in the Scrub Jay, however, because (1) the membership core is an extended rather than a nuclear family, and (2) three, rather than two, generations of breeders and helpers may be present.

During the breeding season, one to three pairs of adults will leave the group flock and engage in nest-building activities. Occasionally, these different pairs interfere with one another, and nest-lining material may be robbed or eggs tossed from nests. Nevertheless, in roughly half of the groups, two (and rarely three) nests will be active simultaneously.

The laying female alone incubates the eggs, but she is fed on her nest both by her mate and by auxiliaries. After the eggs hatch, visits by auxiliary group members increase, and between half and three-quarters of the nestling feedings are performed by birds other than the parents. When more than one nest is active, the auxiliaries will divide their helping activities among them. Interestingly, Brown and Brown (1980) have found that some of the breeding females themselves will temporarily leave their own nestlings during the few days prior to their fledging and visit and feed at other nests belonging to the group. After the young fledge, they intermingle and are fed indiscriminately by all group members.

The social organization of the Mexican Jay is a step more complex than that of its phylogenetic relative, the Scrub Jay. Groups are larger, more generations coexist in the groups, and more than one pair of birds may breed in any given season. The breeding system is one of helpers at the nest with the proviso that when multiple nests are active, auxiliaries serve as helpers to them all. Data on the effects of helpers on reproductive success are not yet available.

Acorn Woodpecker *(Melanerpes formicivorous)*

The Acorn Woodpecker ranges from the western and south-western United States, through Mexico and Central America, and into Colombia. It has been studied most thoroughly in coastal California, where it lives in permanently territorial groups of from 2 to 15 individuals. Such groups consist of from one to four breeding males, one or two breeding females, and up to eight nonbreeding helpers at the nest (generally grown offspring from previous years) (MacRoberts and MacRoberts 1976; Koenig and Pitelka 1979).

The species occupies a specialized ecological niche, being dependent upon acorns for its food resource. These acorns are stored in specially constructed granaries located on the permanent territory. All group members take part in building and defending the granaries, as well as in storing and harvesting the acorns.

During breeding, only a single nest is attended at any one time by a group, and most or all group members help to incubate the eggs and feed and defend the young. Monogamous pairbonds appear to be absent, and several males engage in sexual activity with the nesting female (Stacey 1979b). Furthermore, in one-quarter of the groups studied in California, two females (sisters) laid eggs communally in the single active nest (Koenig and Pitelka 1979; Mumme et al., in press).

The breeding system of the Acorn Woodpecker thus combines elements of both helper-at-the-nest and communal breeding types. Several males, and up to two females, may share in the paternity/maternity of the group clutch. The remaining individuals (roughly half of each group) act as helpers at the nest.

Reproductive success per group increases with group size up to size seven or eight, but reproductive success per bird declines with increasing group size (Koenig 1981a). Communally laid clutches have a slightly lower success than single-female nests (Koenig and Pitelka, in press). This is due in large part to behavioral interference and egg tossing by the communally competing females.

Groove-billed Ani *(Crotophaga sulcirostris)*

Anis are neotropical members of the cuckoo family (Cuculidae). They live in small groups with fairly closed membership and maintain year-round territories that are defended by all. The Groove-billed Ani has been studied extensively by Vehrencamp (1977, 1978) in Costa Rica. In this species, groups consist of from one to four monogamous pairs of unrelated adults plus an occasional unpaired helper. All members of a group contribute to the building of a single nest into which *all* females communally lay their eggs. Incubation and care of the young are shared by the different members of the group.

There is a limit to the number of eggs that can be efficiently incubated. Beyond a certain clutch size, the eggs tend to be buried, fail to be turned regularly, and are unable to receive sufficient heating from incubation to guarantee normal development. Thus, the percentage of eggs that hatch decreases as the communal clutch size increases. Vehrencamp discovered that females engage in several behaviors that increase the probability that *their* eggs will be among those successfully incubated. Foremost among these behaviors is egg tossing, in which a female repeatedly visits the nest and removes eggs already present. No eggs were tossed at nests tended by single pairs, but the proportion of the total eggs laid that were tossed increased with increasing group size.

The Anis exemplify a clear case of the communal breeding strategy. At one time, these birds were thought to represent a pinnacle in the evolution of true cooperation. Vehrencamp's results, however, demonstrate that increased competition and conflict go hand in hand with increased communality.

Analyses of reproductive success are compounded by tendencies of Groove-billed Anis to inhabit a mosaic of two habitat types and for different-sized groups to predominate in each type. Within-habitat analyses for each type show that reproductive success of groups exceeds that of simple pairs and that per capita reproductive success is roughly equal across all group sizes.

White-fronted Bee Eater *(Merops bullockoides)*

Many members of the bee-eater family (Meropidae) are both cooperative and colonial. The White-fronted Bee Eater, studied by Emlen and his co-workers in Kenya (Emlen 1981; Hegner et al. 1982; Emlen and Demong, unpub.), inhabits the savannahs of eastern and southern Africa. These birds live in large colonies of several hundred individuals, but each colony is socially substructured into a number of smaller interacting units termed "clans." It is within these smaller groups that helping behaviors occur. Clans range in size from single pairs to groups of 11 individuals (mean = 7) and have a kin structure based largely upon extended family relationships. Each individual bee eater maintains a feeding home range that may be located a considerable distance from the colony site. Different members of the same clan occupy adjacent and partially overlapping home ranges. Thus, there is a spatial representation of the clan structure in the form of neighborhood clusters of feeding home ranges away from the colony.

Each clan is composed of from one to five monogamous pairs of birds, plus a smaller assortment of single individuals. All pairs are potential reproductives in any given year, but the number of pairs that exercise this option varies greatly across years depending upon local environmental conditions. Individuals that do not breed are likely to serve as helpers at the nest of a clan member that does. Additionally, birds that start a season as breeders, but whose nesting attempts fail, are likely to join as helpers with other clan members at nests where eggs or young are still being tended. Such redirected helping demonstrates that adult *M. bullockoides* shift back and forth between breeder and helper status. As a consequence, reciprocal exchanges of helping between donor and recipient, or donor and the offspring of the original recipient, are not uncommon.

White-fronted Bee Eater society is basically a helper-at-the-nest society, onto which an overlay of colony complexity has been added. Here again cooperation and conflict go hand in hand. Within clans, we find redirected and reciprocal helping exchanges occurring; but we also find dominant breeders actively recruiting subordinate auxiliaries to be helpers (Emlen, in preparation). Colonial living means that 15 to 30 clans associate and interact with one another on a regular basis. Often these interactions are competitive, and nest takeovers, forced extra-pair (extra-clan) copulations, and parasitic egg laying in other nests at the colony all occur (Emlen et al., unpub.).

The activities of helpers significantly increase reproductive success of groups compared with pairs. On a per capita basis, reproductive success is constant across different group sizes (Emlen 1981).

Many other species could be listed. However, the six examples presented provide an idea of the spectrum of social systems found among cooperatively breeding birds. The majority of cooperative breeders exhibit helper-at-the-nest systems, and much of our discussion in the following two sections will emphasize such systems.

Ecological determinants of cooperative breeding

Cooperative breeding is not a rare social system among birds. Approximately 3% (300) of the living avian species are known to exhibit aid-giving behavior during nesting, and the number will increase as more are studied. The list includes species belonging to a wide variety of taxonomic groups, inhabiting a broad range of geographic locations and habitats and filling the spectrum of ecological niches. This diversity has stifled the search for common ecological denominators and has caused some workers to conclude that no common thread underlies the parallel evolution of cooperative breeding in birds. Although agreeing that aid-giving behaviors have undoubtedly evolved independently many times in the class Aves, and that each species can be understood fully only when studied in conjunction with its own, unique, ecology, we nevertheless propose that certain broad ecological generalizations can be made about the necessary and sufficient conditions that have led to the evolution of cooperative breeding systems.

Consider any postbreeding group of birds comprising parents and maturing offspring. There are two behavioral options available to the maturing individuals: They can disperse and attempt to become established as independent breeders in the upcoming nesting season, or they can remain with their parents in their natal groups through the next season. In the latter case, they face a second subset of options: to help or not to help the group's breeders in rearing the next generation of nestlings. Our first question is; Under what conditions will grown offspring postpone dispersal and remain in their natal groups?

There is a large body of literature discussing the gains that accrue to individuals by virtue of living in groups. The two most cited benefits include (1) increased alertness and protection against predators and (2) enhanced efficiency in localization and exploitation of patchy, ephemeral food resources (e.g., Alexander 1974; Bertram 1978).

Surprisingly, neither of these factors usually is implicated in studies of cooperative breeders. Instead, field workers consistently stress the difficulties that younger birds have in finding vacant territories, obtaining mates, or breeding successfully alone. We suggest that the selective pressures favoring group living in most helper-at-the-nest species are fundamentally different from those favoring grouping in other types of avian aggregations, such as nesting colonies, large foraging flocks, or gregarious roosts.

Constraints on the option of independent breeding

If foraging and/or antipredation benefits form the primary reason for gregariousness in cooperative breeders, then the average fitness of individual group members should increase as some function of increasing group size up to an optimum size and decrease thereafter. Koenig (1981a) applied this logic to an analysis of the data available for cooperative breeders. He found that annual, per capita reproductive success showed no consistent pattern with group size. Of 16 cases analyzed, pairs were the most productive units in 7 cases, pairs were roughly equivalent to larger groups in 5 cases, and groups were most successful (on a per capita basis) in only 4 cases. Koenig further predicted that if cooperative groups formed because of inherent benefits of grouping, then the most frequently observed group sizes should coincide with the most productive group size as determined by per capita success. Such was not the case: Although in 12 cases pairs were equally or more productive than larger groups, in all cases the observed mean group size exceeded two. From these analyses, Koenig (1981a) concluded that most cooperatively breeding birds live in groups because they are "forced" to do so by severe ecological constraints that limit the option of younger birds to become established as independent breeders.

The idea is not new; it was first formulated by Selander (1964) during his study of helpers in *Campylorhynchus* wrens. It was generalized (for permanently territorial species) by Brown in 1974 and again in 1978, and recently it has been expanded and developed into more specific models by Woolfenden and Fitzpatrick (1978), Gaston (1978a), Koenig and Pitelka (1981), Koenig (1981a), and Emlen (1981, 1982a). What is new is the realization that the ecological constraints concept can be generalized to encompass all categories of cooperative breeders. Furthermore, data are becoming available that, for the first time, allow a test of the major predictions of the model.

In its most concise form, the model states that when ecological constraints exist that severely limit the possibility of personal, independent reproduction, selection will favor delayed dispersal and continued retention of grown offspring within their natal units. It is the restriction upon independent breeding more than any inherent gain realized by grouping that leads to the formation of cooperative breeding units. Differing proximate factors can be responsible for limiting the option of personal reproduction in different species. To date, three categories of constraining factors have been recognized:

Shortage of territory openings. Many cooperative breeders are permanently territorial species that inhabit stable or regularly predictable environments. Furthermore, many have specific ecological requirements such that suitable habitat is restricted. All available high-quality habitat becomes filled or "saturated." Unoccupied territories are rare, and territory turnovers are few. As the intensity of competition for space increases, fewer and fewer individuals are able to establish themselves on quality territories. Assuming that occupancy of a suitable territory is a prerequisite for reproduction, the option of breeding independently becomes increasingly limited. A new individual can become established as a breeder only (1) when it challenges and defeats a current breeder on an occupied territory, (2) when it competes to fill a vacancy that results from a death of a nearby breeder, or (3) when it buds off or inherits a portion of the parental territory itself (Woolfenden and Fitzpatrick 1978; Gaston 1978a). The nonbreeder must wait until it attains sufficient age, experience, and status to enable it to obtain and defend an independent territory.

The notion of habitat saturation does not explain why auxiliaries do not disperse into more marginal habitats where competition is less. Koenig and Pitelka (1981) propose that a second factor was necessary for the evolution of cooperative breeding in permanently territorial species. Not only must optimal habitat be saturated, but marginal habitat must be rare. When this is the case, a maturing individual is severely constrained either from establishing itself as an independent breeder in the optimal habitat or successfully surviving and breeding in an outlying area. By this model (shown graphically in Figure 4.1), cooperative breeding is predicted to occur in those species whose ecological requirements are sufficiently specialized that marginal habitat is rare or which occupy habitats that are physically restricted or relict in distribution (line A in Figure 4.1).

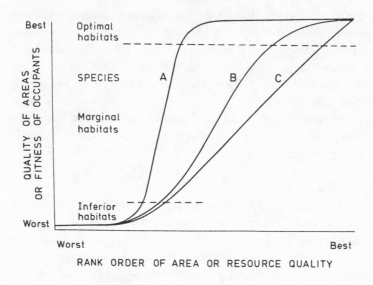

Figure 4.1. Graphic model showing fitness curves for three hypothetical species as a function of habitat quality. When optimal areas are saturated and marginal habitat is rare (as in species A), the option of dispersal and independent breeding is severely constrained. Grown offspring will postpone dispersal and be retained in the parental unit in such circumstances. (Reprinted, by permission, from Koenig and Pitelka, 1981.)

Shortage of sexual partners. A parallel argument can be made to emphasize the shortage of sexual partners rather than spatial territories. Many species of cooperatively breeding birds have a skewed tertiary sex ratio, with an excess of males (e.g., Rowley 1965; Fry 1972; Ridpath 1972; Dow 1977; Reyer 1980). The reason for such skewing is poorly understood, but its effect is to increase competition for mates, leading to a demographic constraint on the option of becoming established as an independent breeder.

Prohibitive costs of reproduction. A major stumbling block to the acceptance of the generality of a breeding constraints model lies in the realization that many species of cooperatively breeding birds reside in areas where the concepts of ecological saturation or shortage of marginal habitat simply do not apply. This is especially true of arid and semiarid environments in Africa and Australia. Not only are cooperative breeders common in such environments, but many are either nomadic or inhabit areas subject to large-scale, unpredictable fluctuations in environmental quality (Rowley 1968, 1976; Harrison 1969; Grimes

1976). The carrying capacity in such environments changes markedly and erratically from year to year. Avian populations, with their relatively low intrinsic rates of increase, cannot track these changes. Consequently, the degree of habitat saturation (if any) changes dramatically across seasons. We cannot speak of any consistent shortage of territory openings as the driving force in the evolution of cooperative breeding. We can, however, still speak of constraints on the option of independent breeding.

In variable and unpredictable environments, erratic change in the carrying capacity create the functional equivalents of breeding openings and closures (Emlen 1982a). As environmental conditions change from year to year, so too does the degree of difficulty associated with successful breeding. In benign seasons, abundant food and cover decrease the costs to younger, less experienced individuals of dispersing from their natal groups and breeding independently. In harsher seasons, the costs associated with such reproductive ventures increase, eventually reaching prohibitive levels. As conditions deteriorate, breeding options become more constrained, and the constraints hit first at the younger, more subordinate individuals. The predicted outcome is the continued retention of such individuals in the breeding groups of older (usually parental) individuals.

Tests of the constraints model

The ecological constraints model predicts that the frequency of occurrence of nonbreeding auxiliaries will vary directly with (1) the degree of difficulty in becoming established as a breeder (for a permanently territorial species residing in saturated habitats), (2) the degree of skew in the population sex ratio (for species facing a shortage of mating partners), and (3) the level of environmental harshness (for species in erratic, unpredictable habitats). Available data for three species of cooperative breeders have been analyzed and plotted in Figures 4.2a–c. Figure 4.2a shows the effect of territory constraints on the retention of offspring in the Acorn Woodpecker. It incorporates data reported by MacRoberts and MacRoberts (1976), Stacey and Bock (1978), and Stacey (1979a), as well as unpublished results kindly provided by P. Stacey and W. Koenig, R. Mumme, and F. Pitelka. Acorn Woodpecker populations from Arizona, New Mexico, and coastal California are pooled in the diagram. The percentage of territories under observation that became vacant (the annual turnover rate) is used as the measure of ecological constraint. Figure 4.2b comes from Rowley's (1965, 1981) studies of cooperative breeding in *Malurus* wrens in Aus-

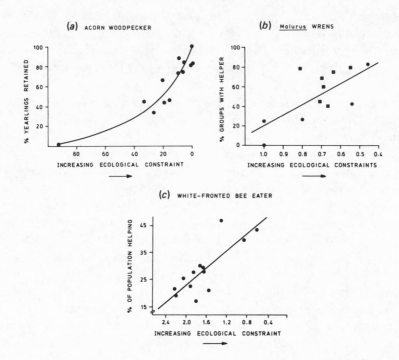

Figure 4.2. The occurrence of helpers plotted as a function of the severity of ecological constraints for three species of cooperative breeders. (a) Acorn Woodpeckers (*Melanerpes formicivorous*); (b) *Malurus* wrens (*M. cyaneus*, circles, and *M. splendens*, squares); (c) White-fronted Bee Eaters (*Merops bullockoides*). See text for details. (a: Data from MacRoberts and MacRoberts, 1976; Stacy and Bock, 1978; Stacy, 1979a; b: redrawn from Rowley, 1981; c: redrawn from Emlen, 1982a.)

tralia and uses the shortage of sexual partners as the index of demographic constraint. Figure 4.2c is taken from Emlen's (1982a) work on the White-fronted Bee Eater (*Merops bullockoides*). This is an example of a colonial, cooperative breeder that inhabits an erratic, unpredictable environment. The data utilize a measure of environmental harshness (log of rainfall occurring in the month preceding breeding) as the index of ecological constraint.

Although the proximate factors responsible for the constraint upon independent breeding differ among the three species, in each case the intensity (magnitude) of the constraint is a good predictor of breeding group size.

The retention of nonbreeding auxiliary individuals in breedings groups is the usual first step in the evolution of cooperative breeding.

By itself, however, it is insufficient to explain the development of actual helping behavior. We must also consider the question, When should a retained auxiliary help in the rearing of the next generation of nestlings?

Genetic advantage of helping

Helping behavior is an intriguing phenomenon from the standpoint of evolutionary theory because of its seemingly altruistic nature. The auxiliary that feeds and defends the nestlings of another breeding pair not only incurs the costs and risks of alloparental care, but it also improves the reproductive success of other individuals in the population relative to its own. Evolutionary theory predicts that helping should only evolve when it actually benefits the individual helper. The focus of much current research on helper-at-the-nest species is on identifying the ways in which helpers benefit from helping.

In formulating a testable hypothesis for investigating the selective advantage of helping, it is necessary to establish an alternative behavioral strategy as a reference point for comparison. The obvious alternative to helping is not helping. However, there are few cases of retained auxiliaries that do not also aid the breeding pair in some way. Rather, nonhelpers typically disperse and attempt to breed independently. Therefore, the alternative strategy for most studies, and the one that the helper species also appear to recognize, is dispersal from the natal territory at the age of sexual maturity and independent breeding thereafter. The evolutionary hypothesis specifically predicts that the fitness of birds that help during their lives will be equal to, or greater than, the fitness of birds that attempt early dispersal and breeding. Such a comparison incorporates the fitness effects of both delayed reproduction and helping. Thus, the question, Why do helpers provide aid at the nest of their parents? is closely tied to the question of why the auxiliaries remain on their natal territory.

Recent studies indicate that the helping phenomenon is a complex one and that there is no single advantage to helping. Each component of helping has its own set of costs and benefits that must be tallied in the final fitness equation. For example, the contribution of the aidgiving component can in theory be separated from the contribution of the delayed reproduction component, and their independent ecological causes and fitness consequences can be investigated. The helping strategy affects not only the production of offspring but also the chances of surviving, so both fertility and survivorship must be incorporated. Finally, since helpers often become breeders later in their lives, short-term effects must be balanced against long-term possibilities. To test quantitatively the evolutionary hypothesis for helping, the investigator must not only identify all components of fitness that differ for helpers and nonhelpers but also combine these into a composite measure of lifetime reproductive success. Long-term data are just now becoming available. The next decade should see the verification or refutation of the adaptationist viewpoint concerning the evolution of helping.

How might helping behavior be adaptive to the helper? Five general types of advantages may accrue to helpers and these will be discussed separately. The first deals with the potential benefit of delayed reproduction, and the second through fifth deals specifically with the question, Why help?

1. Survivorship advantage from delayed breeding. If the prospects for breeding independently are poor, it is to the advantage of a bird to forego reproduction until conditions are better. As discussed in the previous section, such constraints occur when there is (1) a shortage of territory openings, (2) a shortage of sexual partners, or (3) a prohibitive cost of reproduction. These constraints tend to fall more heavily on younger than older birds. By remaining on the parental territory and delaying reproduction, a young bird can ensure its survival through the difficult period and greatly improve its chances of becoming a breeder in the future when conditions are more favorable.

If the short-term cost of not reproducing is more than compensated for by improved longevity, then the survivorship advantage is sufficient to explain evolution of delayed breeding. However, improved survivorship may be only one component of the total benefits that result in a higher lifetime reproductive success for delayed reproducers compared to nondelayers.

2. Breeding experience. Several studies of helpers-at-the-nest species have indicated that experienced breeders are more successful in rearing offspring than inexperienced breeders and that older helpers provide more effective aid than younger ones (e.g., Woolfenden 1975; Lawton and Guindon 1981). These findings suggest that one advantage of helping during the first year or two of life may be improved breeding expertise. In order for breeding experience to be the sole explanation for the helping phenomenon, however, the amount of improvement in a helper's subsequent reproductive success as a breeder would have to outweigh the cost of its not reproducing during the helping years. This will only be true if helping is a more effective learning process than actual breeding. Although helping may be a less risky method of learning, it seems unlikely that it is a more effective one. Additionally, although it is not clear exactly how much experience a bird needs before it can reproduce effectively, the duration of helping in some species greatly exceeds any reasonable notion of what is required.

Consequently, the attainment of breeding experience is unlikely to the sole advantage of delayed breeding and helping. However, it may well be an important contributory factor. The most likely benefit of breeding experience occurs in conjunction with the survival advantage of remaining on the natal territory when ecological conditions from independent breeding are constraining. The breeding experience of helping provides a bonus to auxiliaries who are otherwise forced to remain with their parents. In this case, the improvement in a helper's consequent reproductive success as a breeder need only outweigh the cost of the aid for the helping to be advantageous.

3. Parentally manipulated helping. Helping may function to assure the young bird of continued membership in the group. Youngsters that remain on their natal territory because of ecological constraints on breeding impose a cost on their parents. The young may not only reduce the level of food on the territory and compete with the new generation for food, but they also may disrupt the breeding pair's reproductive efficiency by interfering with nest construction, interrupting copulation and incubation, and attracting predators to the nest (Zahavi 1974). Helping to rear nestlings may be the price the youngsters must pay to be allowed to remain on the parental territory (Gaston 1978a). Parents should be selected to expel young that do not help. Parents are in a position to demand such aid, because it is the youngster that stands to benefit most from retention (Emlen 1982b). If the amount of help a helper provides is determined by parental pressure rather than by the helper itself, then it should be the case that the greater the benefit of retention to the helper, the greater the helping effort the parents can demand.

In most cooperative species, helpers do improve the reproductive success of the breeders (for reviews, see Brown 1978; Emlen 1978). However, helpers usually do not work as hard as the parents themselves, and this is reflected by the fact that per capita reproductive success usually declines with increasing group size (Koenig 1981a). In most species, there is variation in the amount of help given as a function of age, sex, and group size. Proof of the parental manipulation hypothesis for the "advantage" of helping will rest upon its ability to predict these variations in helper effort.

4. Liasons for the future. Given that auxiliaries are temporarily constrained from breeding and are retained on the natal territory as "hopeful reproductives," helping behavior may function in part to speed up the auxiliary's ascendency to breeding status by cementing bonds and creating liasons for future cooperative ventures. There are several mechanisms by which this might work.

a. Helping may heighten the dominance position of the helper relative to other group members and improve its position in the queue for breeding slots on the parental territory or on neighboring territories.

b. Helping may produce potential liasons that later help the auxiliary to attain breeding status. In some helper-at-nest species, coalitions of siblings or half-siblings disperse together. Such groups of dispersers, it has been claimed, have a greater chance of winning territorial disputes than solitary individuals. (Ridpath 1972; Ligon and Ligon 1978a; Gaston 1978b; Koenig 1981b). To the degree that the social bonds formed during helping itself are important for the development of these coalitions, we may conclude that helping in one season may increase a helper's prospects for breeding in future years (Ligon and Ligon 1978a).

c. By helping cooperate in territorial defense, an auxiliary may bring about an expansion of the natal territory. Cases of such expansion being followed by the auxiliary budding off a portion of the natal area as its own have been reported by Woolfenden and Fitzpatrick (1978) and Gaston (1978a). In this way, helpers may inherit part or even all of the parental territory itself.

5. Increased inclusive fitness via collateral kin. All of the previously mentioned benefits of helping increase the fitness of a helper via the personal component of inclusive fitness, that is, lifetime production of offspring. A second type of genetic gain inherent in helping behavior involves that component of inclusive fitness attributable to gene copies through collateral kin. The debate over the importance of kin selection has led to an unfortunate and, in our opinion, unproductive controversy.

It is true that for the vast majority of helper species for which data are available, the helpers are predominantly grown offspring that remain with their parental groups and help to rear full or half-siblings. Thus, the kin component of inclusive fitness is large and could prove to be a major factor in the evolution of helping behavior. Some workers, however, have mistakenly taken this correlation of group relatedness with helping as the sole evidence to build a case for the existence of and importance of kin selection in the evolution of seemingly altruistic behaviors. Other workers have erred in the opposite direction. By finding ways in which individual helpers gain personally through helping (i.e., points 1, 2, and 4), they have concluded that kin selection is either unimportant to understanding helping behavior or that kin selection does not even exist.

Both approaches are too narrow in outlook and miss the point. Evolution is the process of changing gene frequencies in a population through time. Natural selection does not distinguish between a gene copy produced by a direct descendent and one fostered by a collateral kin. All gene copies are tallied, irrespective of who or what was responsible for their survival in the population. The interesting question, then, is not whether kin selection exists or not, but rather whether collateral kin interactions (as opposed to personal offspring production) have been an important or essential component in the evolution of helping behavior.

To take this debate out of the realm of semantic argument and into the realm of quantitative biology, Vehrencamp (1979) has devised a simple index for determing the relative importances of individual and kin selection to the current evolutionary maintenance of a behavioral trait. The index calculates the proportion of the total gain in inclusive fitness that is due to the kin component for any type of cooperation among kin, compared to the noncooperative situation:

$$I_k = \frac{(W_{RA} - W_R)r_{ARy}}{(W_{AR} - W_A)r_{Ay} + (W_{RA} - W_R)r_{ARy}}$$

where $(W_{RA} - W_R)$ is the change in lifetime reproductive success of the recipient, R, when it is aided by the donor A, $(W_{AR} - W_A)$ is the change in lifetime reproductive success of the donor, A, when it provides aid to R, r_{ARy} is the relatedness of A to R's young, and R_{Ay} is the relatedness of A to its own young. The index is applicable only to situations in which the net change in inclusive fitness of A (i.e., the denominator) is positive. When I_k is greater than 1, there is a net cost to A's personal reproduction, and pure kin selection is acting. When I_k is less than 0, than A is manipulating R at a net cost to R's personal reproduction. When I_k is between 0 and 1, both the personal component and the kinship component are increased by the cooperation, and the value of the index gives the proportion that is due to the kin component.

Rowley (1981) has calculated I_k for the Splendid Wren (*Malurus splendens*), an Australian helper-at-the-nest species. He found that the value of I_k ranged from 0.35 to 0.51 for a typical male helper. This suggests that both the kin and personal component are increased via the helping behavior and that the personal component is slightly more important than the kin component. Similar calculations on other cooperative breeders also yield values of I_k of 0.4 to 0.5 (Vehrencamp 1979).

The main point of this section is that helping does not seem to present the evolutionary paradox that it first appeared to do. Severe ecological constraints often eliminate or greatly restrict the option of independent breeding. When viewed against the backdrop of this constraint, a number of possible adaptive functions can be proposed for helping behavior. These range from direct improvements in a helper's lifetime reproductive success to improvements in the kinship component of fitness by aiding close relatives. The differing importance of such factors for each type of cooperative system remains one of the principal challenges of future work on cooperative birds.

The communal breeding perspective: balancing the within-group conflict

Despite the fact that most cooperatively breeding avian species are characterized by one-sided helping behavior, one cannot dismiss the communal breeding systems. In these species, egg laying and/or fertilization are shared, so that the fitness benefits as well as the parental care costs are more equitably distributed among group members. Many workers have conceived of the communal breeders as belonging to a totally separate and distinct category from the helper species. However, communal and helper systems differ in only one key parameter: the degree of bias in current personal reproductive benefits. The degree of bias is a continuously varying characteristic, ranging from no bias (i.e., equal division of benefits) to a very high bias (i.e., zero personal benefits to some members). Intermediate levels of bias are not only theoretically possible, many of the so-called equitably communal breeders in fact show slight to moderate degrees of bias. This perspective on cooperative breeding now forces the investigator of helper-at-the-nest species to address a third question, Why don't helpers reproduce communally along with their parents? Once the communal breeding strategy is established as an alternative hypothesis to helping, the focus of research is shifted away from just the helper. The breeder/helper *relationship* is now viewed as a general dominant/subordinate interaction in which the interests and strategies of the two parties diverge to different degrees. This view forces us to recognize that the cooperatively breeding group is composed of competing individuals with conflicting strategies. Different individuals hold different options and different leverages for imposing their "will" on the group. The social result of these conflicts, in terms of the degree of bias and type of cooperative society, must represent the balance among the various competing interests.

Types of conflicts

Emlen (1982b) has employed this approach in evaluating the nature of behavioral interactions in helper species. In modeling the conflict between helpers and breeders, he used the inclusive fitness equations of Trivers (1974), West-Eberhard (1975), and Vehrencamp (1979) to establish the zones of conflict. The direction of the conflict, that is, whether the breeder or the helper had the greatest leverage, changed depending on two critical variables: (1) the severity of the ecological constraints on younger, subordinate individuals and (2) the magnitude of the advantage (or disadvantage) of group living.

The results of this analysis are summarized graphically in Figure 4.3. When retention and helping benefit both helper and breeder, there is clearly no conflict and helping will evolve (speckled zone). Similarly, when retention and helping are disadvantageous to both parties, helping will not evolve (white zone). Because of certain asymmetries in the inclusive fitness equations, however, there are two zones of breeder/helper conflict over the issue of helping. The size of these conflict zones increases as the degree of relatedness between breeder and helper decreases (compare Figure 4.3a and b). In conflict type I (striped zone), severe ecological constraints effectively prohibit dispersal and independent breeding by yearlings, yet retention of such auxiliaries results in a loss to the personal fitness of the breeder via competition, interference, greater conspicuousness to predators, and so forth. This means that the yearling gains, but the breeder loses, from retention. Assuming that the breeder is dominant over the yearling, the breeder will either expel the subordinate from the group or demand that the subordinate aid in the care of its young so as to increase the fitness of the breeder above the breeder retention tolerance line (labeled B-B). Notice that retention of related auxiliaries can evolve even when there is a cost in personal reproduction to the breeder, since the

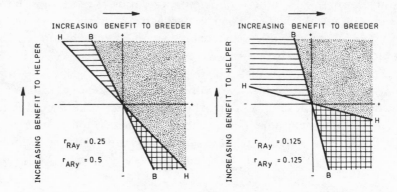

Figure 4.3. Fitness space diagrams for breeders and helpers of varying degrees of relatedness. The inclusive fitness of a breeder is increased through the retention of a helper in all areas to the right of line B–B; the inclusive fitness of an auxiliary is increased by such an association in all areas to the right of line H–H. Speckled areas indicate zones where continued association is mutually beneficial. Striped and crosshatched areas represent zones of conflict, in which one party gains, but the other loses, by the association. Two distinct types of breeder–helper conflict, and the behaviors expected with each, are described in the text. Note that as the degree of relatedness between breeder and helper decreases, the zones of conflict increase. (Modified from Emlen, 1982b.)

improved survival chances of a retained auxiliary can increase the kinship component of the breeders' inclusive fitness. In conflict type II (crosshatched zone), the breeder gains from the presence of the yearling but the yearling loses from retention. This will occur when ecological constraints against yearling breeders are slight, but there is a direct benefit of group living to breeders. Even though the breeder would still benefit from the presence of the helper, the breeder has little leverage to prevent helper dispersal. The only option the breeder has is to entice the helper to stay by "forfeiting" (sensu, Alexander 1974) some of its fitness benefits via shared fertilization, shared egg laying, or reciprocation of helping (Emlen 1982b). The dominant breeder must allow the helper a large enough genetic contribution to the clutch to increase the helper's fitness above its retention tolerance line (labeled H-H). This, of course, leads to a communal breeding system with a lower degree of bias.

This model illustrates the nature of conflicts within groups and shows how the options and leverages of the different group members change as a function of changing environmental conditions. It provides a far more reasonable resolution to the question of why helpers help than the verbal arguments presented in the previous section. Further-more, it specifically shows that the strongly biased helper-at-the-nest system can only evolve when the helper stands to gain more from retention in the long run than the breeder does, whereas lower-biased communal systems evolve when the breeder gains more than the helper.

How much bias?

Following up this conflict approach, we can now pursue the questions of how much nestling care the helper should provide, and how much fitness the dominant must forfit to keep the helper in the group. A short-term model of inclusive fitness costs and benefits is required, and the boundary conditions for such a model are that both the helper and breeder must break even, that is, all solutions must lie within the speckled area of the graph in Figure 4.3. Furthermore, this model assumes that, within the limits set by the subordinate's options, the dominant individual has the greatest leverage to push the fitness advantage in its favor. Given these assumptions, it is a simple matter to solve for the optimal degree of bias (Vehrencamp 1979, 1980, in press).

Take the simplest case of a group of unrelated individuals where grouping benefits all group members on average. The function relating the per capita reproductive success and group size, k, is given by $\bar{W}(k)$ (Figure 4.4a). For each unit of fitness that the dominant can usurp from a subordinate, one unit of fitness accrues to the dominant. For simplicity, we also assume that there is only one dominant in the group and that all subordinates are affected by the same amount. The dominant cannot lower the fitness of any one subordinate below the fitness of a solitary breeder, $\bar{W}(1)$, otherwise subordinates will be selected to leave and breed solitarily. The lowest possible fitness of a subordinate in a group $\bar{W}_\omega(k)$, is therefore depicted as a horizontal line intersecting W (1). The fitness increment of the dominant, $W_\alpha(k)$, is the difference between $\bar{W}(k)$ and $\bar{W}(1)$ times the number of subordinates. The bias in fitness is, therefore, reflected in the degree to which the fitness of dominant and subordinate diverge. Clearly, the greater the average benefit to grouping compared to solitary living, the greater the bias can be.

For groups of related individuals, a similar graph can be derived, but in this case the subordinates leave the group when their *inclusive fitness* would be greater if they bred solitarily (Figure 4.4b). Here, the personal fitness of the subordinate can go well below the personal fitness of a solitary breeder, and the higher the degree of relatedness, the

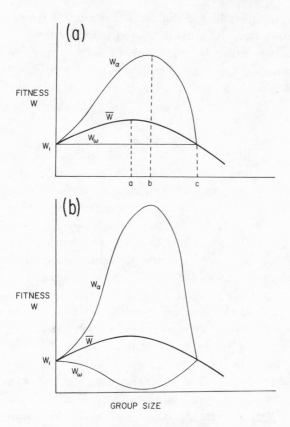

Figure 4.4. The maximum degree of bias in fitness that can be generated between dominants and subordinates in groups containing (a) unrelated individuals and (b) individuals related by a coefficient of relatedness $r = 0.25$. $\bar{W}$ is the average or per capita fitness of group members, W_α is the fitness of the dominant, and W_ω is the fitness of a subordinate. The distance between the W_α and W_ω curves represents the amount of bias possible at that group size. See text for further explanation.

greater the bias can be. As in the unrelated case, grouping and biasing can only occur when there is a net benefit to grouping, $\bar{W}(k)$ $\bar{W}(1)$.

As we saw earlier, there are several situations in which grouping can be advantageous. These conditions are depicted graphically in Figure 4.5. When group breeding leads to a per capita increase in fitness compared to solitary breeding, there is a direct benefit to grouping (Figure 4.5a). In another case, grouping could lead to a net decrease in per capita fitness compared to solitary breeding (Figure 4.5b, solid line). However, if the cost of dispersing to breed independently is high, the intrinsic benefit of independent breeding must be devalued by this

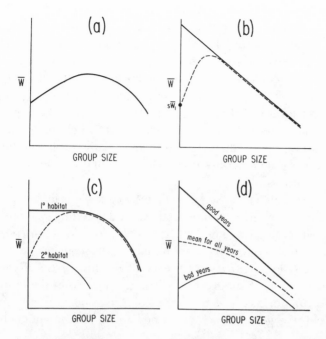

Figure 4.5. Four ways to generate a benefit to grouping: (a) direct group benefit, for example, White-winged Chough; (b) low probability of dispersing to the optimal solitary situation, for example, Scrub Jay; (c) limited primary habitat with no disadvantage due to grouping, for example, Groove-billed Anis; and (d) direct benefit to grouping occurring unpredictably in poor years only, for example, White-fronted Bee Eaters. See text for further explanation.

cost. This leads to an effective "hump" in the $W(k)$ curve, and grown offspring may be "forced" to remain on their natal territories (Figure 4.5b, dashed line). The probability of dispersing successfully, here denoted as s, can be calculated for permanently territorial species as the number of breeding slots available relative to the number of auxiliaries competing for them (see Rowley 1981) and is an inverse measure of the level of habitat saturation or ecological constraint. This condition can only lead to groups of close relatives since the groups form by retention of natal members. Another situation in which grouping might be favored is illustrated in Figure 4.5c. Here, there is a primary habitat that is limited and not greatly detrimental to grouping over some range of group sizes, and a secondary habitat that still supports breeders and is not limited. The two types of habitats must be close by or arranged in a mosaic so that dispersers have a choice between them. Large groups will occur on good territories, small groups or pairs on the poorer territories. This situation is graphically similar to the Verner–Orians polygamy model (Orians 1969) and can lead to groups of nonrelatives. Finally, the conditions for grouping may change unpredictably from year to year (Figure 4.5d). If grouping is disadvantageous in good years and advantageous in bad years, then birds will hedge their bets by attempting to breed solitarily in all years but will switch to groups if conditions deteriorate. If there is only a short time for breeding, some birds will initiate breeding but give up their nest to help others during bad times. The model predicts that only close relatives will be willing to accept this cost *and* be in a position to benefit from groupng via heightened inclusive fitness.

It can be seen that certain sets of ecological conditions lead to groups composed of either relatives or nonrelatives, and breeding systems characterized by either high or low bias. In theory, four kinds of cooperative social systems can occur. To summarize the conclusions of the previous models, let us now take the point of view of the dominant individual in a group of subordinate relatives and, within the limits set by the subordinate, ask which of the following possible alternative strategies yields the greatest inclusive fitness for the dominant (Vehrencamp, in press):

1.   Disperse the relatives and all breed solitarily
2.   Retain the relatives and cooperate equitably (without bias)
3.   Retain the relatives and impose the greatest possible bias
4.   Disperse the relatives and recruit nonrelatives, imposing the greatest possible bias
5.   Disperse the relatives and cooperate equitably with recruited nonrelatives

The critical variables once again are the level of ecological constraints (s in this model) and the effect of grouping on the per capita reproductive success (ratio of $\bar{W}$ (k) to $\bar{W}$ (1)). Solitary breeding is best when s is high (i.e., constraints are slight) and group breeding is disadvantageous. Equitable breeding with relatives can only occur when s is high, grouping is advantageous, and biasing reduces the efficiency of reproduction. Biased breeding with relatives evolves whenever s is low. Strong biasing will result even when grouping is disadvantageous if s is low enough. It is best for the dominant to disperse his relatives, recruit nonrelatives, and impose a bias upon them when s is high and grouping is highly advantageous. This is because the dominant gains a greater inclusive fitness by allowing the relatives to disperse and join other groups where their fitness potential is high. Finally, equitable (unbiased) breeding with nonrelatives is usually the worst strategy for all combinations of the critical variables and will therefore only occur when grouping is advantageous, biasing is impossible, *and* no relatives are available.

A final point from this analysis is worth noting with regard to avian helpers at the nest. In long-term monogamous animals such as many birds, a low s that favors retention of offspring leads to very high degrees of relatedness. The model proposes that this in turn always leads to high degrees of bias. In fact, in monogamous animals, the short-term personal fitness of a retained offspring is always predicted to be zero. As long as there is a net benefit to remaining, an offspring will help without requiring personal reproduction. If either s increases or the benefit the helper can provide to the breeders decreases, then the offspring will leave (Emlen 1978, 1982b). Thus, there are only two possible alternatives to a yearling of monogamous parents when s is low, helping or leaving, and researchers are therefore somewhat justified in only considering these two possibilities. Communal breeding rarely occurs with monogamous parents and their offspring. It has been suggested that this is entirely due to the problems of inbreeding, and this may be the explanation in some cases (Koenig and Pitelka 1979; Stacey 1979b). However, many helper systems have available nonrelatives in the group (i.e., babblers, Zahavi 1974; woodhoopoes, Ligon and Ligon 1978b), some systems inbreed anyway (Pukeko, Craig 1980, pers. comm.), and, in eusocial insects, inbreeding is commonplace (Cowan 1979). Deleterious genes are rapidly selected out in an inbreeding population, and any long-term negative effect probably will be insufficient compared to other sources of juvenile mortality. If equitable breeding with parents were highly advantageous, it would evolve in spite of a slight inbreeding cost. The observed high bias in helper species is therefore more in line with the predictions of the conflict models than with inbreeding avoidance.

Cooperatively breeding avian species are falsely typified as living in harmonious, aid-giving societies. Rather, they represent societies faced with sometimes severe ecological hardships that limit their behavioral options and societies in which competitive conflicts of interest reach extreme limits. Viewing the various forms of cooperative breeding as the balance of interests between conflicting group members, with dominants having greater behavioral leverage, provides a more all-inclusive understanding of each system and the evolutionary forces that shaped it.

Literature cited

Alexander, R. D. 1974. The evolution of social behavior. *Annu. Rev. Ecol. Syst.* *5*:325–383.

Bertram, C. C. R. 1978. Living in groups: predators and prey. In J. Krebs and N. B. Davies, (eds.). *Behavioral ecology: an evolutionary approach.* pp. 64–96. Oxford: Blackwell Scientific Publications.

Brown, J. L. 1963. Social organization and behavior of the Mexican Jay. *Condor 65*:126–153.

– 1970. Cooperative breeding and altruistic behaviour in the Mexican Jay, *Aphelocoma ultramarina. Anim. Behav. 18*:366–378.

– 1974. Alternate routes to sociality in jays – with a theory for the evolution of altruism and communal breeding. *Am. Zool. 14*:63–80.

– 1978. Avian communal breeding systems. *Annu. Rev. Ecol. Syst. 9*:123–156.

Brown, J. L., and E. R. Brown. 1980. Reciprocal aid-giving in a communal bird. *Z. Tierpsychol. 53*:313–324.

– 1981a. Extended family system in a communal bird. *Science 211*:959–960.

– 1981b. Kin selection and individual selection in babblers. In R. D. Alexander and D. Tinkle (eds.). *Natural selection and social behavior: recent research and new theory,* p. 244–256. New York: Chiron Press.

Cowan, D. A. 1979. Sibling matings in a hunting wasp: adaptive inbreeding? *Science 205*:1403–1405.

Craig, J. L. 1980. Pair and group breeding behaviour of a communal gallinule, the Pukeko, *Porphyrio p. melanotus. Anim. Behav. 28*:593–603.

Dow, D. D. 1977. Reproductive behavior of the Noisy Miner, a communally breeding honeyeater. *Living Bird 16*:163–185.

Emlen, S. T. 1978. The evolution of cooperative breeding in birds. In J. R. Krebs and N. B. Davies (ed.). *Behavioral ecology: an evolutionary approach,* pp. 245–281. Oxford: Blackwell Scientific.

– 1981. Altruism, kinship and reciprocity in the White-fronted Bee-eater. In R. D. Alexander and D. Tinkle (eds.) *Natural selection and social behavior: recent research and new theory,* pp. 217–230. New York: Chiron Press.

– 1982a. The evolution of helping. I. An ecological constraints model. *Am. Nat. 119*:29–39.

– 1982b. The evolution of helping. II. The role of behavioral conflict. *Am. Nat. 119*:40–53.

Fry, C. H. 1972. The social organization of bee-eaters (Meropidae) and cooperative breeding in hot-climate birds. *Ibis 114*:1–14.

Gaston, A. J. 1978a. The evolution of group territorial behavior and cooperative breeding. *Am. Nat. 112*:1091–1100.

– 1978b. Demography of the Jungle Babbler *Turdoides striatus. J. Anim. Ecol. 47*:845–879.

Grimes, L. G. 1976. The occurrence of cooperative breeding behavior in African birds. *Ostrich 47*:1–15.

Harrison, C. J. 1969. Helpers at the nest in Australian passerine birds. *Emu 69*:30–40.

Hegner, R. E., S. T. Emlen, and N. J. Demong. 1982. Spatial organization of the White-fronted Bee-eater. *Nature (London) 298*:264–266.

Koenig, W. D. 1981a. Reproductive success, group size, and the evolution of cooperative breeding in the Acorn Woodpecker. *Am. Nat. 117:*421–443.

– 1981b. Space competition in the Acorn Woodpecker: Power struggles in a cooperative breeder. *Anim. Behav. 29:*396–409.

Koenig, W. D., and F. A. Pitelka. 1979. Relatedness and inbreeding avoidance: counterploys in the communally nesting Acorn Woodpecker. *Science 206:*1103–1105.

– 1981. Ecological factors and kin selection in the evolution of cooperative breeding in birds. In R. D. Alexander and D. Tinkle (eds.). *Natural selection and social behavior: recent research and new theory,* pp. 261–280. New York: Chiron Press.

– in press. Female roles in cooperatively breeding Acorn Woodpeckers. In S. W. Wasser (ed.). *Female social strategies.* New York: Academic Press.

Lawton, M. F. and C. F. Guindon. 1981. Flock composition, breeding success, and learning in the Brown Jay. *Condor 83:*27–33.

Ligon, J. D., and S. H. Ligon. 1978a. The communal social system of the Green Woodhoopoe in Kenya. *Living Bird 17:*159–198.

– 1978b. Communal breeding in Green Woodhoopoes as a case for reciprocity. *Nature (London) 276:*496–498.

MacRoberts, M. H., and B. R. MacRoberts. 1976. Social organization and behavior of the Acorn Woodpecker in central coastal California. *Ornithol. Monogr. 21:*1–115

Maynard Smith, J., and M. G. Ridpath. 1972. Wife sharing in the Tasmanian native hen, *Tribonyx mortierii:* a case of kin selection? *Am. Nat. 106:*447–452.

Mumme, R. L., W. D. Koenig, and F. A. Pitelka. in press. Infanticide in the communally nesting Acorn Woodpecker: sisters destroy each other's eggs.

Orians, G. A. 1969. On the evolution of mating systems in birds and mammals. *Am. Nat. 103:*589–603.

Reyer, H.-U. 1980. Flexible helper structure as an ecological adaptation in the Pied Kingfisher (*Ceryle rudis*). *Behav. Ecol. Sociobiol. 6:*219–227.

Ridpath, M. G. 1972. The Tasmanian Native Hen, *Tribonyx mortierrii. CSIRO Wildl. Res. 17:*1–118.

Rowley, I. 1965. The life history of the Superb Blue Wren, *Malarus cyaneus. Emu 64:*251–297.

– 1968. Communal species of Australian birds. *Bonn Zool. Beitr. 19:*362–370.

– 1976. Co-operative breeding in Australian birds. In H. Frith and J. H. Calaby (eds.). *Proceedings of the XVI International Ornithological Congress (Canberra, Australia),* p. 657–666.

– 1978. Communal activities among White-winged Choughs, *Corcorax melanorhamphus. Ibis 120:*178–197.

– 1981. The communal way of life in the Splendid Wren, *Malurus splendens. Z. Tierpsychol. 55:*228–267.

Rowley, I., S. T. Emlen, A. J. Gaston, and G. E. Woolfenden. 1979. A definition of "group." *Ibis 121:*231.

Selander, R. K. 1964. Speciation in wrens of the genus Campylorhynchus. *Univ. Calif. Berkeley Publ. Zool. 74:*1–224.

Stacey, P. B. 1979a. Habitat saturation and communal breeding in the Acorn Woodpecker. *Anim. Behav. 27:*1153–1166.

– 1979b. Kinship, promiscuity, and communal breeding in the Acorn Wood-pecker. *Behav. Ecol. Sociobiol. 6:*53–66.

Stacey, P. B., and C. E. Bock. 1978. Social plasticity in the Acorn Wood-pecker. *Science 202:*1297–1300.

Trail, P. W., S. D. Strahl, and J. L. Brown. 1981. Infanticide in relation to individual and flock histories in a communally breeding bird, the Mexican Jay (*Aphelocoma ultramarina*). *Am. Nat. 118:*72–82.

Trivers, R. L. 1974. Parent-offspring conflict. *Am. Zool. 14:*249–264.

Vehrencamp, S. L. 1977. Relative fecundity and parental effort in commu-nally nesting anis, *Crotophaga sulcirostris. Science 197:*403–405.

– 1978. The adaptive significance of communal nesting in Groove-billed Anis (*Crotophaga sulcirostris*). *Behav. Ecol. Sociobiol. 4:*1–33.

– 1979. The roles of individual, kin, and group selection in the evolution of sociality. In P. Marler and J. G. Vandenbergh (eds.). *Handbook of behavioral neurobiology,* pp. 351–394. Vol. 3. New York: Plenum.

– 1980. To skew or not to skew? In *Proceedings of the XII International Ornitho-logical Congress (Berlin, W. Germany),* Vol. 2, pp. 869–874.

– in press. A model for the evolution of despotic versus egalitarian societies. *Anim Behav.*

West-Eberhard, M. J. 1975. The evolution of social behavior by kin-selection. *Q. Rev. Biol. 50:*1–33.

Wilson, E.O. 1971. *The insect societies.* Cambridge, Mass.: Belknap Press.

Woolfenden, G. E. 1973. Nesting and survival in a population of Florida Scrub Jays. *Living Bird 12:*25–49.

– 1975. Florida Scrub Jay helpers at the nest. *Auk 92:*1–15.

– 1981. Selfish behavior by Florida Scrub Jay helpers. In R. D. Alexander and D. Tinkle eds.) *Natural selection and social behavior: recent research and new theory,* pp. 257–260. New York: Chiron Press.

Woolfenden, G. E., and J. W. Fitzpatrick. 1978. The inheritance of territory in group-breeding birds. *Bioscience 28:*104–108.

Zahavi, A. 1974. Communal nesting by the Arabian Babbler: a case of indi-vidual selection. *Ibis 116:*84–87.

Commentary

J. DAVID LIGON

This chapter provides a fine synthetic overview of many of the most important aspects of cooperative breeding. The emphasis that Emlen and Vehrencamp place on within-group conflict and competition for reproductive success probably results from insights gained via their studies on White-fronted Bee Eaters (*Merops bullockoides*) and Groove-billed Anis (*Crotophaga ani*), respectively, and is justified for those and a good many other cooperative breeders. However, many other species are characterized by social units composed of a single monogamous breeding pair, plus a variable number of younger and subordinate helpers that ordinarily do not participate at all in personal reproductive activities. Competition for copulations or placement of eggs in the nest as occurs in the bee eaters and anis is very rare or essentially nonexistent in those species (e.g., Florida Scrub Jay, *Aphelocoma c. coerulescens;* Woolfenden 1975).

In their concluding remarks, Emlen and Vehrencamp seem to deny that apparently "harmonious, aid-giving" aspects of cooperative breeding are real. I suggest that some cooperative breeders do live in societies as harmonious and aid giving as those of any other vetebrate. This, of course, is not to say that conflict does not occur or that individuals are not behaving in manners designed to promote their own fitness. Rather, my point is that under certain social and demographic conditions complex levels of cooperation and reciprocity (Alexander 1974) can evolve in birds. In this commentary, I will describe some examples supporting this contention.

Reciprocity appears to be a plausible explanation for cooperative behaviors exhibited within closed social systems between individuals that are both genetically unrelated and of the same sex (e.g., Packer 1977); that is, the behaviors cannot be associated with pairbonding. Such species should prove to be especially informative, for it is only in these that kinship considerations and personal gain strategies can be empirically teased apart. This separation is critical to the eventual understanding of cooperative breeding and kinship theory in general. As Emlen and Vehrencamp point out, the significant question is, Are kinship ties fundamental to the evolution of cooperative breeding? (See Koenig and Pitelka 1981.)

Nest helpers unrelated to either breeder occur uncommonly but regularly (8–11% of helper population each year, 1978–1981) in Green Woodhoopoes (*Phoeniculus purpureus*) in central Kenya (Ligon and Ligon 1978a,b; Ligon, unpub.) and commonly in Pied Kingfishers (*Ceryle rudis*) on Lake Victoria, Kenya (Reyer 1980). In addition to nest helpers, cooperative breeding occurs in the context of probably unrelated male–male "teams" that compete with other groups of males for attraction of females. This is best known in manakins of the genus *Chiroxiphia* (Snow 1977; Foster 1977, 1981). I will briefly consider aspects of the biology of these birds that might have been critical to the evolution of cooperation between individuals that either are not or not known to be at all closely related genetically.

Cooperation involving nest helpers

Green Woodhoopoe

Our studies on Green Woodhoopoes in Kenya over 7 years have impressed us (Ligon and Ligon 1978a,b, unpub.) with the importance to these birds of personal alliances characterized by mutually beneficial behaviors. Although intragroup conflict of interest and aggression certainly exists within woodhoopoe social units, these generally are uncommon, and we believe usually less productive for each group member than cooperative interactions, even when individuals within the same group are not genetically related.

Social units are composed of a single breeding pair that almost always have a variable number of helpers of one or both sexes. As is the case for many cooperative breeders, each flock is highly territorial, with the most suitable habitat owned by one group or another. Thus, establishment of new territories is rare. Unassisted pairs cannot establish a territory in areas of good habitat because of the aggression of neighboring flocks; all of the few new territories originating in good habitat during the course of our studies developed over a prolonged period and involved from three to six birds. In about half of these, unrelated birds of the same sex merged, forming a social unit large enough to defend a territory (Ligon and Ligon, unpub.), with the oldest and dominant bird of each sex eventually becoming a breeder. Most procurement of new territorial space, however, occurs via movement of a unisexual group of nonbreeding woodhoopoes from one territory into a nearby established territory after most or all previous flock members of their sex have disappeared (Ligon and Ligon 1978b). Again, it is the oldest birds that breed. In both situations, groups of two or more individuals of each sex are ordinarily involved in acquisition and retention of a new territory (also see Koenig 1981).

A demographic variable – death rate – appears to have been critical to the origin of cooperative breeding. Mortality of adult woodhoopoes is high as compared to some other cooperative breeders (e.g., 30–45% year^{-1} versus 17–20% year^{-1} for adult Florida Scrub Jays; Woolfenden 1973, pers. comm.) and is primarily a result of predation at the roost cavity (Ligon and Ligon 1978b; Ligon 1981). Parents allow their grown offspring to remain in the safety of their territory until a vacancy appears in a nearby territory (Ligon 1981). These young birds later become assistants to the parents and give them a competitive edge over mated pairs with no helpers. This differential between the success of breeding pairs that allowed their grown offspring to remain at home and those that did not probably led to the group-based population structure of present-day Green Woodhoopoes.

Because most woodhoopoe social units consist of breeders plus their subordinate allies (1–10 or more), usually relatives, a single individual or lone pair is placed at a great competitive disadvantage for space vis-a-vis the larger groups. However, lone woodhoopoes without supportive kin do occur regularly as a result of the high mortality and subsequent displacement of a surviving singleton from a territory by a group or, in the case of some individual females in particular, dispersal (Ligon 1981). Because of the group-based population structure, initially based on parent–offspring relationships, the best option open to most singletons, if they are ever to gain territory ownership and eventually breed, is to procure one or more allies of their sex. Often an older and dominant (alpha) bird allows a younger unrelated subordinate (beta) woodhoopoe of its sex that is unaffiliated with another flock to join it. Later, beta becomes an active helper to the nestlings of alpha. Eventually, depending on the pattern of mortality, some of the offspring of alpha that beta had earlier provided care for assist beta in obtaining and retaining breeding status and then become helpers for beta's offspring (Ligon, unpub.).

Pied Kingfisher

At Lake Victoria, in Kenya, 27 of 41 breeding pairs (66%) of Pied Kingfishers had one or more male helpers. About half of the helper population consisted of individuals unrelated to either breeder, with other helpers, also males, known or thought to be yearling sons of at least one of the breeding pair. In contrast, at Lake Naivasha, Kenya, helpers assisted only 5 of 18 breeding pairs (28%), and all were thought to be related to at least one of the breeders (Reyer 1980). This striking difference both in frequency and origins of helpers may basically be due to differences in their effects on the reproductive success of breeders at the two sites. At Lake Victoria, where the rearing of young is difficult, a single helper doubles the number of young kingfishers fledged per nest (1.8 vs. 3.6); at Lake Naivasha, the effect of a helper was much less (3.7 vs. 4.3 young per nest). I suggest that at Lake Victoria, where the lack of a helper is extremely costly to the reproductive success of breeders as compared to those pairs with a helper, natural selection has favored breeders that acquire one or more helpers, either by recruiting a matured offspring or, failing to do so, by allowing an unrelated male to join them, *after* the young hatch (no risk of shared paternity to the male breeder).

Reyer (1980) suggests two ways by which secondary helpers may be repaid for their help. First, acquisition of knowledge of a colony in 1 year should improve reproductive success in future years via early breeding and recruitment of secondary helpers. Second, a male helper may mate with the breeding female the next year and may, moreover, acquire a primary helper (his mate's offspring that he has helped to rear) for his own subsequent reproductive efforts. Reyer presents a few examples illustrating these points.

In some important respects, the biology of the nonterritorial, colonial Pied Kingfisher resembles that of the highly territorial, noncolonial Green Woodhoopoe. In both species, helpers clearly enhance overall reproductive success in one or more ways and annual turnover of adults is high, which leads to frequent breeding vacancies for the helpers. The kingfisher and woodhoopoe systems also are similar in that unrelated individuals are allowed to join a dominant mated pair *only* when the dominant birds stand to obtain a direct benefit at little cost. For breeding kingfishers at Lake Victoria, the benefit is readily seen – a doubling of reproductive success with only one helper. Direct reproductive gains via helpers are less obvious in the woodhoopoes, but, as described earlier, helpers probably are even more critical to potential breeders in that an unassisted pair has an extremely low probability of breeding at all, due to the difficulty of gaining or holding a territory. In both the kingfishers at Lake Victoria and the woodhoopoes, the "preadaptation" of related helpers together with the great disparity in reproductive success between breeders with and without such helpers may have led to selection for behavioral traits of potential breeders that provide them with allies, related or not. The same selective pressures later repay the unrelated helpers for their beneficence as they ascend to breeding status.

Cooperative displays of male manakins

Male manakins of the genus *Chiroxiphia* exhibit a striking form of teamwork in their courtship patterns, with two or three birds engaging in a precopulatory display that leads to matings by only the dominant male of the group (Snow 1977; Foster 1977, 1981). At present, there is no evidence to suggest that such cooperating males are closely related and some sound reasons for suspecting that they are not (Foster 1977, 1981). If not, their cooperation must be based on factors other than kinship ties. Display sites are limited and may persist longer than individual birds (Snow 1977); although a subordinate young male receives few or no copulations as compensation for his participation in the displays, he stands eventually to inherit both the site and dominance status (Foster 1977, 1981), along with any still younger males attracted to the site, and to obtain the great majority of copulations with females that come to the display area.

The social system of these male *Chiroxiphia* manakins appears to be analogous in some important ways to those of Green Woodhoopoes and perhaps Pied Kingfishers. First, younger, subordinate birds provide critically important aid to older, socially dominant individuals. Second, they will be repaid by eventual inheritance of ownership of the territory or display site plus any still younger birds present (true only for males in Pied Kingfishers and *Chiroxiphia* manakins and for both sexes in Green Woodhoopoes) as their assistants, provided that they live long enough. Among emigrant groups of male woodhoopoes, the beta individual does live significantly longer than the alpha bird (Ligon, unpub.), and this may be true for manakins as well (Foster 1981).

Concluding remark

This brief commentary has dealt with tropical birds. Complex, apparently cooperative, behaviors also occur in a variety of temperate zone species between individuals known or thought to be unrelated. In North America, this is recorded for turkeys, *Meleagris gallopavo* (Watts and Stokes 1971; see Balph et al. 1980); Western and Herring gulls, *Larus occidentalis* and *Larus argentatus*, respectively (Pierotti 1980, 1982); Chimney Swifts, *Chaetura pelagica* (Dexter 1952, 1981); Yellow-bellied Sapsuckers, *Sphyrapicus varius* (Kilham 1977, Pinkowski 1978); Barn Swallows, *Hirundo rustica* (Myers and Waller 1977); Eastern and Mountain Bluebirds, *Sialia sialis* and *Sialia currucoides*, respectively (Power 1975; Pinkowski 1978); and House Sparrows, *Passer domesticus* (Sappington 1975). With very few exceptions (e.g., Pierotti 1980, 1982), the possible adaptive significance of these behaviors is not at all well understood; they provide an exciting challenge to students of avian social systems.

Literature cited

Alexander, R. D. 1974. The evolution of social behavior. *Annu. Rev. Ecol. Syst.* 5:325–383.

Balph, D. F., G. S. Imis, and M. H. Balph. 1980. Kin selection in Rio Grande turkeys: a critical assessment. *Auk 97*:854–860.

Dexter, R. W. 1952. Extra-parental cooperation in the nesting of Chimney Swifts. *Wilson Bull. 64*:133–139.

– 1981. Nesting success of Chimney Swifts related to age and the number of adults at the nest, and the subsequent fate of the visitors. *J. Field Ornithol. 52*:228–232.

Foster, M. S. 1977. Odd couples in manakins: a study of social organization and cooperative breeding in *Chiroxiphia linearis*. *Am. Nat. 111*:845–853.

– 1981. Cooperative behavior and social organization of the Swallow-tailed Manakin (*Chiroxiphia caudata*). *Behav. Ecol. Sociobiol. 9*:167–177.

Kilham, L. 1977. Altruism in nesting Yellow-bellied Sapsucker. *Auk 94*:613–614.

Koenig, W. D. 1981. Space competition in the Acorn Woodpecker: power struggles in a cooperative breeder. *Anim. Behav. 29*:396–409.

Koenig, W. D., and F. A. Pitelka. 1981. Ecological factors and the role of kin selection in the evolution of cooperative breeding in birds. In R. D. Alexander and D. W. Tinkle (eds.). *Natural selection and social behavior: recent research and new theory*, pp. 261–280. New York: Chiron Press.

Ligon, J. D. 1981. Demographic patterns and communal breeding in the Green Woodhoopoe (*Phoeniculus purpureus*). In R. D. Alexander and D. W. Tinkle (eds.). *Natural selection and social behavior: recent research and new theory*, pp. 231–243. New York: Chiron Press.

Ligon, J. D., and S. H. Ligon. 1978a. Communal breeding in the Green Woodhoopoe as a case of reciprocity. *Nature (London) 176*:496–498.

– 1978b. The communal social system of the Green Woodhoopoe in Kenya. *Living Bird 17*:159–197.

Myers, G. R., and D. W. Waller. 1977. Helpers at the nest in Barn Swallows. *Auk 94*:596.

Packer, C. 1977. Reciprocal altruism in Olive Baboons. *Nature (London) 265*:441–443.

Pierotti, R. 1980. Spite and altruism in gulls. *Am. Nat. 115*:290–300.

– 1982. Spite, altruism, and semantics: a reply to Waltz. *Am. Nat. 119*:116–120.

Pinkowski, B. C. 1978. Two successive male Eastern Bluebirds tending the same nest. *Auk 95*:606–608.

Power, H. W. 1975. Mountain Bluebirds: experimental evidence against altruism. *Science 189*:142–143.

Reyer, H. V. 1980. Flexible helper structure as an ecological adaptation in the Pied Kingfisher (*Ceryle rudis rudis* L.). *Behav. Ecol. Sociobiol. 6*:219–227.

Sappington, J. N. 1975. Cooperative breeding in the House Sparrow (*Passer domesticus*). Ph.D. thesis, Mississippi State University.

Snow, D. W. 1977. Duetting and other synchronized displays of the blue-back manakins, *Chiroxiphia* spp. In B. Stonehouse and C. Perrins (eds), *Evolutionary ecology*, pp. 239–251. Baltimore, Md.: University Park Press.

Watts, C. R., and Stokes, A. W. 1971. The social order of turkeys. *Sci. Am. 224*:112–118.

Woolfenden, G. E. 1973. Nesting and survival in a population of Florida
  Scrub Jays. *Living Bird 12*:25–49.
– 1975. Florida Scrub Jay helpers at the nest. *Auk 92*:1–15.

Commentary

IAN ROWLEY

By the early 1950s, two new disciplines, ecology and ethology, dominated field biological research. Interest in the former at first centered
around population regulation while ethology employed the classical
comparative approach of the evolutionary zoologist to study animal
behavior. The importance of individual and social behavior and their
plasticity within groups of animals led to the recognition that such traits
were just as characteristic of a species as were differences in morphology or plumage and that frequently the behavior of a species was a
response to its ecological environment.

Wynne-Edwards (1962) presented a theory about the evolution of
social behavior that linked together populations and behavior and depended on the idea that evolution could operate through selection at
the level of the social group as well as at the individual level. This
provoked a storm of protest from not only ethologists and ecologists
but also from geneticists, all of whom pointed to the altruistic nature of
such evolutionary selection via the group and to the incompatability
between altruism and the Darwinian theory of natural selection. Hamilton (1964), however, showed how selection could perpetuate altruistic
traits provided that the altruism was directed toward relatives carrying
the same genes as the altruist. These two processes that "could cause
the evolution of characteristics which favoured the survival, not of the
individual, but of other members of the species" were dubbed *kin selection* and *group selection* (Maynard Smith 1964).

Although the classic cases of altruism had been described in insect
societies, a review of cooperative behavior in over 130 species of bird
(Skutch 1961) drew attention to the possibility that altruism might occur in larger animals as well. However, Skutch's paper lacked quantitative data, and many people still tended to regard the behavior as aberrant, especially because it rarely occurred in the northern-temperate
regions where most ornithologists lived. It was not until the first studies

of color-banded birds breeding cooperatively were made in Australia (Carrick 1963; Rowley 1965; Ridpath 1972) that quantitative data became available. The first such American study was that of Brown (1970) closely followed by Woolfenden (1973) and MacRoberts and MacRoberts (1976), three long-term projects that are still continuing. Reviews such as those by Skutch (1961) and Lack (1968) showed the need for both quantitative and qualitative data on birds that bred cooperatively, and in the 1970s this stimulated people to look at the problem in many different parts of the world.

By 1974, there was sufficient interest for an international symposium to review the incidence of cooperative breeding on a continental basis (see Rowley 1976). By now it was obvious that this behavior was not only widespread in many different families but that it was no rare happening and was an important aspect of the species' life history. The data from these, mainly short-term, studies tended to maintain the emphasis on altruism and to explain its evolution through kin selection, because most cooperatively breeding groups consisted of close relatives, the "helpers" usually being progeny from an earlier nesting retained in the family after reaching independence. Not until data from long-term studies such as that for the Florida Scrub Jay, *Aphelocoma coerulescens* (Woolfenden and Fitzpatrick 1978) became available did people begin to realize that what superficially appeared to be altruistic behavior toward kin was in reality sound tactics ensuring a prime position from which to seize a breeding opportunity should a vacancy occur in the family group or nearby. This change of thinking can be traced through the reviews and theoretical papers that appeared in the late 1970s (Brown 1978; Emlen 1978; and others) to Vehrencamp's (1979) paper that suggested how *both* individual and kinship components of fitness could be assessed in the evolution of cooperative breeding. Papers presented recently in books edited by Markl (1980) and Alexander and Tinkle (1981) continued to develop this theme, which culminated in the present chapter.

Much of science is characterized by the search for unitary answers, and this has tended to polarize people into holding one or the other of two opposite views, whereas in many cases the answer lies between the two extremes. The study of cooperative breeding has tended to show this sort of polarization with the protagonists striving to fit the limited data into their particular solution – individual selection, kin selection, or altruism. A refreshing aspect of Emlen and Vehrencamp's chapter is their recognition that no single answer will explain the whys and wherefore of cooperative breeding in all species; the authors present a series of elegantly simple models, which they explain with a minimum of algebra. By showing how each species can be evaluated in terms of costs and benefits, they set the stage for valid comparisons to be made and meaningful conclusions to be drawn in a standard way.

One aspect of cooperative breeding that does not seem to have received the attention it deserves is the fact that most species behaving in this way are long-lived (but see the previous commentary). Ornithological theory has been dominated by northern hemisphere institutions and data for obvious reasons of history and geography. As a consequence, the "normal" life history pattern in the minds of most people is that of a short-lived passerine faced with all the hazards of either a severe northern winter or a lengthy migratory journey. It has therefore come as a surprise to many that tropical, subtropical, and southern-temperate species tend to live for much longer than their northern-temperate counterparts (Rydzewski 1978; Fry 1980). It is these same faunas that provide most of the species that breed cooperatively. Most cooperative species therefore are long-lived and I disagree with Koenig and Pitelka (1981), who suggest that longevity in these species is the result of, rather than a cause of, cooperative breeding.

In considering the costs and benefits of cooperation, the longevity of the actors is very important. On the one hand, individuals likely to live for several breeding seasons can "afford" to wait as helpers until that scarce commodity, a quality breeding territory, becomes available. On the other hand, bearing in mind the criteria of lifetime inclusive fitness measured in terms of progeny surviving to reproduce, long-lived parents can afford the luxury of prolonged parental care in the form of progeny retained in the family group as helpers until a quality opportunity for them to reproduce on their own occurs. For example, a Great Tit (*Parus major*) with only a 50% chance of surviving to breed another year can be expected to put all its efforts into one reproductive effort, whereas a Splendid Wren (*Malurus splendens*), only half the size but with twice as good a chance of survival, can afford to wait a year to obtain a prime breeding site or a mate rather than "disperse in desperation" and risk the hazards of unfamiliar surroundings.

Longevity not only affects the costs of current reproductive efforts but is also reflected in the age structure of the population and in the proportion of experienced:inexperienced breeders. It is the *lifetime* inclusive fitness of an individual that is selected for in the evolutionary process. High adult survival means that a successful individual probably has several opportunities to reproduce, which suggests that perhaps we should pay more regard to the size of the contribution made by the successful "top 10%" rather than think in terms of the average animal.

Such an approach emphasizes the value of long-term studies of individually recognizable and pedigreed individuals. Only then can the real identity and background of the participants enable the present demographic situation to be viewed as a consequence of past ecological events. I do not want to raise the complex subject of population cycles, but it is important to recognize that all demes have their ups and downs and that relative demographic stability is a characteristic of long-lived animals. Instead of regarding cooperative breeding as a behavioral aberration to be studied on its own, I suggest that it would be more profitable to think in terms of life history strategies and to regard cooperative breeding as just one other technique available for tactical employment to achieve the ultimate goal of successful reproduction under a variety of ecological situations. The incidence of cooperative breeding within a species is just as subject to variation as any other character and, therefore, should be regarded as an extra chance for heterogeneity to manifest itself and render the species more adaptable to change. In contrast, monogamous simple pairs may be regarded as restricted and stereotyped in their social behavior.

The importance of looking at the long-term view when considering life history strategies is self-evident. Once one relegates cooperative breeding to its proper role as one of many available tactics, it is easier to recognize its value in certain situations and to appreciate that such situations are not necessarily the outcome of present environmental conditions but may stem from past events, for example, good or bad previous recruitment, 12 months or more earlier. Recently Emlen (1981) has shown that *Merops bullockoides* individuals can oscillate from helping, to breeding, to helping, and back to breeding again. Emlen and Vehrencamp have drawn attention also to the different responses shown by *Melanerpes formicivorus* under increasingly severe conditions of ecological restraint. Reyer (1980) found different responses in two populations of Pied Kingfisher (*Ceryle rudis*) in Africa.

Another example of the importance of maintaining long-term studies comes from a study of the cooperatively breeding malurid wren, *Malurus splendens* (Rowley 1981). That paper described the cycles of a population over 7 years; an eighth year passed before publication and that confirmed the previous findings. Then came the first wet winter for many years followed by a long spring and a mild summer–climatic circumstances that had not occurred in the previous decade. Added to this it was 4 years since the last bushfire, and the vegetation had responded accordingly. Whatever the proximate cause, the result was that although the productivity of the 1980–1981 season was average, twice as many progeny survived to adulthood and were available to start the 1981–1982 season after other winters. For the first time in 9 years, there were more females than males (27 males, 28 females) and the population responded in ways that had not occurred previously. Six new territories were set up, each staffed by a simple pair, five of them with females that had previously been helpers of known origin in other groups while the sixth contained a pair of siblings that budded off (sensu Woolfenden and Fitzpatrick 1978) part of the parental territory. The effect of this unusually large number of yearlings did not end there; two groups, in both of which three siblings had survived from the final brood of the previous year, had two simultaneous nests in the one territory, one of them doing so twice in the season. Whether either of these groups practiced polygamy must remain unknown, because I did not witness copulation; I prefer to regard these as instances of budding off that failed to consolidate when the subgroups maintaining separate concurrent nests coalesced after fledging young.

The importance of continuing long-term studies so that a full range of seasonal and demographic features can be allowed their cyclical expression in terms of different population densities, sex ratios, productivities, and mortalities should be obvious. Such work will inevitably provide data for an interpretation of the genetics of helping or communal breeding and will allow a more precise assessment of the frequency and effects of inbreeding although paternity might be uncertain when more than one male is present in the breeding group. Such long-term studies of individually marked birds will emphasize the importance of slow maturity, delayed onset of breeding, and, ultimately, the large contribution made by relatively few successful long-lived individuals. It will also provide data to fit the models provided in Chapter 4.

Further advances in understanding cooperative breeding will come from monitoring hormone levels and other physiological parameters within the various constituent castes of group-living species. How effective and how long-lasting is the psychological castration imposed on a subordinate member? How severe is the stress of breeding? Also, how great is the relief afforded by helpers? After measuring these facets of life history in the field, manipulation and monitoring an experimental population of known age and origin would confirm the ecological importance of helping as yet another variable feature in an individual's repertoire, enabling it to cope with a continually varying environment. For my part, I hope to continue a series of long-term studies within the same genus (*Malurus*) to show how cooperative breeding functions under a wide range of climate and habitat.

Literature cited

Alexander, R. D., and D. W. Tinkle. 1981. *Natural selection and social behavior.* New York: Chiron Press.

Brown, J. L. 1970. Cooperative breeding and altruistic behaviour in the Mexican Jay, *Aphelocoma ultramarina. Anim. Behav. 18:*366–378.

– 1978. Avian communal breeding systems. *Annu. Rev. Ecol. Syst. 9:*123–155.

Carrick, R. 1963. Ecological significance of territory in the Australian Magpie. In *Proceedings of the XIII International Ornithological Congress,* pp. 740–753.

Emlen, S. T. 1978. The evolution of cooperative breeding in birds. In J. B. Krebs and N. B. Davies (eds.). *Behavioural ecology,* pp. 245–281. Oxford: Blackwell Publisher.

–1981.Altruism, kinship, and reciprocity in the White-fronted Bee-eater. In R. D. Alexander and D. W. Tinkle (eds.). *Natural selection and social behavior,* pp. 217–230. New York: Chiron Press.

Fry, C. H. 1980. Survival and longevity among tropical land birds. In *Proceedings of the IV Pan-African Ornithogical Congress,* pp. 333–343.

Hamilton, W. D. 1964. The genetical evolution of social behaviour, I and II. *J. Theor. Biol. 7:*1–52.

Koenig, W. D., and F. A. Pitelka. 1981. Ecological factors and the role of kin selection in the evolution of cooperative breeding in birds. In R. D. Alexander and D. W. Tinkle (eds.). *Natural selection and social behavior,* pp. 261–280. New York: Chiron Press.

Lack, D. 1968. *Ecological adaptations for breeding in birds.* London: Methuen & Co.

MacRoberts, M. H., and B. R. MacRoberts. 1976. Social organization and behavior of the Acorn Woodpecker in central coastal California. *Ornithol. Monogr. 21:*1–115.

Markl, H. 1980. *Evolution of social behaviour: hypotheses and empirical tests.* Basel: Verlag Chemie.

Maynard-Smith, J. 1964. Group selection and kin selection. *Nature (London) 201:*1145–1147.

Reyer, H. V. 1980. Flexible helper structure as an ecological adaptation in the Pied Kingfisher (*Ceryle rudis rudis* L.). *Behav. Ecol. Sociobiol. 6:*219–227.

Ridpath, M. G. 1972. The Tasmanian native hen, *Tribonyx mortierii. CSIRO Wildl. Res. 17:*1–118.

Rowley, I. 1965. The life history of the Superb Blue Wren *Malurus cyaneus. Emu 64:*251–297.

– 1976. Co-operative breeding in Australian birds. In H. J. Frith and J. H. Calaby (eds.). *Proceedings of the XVI International Ornithological Congress, (1974), Canberra,* pp. 657–666.

–1981.The communal way of life in the Splendid Wren, *Malurus splendens. Z. Tierpsychol. 55:*228–267.

Rydzewski, W. 1978. The longevity of ringed birds. *The Ring 8:*218–262.

Skutch, A. F. 1961. Helpers among birds. *Condor 63:*198–226.

Vehrencamp, S. L. 1979. The roles of individual kin, and group selection in the evolution of sociality. In P. Marler and J. C. Vandenbergh (eds.). *Handbook of behavioral neurobiology;* Vol. 3, pp. 351–394. New York: Plenum.

Woolfenden, G. E. 1973. Nesting and survival in a population of Florida Scrub Jays. *Living Bird 12:*25–49.

Woolfenden, G. E., and J. W. Fitzpatrick. 1978. The inheritance of territory in group-breeding birds. *BioScience 28:*104–108.

Wynne-Edwards, V. C. 1962. *Animal dispersion in relation to social behaviour.* Edinburgh: Oliver & Boyd.

5 Ecological energetics: what are the questions?

GLENN E. WALSBERG

A major basis of modern biology is the recognition that organisms represent temporary and complex associations of molecules that are maintained and reproduced only by expenditure of energy. Associated with this power consumption, organisms are involved continuously in internal energy transformations and exchange of energy with their environment. Two biologically important classes of energy are chemical potential energy, particularly that liberated by oxidation of carbon or hydrogen, and random kinetic energy at the molecular level, which is equal to heat content and proportional to temperature. These types of energy are interrelated in animals in complex ways. All work accomplished by conversion of chemical potential energy necessarily is inefficient and results in metabolic heat production. All chemical reactions, including the biochemical reactions of catabolism and synthesis, produce changes in the heat content of a system. In turn, heat content is proportional to organismal temperature, which is a critical determinant of metabolic performance. Therefore, it is impossible to segregate neatly problems of thermal balance from those related to the acquisition and allocation of chemical potential energy. For the purposes of the following discussion, however, I have attempted at least a rough division. I will initially discuss questions related to thermal energy as an important property of the physical environment; such questions are most directly pertinent to regulation of the bird's body temperature. Second, I will discuss questions related to the acquisition and allocation of chemical potential energy as a vital resource.

Energy as an environmental property

Questions
The fundamental question in avian thermal ecology is, How do thermal relations affect an organism's survival, reproduction, and con-

135

sequent contribution to succeeding generations? This most basic question can be addressed through a series of subsidiary questions.

1. To what degree are free-living birds thermally stressed, and what is the nature of this stress? Such stress may be direct physiological stress, that is, the avoidance of potentially lethal hyper- or hypothermia by physiological mechanisms requiring allocation of energy or mass (e.g., water). It also could be an "indirect" stress, such as manifested by disturbances of vital components of the time budget.

2. How do birds cope with or minimize this thermal stress? Physiological and morphological mechanisms used to cope with thermal stress are relatively well known, although their importance for birds in nature compared to behavioral mechanisms is virtually unknown. Behavioral mechanisms for coping with thermal stress are perhaps of major importance but have received relatively little attention. Such behavioral adjustments encompass both habitat selection and adjustments within a single habitat. The latter, for example, includes thermally adaptive postural shifts, such as described for Herring Gulls (*Larus argentatus*) by Lustick et al. (1978). Microhabitat selection in the special case of the nest site has received substantial attention (e.g., Calder 1971, 1973; Balda and Bateman 1973; Smith et al. 1974; Southwick and Gates 1975; Walsberg 1981), and a few studies have examined the thermal consequences of nocturnal roost-site selection for birds during the winter (e.g., Kendeigh 1961; Caccamise and Weathers 1977; Gyllin et al. 1977; Kelty and Lustick 1977; Yom-Tov et al. 1977; Walsberg and King 1980). Microclimate selection during other portions of annual or daily cycles has yet to be examined extensively.

3. What are the costs associated with coping with thermal stress and how do these costs affect fitness? Even in cases in which mechanisms such as microclimate selection have been analyzed, few authors have quantified the costs of such adjustments in terms of time, energy, materials, or other quantities.

4. How does the degree of stress and the mechanisms used to cope with it change with variation in important attributes of avian life histories? Such variation over the annual cycle or variation associated with differing mating systems, foraging techniques, or modes of resource defense has received almost no attention (for an important exception, see Mugaas and King 1981).

Suggested emphases

Current knowledge of the adaptations of birds to their thermal environments comes primarily from two sources: laboratory studies of thermoregulatory physiology and field studies describing the behavioral reactions of free-living individuals to thermal stress (reviewed by Dawson and Hudson 1970; Calder and King 1974). Although these analyses are of fundamental importance, progress in understanding the adaptations of animals to their thermal environments has been inhibited by the hiatus between such field and laboratory studies. This gap primarily reflects the paucity of information describing in physiological terms the degree and mode of thermal stress experienced by free-living animals. The absence of such information makes it difficult to evaluate the ecological significance of either physiological responses in the well-defined, but artificial, environment produced in the laboratory or behavioral responses in the complex and poorly defined natural environment. Acquisition of such data has been inhibited by the technical difficulty of analyzing all ecologically significant avenues of heat transfer (e.g., convection and radiation) in a particular microhabitat, as well as the theoretical problems involved in integrating these different modes of heat flux in a manner so that physiological responses of a particular species may be predicted reliably. The recent development of the *operative environmental temperature* (= equivalent black-body temperature) theory and associated techniques shows great promise for providing this integration (for recent theoretical developments and laboratory analyses, see Bakken 1976, 1980; Robinson et al. 1976; Campbell 1977; Mahoney and King 1977; Walsberg et al. 1978; Bakken et al. 1981; for examples of recent applications of these techniques to birds in nature, see Mahoney 1976; Walsberg and King 1978a,b, 1980; DeWoskin 1980; Mugaas and King 1981). This body of theory allows incorporation of all major modes of heat transfer into a single thermal index that can be used together with laboratory analyses to predict physiological responses of animals to complex natural environments. Such analyses usually employ one of two techniques. The most common to date is computation of operative environmental temperature based upon micrometeorological information and data describing animal surface properties (e.g., Morhardt and Gates 1969; Mahoney and King 1977; DeWoskin 1980; Mugaas and King 1981; Walsberg 1981). Alternatively, operative temperative may be measured in a particular microclimate using a metal cast of the bird's body covered with the animal's integument by taxidermy techniques (Bakken and Gates 1975; Bakken 1976; Bakken et al. 1981). I suggest that refinement and extensive application of these techniques would be a particularly productive avenue of research, since this approach potentially allows accurate quantification of the thermal stresses to which birds are exposed in nature. Establishment of such a data base would facilitate initial analyses of the major questions previously stated.

A second major emphasis in avian thermal ecology should be the thermoregulatory physiology of birds engaged in activities other than resting, such as locomotion and intense synthetic activity. Current information suggests that approximately 40–60% of a free-living bird's total energy expenditure is devoted to basal and thermostatic demands; in some species, variation in these demands may determine annual variation in total expenditure (Walsberg, in press). However, current analyses rely largely upon physiological studies of inactive, postabsorptive birds, a condition that is likely to be characteristic of free-living birds only during a portion of the inactive phase of their daily cycle. The thermostatic physiology of birds engaged in normal levels of activity is not well known and ecologically important questions in this area remain unresolved in spite of substantial effort in the last 10–15 years (e.g., Tucker 1968; Pohl 1969; Taylor et al. 1971; Berger and Hart 1972; Pohl and West 1973; Torre-Bueno 1976; Paladino 1979). An example concerns the relationship between the waste heat of muscular activity and the thermostatic power requirement of a bird active at low environmental temperatures. If the muscular production of heat that is a by-product of activity reduces or eliminates the necessity for a bird to use shivering themogenesis, then under cold conditions there might be little difference in power consumption between an active bird and one that is immobile but intensively shivering. The ecological significance of this possibility is striking, because under these conditions there might be little or no effective net energy cost for activities such as foraging or territorial defense.

Energy as a vital resource

Birds lie at one extreme of the spectrum of patterns of animal energy consumption. When averaged over extended periods of time, only a small part of the chemical potential energy they acquire is incorporated into new protoplasm (e.g., about 1–2% in four passerine species studied by Holmes et al., 1979). Little somatic growth is shown by adults and energy represented in the tissues of the offspring is small

compared with the power consumed in locomotor activity and maintenance of high body temperatures. In general, avian rates of power
consumption also are high compared with their capacity for storage of
chemical potential energy. Although ectothermic species commonly use
stored energy reserves to survive major parts of the annual cycle, this
apparently is characteristic of only the few avian species that exhibit
seasonal hibernation or that store large amounts of energy outside of
the body. The epitome of avian reliance upon lipid stores is that of
long-range migrants, which in the extreme may store fat approximately
equal to their lean body mass (Blem 1980). If power consumption
during flight is estimated using the equation of Hart and Berger (1972)
and stored lipids have an average energy density of 40 kJ/g (Kleiber
1961), then 25 g of stored fat would maintain a bird of 25 g lean body
mass in continuous migratory flight for about 3.2 days. This probably is
an overestimate, because doubling body mass undoubtedly results in an
increase in the rate of power consumption during flight and because
not all lipid in the body can be catabolized without damaging vital
functions. Outside of migration, the most conspicuous energy storage
by small birds is winter fattening. The primary function of increased
lipid reserves during the winter apparently is to allow the bird to survive its nocturnal fast and perhaps short periods of inclement weather
that would prevent foraging (King 1971; Ketterson and King 1977).
Studies of 10 passerine species summarized by King (1972) indicate
that when exposed to normal winter conditions, few of the birds
studied could survive on stored energy for more than one night and
part of the following day. Thus, the chemical potential energy budget
of a typical bird probably is regulated over short periods of 1 or a few
days. Exceptions to this generalization include the use of lipid stores in
supplying the energy required for ovogenesis in several species (e.g.,
Weller 1957; Barry 1962; Krapu 1974, 1981; Jones and Ward 1976;
Korschgen 1977; Ankney and MacInnes 1978, Raveling 1979; Drobney
1980).

A high rate of power consumption compared to capacity for energy
storage has important consequences for ecological analyses. An obvious
example concerns restrictions placed on the variety of foraging specializations exhibited by birds. Small birds typically rely upon food sources
of dependable daily occurrence. In some cases, species such as swifts
for a few days during inclement weather may survive through faculta-
tive hypothermia (Bartholomew et al. 1957). This contrasts with ectothermic species that may specialize on very unpredictable or rare types

of food. An extreme example is the spadefoot toad *Scaphiopus couchi*, which inhabits the Sonoran Desert and feeds primarily on alate termites. Major flights of these reproductive termites occur one to three times per year in a given locality and during each event the alates are vulnerable to toad predation for only about one-half hour (Dimmit and Ruibal 1980). However, a spadefoot toad can acquire sufficient energy in a single feeding to produce fat reserves that enable it to survive for more than a year.

A less obvious example of the importance of this limited capacity for storage of energy in birds compared to their rates of expenditure occurs in the application of parental investment theory (sensu, Trivers 1972). Parental investment is defined as any investment made by a parent in one reproduction at the expense of its ability to invest in future reproductions (Trivers 1972). This investment may subsume a variety of factors, such as time expenditure, energy allocation, or exposure to predation. Equating these modes of investment to a common currency is a major problem in applying this body of theory, and several authors have suggested that energy allocation be used as a general index of parental investment (e.g., Trivers 1972; Gladstone 1979). For many bird species, however, it is questionable as to what fraction of the energy devoted to reproduction may be correctly classified as parental investment, because the parent may not effectively possess the option of storing that energy for allocation to future reproduction. For example, energy devoted to egg production generally is assumed to be an important part of parental investment, yet I have calculated that for five small (under 25 g) passerine species the clutch contents represent an average of only 2% of the female's total energy expenditure during a single reproductive event (Walsberg, in press). This indicates that changes in allocation to egg production may not significantly affect the female's energy balance at the start of a subsequent reproduction. Indeed, rates of power consumption are so high in these species compared to their capacity for energy storage that a female that elected to reduce her expenditure 25% during one breeding event and store that energy to invest in a subsequent reproduction would be required to retain fat equal to 56–85% of normal body mass ($\bar{x} = 72\%$) (Walsberg, in press). For such small, altricial species, it thus seems likely that the only energy devoted to reproduction that can be appropriately classified as parental investment is that expended near the end of one reproductive event and immediately prior to the start of a second breeding attempt (e.g., care of fledged young immediately prior to laying a second clutch).

Questions

The fundamental question when evaluating the role of energy as a vital resource to birds is, How do the demands related to energy acquisition and allocation affect a bird's contributions to succeeding generations and consequently influence the patterns observed in avian biology? Either on a practical or philosophical basis, this is an exquisitely difficult problem to analyze. To approach this fundamental question, we must first deal with important secondary questions such as the following.

1. What are the energetic consequences of activities important for birds in nature? Relatively little is known regarding the energetic correlates of a host of activities usually considered to be of major ecological importance, such as resource defense, mating behavior, or migration.

2. Do birds approach energetic limitations on such activities, and, if so, how are these limits expressed? Resolving this question for a particular species and activity is complex, because such an energetic limit could result from, among other causes, (a) limits on the rate of power consumption at the tissue level, (b) a simple insufficiency of food in the environment, or (c) restrictions on foraging behavior due to conflicts in time allocation to other functions.

3. By what mechanisms do animals cope with potential energy limits?

4. What are the evolutionarily significant costs associated with the use of these mechanisms?

5. How do the stresses, mechanisms, and costs vary with important aspects of the life-styles of birds?

Areas most conspicuously in need of study

There are truthfully no areas of ecological energetics that are particularly strong, except for the laboratory-based knowledge of the physiology of maintenance metabolism and thermoregulation in inactive birds. The following survey merely examines activities likely to be of particular ecological importance and notes the most obvious problems that have yet to be resolved regarding the allocation of energy as a resource.

Resource acquisition and defense. Within recent years, studies of resource acquisition have been dominated by theoretical or speculative analyses that fall under the general heading of "optimal foraging theory" (e.g., Schoener 1971; MacArthur 1972; Pyke et al. 1977; Chapter 6). Such analyses typically predict that foraging behavior should be organized so as to maximize the net rate of energy intake. Empirical studies that allow quantitative tests and refinement of optimal foraging theory have been rare, but this situation currently is improving. In the laboratory, at least some species have been shown to be capable of modifying their foraging in a manner that maximizes energy intake (e.g., Krebs et al. 1977). A limited number of studies of birds in nature have supported such predictions of maximization of energy intake rates (e.g., Gill and Wolf 1975; Goss-Custard 1977a; Zach 1979). Most such analyses have dealt with extremely simple situations, such as that of a nonbreeding bird feeding on a single food type. In more complex situations, at least two workers have found that birds did not forage in a manner that maximized their rate of net energy gain (Goss-Custard 1977b; Kushlan 1978). Relatively simple foraging systems have provided the ease of analysis necessary for pioneering studies and continued attention to such systems undoubtedly will allow test and refinement of the major assumptions and hypotheses of optimal foraging theory. Clearly, an important goal of workers in this area must be to examine the usefulness of current theory in predicting foraging behavior of animals in the complex environments to which they typically occur. To what extent do phenomena such as the predicted maximization of net rates of energy acquisition exist when animals encounter conflicting selective pressures, such as those involving social dominance, predation, competition, thermal stress, water balance, and nutrient (e.g., protein) requirements? (For examples of such analyses of mammalian systems, see Belovsky 1981a,b.)

Closely related to the problem of foraging is that of resource defense. Some major initial questions regarding defense of resources by birds are: How much energy is expended in resource defense? What are the relative energy costs and benefits of different types of resource defense? To what degree do energetic costs constrain defense? In spite of widespread interest in the ecology of resource defense in birds, few data are available to address even these most basic questions.

It is commonly supposed that resource defense is a major energy drain on birds, but data for six species suggest that these birds typically devote less than 8% of total daily energy expenditure to territorial advertisement and defense (Walsberg, in press). It is clear, however, that such costs may be quite variable within a species. For example, Ewald and Carpenter (1978) demonstrated for *Calypte anna* that energy expenditure for chasing intruders declined from about 4% of total daily energy expenditure when territories rich in food were defended to about 0.3% of daily expenditure when food was absent from the territories. Few studies have examined the energetic cost of initially establishing rather than simply maintaining territories or considered the consequences of major shifts in resource defense. The few analyses of major shifts in patterns of resource defense, such as the change from nonterritorial behavior to defense of a feeding territory, have indicated that such shifts are associated with major changes in daily energy expenditure (i.e., 17–46%; Carpenter and MacMillen 1976a,b; Carpenter 1976; Walsberg 1977). Thus, the limited available data show a contrast between the amount of energy devoted to territorial defense and the variation in energy expenditure associated with large changes in territorial behavior. For example, daily energy expenditure is elevated 17% in territorial Hawaiian honeycreepers *Vestiaria coccinea* compared with nonterritorial individuals, but less than one-third of this increase can be attributed to the costs of active territorial defense (Carpenter and MacMillen 1976a). Most of the increase is due to shifts in movement patterns that may increase the conspicuousness of the territorial bird and, therefore, function as advertisement. Thus, the major energy costs associated with territorial defense in at least some species may be due to much more subtle phenomena than simply active expulsion of intruders.

Gametogenesis. The analyses of King (1973), Ricklefs (1974), and Walsberg (in press) indicate that egg production is the only form of gametogenesis likely to be of major energetic significance. The estimated peak daily energy cost of egg production ranges from 37–55% of basal metabolism (BM) in some passerine species to 160–216% of BM in ducks and the Brown Kiwi (data summarized in Walsberg, in press). The degree to which such relatively large costs actually produce an increase in the daily energy expenditure of free-living birds is unknown. The manner in which energy is acquired for ovogenesis also is poorly known. Such energy can be supplied by increasing dietary intake, use of internal stores, or reducing allocation to other activities. Each of these methods is used in at least some species (Walsberg, in press); the most extensively documented is the use of fat stores during ovogenesis in various ducks and geese (e.g., Barry 1962; Hanson 1962; Krapu 1974, 1981; Ankney and MacInnes 1978; Raveling 1979; Drobney 1980). Such energy storage in some anseriform species may supply most of the energy for egg production (Drobney 1980; Krapu 1981). Data for other taxa are sparse (Jones and Ward 1976; Ankney and Scott 1980). For no species, however, is the relative importance known for all three avenues by which energy may be acquired for reproduction (storage, reduced activity, and increased dietary intake). The degree to which birds compensate for energy costs of egg production by behavioral shifts in activity, such as foraging, must be quantified before the ecological significance of such costs can be appreciated.

Incubation. There has been controversy as to whether maintenance of appropriate egg temperatures requires a significant increase in adult energy expenditure. For birds incubating small to medium-sized clutches (three to five eggs), the most complete and direct analyses indicate that power consumption during incubation increases about 20–30% over that of nonbreeding birds roosting in the nest cavity (Biebach 1977, 1979; Vleck 1981). Other analyses commonly indicate a similar or smaller increase, if any (e.g., Gessaman and Findell 1979). I question the usefulness of such comparisons for ecological analyses, however, because nonbreeding birds rarely occupy nests. Thus, a more suitable comparison is that of the power consumption of an incubating bird to that of one in its normal microclimate outside the nest. Such a comparison can be made for only a few species but does indicate that restriction in heat loss due to the addition of the nest's insulation more than compensates for heat lost by the adult to the clutch. Thus, attending the nest and incubating eggs typically entails a 15–20% decrease in resting energy expenditure compared to that of an adult outside of the nest (Walsberg and King 1978a,b; Vleck 1981; Walsberg, in press).

This controversy as to whether incubation requires an increase in resting energy expenditure is vitiated, however, by the small size of the predicted or measured changes in power consumption. At realistic environmental temperatures for the species in question, the change in resting energy expenditure attributed to incuation typically is equal to 30% or less of BM (see, for example, the analyses of Kendeigh 1963; El-Wailly 1966; Biebach 1977, 1979; Hubbard 1978; Walsberg and King 1978a,b; Vleck 1981). This is minor compared to that produced by the greatly reduced activity characteristic of incubating birds. For example, flight requires power consumption at a rate averaging about 9.5 times BM (Hart and Berger 1972). If incubation entailed a change in resting metabolism equal to 0.3 times BM, the change in energy expenditure produced during 1 hour of incubation is equal to the power consumed in less than 2 minutes of flight. This overwhelming effect of reduced locomotor activity is reflected in the daily energy budgets of a variety of species (e.g., *Empidonax traillii*, Ettinger and King 1980; *Empidonax minimus*, Holmes et al. 1979; *Petrochelidon pyrrhonota*, Withers 1977; *Pica pica*, Mugaas and King 1981; *Calcarius lapponicus*, Custer 1974; *Phainopepla nitens*, Walsberg 1977, 1978; *Vireo olivaceous, Dendroica caerulescens*, Holmes et al. 1979; *Zonotrichia leucophrys*, Hubbard 1978). With the exception of *Zonotrichia*, total daily energy expenditure in these species by the female during incubation was estimated at 10–37% below the egg-laying period and 5–30% below the nestling period. Thus, energy expenditure during incubation usually is low compared to other portions of the annual cycle. Major energetic constraints might occur, however, due to limits on time available for foraging because of the large fractions of the day allocated to nest attendance. Thus, time available to acquire energy in the form of food may be reduced even more than is the animal's energy requirement. I have calculated the net effect of these shifts in energy requirements and time available for foraging by estimating an average foraging rate (Walsberg, in press). This is the average rate at which a bird must acquire energy while foraging in order to maintain energy balance over a 24 hour period. It is estimated as the animal's daily energy expenditure divided by time available for foraging during a day. This computation is an underestimate because it does not consider inefficiency of assimilation. During reproduction, time available for foraging may be approximated crudely as the animal's daily active period minus time devoted to nest attendance. Estimates for the five species in which appropriate data are available indicate that for the parent that takes the major role in nest attendance, required foraging rates during the incubation period are the highest reached during reproduction in three species (*Actitus macularia, E. traillii,* and *P. pica*) and only slightly below the maximum in two other species (*P. nitens* and *Z. leucophrys*) (for calculations and original data sources, see Walsberg, in press). Although these estimates are inexact, they indicate that restrictions in foraging time during incubation could significantly limit reproductive performance. Investigations of the consequences of these behavioral limits should increase substantially our understanding of energetic constraints during incubation. An example of such an approach is the analysis of Sherry et al. (1980) of loss in body mass by incubating Jungle Fowl (*Gallus gallus*). They demonstrated that such weight loss does not reflect partial starvation as has been suggested previously but rather a change in the level at which body mass is regulated. This loss in mass and consequent reduction in food requirements apparently is an adaptation that allows the incubating adult to resolve conflicts in time allocation between foraging and nest attendance.

Care of nestlings and fledglings. In spite of the importance of this phase of the life history, the energetics of adults caring for their young is one of the most conspicuously neglected areas in avian energetics. Although substantial attention has been devoted to nestling growth per se (for recent reviews, see Ricklefs 1979; O'Connor 1980), the energetics of supplying food to nestlings or fledglings has been little studied.

The data that describe parental energetics during the period of nestling care generally are descriptive; comparative or experimental analyses (e.g., Hails and Bryant 1979) are rare. Although the energy requirements of nestlings have been quantified more frequently, this rarely has been accomplished in conjunction with analyses of the parental portion of the adult/nestling energetic complex. Previously, I summarized data from several studies of nestling energy expenditure and estimated simultaneous parental energy expenditure using a general relation between body mass and adult power consumption (Walsberg, in press). This analysis suggested that even if both parents contribute equally to care of the nestlings, supplying food to broods of typical size requires approximately a two- to threefold increase in the peak amount of energy that a parent must acquire in its daily foraging. A three- to fivefold increase is common if only a single parent feeds the brood. This supports the general belief that the nestling period is characterized by particularly great demands on parental foraging. This analysis also suggested that feeding a typical brood may require a greater proportional increase in parental foraging for small species compared to larger forms (Walsberg, in press). Few detailed studies of parental energetics are available, however, although some authors have examined the relation between parental power consumption and brood size (e.g., Hails and Bryant 1979). In view of the potential importance of such relations, this dearth of information is striking. In a field of such primitive development, only the most basic questions can be identified. It is clear, however, that the following deserve substantial attention.

1.   How much do the young add to the parents' foraging burden, and how does this vary during development?
2.   What are the consequences of variation in brood size or hatching asynchrony for parental energy requirements or energy supply to the offspring?
3.   What are the energetic characteristics of the postfledging period?

Molt. Seasonal replacement of feathers is the most easily identified event of somatic production in adult birds and apparently involves major changes in the bird's energy budget (King 1980). The magnitude and source of these energetic shifts represent an unresolved controversy. Technical difficulties in analyzing molt energetics notwithstanding, it is notable that a variety of workers using several methods have produced generally similar results that indicate that a complete body molt requires average energy expenditure equal to 20–40% of BM (King 1980). Over the entire molting period, this represents about 20 times more energy than can be accounted for by the potential energy content of the new plumage (King 1980). This strongly suggests that molt involves energetically expensive processes other than keratin synthesis, but such processes have yet to be identified. The ecological significance of such expenditures is unknown, because the energetics of free-living birds during molt has received almost no attention (for a significant exception, see Ankney 1979). Although laboratory studies have shown that molt expenditures are potentially large, such costs need to be placed in the context of the free-living bird's total energy budget. In both captive and free-living birds, molt often accompanies marked reductions in activity that largely could compensate for the energetic costs of feather production. Thus, a particularly useful approach to the ecological energetics of molt would be to examine the combined effect of feather replacement and associated changes in activity upon the energy budget of birds in nature.

Migration. Migration perhaps is the most dramatic activity performed by birds as individuals and is a period in which energy relations may be particularly important. Demands upon the energy budget may be critical at this time because (1) energy requirements probably are high, (2) migration often occurs during seasons when food supply is low or unpredictable, and (3) time available for foraging may be restricted due to conflicting allocations to migratory flight. Unfortunately, the energetics of migration has received little attention, with the notable exceptions of laboratory analyses of flight metabolism and field studies of changes in body mass experienced by migrating birds (see review by Blem 1980). Because of the expected simultaneous and conflicting demands on time and energy allocation, I suggest that intensive studies of foraging would be a particularly useful approach to the energetics of migration. Important questions in this area include: (1) In supplying energy during migration, what are the relative roles of foraging and fat storage? (2) What special adaptations facilitate the intensive foraging that may be required during migration? (3) To what extent does foraging ability or fat storage limit migratory performance?

Questions related to other aspects of the nonbreeding period. One important question is, To what degree is thermoregulation a major expense for birds that winter in temperate or cold climates? Thermoregulatory physiology historically has been a major focus for physiological ecologists. This has produced an extensive knowledge of the physiological responses of birds to cold stress, but the relationship of such thermostatic expenditures to a free-living bird's total energy budget is largely unknown. However, the few data describing annual cycles of power consumption suggest that winter may not be a period of comparatively intense energy expenditure, as is commonly assumed. In the Andean Hillstar hummingbird (*Oreotrochilus estella*), the lowest rate of daily energy expenditure over the annual cycle occurs in the winter (Carpenter 1976). In *Phainopepla nitens,* winter levels of power consumption average only about 7% above the lowest daily value that occurs during the annual cycle (Walsberg 1977). In contrast, the female Black-billed Magpie (*Pica pica*) in early winter reaches her annual maximum level of daily energy expenditure, and the male expends energy at a level only about 6% below his annual maximum (Mugaas and King 1981). However, most of the variation in the magpie's energy expenditure over the annual cycle is due to changes in activity patterns and not variation in the thermostatic demand. During the winter period of maximum energy expenditure, 58% of the female's power consumption is devoted to basal and thermostatic demands. This value is similar to that during the summer, which typically ranges from 55 to 61%. Thus, there are few data to support the idea that winter thermostatic demands require substantial increases in total power consumption.

A second question is, For species that inhabit high latitudes during winter, how important are restrictions on foraging time due to shortened day lengths? Most birds species are diurnal, and winter brings a reduction in the time that they are active each day and can acquire energy. It is therefore possible that a bird is energetically stressed in winter primarily because it has relatively little time to acquire its perhaps moderate energy requirements. Unfortunately, few data exist to evaluate this possibility (Walsberg, in press).

Obstacles to resolving questions about energy as a vital resource

The most conspicuous factor retarding successful approaches to the questions presented earlier is the technical difficulty of estimating the energy expenditure of free-living birds engaged in normal activities. I will discuss the four most commonly used or advocated techniques for making such estimates.

Food consumption by caged birds. Historically, large bodies of data on avian energetics have been collected by analyzing the "metabolized" energy of caged birds, that is, the net difference between the energy content of their ingested food and their excreta (see review by Kendeigh et al. 1977). Such data are used extensively, but extreme care must be taken when attempting to use studies of caged birds to analyze the energetics in nature. The mode and intensity of the activity of captives often differ greatly from that of free-living birds. Current data indicate that variation in allocation to activity contributes significantly to temporal variation in total energy expenditure. Indeed, in the majority of the species for which data are available, such variation in activity is the single most important determinant of phasic variation in total power consumption (Walsberg, in press). Thus, any technique that does not account for the behavioral flexibility characteristic of birds in nature is unlikely to produce acceptable estimates of their patterns of power consumption.

Telemetry of heart rate. Of all physiological rate functions, heart rate is among the easiest to measure on unrestrained animals because of the relative ease by which it can be monitored by radiotelemetry. Several workers have exploited this ability to measure heart rate in attempts to estimate energy expenditure in free-ranging birds. This is based upon the principle that avian power consumption is reflected by rates of oxygen consumption ($\dot{V}_{02}$) and that $\dot{V}_{02} = HR \times SV \times \Delta A\text{-}V_{02}$, where HR is heart rate, SV is stroke volume, and $\Delta A\text{-}V_{02}$ is the difference in oxygen content between arterial and venous blood. This equation illustrates the major difficulty in such analyses: Heart rate is only one of three factors that can vary with oxygen consumption. A single rate of oxygen consumption can be associated with a range of heart rates, and one thus would expect only a loose correlation between heart rate and metabolism. Indeed, typically under controlled conditions as little as 44% of the variance in metabolic rate is explained by variation in heart rate (Wooley and Owen 1977), although values for individual birds may be much higher (e.g., up to 94% in *Columba livia;* Flynn and Gessaman 1979). However, modest changes in experimental conditions can alter the heart rate–metabolism relationship. For example, Flynn and Gessaman (1979) measured heart rate and metabolic rate of pigeons (*C. livia*) in a darkened metabolism chamber. The correlation of heart rate and metabolic rate derived from this study was combined with telemetered data on average daily heart rate to predict the daily energy expenditure of birds housed outdoors in a coop. Values for daily energy expenditure predicted from monitoring heart rates averaged 42% higher than those measured on the basis of food consumption. Thus, Gessaman (1980) concludes that "the heart rate method provides a good index of daily metabolism (i.e., less than 10% error) only in some individuals." I concur and conclude by noting that this technique contains theoretical flaws that are reflected in substantial and unpredictable changes in the heart rate–metabolism relationship. Clearly, this militates against its usefulness as a tool in ecological energetics.

Doubly labeled water. If correct procedures are followed, measurement of CO_2 production by the doubly labeled water ($^3H_2{}^{18}O$) method is potentially the most accurate technique available for measuring the average metabolic rate of free-living animals (Nagy 1980). This technique involves labeling the animal's body water with a known quantity of oxygen-18 (^{18}O) and tritium or deuterium (^3H or ^2H). The rate of decline in ^{18}O concentration in body water is a function of both CO_2 and H_2O loss from the body, whereas the rate of decline of labeled hydrogen is a function of simple water loss. Therefore, CO_2 production and loss can be measured by analyzing the difference in the turnover rates of labeled oxygen and hydrogen (Lifson and McClintock 1966). Unfortunately, this technique is expensive and requires technical expertise that apparently is possessed by few ecologists. For example, most recent field studies of animals using $^3H_2{}^{18}O$ have been accomplished by, or with the assistance of, K. A. Nagy of the Laboratory of Biomedical and Environmental Sciences at the University of California, Los Angeles (e.g., King and Hadley 1979; Nagy and Milton 1979; Nagy and Montgomery 1980; Weathers and Nagy 1980). Other workers have been discouraged by serious technical difficulties. For example, Gessaman (1980) states that about 40% of the values obtained from $^2H_2{}^{18}O$ turnover rates in American Kestrels (*Falco sparverius*) were "totally ridiculous" and also reports personal communications from other biologists who encountered similar problems. Difficulties encountered by these workers apparently were due to unreliable analyses by a laboratory outside their respective universities where they sent samples for analysis of isotope ratios. In contrast, workers who performed their own analyses of isotope ratios have obtained acceptable results (e.g., Mullen 1970, 1971). Thus, the doubly labeled water technique is not flawed in itself, but the need for specialized and very expensive equipment (e.g., a cyclotron) limits its utility.

In addition, this technique requires the initial capture of a bird for isotope injection and recapture over daily or multiday periods. This may be difficult, and such disturbance might affect the bird's behavior and, consequently, its energy budget. In a recent analysis using this

technique, for example, difficulties in the field forced workers to restrict study to birds held in aviaries (Weathers and Nagy 1980). This
obviously limits the usefulness of such data for ecological analyses.

Time-budget analyses. Currently, the most popular technique
for estimating the daily energy expenditure of free-living birds is to
record the time allocated by the animal to various activities throughout
the day and multiply those time values by estimated metabolic costs of
each activity. Such estimates of metabolic costs for activities are derived
from laboratory analyses. This technique requires minimal field equipment (i.e., stopwatches), and data collection by a careful observer generally should not have an important effect on the bird's behavior. Unfortunately, the accuracy of such time-budget estimates is poorly
known. A few authors (e.g., Ettinger and King 1980) have used sensitivity analyses to demonstrate that their estimates probably are sufficiently accurate. Such analyses, however, are rare. More direct tests of
this technique by comparing estimates derived from time budgets and
those produced from independent measures of energy metabolism also
are rare and have produced mixed results. Several analyses have indicated that particular versions of the time-budget method are acceptably
accurate (e.g., Utter and LeFebvre 1973; Koplin et al. 1980; Ettinger
and King 1980), whereas at least one analysis has indicated substantial
error in one version (Weathers and Nagy 1980).

How should this set of technical difficulties in quantifying the power
consumption of free-living birds best be resolved? I suggest that substantial effort be devoted to developing time-budget techniques that
are demonstrably accurate to an acceptable degree. A major advantage
of such techniques is that they simultaneously quantify power consumption and behavior. Current information indicates that behavioral
variation and consequent changes in time allocation to various activities
commonly is the major determinant of changes in a bird's energy budget. Thus, variation in energy budgets cannot be understood without
behavioral analyses or, conversely, useful analyses of the ecology and
evolution of behavioral patterns may require examination of their energetic consequences. Measurement of a purely physiological variable
such as an average rate of power consumption is of very limited use for
ecological analyses unless set in the context of its interaction with major
selective pressures acting on the animal in nature. Thus, workers analyzing daily energy expenditure using such techniques as the doubly
labeled water method probably will find it useful to quantify simultaneously the animal's behavior. Such behavioral analyses may be compli-

cated by irregularities in the bird's time and energy budget associated with the capture, manipulation, and release of the animal that is required by the doubly labeled water technique. In contrast, the time-budget method potentially allows the simultaneous quantification of both behavioral and energetic patterns with minimal disturbance. Improvement in this approach must involve not only tests of such techniques (e.g., Weathers and Nagy 1980) but also efforts to identify and correct sources of error. Such a constructive approach has not been characteristic of any tests of this approach.

Concluding comments

In this analysis, I have isolated particular phases of avian life cycles and considered the energetics of an individual phase for a variety of species. Such a comparative approach is useful in clarifying patterns and problems characteristic of these portions of avian life histories and also reflects the general divisions used by students of avian energetics. However, this approach tends to overemphasize the discreteness of these phases and ignores their fundamental integration in the life history. Natural selection should favor the evolution of integrated life history patterns that tend to maximize fitness. It is important to recognize that this could produce selection for variations in a particular temporal phase that only can be comprehended within the context of the entire life history. Obviously, the energetic characteristics of one portion of the life cycle can significantly affect other portions. The comparatively small fraction of energy devoted to storage or somatic growth in adult birds may restrict the range of such interphasic effects compared to ectothermic taxa, but clearly there remains a host of possibly important influences. For example, variable allocation of energy to nest construction and consequent variation in nest insulation will affect the heat budget of eggs and young, as well as the adult's time and energy budgets, throughout the reproductive cycle. Ideally, analyses of ecological energetics should consider the entire annual cycle, the major iterative unit of avian life histories. If such analyses are impractical, it is worthwhile to attempt to deal with major fractions of the annual period, such as the complete breeding cycle. Such integrated analyses ultimately will be necessary for an adequate understanding of the role that selection pressures related to energy have played in the evolution of the patterns observed in avian biology.

Acknowledgments

This analysis was supported by National Science Foundation Grant DEB 80-04266 and an Arizona State University Faculty grant-in-aid. I thank J. Alcock, J. R. Hazel, J. R. King, R. L. Rutowski, and A. T. Smith for stimulating discussions.

Literature cited

Ankney, C. D. 1979. Does the wing molt cause nutritional stress in Lesser Snow Geese? *Auk 96:*68–72.

Ankney, C. D., and D. D. MacInnes. 1978. Nutrient reserves and reproductive performance of female Lesser Snow Geese. *Auk 95:*459–471.

Ankney, C. D., and D. M. Scott. 1980. Changes in nutrient reserves and diet of breeding Brown-headed Cowbirds. *Auk 97:*684–696.

Bakken, G. S. 1976. A heat-transfer analysis of animals: unifying concepts and the application of metabolic chamber data to field ecology. *J. Theor. Biol. 60:*337–384.

– 1980. The use of standard operative temperature in the thermal energetics of birds. *Physiol. Zool. 53:*108–119.

Bakken, G. S., W. A. Buttemer, and D. M. Gates. 1981. Heated taxodermic mounts: a means of measuring the standard operative temperature affecting small animals. *Ecology 62:*311–318.

Bakken, G. S., and D. M. Gates. 1975. Heat-transfer analysis of animals: some implications for field ecology, physiology, and evolution. In D. M. Gates and R. B. Schmerl (eds.). *Perspectives in biophysical ecology,* pp. 255–290. New York: Springer-Verlag.

Balda, R. P., and G. C. Bateman. 1973. The breeding biology of the Pinyon Jay. *Living Bird 11:*5–12.

Barry, T. W. 1962. Effect of late seasons on Atlantic Brant reproduction. *J. Wildl. Manage. 26:*19–26.

Bartholomew, G. A., T. R. Howell, and T. J. Cade. 1957. Torpidity in the White-throated Swift, Anna Hummingbird, and Poor-will. *Condor 59:*145–155.

Belovsky, G. E. 1981a. Food plant selection by a generalist herbivore: the moose. *Ecology 62:*1020–1030.

– 1981b. Optimal activity times and habitat choice by moose. *Oecologia 48:*22–30.

Berger, M., and J. S. Hart. 1972. Die Atmung beim Kolibri *Amazilia fimbriata* wahrend des Schwirrfluges bei verschiedenen Umgebungstemperaturen. *J. Comp. Physiol. 81:*363–380.

Biebach, H. 1977. Der Energieaufwand für des Bruten beim Star. *Naturwissenschaften 64:*343.

– 1979. Energetik des Brutens beim Star (*Sturnus vulgaris*). *J. Ornithol. 120:* 121–138.

Blem, C. R. 1980. The energetics of migration. In S. A. Gauthreaux, Jr. (ed.).

Animal migration, orientation, and navigation, pp. 175–224. New York: Academic Press.

Caccamise, D. F., and W. W. Weathers. 1977. Winter-nest microclimate of Monk Parakeets. *Wilson Bull. 89:*346–349.

Calder, W. A. 1971. Temperature relationships and nesting of the Calliope Hummingbird. *Condor 73:*314–321.

– 1973. Microhabitat selection during nesting of hummingbirds in the Rocky Mountains. *Ecology 54:*127–134.

Calder, W. A., and J. R. King. 1974. Thermal and caloric relations of birds. In D. S. Farner and J. R. King (eds.). *Avian biology,* pp. 259–413. New York: Academic Press.

Campbell, G. S. 1977. *An introduction to environmental biophysics.* New York: Springer-Verlag.

Carpenter, F. L. 1976. Ecology and evolution of an Andean hummingbird (*Oreotrochilus estella*). *Univ. Calif. Berkeley Publ. Zool. 106:*1–74.

Carpenter, F. L., and R. E. MacMillen. 1976a. Energetic cost of feeding territories in an Hawaiian honeycreeper. *Oecologia 26:*213–223.

– 1976b. Threshold model of feeding territoriality and test with a Hawaiian honeycreeper. *Science 194:*639–642.

Custer, T. W. 1974. Population ecology and bioenergetics of the Lapland Longspur (*Calcarius lapponicus*) near Barrow, Alaska. Ph.D. thesis, University of California, Berkeley.

Dawson, W. R., and J. W. Hudson. 1970. Birds. In G. C. Whittow (ed.). *Comparative physiology of thermoregulation,* Vol. 1., pp. 223–310. New York: Academic Press.

DeWoskin, R. 1980. Heat exchange influence on foraging behavior of *Zonotrichia* flocks. *Ecology 61:*30–36.

Dimmit, M. A., and R. Ruibal. 1980. Exploitation of food resources by spadefoot toads (*Scaphiopus*). *Copeia 1980:*854–862.

Drobney, R. D. 1980. Reproductive bioenergetics of Wood Ducks. *Auk 97:* 480–490.

El-Wailly, A. J. 1966. Energy requirements for egg-laying and incubation in the Zebra Finch, *Taeniopygia castanotis. Condor 68:*582–594.

Ettinger, A. O., and J. R. King. 1980. Time and energy budgets of the Willow Flycatcher (*Empidonax traillii*) during the breeding season. *Auk 97:*535–546.

Ewald, P. W., and F. L. Carpenter. 1978. Territorial responses to energy manipulations in the Anna Hummingbird. *Oecologia 31:*277–292.

Flynn, R. K., and J. A. Gessaman. 1979. An evaluation of heart rate as a measure of daily metabolism in pigeons (*Columba livia*). *Comp. Biochem. Physiol. 63A:*511–514.

Gessaman, J. A. 1980. An evaluation of heart rate as an indirect measure of daily energy metabolism of the American Kestrel. *Comp. Biochem. Physiol. 65A:*273–289.

Gessaman, J. A., and P. R. Findell. 1979. Energy cost of incubation in the American Kestrel. *Comp. Biochem. Physiol. 63A:*57–62.

Gill, F. B., and L. L. Wolf. 1975. Foraging strategies and energetics of East African sunbirds at mistletoe flowers. *Am. Nat. 109:*491–510.

Gladstone, D. E. 1979. Promiscuity in monogamous colonial birds. *Am. Nat.* *114:*545–557.

Goss-Custard, J. D. 1977a. Optimal foraging and size selection of worms by the Redshank, *Tringa totanus*, in the field. *Anim. Behav. 25:*10–29.

– 1977b. The energetics of prey selection by Redshank, *Tringa totanus*, in relation to prey density. *J. Anim. Ecol. 46:*1–19.

Gyllin, R., H. Kallander, and M. Sylven. 1977. The microclimate explanation of town centre roosts of Jackdaws (*Corvus monedula*). *Ibis 119:*358–361.

Hails, C. J., and D. M. Bryant. 1979. Reproductive energetics of a free-living bird. *J. Anim. Ecol. 48:*471–482.

Hanson, H. C. 1962. The dynamics of condition factors in Canada Geese and their relation to seasonal stresses. *Arct. Inst. North Am. Tech. Bull. 12:*1–68.

Hart, J. S., and M. Berger. 1972. Energetics, water economy and temperature regulation during flight. In *Proceedings of the 15th International Ornithogical Congress*, pp. 189–199.

Holmes, R. T., C. P. Black, and T. W. Sherry. 1979. Comparative population bioenergetics of three insectivorous passerines in a deciduous forest. *Condor 81:*9–20.

Hubbard, J. D. 1978. Breeding biology and reproductive energetics of Mountain White-crowned Sparrows in Colorado. Ph.D. dissertation, University of Colorado.

Jones, P. J., and P. Ward. 1976. The level of reserve protein as the proximate factor controlling the timing of breeding and clutch-size in the Red-billed Quelea, *Quelea quelea. Ibis 118:*547–573.

Kelty, M. P., and S. B. Lustick. 1977. Energetics of the starling (*Sturnus vulgaris*) in a pine woods. *Ecology 58:*1181–1185.

Kendeigh, S. C. 1961. Energy of birds conserved by roosting in cavities. *Wilson Bull. 73:*140–147.

– 1963. Thermodynamics of incubation in the House Wren, *Troglodytes aedon.* In *Proceedings of the 13th International Ornithological Congress*, pp. 884–904.

Kendeigh, S. C., V. R. Dolnik, and V. M. Gavrilov. 1977. Avian energetics. In J. Pinowski and S. C. Kendeigh (eds.). *International Biological Programme*, Vol. 12, *Granivorous birds in ecosystems*, pp. 127–204. Cambridge: Cambridge University Press.

Ketterson, E. D., and J. R. King. 1977. Metabolic and behavioral responses to fasting in the White-crowned Sparrow (*Zonotrichia leucophrys gambelii*). *Physiol. Zool. 50:*115–129.

King, J. R. 1972. Adaptive periodic fat storage in birds. In *Proceedings of the 15th International Ornithological Congress*, pp. 200–217.

– 1973. Energetics of reproduction in birds. In D. S. Farner (ed.). *Breeding Biology of birds*. Washington, D.C.: National Academy of Science.

– 1980. Energetics of avian moult. In R. Nöhring (ed.). *Acta XVII Congressus Internationalis Ornitholgici*, pp. 312–317. Berlin: Verlag der Deutschen Ornithologen-Gesellschaft.

King, W. W., and N. F. Hadley. 1979. Water flux and metabolic rates of free-roaming scorpions using the doubly labeled water technique. *Physiol. Zool. 52:*176–189.

Kleiber, M. 1961. *The fire of life.* New York: Wiley.

Koplin, J. R., M. W. Collopy, A. R. Bammann, and H. Levenson. 1980. Energetics of two wintering raptors. *Auk 97:*795–806.

Korschgen, C. E. 1977. Breeding stress of female eiders in Maine. *J. Wildl. Manage. 41:*360–373.

Krapu, G. L. 1974. Feeding ecology of Pintail hens during reproduction. *Auk 91:*278–290.

– 1981. The role of nutrient reserves in Mallard reproduction. *Auk 98:*29–38.

Krebs, J. R., J. Ryan, and E. L. Charnov. 1977. Optimal prey selection in the Great Tit *Parus major. Anim. Behav. 25:*30–38.

Kushlan, S. A. 1978. Nonrigorous foraging by robbing egrets. *Ecology 59:*649–653.

Lifson, N., and R. McClintock. 1966. Theory and use of the turnover rates of body water for measuring energy and material balance. *J. Theor. Biol. 12:*46–74.

Lustick, S., B. Battersby, and M. P. Kelty. 1978. Behavioral thermoregulation: orientation toward the sun in Herring Gulls. *Science 200:*81–83.

MacArthur, R. H. 1972. *Geographical ecology.* New York: Harper & Row.

Mahoney, S. A. 1976. Thermal and ecological energetics of the White-crowned Sparrow (*Zonotrichia leucophrys*) using the equivalent black-body temperature. Ph.D. dissertation, Washington State University.

Mahoney, S. A., and J. R. King. 1977. The use of the equivalent black-body temperature in the thermal energetics of small birds. *J. Therm. Biol. 2:*115–120.

Morhardt, S. S., and D. M. Gates. 1969. Energy exchange analysis of the Belding ground squirrel and its habitat. *Ecol. Monogr. 44:*17–44.

Mugaas, J. N., and J. R. King. 1981. The annual variation in daily energy expenditure of the Black-billed Magpie: a study of thermal and behavioral energetics. *Stud. Avian Biol. 5:*1–78.

Mullen, R. K. 1970. Respiratory metabolism and body water turnover rates of *Perognathus formosus* in its natural environment. *Comp. Biochem. Physiol. 32:*259–265.

– 1971. Energy metabolism and body water turnover rates of two species of free-living kangaroo rats, *Dipodomys merriami* and *Dipodomys microps. Comp. Biochem. Physiol. 39A:*379–380.

Nagy, K. A. 1980. CO_2 production in animals: analysis of potential errors in the doubly labeled water method. *Am. J. Physiol. 238:*R466–R473.

Nagy, K. A., and K. Milton. 1979. Energy metabolism and food consumption by wild howler monkeys (*Alouatta palliata*). *Ecology 60:*475–480.

Nagy, K. A., and G. F. Montgomery. 1980. Field metabolic rate, water flux, and food consumption in three-toed sloths (*Bradypus variegatus*). *J. Mammal. 61:*465–472.

O'Connor, R. J. 1980. Energetics of reproduction in birds. In R. Nöhring (ed.). *Acta XVII Congressus Internationalis Orthithologici,* pp. 306–311. Berlin: Verlag der Deutschen Ornithologen-Gesellschaft.

Paladino, F. V. 1979. Energetics of terrestrial locomotion in White-crowned Sparrows (*Zonotrichia leucophrys gambelii*). Ph.D. thesis, Washington State University.

Pohl, H. 1969. Some factors influencing the metabolic response to cold in birds. *Fed. Proc. 28:*1059–1064.

Pohl, H., and G. C. West. 1973. Daily and seasonal variation in metabolic response to cold during rest and forced exercise in the Common Redpoll. *Comp. Biochem. Physiol. 45A:*851–867.

Pyke, G. H., H. R. Pulliam, and E. L. Charnov. 1977. Optimal foraging: a selective review of theory and tests. *Q. Rev. Biol. 52:*137–154.

– 1979. The annual cycle of body composition of Canada Geese with special reference to control of reproduction. *Auk 96:*234–252.

Ricklefs, R. E. 1974. Energetics of reproduction in birds. In R. A. Paynter (ed.). *Avian energetics*, pp. 152–292. Cambridge, Mass.: Nuttall Ornithological Club.

– 1979. Adaptation, constraint and compromise in avian postnatal development. *Biol. Rev. 54:*269–290.

Robinson, D. E., G. S. Campbell, and J. R. King. 1976. An evaluation of heat exchange in small birds. *J. Comp. Physiol. 105:*153–166.

Schoener, T. W. 1971. The theory of foraging strategies. *Annu. Rev. Ecol. Syst. 2:*369–404.

Sherry, D. F., N. Mrosovsky, and J. A. Hogan. 1980. Weight loss and anorexia during incubation in birds. *J. Comp. Physiol. Psychol. 94:*89–98.

Smith, W. K., S. W. Roberts, and P. C. Miller. 1974. Calculating the nocturnal energy expenditure of an incubating Anna's Hummingbird. *Condor 76:*176–183.

Southwick, E. E., and D. M. Gates. 1975. Energetics of occupied hummingbird nests. In D. M. Gates and R. B. Schmerl (eds.). *Perspectives in biophysical ecology*, pp. 417–430. Berlin: Springer.

Taylor, C. R., R. Dmi'el, M. Fedak, and K. Schmidt-Nielsen. 1971. Energetic cost of running and heat balance in a large bird, the Rhea. *Am. J. Physiol. 221:*597–601.

Torre-Bueno, J. R. 1976. Temperature regulation and heat dissipation during flight in birds. *J. Exp. Biol. 65:*471–482.

Trivers, R. L. 1972. Parental investment and sexual selection. In B. Campbell (ed.). *Sexual selection and the descent of man, 1871–1971*, pp. 136–179. Chicago: Aldine.

Tucker, V. A. 1968. Respiratory exchange and evaporative water loss in the flying Budgerigar. *J. Exp. Biol. 48:*67–87.

Utter, J. M., and E. A. LeFebvre. 1973. Daily energy expenditure of Purple Martins (*Progne subis*) during the breeding season: estimates using D_2O^{18} and time methods. *Ecology 54:*597–604.

Vleck, C. M. 1981. Energetic cost of incubation in the Zebra Finch. *Condor 83:*229–237.

Walsberg, G. E. 1977. Ecology and enegetics of contrasting social systems in *Phainopepla nitens* (Aves: Ptilogonatidae). *Univ. Calif. Berkeley Publ. Zool. 108:*1–63.

– 1978. Brood size and the use of time and energy by the Phainopepla. *Ecology 59:*147–153.

– 1981. Nest-site selection and the radiative environment of the Warbling Vireo. *Condor 83:*86–88.

— in press. Avian ecological energetics. In D. S. Farner and J. R. King (eds.). *Avian biology*, Vol. 7. New York: Academic Press.

Walsberg, G. E., G. S. Campbell, and J. R. King. 1978. Animal coat color and radiative heat gain: a re-evaluation. *J. Comp. Physiol. 126*:211–222.

Walsberg, G. E., and J. R. King. 1978a. The heat budget of incubating Mountain White-crowned Sparrows (*Zonotrichia leucophrys oriantha*) in Oregon. *Physiol. Zool. 51*:92–103.

— 1978b. The energetic consequences of incubation for two passerine species. *Auk 95*:644–655.

— 1980. The thermoregulatory significance of the winter roost-sites selected by robins in eastern Washington. *Wilson Bull. 92*:33–39.

Weathers, W. W., and K. A. Nagy. 1980. Simultaneous doubly labeled water (^{3}HH^{18}O) and time budget estimates of daily energy expenditure in *Phainopepla nitens. Auk. 97*:861–867.

Weller, M. W. 1957. Growth, weights, and plumages of the Redhead (*Aythya americana*). *Wilson Bull. 69*:5–38.

Withers, P. C. 1977. Energetic aspects of reproduction by the Cliff Swallow. *Auk 94*:718–725.

Wooley, J. B., and R. B. Owen. 1977. Energy costs of activity and daily energy expenditure in the Black Duck. *J. Wildl. Manage. 42*:739–745.

Yom-Tov, Y., A. Imber, and J. Otterman. 1977. The microclimate of winter roosts of the starling *Sturnus vulgaris. Ibis 119*:366–368.

Zach, R. 1979. Shell-dropping: decision-making and optimal foraging in Northwestern Crows. *Behaviour 68*:106–117.

Commentary

WILLIAM A. CALDER III

The voluminous literature on standard metabolic rates and metabolism curves that followed the "seeding" of the field by Scholander et al. (1950) and Dawson (1954) attests to the value of birds for the study of energetics. As the oxygen analyzer's life list grew from the original Snow Buntings, manakins, and towhees, generalizations about *The Bird* and its function were in order. King and Farner (1961), Lasiewski and Dawson (1967), and Aschoff and Pohl (1970) used allometry to generalize or summarize. Although the "splitters" tended to be more concerned about the variety of points that did not fall on the line, body size did account for the major portion of the variability in basal metabolic rates. Faced with the variety of nine thousand species of birds, many are more grateful for the body mass to the $\frac{3}{4}$ power ($M^{3/4}$) that describes three-quarters or more of the variance than they are disturbed by the adaptive departures from such a generalization.

However, the study of ecological energetics cannot be confined to the metabolic chamber. To find biological meaning and broad interest, our attention turns to the "real world." Walsberg has given an excellent inventory of the relevant questions about energetics in natural environments, the questions that we should attempt to answer. In addition to those conceptual questions, there are questions of procedure: What species should we follow? What will we measure? How will we relate the data? Can we extrapolate from laboratory successes to field conditions? In a time of diminished support for basic research, how can we gain the maximum insight with the least dead ends? If you attempt to answer Walsberg's questions 1 with a 30-g yellow bird from the Great Plains and I go after question 2 with a 100-g brown bird from a semitropical forest, can we get together in a synthesis for the next symposium?

The splitters have a point—the unusual or atypical varieties are what makes the study of evolutionary adaptation interesting, but until we have a common baseline for comparisons, the interesting adaptations are only subjective. The metabolism chamber data showed the predominant importance of size, and no one would doubt that a goose or an ostrich needs more food than a hummingbird and could not nest on the same branch or feed from the same gnat swarm or flower patch. Size must be the primary consideration in nature as it was in the flat-blacked paint can. The best baseline is the one that absorbs the greatest proportion of the variance in the data. Consequently, I feel that we will start answering Walsberg's questions and finish answering them with the recognition of body size as the best "handle" for progress in ornithological energetics. I will therefore begin to anticipate an allometric basis for ecological energetics by examining some of Walsberg's questions.

1. *To what degree are free-living birds thermally stressed, and what is the nature of this stress? To what degree is thermoregulation a major expense for birds that winter in cold?* The available allometric generalizations that are relevant to these questions appear in Table 5.1. Actual energetic costs per bird per day ($\dot{E}_{total}$) would be somewhere between the minimum and the maximum capacities of the birds, that is, between standard or basal power ($\dot{E}_{sm}$) while resting in a thermoneutral environment and the rate during flight ($\dot{E}_{fly}$) or during inactive exposure to an extremely cold environment ($\dot{E}_{cold}$).

The ratio $\dot{E}_{fly}/\dot{E}_{sm}$ is between 7 and 12 and size independent ($\propto M^{\sim 0}$), depending upon whether comparison is made with the typical homeothermic $\dot{E}_{sm}$ of nonpasserines or the higher $\dot{E}_{sm}$ of passerines. The ratios $\dot{E}_{o}/\dot{E}_{sm}$, comparing exposure to O° and thermoneutrality, are lower, about 5, and size dependent with body mass exponent thought to reflect size differences in the lower critical temperature at which the

Table 5.1. *Allometric equations for avian energetics* (m = body mass in grams)

	Milliwatts	Kilojoules per day	Relative to standard power[a]	Percentage	Reference
Flight metabolic rate, all birds	$341.4m^{0.73}$	–	$7.32\ m^{0.01}$ (p) $11.91\ m^{0.00}$ (np)		Calder (1974)
Percentage of active day flying				$44.3m^{-0.603}$	Walsberg (in press)
Estimated total energy per day, all birds	–	$13.05m^{0.605}$	$3.24m^{-0.10}$ (p) $5.26m^{-0.11}$ (np)		Walsberg (in press)
Flight foragers	–	$14.17m^{0.607}$	$3.52m^{-0.10}$ (p) $5.71m^{-0.11}$ (np)		Walsberg (in press)
Non-flight foragers	–	$12.84m^{0.610}$	$3.19m^{-0.10}$ (p) $5.18m^{-0.11}$ (np)		Walsberg (in press)
Resting metabolic power at 0°					
Passerines	$230.9m^{0.42}$	$19.95m^{0.42}$	$4.95m^{-0.30}$		Calder (1974)[b]
Nonpasserines	$161.8m^{0.53}$	$13.98m^{0.53}$	$5.64m^{-0.20}$		Calder (1974)[b]
Standard (basal) metabolic power					
Passerines	$46.63m^{0.72}$	$4.03m^{0.72}$	$1.0m^{0}$		Calder (1974)[b]
Nonpasserines	$28.67m^{0.73}$	$2.48m^{0.73}$	$1.0m^{0}$		Calder (1974)[b]

[a]P, Passerine; np, nonpasserine.
[b]Equations appear in review by Calder (1974), converted from works of Lasiewski and Dawson, Kendeigh, Hart, and Berger, cited therein.

$\dot{E}_{sm}$ is adequate (i.e., exposure to O° is a greater departure from thermoneutrality for the smaller bird). This allometric ratio is in agreement with the specific determination of a factor of 5.5 for $\dot{E}_{max}$ of cold-stressed cardueline finches (Dawson and Carey 1976).

In another review, Walsberg (in press) has analyzed the estimated total daily energy requirement ($\dot{E}_{total}$) as a function of size (Table 5.1), which can be translated into approximately allometric ratios that show size dependencies intermediate between those of $\dot{E}_{sm}$ and $\dot{E}_{0°}$. Does this indicate an average level of thermal stress intermediate between $\dot{E}_{sm}$ and $\dot{E}_{O}°$? Not necessarily, because both consider only the resting bird.

What happens in activity? We can use the equations in Table 5.1 to generate values for $\dot{E}_{total}$ of seven species of hypothetical birds (four passerines bracketed in size by three nonpasserines; Table 5.2), using Walsberg's (in press) equation for the portion of active day spent in flight. Because this portion is inversely related to size, the larger the bird, the greater the time spent at the lower resting rate. A regression of these estimates is

Table 5.2. *Estimated total daily energy requirements for hypothetical birds*

	Humming-bird	Warbler	Finch	Plover	Thrush	Raven	Goose
Body mass	3 g	10 g	40 g	50 g	75 g	1,000 g	5,000 g
Standard metabolism (mWatt)	63.9	245	664	499	1,044	6,605	15,870
Flight metabolism (J hour^{-1})	0.761	6.60	18.16	21.37	28.73	190.4	616.3
Flight (% of active period)	22.8	11.1	4.79	4.19	3.28	0.69	0.26
Flight time in 13-hr day	2.97	1.44	0.62	0.54	0.43	0.09	0.03
Total flight energy kJ)	8.14	9.50	11.26	11.54	12.35	17.13	20.88
Resting time (hours)	21.03	22.56	23.38	23.46	23.57	23.91	23.97
Basal energy (kJ)	4.84	19.88	65.88	42.10	88.60	956.7	1,369
Total energy (kJ)	12.98	29.38	77.14	53.64	101.0	973.8	1,390

$$\dot{E}_{\text{total}} = 5.87m^{0.67}; \qquad r^2 = 0.970 \tag{5.1}$$

The estimates used in this calculation give a low value, because there is no allowance for the fact that metabolic rates while perched, singing, feeding, and other nonflying activities are probably above the basal level. However, they can account for a scaling of $\dot{E}_{\text{total}}$ that is of smaller exponent than resting or flying rates, without invoking cold stress. From this, one can see the possibility that the increased heat-generation during activity, diurnal solar radiation, and nocturnal microclimate selection might limit cold stress to relatively rare emergency situations.

Heat stress may be another matter. Regal (1975) in fact hypothesized that the evolution of insulation on birds was a response to the threat of excess heat influx to diurnal birds, whereas the pelage of nocturnal mammals served to conserve heat. The allometric analysis of heat stress has only begun with the significant contribution of Weathers (1981).

Although one can do little more than speculate with the present-day energetics allometry, it is clear from the different exponents that the questions cannot be answered without considering the importance of size.

2. *What are the energetic consequences of activities important for birds in nature? How much energy is expended in resource defense? What are the relative energy costs and benefits of different types of resource defense? To what degree do energetic costs constrain defense?* Before these questions can be

dealt with effectively (quantitatively), we need a better quantitative framework that describes the resources being defended.

Schoener (1968) showed that territory size for 61 species of birds (not separated into feeding types) is proportional to $m^{1.09 \pm 0.11}$. Why should territory size be scaled in this fashion? For simplification, we can treat the territory as if it were circular or square in which case the diameter or distance across the middle as well as the perimeter would be proportional to $(m^{1.09})^{1/2}$ or $m^{0.55}$. Pennycuick (1969) approximated the speed ranges of flying birds as $m^{1/6}$ or $m^{0.17}$. The time required for birds to fly across or around the territory should then scale as $m^{0.55}/m^{0.17} = m^{0.38}$. Because most other physiological times are functions of $m^{-0.25}$ (Lindstedt and Calder 1981), this suggests that the larger the bird, the fewer defense trips per day.

What is being defended is a resource for reproduction. T. Marr (unpub.) has calculated the following relationships:

mean number of feeding trips to brood per day = $1,430m^{-0.57}$
($r = 0.91$, $t = 4.9$, $p = 0.002$)   (5.2)

average fresh mass of food (g trip^{-1}) = $0.00606m^{1.14}$
($r = 0.97$, $t = 7.98$, $p = 0.0003$)   (5.3)

These can be combined to give

food mass to nest (g day^{-1}) = $(1,430m^{-0.97})$ $(0.00606m^{1.14})$ = $8.67m^{0.57}$ (5.4)

This mass of food transported appears to have an essentially size-independent relationship to total energy costs of an adult per day (combining Equation 5.4 and $\dot{E}_{tot}$ from Tabl .1):

$$\dot{M}_{food}/\dot{E}_{tot} \propto m^{0.57}/m^{0.61} \propto m^{-0.04}$$ (5.5)

Marr (unpub.) also calculated the mean distances ($l_{0.2E_{tot}}$) from foraging site to nest for which the daily round-trip cost would amount to 20% of $\dot{E}_{sm}$:

$$l_{0.2E_{tot}} = 10.45m^{0.52}$$ (5.6)

If this is the characteristic linear dimension of the territory in which the bird feeds, then the area would be proportional to

$$l^2 \propto m^{1.04}$$ (5.7)

As Marr points out, this is statistically indistinguishable from the exponent for territory size. Note also the similarity in exponents for percentage of active day spent in flight (Table 5.1) and for number of feeding trips to brood per day. If the active time per day is independent of body size, this means that the ratio of time per day in flight (t_{fly}) to trips per day is essentially a size-independent factor for time per trip:

$$t_{fly} \, day^{-1}/trips \, day^{-1} \propto m^{-0.60}/m^{-0.57} \propto m^{0.03} \tag{5.8}$$

From Equation 7 of Tucker (1970) the energy cost of travel for birds is

$$J_{km}^{-1} = 25.1 m^{0.77} \tag{5.9}$$

The product of this cost per distance and the distance across the territory is the cost per trip:

$$\propto m^{0.77} \times m^{0.55} = m^{1.32} \tag{5.10}$$

The travel cost of territorial or patrol per day should be the product of trip cost and number of trips. What is needed to fill in this chink is an allometry of territorial defense or patrolling frequency, usually by the male. If his trip frequency were scaled as is that for the female feeding her brood, or if both sexes share in these responsibilities, the territorial travel cost would be proportional to

$$m^{1.32} \times m^{-0.57} = m^{0.75} \tag{5.11}$$

we would have a scaling more nearly like that for $\dot{E}_{sm}$ than for $\dot{E}_{tot}$.

It is necessary to emphasize that these speculations are made on the basis of some equations with limited data bases. Their virture is in telling us what sorts of basic data and analysis are needed to complete the answer to How do stresses, mechanisms, and costs vary with important aspects of the life-styles of birds? The term "life-style" implies a specialized departure from the general or typical, and the only way a departure can be measured quantitatively is by comparison to a baseline. Hopefully, this preliminary effort has demonstrated the potential for allometric baselines in the ecological energetics of the future.

Acknowledgment

The work described here has been supported in part by NSF Grant DEB 79-03689.

Literature cited

Aschoff, J., and H. Pohl. 1970. Rhythmic variations in energy metabolism. *Fed. Proc. 29*:1541–1552.
Calder, W. A., III. 1974. Consequences of body size for avian energetics. In R. A. Paynter, Jr. (ed.). *Avian energetics. Publ. Nuttall Ornithol. Club 15*:86–151.

Dawson, W. R. 1954. Temperature regulation and water requirements of the Brown and Abert Towhees, *Pipilo fuscus* and *Pipilo aberti*. *Univ. Calif. Berkeley Publ. Zool.* 59:81–124.

Dawson, W. R., and C. Carey. 1976. Seasonal acclimatization to temperature in cardueline finches. I. Insulative and metabolic adjustments. *J. Comp. Physiol.* 112:317–333.

King, J. R., and D. S. Farner. 1961. Energy metabolism, thermoregulation, and body temperature. *In* A. J. Marshall (ed). *Biology and comparative physiology of birds*, Vol. 2, pp. 215–288. New York: Academic Press.

Lasiewski, R. C., and W. R. Dawson. 1967. A re-examination of the relation between standard metabolic rate and body weight in birds. *Condor* 69:13–23.

Lindstedt, S. L., and W. A. Calder III. 1981. Body size, physiological time, and longevity of homeothermic animals. *Q. Rev. Biol.* 56:1–16.

Pennycuick, C. J. 1969. Mechanics of bird migration. *Ibis* 111:525–556.

Regal, P. J. 1975. The evolutionary origin of feathers. *Q. Rev. Biol.* 50:35–66.

Schoener, T. W. 1968. Sizes of feeding territories among birds. *Ecology* 49:123–141.

Scholander, P. F., R. Hock, V. Walters, F. Johnson, and L. Irving. 1950. Heat regulation in some arctic and tropical mammals and birds. *Biol. Bull.* 99:237–258.

Tucker, V. A. 1970. Energetic cost of locomotion in animals. *Comp. Biochem. Physiol.* 34:841–846.

Walsberg, G. E. in press. Avian ecological energetics. In D. S. Farner, and J. R. King (eds.). *Avian biology*, Vol. 7. New York: Academic Press.

Weathers, W. W. 1981. Physiological thermoregulation in heat-stressed birds: consequences of body size. *Physiol. Zool.* 54:345–361.

6   Perspectives in optimal foraging

JOHN R. KREBS, DAVID W. STEPHENS, AND
WILLIAM J. SUTHERLAND

Optimal foraging theory (OFT) is one of the few areas of study in behavior and ecology in which mathematical models derived from first principles have been seriously tested in the laboratory and field. The balance between theory and data has remained good, unlike, for example, the field of community ecology in the 1960s and 1970s, where arcane models of baroque complexity were generally matched only with qualitative and unconvincing tests. Part of the success of OFT lies in the fact that although ecological in origin, the models have been tested with both ecological methods and the methods developed by ethologists and comparative psychologists (Pyke et al. 1977; Krebs 1978, Staddon 1980; Hughes and Townsend 1981; Kamil and Yoerg 1982). In 1966, the first two papers published on OFT (MacArthur and Pianka 1966; Emlen 1966) amounted to 0.5% of the articles in *American Naturalist, Ecology, Journal of Animal Ecology,* and *Animal Behavior.* The proportion of papers on OFT in just these journals had quadrupled to 2% by 1974 and to 8% in 1981.

In this chapter, we will start with a brief general comment on optimal foraging theory, then we review the evidence relating to "classical" foraging models. This is followed by two more detailed discussions; the first considers the relationship between classical models and two more recent developments, models of "rules of thumb" and stochastic models, and the second looks at some implications of the traditional models for population interactions of predators and prey.

What is OFT?

OFT is an attempt to find out if there are any general rules about what animals feed on, where they go to feed, and how they search for food. One convenient way to think about these questions is

165

in terms of "decision rules": The animal "decides" to eat this food item and not that one, it decides to hunt here and not there. In this terminology, OFT can be said to be an attempt to understand the decision rules of foraging animals. The rationale of OFT is that these decision rules have been shaped by natural selection to allow the animal to perform as efficiently as possible, and the aim of a foraging model is to make an informed guess about the meaning of "efficiency" and the constraints that limit efficiency. When these guesses have been built into a model, they can be examined by testing the predictions or assumptions of the model against real data. For example, many foraging models hypothesize that efficiency can be equated with "net rate of energy intake" and that constraints on maximizing net rate of energy intake include such things as handling time, time needed to recognize cryptic prey (Hughes 1979), the requirement for certain nutrients (Belovsky 1978), and the ability to discriminate between prey types (Rechten et al., in press).

Two common misunderstandings about this approach are illustrated by such often-heard quotes as "aardvarks [or whatever animal is under discussion] do not seem to forage optimally" and "optimal foraging models are all right in simple laboratory environments but they cannot handle the complexities of nature, where predation risk etc. impinge on foraging behavior" (Morse 1980; Zach and Smith 1981; Schluter 1981). The first statement reflects a misunderstanding of the use of optimality models in biology. They cannot be used to test the proposition that animals are (or are not) optimal but only the proposition that one particular hypothesis, for example, maximizing net rate of intake subject to constraints a, b, and c, describes the animal's foraging behavior. The apocryphal aardvark connoisseur probably meant that aardvarks do not eat prey or visit foraging sites that maximize net gain rate. This is not, however, to say that there is no conceivable maximization model that could predict aardvark foraging behavior. An actual example to illustrate the point is the finding (Krebs et al. 1977) that Great Tits in cages selectively ate the profitable prey in a way that was qualitatively but not quantitatively consistent with an intake maximizing model of prey choice. Rechten et al. (in press) subsequently showed that part of the discrepancy between predicted and observed behavior could be explained by the fact that the birds were not able to make perfect discriminations between prey sizes. This meant that the model of prey choice had to be reformulated to incorporate the constraint of discrimination errors; it did not and could not show that Great Tits were failing to forage optimally.

The second misunderstanding arises because for some ecologists and ethologists OFT is identified only with energy (or intake) maximizing models. However, as discussed later, OFT is not inherently limited to such models (although they form the bulk of the first generation of models), and it is quite feasible to formulate models that include trade-offs between, for example, foraging and vigilance (Milinski and Heller 1978) or foraging and territorial defense (Kacelnik et al. 1981; Davies and Houston 1981; Martindale 1982). The first OFT model restricted themselves to simple optimality criteria, such as intake maximizing, because these were more tractable both theoretically and experimentally. The success of these early models in spite of their simplicity speaks for the advisability of a piecemeal approach, building from simple to complex rather than a global all-inclusive model from the start (e.g., Sibly and MacFarland 1976). In a parallel discussion of different approaches to community ecology, Harper (1982) has expressed the following view, with which we concur:

For those who make it an act of faith that the whole is more than the sum of its parts plus their interactions, the behavior of deliberately simplified systems is irrelevant to understanding. . . . If we accept . . . that the activities of communities are no more than the activities of their parts plus their interactions it becomes appropriate to break down the whole into parts and study them separately.

OFT is just one example of the application of optimality models in behavioral and evolutionary ecology, and the distinctions between models of foraging, life histories (Horn 1978), mating strategies (Parker 1978), territorial defense (Pyke 1979b; Myers et al. 1981), and so on will gradually dissolve as a result of piecemeal extension of OFT.

The main aim of early papers on OFT (MacArthur and Pianka 1966; Emlen 1966; Royama 1970; Schoener 1971) was ecological: to understand the determinants of diet breadth, changes in diet, and use of feeding habitats. The initial hope was that much of the variation in feeding ecology between and within species and some of the consequences of competition for food could be understood in the framework of energy maximizing. In this approach, OFT was used to make qualitative predictions (e.g., diet breadth should increase as food abundance decreases or competition increases), and, on the whole, the success of these predictions does not seem to have been high (Smith et al. 1978; Schluter 1981). However, it is likely that diet breadth or habitat use in the wild are at least partly related to energetic returns of foraging, because the experimental evidence for the sensitivity of animals to energy gain in more detailed studies is quite good (Table 6.1). In part, the apparent failure of energy maximizing models to describe aspects of community organization is because the data used to test the models were insufficiently detailed (Schluter 1981). For example, diet changes are sometimes inferred from stomach analyses, and it is not known whether they reflect changes in prey choice or habitat use. This distinction can be crucial in testing OFT models, because the prediction about the effects of competition and changing food abundance are opposite for patches (greater restriction of diet with increased competition and/or decreased food abundance) and prey choice within a patch (Schoener 1974). A second point about using energy maximizing models to understand community foraging patterns is that feeding ecology in the wild may often be constrained by factors such as predation risk (Sih 1980a; Werner and Mittelbach 1981) and accessibility of prey (Moermond and Denslow, in press). For example, Werner and Mittelbach (1981; see also Mittelbach 1981) found that bluegill sunfish choose prey sizes that maximize energy gain (Werner and Hall 1974; Gardner 1981) and that switches in habitat use can also generally be explained in terms of energy maximizing, with the exception of small fish that avoid energetically profitable feeding sites because of the risk of being eaten by larger fish.

Prey, patch, and related models: assessment of the evidence

In this section, we summarize and assess the evidence relating to the first models of optimal foraging.

Prey and patch models

MacArthur and Pianka (1966) distinguished between foraging for patchily distributed food and foraging for randomly encountered prey within a patch. This distinction between patch and prey choice has proved useful in analyzing foraging behavior and has led to a substantial literature providing models of prey and patch choice and to tests. Many of the tests have been done in the laboratory or in simple field situations. As stressed earlier, this is not a weakness of OFT but reflects the preference of those testing the models for precision at the expense of some realism. Generally speaking, the prey and patch models predict choice of diet and allocation of foraging time, respectively, under the hypothesis that predators aim to maximize net intake with no nutrient constraints, with perfect knowledge of prey and patch qualities, and no trade-offs with other kinds of cost or benefit. These models are now generally seen as the first generation or classical OFT models. The literature surrounding them has been reviewed several times recently (Pyke et al. 1977; Krebs 1978; Hughes and Townsend 1981). Table 6.1 lists most of the published tests.

A number of general points can be made about Table 6.1. First, many studies purporting to "test the predictions of optimal foraging models" do no such thing. Instead, they report observations that are more or less consistent with some of the assumptions of patch and prey models; for example, predators prefer profitable prey – although in some cases this is asserted without measuring E/h (Furnass 1979; Ebersole and Wilson 1980), or predators spend more time in more profitable patches. Although suggesting that energetic gain may be important to foraging animals, these kinds of data do not provide a test of any of the patch or prey models. Other reports are hard to assess because they make predictions (e.g., predators are not selective because profitable prey are scarce; Pastorak 1980) that are also consistent with a null hypothesis of random capture or time allocation. A further problem in some cases is that the models have been misunderstood. For example, several authors have presented their data in the context of the marginal value model, which assumes patch depletion, even though they have clearly studied nondepleting patches (e.g., Zach and Falls 1976; Townsend and Hildrew 1980; Bond 1981). Similarly, both in theoretical (Templeton and Lawlor 1981; Janetos and Cole 1981) and experimental (Howell and Hartl 1980) studies of patch, it has sometimes been assumed that maximizing gain (or gain/cost) per patch is an appropriate optimization criterion. Although there may be some instances where this is appropriate, for most predators that have the option of visiting more than one patch, maximizing gain rate, as the traditional models propose, seems more sensible (Gilliam et al. 1982; Stephens and Charnov 1982; Turelli et al. 1982).

Another error in some analyses of patch depletion is illustrated by the reports of Pyke (1978b) and Hodges (1981). These authors consider a so-called discrete analog of Charnov's marginal value model in which the predator gains food in a series of discrete steps as a function of time in a patch instead of in the smooth curve assumed by Charnov's model. In their analyses of bumblebees and hummingbirds, respectively, Hodges and Pyke show that the animals leave an inflorescence when their expected gain rate from the next flower in that inflorescence is less than their expected gain rate for moving to a new inflorescence. Although this is a necessary condition for energy maximization, it is not sufficient, because it does not eliminate the possibility that

Table 6.1. *Summary of evidence relating to models of prey and patch choice and other related models*

Reference	Species	Result	Laboratory (L) or field (F) test	Agreement with model
Prey/diet model				
Barnard and Brown (1981)	Shrews	Prefer larger of two sizes of mealworm even though E/h smaller; greater preference at higher encounter rates	L	Preference based on size not profitability
Barnard and Stephens (1981)	Lapwing	Select profitable size classes of earthworm; more selective at higher prey density	F	Qualitative agreement with diet model
Belovsky (1978)[a]	Moose	Maximize rate of energy intake subject to sodium constraint	F	Quantitative agreement with nutrient constraint model
Davidson (1978)	Harvester ants	Specialization increases with seed density	F	Consistent with diet model
Davies (1977a,b)	(a)Pied and Yellow Wagtails (b)Spotted Flycatcher	Select more profitable size classes of insects; more selective when profitable prey more abundant; no effect of abundance of less profitable sizes	F	Qualitative agreement with diet model
Ebersole and Wilson (1980)	Mice	No change in diet diversity with seed density; satiation greater handling time increases specialization	L	Relevance to OFT cannot be judged without more data; no measure of E/h presented
Elner and Hughes (1978)	Shore crab	Choose close to most profitable size mussels; take less profitable prey after run of bad luck; selectivity greater at higher densities; density of less profitable sizes affects selectivity	L	Qualitative agreement with recognition time model; but partial preferences

Reference	Predator	Observations	L/F	Conclusion
Erichsen et al. (1980); (see also Rechten et al. (1981)[a])	Great Tit	Choose more profitable of two prey sizes of mealworm, including effects of crypsis; selectivity changes with encounter rates	L	Quantitative agreement with recognition time model; but partial preferences
Gardner (1981)[a]	Bluegill sunfish	Rejection of small *Daphnia* not due to apparent size effects	L	Qualitative agreement with diet model
Gibson (1980)[a]	3-Spined stickleback	Rejection of small *Daphnia* not due to apparent size effects	L	Qualitative agreement with diet model
Goss-Custard (1977a)[a]	Redshank	Prefer most profitable sizes of polychaetes; selectivity increases with encounter rate with profitable prey; no effect of less profitable sizes	F	Quantitative agreement; but partial preferences
Goss-Custard (1977b)	Redshank	*Corophium* preferred to polychaetes, although lower E/h	F	Not consistent with energy maximizing
Hames and Vickers (unpub.)	Amazonian Indians	Lower ranking prey added to diet as prey stocks depleted	F	Consistent with diet model
Houston et al. (1980); see also Rechten et al. (1981)[a]	Great Tits	Test effects of making prey hard to discriminate	L	Qualitative agreement in some treatments and individuals but not others
Hughes and Elner (1979)	Shore crabs feeding on dogwhelks	No selectivity in relation to profitability, when prey presented at artificially high densities	L	Not in agreement with diet model
Hughes and Seed (1981)	Shore crab	Choose smallest mussels that also have highest E/h	L	Agreement with preference based on size or E/h
Jaeger and Barnard (1981)	Salamander	Increased specialization in large *Drosophila* with increasing density; little effect of altering density of small *Drosophila*	L	Gross energy gain would have been higher with no selection, but may have been passage time constraint

Table 6.1 (*cont.*)

Reference	Species	Result	Laboratory (L) or field (F) test	Agreement with model
Kaufman and Collier (1981)	Rats	Prefer husked seeds over unhusked seeds when both provided ad lib.	L	Consistent with diet model
Kislalioglu and Gibson (1976)	15-Spined stickleback	Prefer most profitable (E/h) sizes of prey	L	Consistent with diet model
Krebs et al. (1977); see also Rechten et al. (1981)[a]	Great Tits	Selection for more profitable sizes of mealworms increases with increasing encounter rate; little effect of encounter rate with less profitable prey	L	Quantitative agreement for some individuals; but partial preferences
Lea (1979)[a]	Rock Doves	Pigeons in Skinner box prefer schedule with higher E/h; preference increases with decreasing "search time," immediate small reward preferred to larger reward after delay, even if E/h the same for both	L	Qualitative agreement with diet model; but partial preferences; failure to show "self-control" not consistent with model
Lewis (1980)	Gray squirrel	Prefer high-profitability (E/h) nut species within a patch	F	Consistent with diet model
Lobel and Ogden (1981)	Parrotfish	Prefer algal species with high net E/h except for one toxic species; survival higher on diet of more than one species	L/F	Consistent with prey choice influenced by energy, toxins and nutrients; not consistent with simple energy maximizing
Milton (1979)	Howler monkeys	Preference for young leaves with high protein/fiber ratio	L/F	Consistent with quality ranking; not energy maximizing

172

Reference	Species	Finding	L/F	Conclusion
Mittelbach (1981); see also Werner and Mittelbach (1981)[a]	Bluegill sunfish	Preference for more profitable sizes of *Daphnia* increases with prey density	L/F	Quantitative agreement; but partial preferences
Owen-Smith and Novellie (1982)	Kudu	Selectivity not maximizing protein intake	F	Inconsistent with protein maximizing; no information on energy maximizing
Palmer (1979)	Dogwhelk	Prefer prey species/sizes with highest E/h (and yielding highest growth rate); more selective at high prey densities	F	Quantitative agreement with diet model
Pastorok (1980)	*Chaoborus*	More selectivity when less hungry	L	Consistent with diet model
Pleasants (1981)	Bumblebees	Preference for flowers/patches with high profitability (E/h)	F	Consistent with diet model
Pulliam (1980)[a]	Chipping Sparrow	Preference for high-profitability (E/h) seeds with some notable exceptions; more selective when density high	F	Partial agreement only with energy maximizing
Rapport (1980)	*Stentor*	Prefer mixed diet; fitness higher when eating mixed diet	L	Consistent with complementary resources model; but not with energy maximizing
Ringler (1979)[a]	Trout	Prefer most profitable (E/h) sizes of prey	L	Apparent agreement with diet model
Stein (1977)	Smallmouth Bass	Select crayfish with highest E/h (h includes wasted pursuit time in unsuccessful attempts)	L	Consistent with diet model
Tinbergen (1981)	European Starling	Sometimes prefer prey species which does not have highest E/h	F	Energetic value of prey less important than prey quality
Vadas (1977)	Sea urchins	Preference for species of algae that yield highest reproductive success and have highest assimilation efficiency, but not highest calories	F	Consistent with maximizing assimilation efficiency, but not energy intake

Table 6.1 (cont.)

Reference	Species	Result	Laboratory (L) or field (F) test	Agreement with model
Waddington and Holden (1979)[a]	Honeybees	Preference for more profitable (E/h) depends on relative abundance	F	Quantitative agreement with simultaneous encounter model in average results; individuals vary
Werner and Hall (1974)[a]	Bluegill sunfish	Preference for more profitable (E/h) size classes of Daphnia increases with their density	L	Qualitative agreement with diet model
Winterhalder (1981)	Cree Indians	Selectivity for profitable (E/h) prey species increased with increasing encounter rate (due to acquisition of snowmobiles)	F	Qualitative agreement with diet model
Patch use model				
Best and Bierzychudek (1982); see also Pyke (1978a)[a]	Bumblebee on foxglove	Within florescence diminishing returns from bottom to top; departure rule consistent with energy maximizing	F	Quantitative agreement with patch depression model
Bond (1980, 1981)	Green lacewing	Giving-up time (GUT) increases with hunger level	L	Relevance to patch depression model not clear (apparently no depressing of patches)
Cook and Cockrell (1978)	Notonecta and Ladybird larvae	Prey as patches; food extracted per prey declines with increasing prey density	L	Qualitative agreement with mvt but curve relating patch time and travel time wrong shape

Reference	Subject	Observation		Conclusion
Corbet et al. (1981); see also Waddington and Heinrich (1979)	Bumblebees on *Scrophularia*	Bees move upward on vertical inflorescence even though nectar rewards show no clear pattern of vertical distribution	F	Vertical movement rule not related to maximizing energetic gain
Cowie (1977)[a]	Great Tit	Patch time changes when travel time altered	L	Quantitative agreement with mvt when energy costs included
Giller (1980)	*Notonecta*	As for Cook and Cockrell (1978), but time per prey related to previous intercatch interval	L	Supports gut-filling model rather than mvt
Hartling and Plowright (1979)	Bumblebee	More persistence within a patch with longer travel distances between patches	L	Qualitative agreement with mvt
Heinrich (1979b)	(a) Bumblebee on *Trifolium*	On *Trifolium* visit more flowers per head at low head densities	F	Relevance to mvt not clear since patch depletion not measured
	(b) Bumblebee on *Aconitum*	On *Aconitum* move from bottom on top even though nectar rewards higher at top	F	Bee's movement rule does not maximize energetic gain; may reduce revisiting
Hodges (1981)[a]	Bumblebee	Departure rule from *Delphinium* plants based on rate of intake	F	Mechanism produces results in agreement with energy maximizing
Hodges and Wolf (1981)[a]	Bumblebees on *Delphinium*	Nectar extracted from individual flowers related to standing crop: more nectar left behind in rich sites	F	Quantitative agreement with mvt

Table 6.1 (*cont.*)

Reference	Species	Result	Laboratory (L) or field (F) test	Agreement with model
Howell and Hartl (1980)[a]	Glossophagine bat	Nectar left behind in artificial inflorescences consistent with rule that maximizes benefit/cost per patch	L	Results not consistent with mvt
Hubbard and Cook (1978)[a]	*Nemeritis* (ichneumonid wasp)	Allocation of time to host patches of different density similar to that which would maximize encounter rate with unparasitized hosts	L	Quantitative agreement with patch use model
Larkin (1981)[a]	Barbary Dove	On progressive interval schedules with reset after switching and variable change over delay, birds maximize food or water intake, energetic costs influenced switching as predicted	L	Quantitative agreement with mvt
Lewis (1980)	Gray squirrel	Prefer to feed in patches with highest intake per unit time	F	Qualitative agreement with patch use model
Parker (1978); see also Parker and Stuart (1976)[a]	Male dungfly	Duration of copula and guarding consistent with maximizing fertilization rate	F	Quantitative agreement with mvt

Reference	Organism	Description		Comment
Pyke (1978a); see also Pyke (1979a)	Bumblebee on *Delphinium*	Nectar rewards diminish from bottom to top: bee starts at bottom and works upward; departs before top	F	Qualitatively consistent with mvt
Pyke (1978b)	Hummingbirds on Scarlet Gilia	Number of flowers visited per inflorescence increases with number available; consistent with departure rule based on expected gain	F	Qualitatively consistent with depleting patch depression model
Sih (1980b)	*Notonecta*	Similar to Cook and Cockrell (1978)	L	Qualitative agreement with mvt
Townsend and Hildrew (1980)	Caddis fly larvae	GUT equal to average capture rate in field; no variation in GUT between habitats	L F	Relevance to mvt cannot be assessed without more details of resource depression
Waage (1979)[a]	*Nemeritis* (ichneumonid wasp)	Patch-leaving rule is a result of interaction between habituation to host odor and incremental effect of successful oviposition	L	Simple rule produces results qualitatively consistent with OFT
Waddington and Heinrich (1979)	Bumblebee	Bees move upward on vertical artificial inflorescence regardless of vertical distribution of nectar rewards	L	ROT used by bee not sensitive to variation in nectar rewards
Whitman (1977)[a]	Bumblebee on Desert Willow	Amount of nectar left behind in flowers decreases as general level of availability in habitat decreases	F	Qualitative agreement with mvt

Table 6.1 (*cont.*)

Reference	Species	Result	Laboratory (L) or field (F) test	Agreement with model
Williams (1982)[a]	White rat	On progressive ratio schedules with reset key to simulate travel time "switch between patches" in a way predicted from OFT	L	Quantitative agreement with mvt
Ydenberg (1982)[a]	Great Tit	Similar to Williams (1982)	L	Quantitative agreement with mvt
Zimmerman (1981)	Bumblebee on *Polemonium*	When travel time between inflorescences increases new flower species included in diet; no clear increase in number of flowers visited per inflorescence	F	Significance for mvt not clear since gain curves not measured
Central place foraging model				
Andersson (1981)	Whinchat	Search effort inversely related to distance from nest; adding food to territory caused search effort to shift toward nest	F	Qualitative agreement with model of allocation of search time
Brooke (1981)	Wheatear	Similar to Carlson and Moreno (1981)	F	Apparent agreement with load-size predictions
Carlson and Moreno (1981)	Wheatear	Evidence for loading curve; load size increases with distance; time allocation different to patches at some curves not as predicted	F	Qualitative agreement with load-size predictions

Reference	Species	Finding		Comments
Davidson (1978)	Harvester ants	Take larger prey at greater distances from nest	F	Consistent with single prey loader model
Evans (1982)	Black-billed gull	Intensity of use of feeding site inversely related to distance from central place	F	Qualitative agreement with model of allocation search time
Giraldeau and Kramer (1982)[a]	Eastern Chipmunk	Load size increases with distance	F	Qualitative but not quantitative fit with multiple-prey loader model
Hegner (1982)[a]	White-fronted Bee Eater	Selectivity of parents for large prey to nest increases with distance from nest to feeding site	F	Qualitative agreement with single-prey loader model
Heithaus and Fleming (1978)	Frugivorous bats	Search intensity decreases as a function of distance from roost	F	See Evans (1982)
Jenkins (1980)	Beaver	Select smaller trees at greater distances	F	Appears to conflict with single-prey loader model, but travel cost and "pursuit time" in relation to tree size not measured
Kacelnik (unpub.)	European Starling	Load size increases with distance from nest in experimental patches with progressive interval schedules	F	Quantitative agreement with multiple-prey loader models
Kramer and Nowell (1982)	Eastern Chipmunk	Load size increases with distance	F	Qualitative agreement multiple-prey loader model
Krebs and Avery (in press)[a]	European Bee Eater	Selectivity of parents (for large prey) bringing prey to nest increases with travel distance; not due to changes in prey availability	F	Quantitative agreement with single-prey loader model

Table 6.1 (cont.)

Reference	Species	Result	Laboratory (L) or field (F) test	Agreement with model
Killeen et al. (1981)[a]	Norway rat	Rats accumulate larger "load" of pellets behind a door when "travel time" increased (delay before access to food)	L	Apparent agreement with mvt but not clear whether there is a depression effect
Timbergen (1981)	European Starling	Demonstration of loading curve; load size of parents increases with distance to nest	F	Qualitative agreement with multiple prey-loader model
Other models: trade-offs				
Davies and Houston (1981)[a]	Pied Wagtail	Exclusion or tolerance of satellites on feeding territories varies with prey density as predicted by intake maximizing model	F	Quantitative agreement with intake maximizing subject to constraints of defense time and competition for food
Dunstone and O'Connor (1979)	Mink	Animals maximize capture rate subject to constraint of oxygen reserves while hunting under water	L	Qualitative agreements in predictions of pursuit time, giving up time and dive frequency
Heller and Milinski (1979); see also Milinksi and Heller (1978)	3-Spined stickleback	At high hunger levels fish choose to attack high densities of *Daphnia*; at low hunger levels they prefer lower prey densities; increased predation risk causes hungry animals to behave like satiated ones	L	Qualitative agreement with dynamic optimization trade-off between feeding and vigilance

Reference	Species		Description
Kacelnik et al. (1981)	Great Tit	L	Birds sacrifice foraging efficiency for territorial vigilance	Qualitative agreement with trade-off between feeding and vigilance
Krebs (1980)	Great Tit	L	Handling time increases with cumulative food intake, due to increased vigilance while handling prey	Qualitative agreement with dynamic optimization trade-off between hunger and vigilance
Martindale (1982)	Gila Woodpecker	F	Parents make shorter trips (smaller loads closer to nest) delivering food to young after presentation of stuffed intruder near nest	Qualitative support for trade-off between foraging and nest defense
Mittelbach (1981)	Bluegill sunfish	F,L	Small fish (but not large ones) avoid feeding in profitable sites perhaps because of predation risk	Qualitatively consistent with trade-off between feeding and predation risk
Sih (1980a)	*Notonecta*	F,L	Small individuals (but not large ones) avoid profitable feeding areas, because of predation risk	See Mittelbach (1981)

Other models: energetic costs of foraging and morphological constraints

Reference	Species		Description
DeBenedictis et al. (1978)[a]	Hummingbirds	F,L	Meal size can be interpreted as trade-off between cost of carrying nectar in crop and cost of traveling to flowers	Quantitative agreement
Pyke (1981)	Hummingbirds and honeyeaters	F,L	Difference in feeding made (hovering vs. perching) can be explained in terms of energetic costs and movement speed	Qualitative account of interspecific difference

Table 6.1 (*cont.*)

Reference	Species	Result	Laboratory (L) or field (F) test	Agreement with model
Wilson (1980a,b)[a]	Leaf-cutting ants	Size at which workers become leaf harvesters maximizes energy gain/cost for colony	L/F	Consistent with colony energy-efficiency model
Zach (1979)[a]	Northwestern Crow	Height from which crow drops mollusk shells to break them minimizes cost (in upward flight) per shell	F	Quantitative agreement with cost minimizing

Note: Not listed in the table, but relevant to OFT are tests of the "ideal free distribution" – essentially a model of OFT with competition. The main references are Davies and Halliday (1979), Milinksi (1979), Pyke (1980), Whitham (1980), Zwarts and Drent (1981), Harper (1982), Sutherland (1982a).

[a]A quantitative test of the model.

departure at an earlier moment would result in a higher gain rate (Houston, pers. comm.).

Considering just the relatively few studies in Table 6.1 that correctly test quantitative predictions of the prey and patch models, the conclusion is that most have provided qualitative, but sometimes quantitative, support for the models. This is perhaps surprising in light of the extreme simplifications involved in the models. Although the models were originally viewed as tools for thought rather than predictive analyses (Charnov, pers. comm.), they have survived experimental and observational tests on a wide taxonomic range of animals. Any lack of precision in their predictions is compensated for by their apparent generality.

The degree of congruence between predictions and data ranges from precise quantitative fits (e.g., Cowie 1977; Hodges and Wolf 1981) to qualitative fits where the general direction of a relationship is as predicted, but the form of the relationship is wrong (e.g., Cook and Cockrell 1978). Even in the latter instances, some of the variance in behavior is explained by the model, which therefore has captured part of the problem (although one should bear in mind that the weaker the prediction, the more likely it is that an alternative model will account for it). Perhaps, more importantly, formulating the model in the first place both stimulated the research and by making quantitative predictions demonstrated that there still remains some unexplained variance. This could lead to further experimentation; for example, Giraldeau and Kramer (1982) found that load size of chipmunks carrying seeds back to a central cache was consistently smaller than predicted by an energy maximizing model, although the change in load size with distance was qualitatively as predicted. This suggests that the energy maximizing model explains part of the chipmunks' foraging behavior but that other factors are involved. One possibility is the need for territorial defense, which has been shown experimentally to reduce load size in one species of central place forager (Martindale 1982).

In tests of diet models, a frequently observed deviation from the predictions of OFT is that animals show sigmoid changes in selectivity with prey density and not step changes (Table 6.1). Should this be taken as evidence against the prey selection model? When data are averaged over several individuals and/or observation periods, a step change would not be expected, because even if each individual at a particular time had a threshold, variation in the position of the threshold (due to environmental variation or interindividual differences in search speed, handling time, etc.) would produce a sigmoid curve in the averaged data. The average position of the threshold could be estimated from the sigmoid curve by standard statistical techniques. Sigmoid curves are also seen, however, in studies of individual animals (e.g., Krebs et al. 1977; Lea 1979), and they may reflect an analogous effect. Thresholds vary from moment to moment within an individual. Evidence for this view comes from psychological studies of time perception in animals (appropriate in this context because it is likely that foraging animals can assess interprey intervals or intake rates) that show that rats and pigeons can accurately recognize time differences of the order of a few seconds but that there is intraindividual variance in perception (Gibbon and Church 1981; Gibbon, in press). Gibbon (in press) shows that this variation can be accounted for by a model in which the animal's clock (pulse generator) changes its mean rate slowly over time. If this kind of mechanism underlies the sigmoid curves seen in foraging studies, the curves should not be viewed as evidence against a quantitative fit to the model, but rather the curve should be used to estimate whether or not the threshold occurs in the predicted place.

Among the refinements and modifications of the classical models that have been published are the inclusion of recognition as well as handling times (Hughes 1979), simultaneous encounters with prey (Waddington and Holden 1979; McNair, in press), and mixed prey and patch models (Heller 1980). For herbivores, in particular, energetic intake may be less crucial in dictating behavior than intake of certain nutrients or combinations of nutrients (Owen-Smith and Novellie 1982). Two approaches to modeling this have been to consider nutrient requirements as constraints on maximizing energy intake (Pulliam 1975; Belovsky 1978) and to treat different components as complementary resources, that is, a mixture of A and B gives a higher benefit than A or B alone (Greenstone 1979; Rapport 1980).

Other foraging models

In addition to the literature on prey and patch choice, a number of other aspects of foraging behavior have been modeled within the framework of OFT, usually in terms of maximizing rate of intake.

Central place foraging. Orians and Pearson (1979) and Schoener (1979) were the first to formulate a model for animals harvesting food and bringing it back to a central place, such as a nest or cache. Although applicable in many other contexts, Orians and Pearson's model was originally conceived in relation to parent birds feeding their young. Some birds bring one prey at a time (single-prey loaders), whereas others bring many items per trip (multiple-prey loaders). Orians and Pearson's model predicts prey size and load size as a function of distance for single-prey loaders and multiple-prey loaders, respectively. The superficial similarity of the graphical solutions to these models presented by Orians and Pearson led them to view them as being essentially similar. However, as Lessells and Stephens (in press) have pointed out, this is incorrect. The model for multiple-prey loaders is an extension of Charnov's (1976a) marginal value model and predicts that within-patch search time and load size should increase with travel distance. For single-prey loaders, a decision rule in terms of optimal persistence time within a patch makes no sense (imagine the rule "search for 15 seconds then leave to go to the nest, with or without prey"). Instead, the single-prey loader's decision is about which prey sizes are economically worth transporting to the nest, so the rule is of the form "search until a prey of at least such and such size is captured." The model predicts greater selectivity at great distances. The model for multiple loaders is based on the assumption that as the parent gathers more and more prey in its beak, the capture rate for the next prey diminishes as a result of the encumbrance of prey in the beak. This assumption of the *loading curve* has been tested for several species, as has the qualitative prediction that load size changes with distance (this has also been tested quantitatively in two cases, Table 6.1). There has been less work on single-prey loaders, but work on two species of bee-eaters shows a rather good fit to the predictions (Table 6.1).

A different aspect of central place foraging has been modeled by Andersson (1978), who considers the optimal allocation of search effort in relation to distance from the nest or other central place (see also Evans 1982). His test of the model (Andersson 1981) was largely in qualitative agreement with its predictions (see also Heithaus and Fleming 1978; Evans 1982); but as Aronson and Givnish (in press) point out, most of these predictions could equally well be accounted for by a model of random search from a central place. This illustrates an important general point: Wherever possible, predictions of OFT models (especially qualitative predictions) should be compared with other models including the null hypothesis of random choice.

Search paths. The problem of optimal search paths for foraging animals has not so far proved analytically very tractable for biologically realistic general cases. Simulation models of search for scattered or clumped prey (Cody 1971; Smith 1974; Pyke 1978) show that some degree of directionality (forward momentum) increases search efficiency. An interesting model of a more specific search problem is Wehner and Srinivasan's (1981) analysis of searching in the desert ant *Cataglyphis*. These ants normally return from foraging trips to their nest site by vector navigation. If they are displaced on their return journey, they continue on the same vector in a straight line until they reach the expected nest site position, when they switch to searching. They search for the missing nest entrance by traveling out and back on a series of loops from the starting point (where the nest ought to be). The observed search path is very similar to an optimal one derived analytically and by simulation. The principle behind the model is to set up a subjective probability density function of the nest's position and search in the area of highest probability, updating the distribution as a consequence of unsuccessful search.

Energetic costs of foraging. The classical prey and patch models implicitly assume that search, travel, and handling costs of foraging are important, because they hypothesize that animals maximize their net rate of energy gain. In some cases, comparisons of gross and net energy gain versions of the classical models show that animals are, in fact, sensitive to energy costs (Cowie 1977; Williams 1982). Other models have tackled the question of energy cost more explicitly and considered how variations in search costs might alter the search process itself. Included in the effects that have been studied are the influence of search costs on flight speed (Pyke 1981b), variation in body weight (Norberg 1981), and meal size (DeBenedictis et al. 1978).

Trade-offs with competing activities

Often it is clear that animals do not maximize food intake during foraging. A well-documented case from laboratory studies is the typical *satiation curve* of hungry animals provided with ad libitum food. There is a gradual exponential decline in rate of food intake with increasing cumulative intake; which indicates that animals do not always forage as fast as they possibly can. Even though the relevance of satiation experiments to the natural situation may be questioned, the results illustrate the point that maximizing rate of intake is an inadequate hypothesis to explain some patterns of foraging. However, satiation curves can nevertheless be treated as a problem in optimal foraging. As McCleery (1977) has shown, they can be viewed as the outcome of a trade-off between foraging and doing other activities. When the animal is hungry, reducing food deficit is given high priority (because of the relatively high risk of death by starvation), and the animal devotes all its attention to feeding. When less hungry, the value of reducing deficit relative to, for example, scanning for predators and mates, is lower so the animal divides its time between feeding and other activities, producing the observed decrease in feeding rate (e.g., Milinski and Heller 1978). McCleery's and Milinski and Heller's models of this problem are dynamic optimization models, in which the animal's internal state changes as a function of its behavior. This is in contrast to the classical prey and patch models, which do not include changes in state. As discussed by Houston (1980) and Krebs et al. (1981), the dynamic models simplify to produce predictions identical to the static case when the cost of foraging is independent of foraging rate. Here "cost" refers to the individual cost function that is optimized, either minimized or maximized. This is, of course, not a very realistic assumption, so static models are virtually always likely to be simplifications. Hunger-related changes in foraging, such as changes in handling time (Krebs 1980) and in selectivity for profitable prey (Rechten et al., in press), may be understood within the framework of a dynamic but not a static optimization model.

Another approach that has been taken to the analysis of trade-offs between competing activities is to treat the competing demands as constraints on the foraging animal's time budget in a static optimization model (Pulliam 1976; Sih 1980a; Davies and Houston 1981; Mittelbach 1981). In this approach, the difficult problem of measuring a common currency for the different activities is circumvented by considering how the time spent not feeding decreases energy gains. Another, still controversial, method of calibrating the costs and benefits of different activities against one another is the inverse optimality approach advocated by McFarland (1977). This method involves observing the animal's decision rules and inferring backward what function must have been maximized to produce the observed behavior. Although it has been suggested that this technique does no more than redescribe the data (Maynard-Smith 1978), it has proved useful in one case for understanding individual differences in behavior (Kacelnik et al. 1981) and may be a way of generating predictions about the quantitative trade-offs between different kinds of activity.

Rules of thumb: extension of traditional models

It is often said that OFT is about ultimate explanations of behavior (e.g., Krebs and Davies 1981), describing how animals ought to behave if they are designed by natural selection according to the criteria embodied in the model, without commenting on the proximate mechanisms used by any particular species to solve its foraging problems. By omitting details of proximate mechanisms, OFT achieves generality, at the expense of some precision. In the last few years a number of authors have begun to refer to more precise but parochial models based in the constraints of individual species (e.g., Cowie and Krebs 1979; Janetos and Cole 1981), and these models have become known as models of "rules of thumb" used by the animal to solve its foraging problems. In this section, we will describe two ways of looking at the relationship between traditional OFT models and rules-of-thumb (ROT) models.

Different levels of explanation

According to this view, the distinction between OFT models and ROT models is roughly equivalent to the distinction between ultimate and proximate explanations of behavior (implied in the previous paragraph; Janetos and Cole 1981). The two reflect different levels of explanation: Any ultimate account must have its proximate counterpart (*not alternative*, as implied by Ollason, 1980). When animals behave as if they are maximizing food intake, or solving any other complex problem, such as designing a nest, they achieve this end by using simple ROT (Cowie and Krebs 1979; Janetos and Cole 1981). An excellent example is the work of Hubbard and Cook (1978) and Waage (1979) on the hymenopteran parasitoid *Nemeritis canescens*. The former authors show that *Nemeritis* allocates its searching time in relation to host density approximately as predicted by an optimal foraging model, whereas Waage shows that the decision rule used by *Nemeritis* in allocating its time to a patch is a simple mechanism based on habituation to host scent—a far cry from the Lagrange multipliers and Newton's iterative approximations used by the theorist to solve the problem! The proximate mechanisms used by animals are presumably themselves subject to selection and have evolved to work well in the natural environment. However, when the animal is brought into the laboratory and set to an artificial task, its rules may yield results that do not maximize gain rate (e.g., Mazur 1981).

Continuum of strategy sets

Another way to review the relationship between ROT and traditional OFT models is to consider the two as representing different points on a continuum of strategy sets. The aim of an OFT model is to gain some insight into the currencies and constraints influencing foraging behavior, and the difference between a ROT and a classical OFT model lies in the assumptions about constraints. A ROT model is one in which the animal is constrained in its choice or assessment of the environment. Put another way, the difference between the two kinds of model is in the strategy set available to the animal. In this context it is incorrect to refer to a ROT model as "suboptimal" or "imperfectly optimal," because the animal may still be maximizing payoff within the context of its strategy set.

An hypothetical example illustrates our point. A classical optimal diet model predicts that predators should prefer prey with the highest profitability (energy yield per unit handling time). This assumes that predators are able to recognize profitabilities of prey they encounter. However, suppose a particular species could only categorize prey by size, often a good correlate of profitability and, therefore, a useful ROT for ranking prey. In modeling prey choice in this species, we would have to rework the optimal diet model to include the constraint that size is the only cue used by the predator to distinguish between prey. Predictions such as "take only the three most profitable prey types" may become "take only the three largest prey types." The predator using this rule is not suboptimal but optimizes within a different strategy set from the classical optimal forager.

Accepting the view that ROT models are on a continuum with classical OFT models, one can ask how to choose the strategy sets for a particular model. Janetos and Cole (1981) plead for biologically realistic assumptions about constraints rather than assumptions included for mathematical convenience, a view with which we concur. However, intuition about biologically reasonable constraints may not always lead to the right answer. Parallel to our hypothetical example of size selection earlier, O'Brien et al. (1976) suggested that bluegill sunfish (*Lepomis macrochirus*) may use the rule "take the prey of largest apparent size in the visual field" when selecting *Daphnia*. This seems to be an intuitively reasonable guess at the strategy set, but in later work Gardner (1981; see also Gibson 1980) showed by manipulating the turbidity of the water that sunfish do not in fact base their size selection on this rule alone.

It is possible to do an optimality analysis of the ROT themselves. For example, Breck (1978) and McNair (1982) examined the efficiency of various patch-leaving rules in approximating the solution of maximizing energy gain. There are special cases where a "giving-up" time rule does best and others where a time (Krebs 1973) or a number expectation (Gibb 1958) rule has the highest payoff. In one detailed study of patch-leaving rules of bumblebees, Hodges (1981) found that the bees did not appear to use either a time or a quantity expectation but a rule based on rates of intake. McNair (1982) also points to a confusion in the literature about the concept of giving-up time. For some authors, the predictions of Charnov's (1976a) Marginal Value Theorem (MVT) have become identified with measurements of giving-up times (e.g., Townsend & Hildrew 1980; Bond 1981), but because the MVT deals with continuous food intake in a patch and the concept of giving-up time is only appropriate for discrete captures, this assumption is incorrect. However, what McNair does not indicate is that different authors have used the concept of giving-up time in different ways. Some have used it to refer to a possible ROT of the animal (e.g., Waage 1979); others have taken it as a way for the experimenter to estimate marginal gain rates (e.g., Krebs et al. 1974), whereas others have (incorrectly) taken it as a direct prediction of MVT (Bond 1981).

In some cases, it can be shown that the animal's ROT achieves a lower payoff (e.g., rate of food intake) than alternative possibilities. Janetos and Cole (1981) suggest that the cost of improving or maintaining the physiological machinery to implement complex rules may exceed the extra benefits gained by doing so. In this sense, animal using ROT might be said to be "satisficing" (Simon 1956), but in view of our comments earlier, this terminology seems to be of limited value in relation to foraging theory.

Stochasticity: an extension of traditional models

Most behavioral ecologists have a general feeling that "the world is essentially stochastic" and that, therefore, stochastic models are superior to their deterministic counterparts. This worry has been exacerbated by some theoreticians who have suggested that deterministic models may be wrong and not worth the fieldworker's attention (Oaten 1977). The situation is further complicated by the fact that there are two parts to the problem of stochasticity in the foraging literature. One is the problem of *information* about an unknown environment; the other, the problem of *risk*, variance in payoff from an environment (Stephens and Charnov 1982).

Risk

The problem of risk, how a predator ought to choose between known but probablistic alternatives, has often been treated glibly. Assuming that if in a deterministic model animals maximize their rate of energy gain, then in a stochastic case they should maximize their mean rate of energy gain (see Pyke et al. 1977). This *mean maximization hypothesis* assumes that foragers are indifferent to variability in food reward, but there are examples where it appears intuitively that animals should be sensitive to variance. Suppose that an animal requires 10 units of food to survive the day, and it has a choice between two alternative patches, one offering a fixed payoff of 9.5 units and the other offering a normal distribution about a mean of 9.0. The fixed option has a higher mean but the variable option offers a higher chance (>0) of getting enough food to survive the day; the animal, therefore, ought to prefer the latter.

A number of experiments along the lines of this hypothetical example have been performed to find out whether animals are indifferent to variance. The animal is usually taught that two equivalent behaviors (choices of keys to peck or sides of the aviary in which to feed) produce different probability distributions of reward. Distribution C provides c units of food with probability 1, distribution L is a lottery providing l_1 units with probability 0.5 and l_2 units with probability 0.5. Where c, l_1 and l_2 are chosen so that the average gain is the same in both C and L, an animal might show two kinds of preference. If it consistently chooses the certain alternative, it is said to be risk averse, whereas if it prefers the lottery it is risk prone. Experimental results show that animals may be either risk averse or risk prone. Furthermore, operant psychologists have tended to find risk proneness, and behavioral ecologist have tended to find risk aversion (Table 6.2).

Only Caraco et al. (1980a) and Caraco (1981) have found, working with two species of junco, both risk aversion and risk proneness in the same animals. More importantly, they have found conditions that influence risk sensitivity. Juncos were risk averse when provided with positive expected daily budgets (i.e., when the birds could expect to meet their daily energy requirements) and risk prone when given negative expected daily energy budgets. This is now referred to as the expected-energy-budget rule: "Be risk averse, if your expected energy budget is positive; be risk prone, if your expected energy budget is negative."

Clearly, there are at least some cases in which mean maximization must be abandoned. With what might we replace it? Two simple alternatives to mean maximization are variance discounting and models based on the Z-score, a measure of the standard normal distribution, or simply, Z-score models.

Table 6.2 *Summary of the evidence relating to risk sensitivity*

Reference	Species	Quantity varied	Result
Evidence from operant psychologists			
Davison (1969)	Pigeons	Delay time	Risk prone
Herrnstein (1964)	Pigeons	Delay time	Risk prone
Leventhal et al. (1959)	Laboratory rats	Number of food pellets	Risk prone, but tendency decreased as mean amount increased
Pubols (1962)	Laboratory rats	Delay time before reward	Risk prone, but tendency decreased as mean delay decreased
Evidence from ethologists			
Caraco (1981)	Dark-eyed Juncos	No. of seeds	Risk averse when energy budgets positive; risk prone when energy budgets negative
Caraco et al. (1980a)	Yellow-eyed Juncos	No. of seeds	Risk averse when energy budgets positive; risk prone when energy budgets negative
Real (1981)	Bumblebees and wasps	Nectar reward in artificial flowers	Risk averse
Stephens and Ydenberg (unpub.)	Great Tits	No. of fly pupae	Risk averse

Variance discounting. Many authors have proposed that mean maximization might be replaced (in some cases) by maximizing the mean discounted by a certain amount for variance (Oster and Wilson 1978; Real 1980a,b; Caraco 1980). That is, if μ is the mean energy gain over some fixed period and σ is the variance, it is proposed to maximize $\mu - k\sigma$, where k is a constant (usually positive) that measures the "undesirability" of variance. If k is positive, variance discounting predicts risk aversion, and if k is negative, it predicts risk proneness. However, there seems to be no a priori reason to imagine that k should be positive or negative. Usually, it is assumed that (as does Real, 1980a,b), contrary to the evidence, animals are simply always risk averse and that k is always positive.

Z-Score models. Several authors have proposed that animal risk sensitivity might be understood in terms of the probability of starvation due to energetic shortfall (Caraco 1980; Stephens 1981; Stephens and Charnov 1982; McNamara and Houston, in press). Stephens (1981) and Stephens and Charnov (1982) have investigated the implications of this hypothesis when daily food reward follows a normal distribution. In this case, the probability of having some critical level of energy requirement (R), that is, the probability of survival, can be shown to decrease in a one-to-one way, with the conventional Z-score of the standard normal distribution. Stephens (1981) and Stephens and Charnov (1982) propose a model in which Z is minimized where

$$Z = (R - \mu)/\sigma$$

The advantage of this formulation is that it can predict both risk aversion and risk proneness. More importantly, it predicts the expected energy budget rule, shown empirically by Caraco et al. (1980a) and Caraco (1981).

The Z-score currency of minimization fits well with the variable risk sensitivity of foraging juncos found by Caraco et al. (1980a) and Caraco (1981). However, can it explain all the preferences over variance in Table 6.2? The short answer is that we don't know, because only Caraco et al. (1980a) and Caraco (1981) have manipulated energy budgets (the relative sizes of μ and R). However, it does seem reasonable that those animals tested in the field had positive energy budgets, and so we would expect them to be risk averse. However, the consistent risk proneness found by operant psychologists is more troublesome. There is no evidence that these rats or pigeons were on negative energy budgets, although they were all food deprived (mostly to 80% of ad libitum body weight). There is also evidence that as the mean reward size ($c = (l_1 + l_1)/2$) is increased, the tendency for risk proneness is reduced (Leventhal et al. 1959; Pubols 1962). Until more energy budgets are measured, the operant evidence cannot be fully assessed.

One potential problem for the Z-score is that for an animal to be risk prone, the probability of death must be greater than 0.5. In the Z-score model, the probability of starvation is estimated as $\phi(Z)$, where ϕ is the tabulated cumulative distribution function of the standard ($\mu = 0$, $\sigma^2 = 1$) normal distribution. When risk proneness is predicted, Z is positive, and a quick look at a normal probability table will show that the probability of death must be greater than 0.5. Why then aren't one-half of the risk prone foragers listed in Table 6.2 dead? The answer, according to McNamara (in press), is that the Z-score model assumes that a forager makes only one decision about whether to chose high or low variance. The fixed, one-decision case underestimates the probability of survival, because it ignores the possibility that a risky decision will result in a gain large enough to allow the forager to become conservative (risk averse). The implications of this sequential approach to risk taking are fascinating and have been studied by Houston and McNamara (1982) and McNamara (in press, unpub.).

Information

Deterministic models have normally assumed that when a patch or a prey item is encountered it is recognized by the forager as belonging to a unique and unambiguous type. However, it may often be more realistic to assume that the predator can recognize some categories (which we call "types") and not other divisions within these types (which we call "sub-types"). For example, it might be able to recognize prey from nonprey, but be unable to distinguish between different species of prey, or it might recognize species of prey but not size classes within a species. The problem of information is to ask how the predator might tackle a problem in which some categories are recognized but within these categories classes cannot be distinguished; this is referred to as the "type, subtype" problem.

Recognition. Hughes (1979) and Elner and Hughes (1978) have described how shore crabs can recognize different size classes of mussels only by paying a cost in *recognition time* (the time required to pick up a mussel and assess its weight). Mathematically, we can ask how much should a forager be willing to pay (in the currency of rate of intake in a "conventional" model) to perfectly recognize sizes or subtypes? Just such an analysis has been done by Gould (1974) in the context of microeconomics. Gould provides a definition of the value of perfect information, analogous to the value of recognition in foraging. At first, this may not seem like a stochastic problem at all, but, as Gould shows, the value of recognition is critically dependent on the probability distribution of subtypes, referred to as the "mixing distribution."

Gould argues that the value of recognition should be defined in terms of the decision problem at hand. The important point is that information must be valued in terms of how it can be put to use and that this value is limited. For example, there will certainly be cases where a type of patch or prey item should be treated in the same way regardless of subtype, and, in such cases, the value of recognition will be nil. At the other extreme, there must also be a limit to the value of recognition; it cannot be infinite.

Patch sampling. The problem of recognition applies equally to patches or prey items. The second way that information may be important in OFT applies most directly to patches. If recognition of subtypes is impossible (regardless of cost), the forager is faced with the problem of judiciously exploiting a patch it recognizes only as belonging to any one of several subtypes. This is the problem of patch sampling. The forager can use information gained while exploiting the patch to modify its assessment of the patch's subtype. This problem was first studied by Oaten (1977) and subsequently by Green (1980), Iwasa et al. (1981), and McNamara (1982).

There has been much controversy over the extent to which these patch-sampling models may be taken as a refutation of Charnov's (1976a) deterministic, *perfect patch recognition* model of patch use (cf. Stephens and Charnov 1982). Perhaps the most significant result of these comparisons has been that where variance in the mixing distribution of patch subtypes is greatest, the difference between sampling and nonsampling models is also greatest (Green 1980; Iwasa et al. 1981; McNamara 1982).

Thus, in the terms similar to those used by Gould, the value of sampling is greatest when the indistinguishable patch types are very different in their probability distributions of reward. However, under these conditions, it is relatively easy to recognize the patch types by sampling, and a fairly simple rule may be sufficient to distinguish between them as effectively as a more complicated procedure (Stephens 1982). In other words, when the value of sampling is highest, the need for elaborate sampling procedures is low. Conversely, when the variance in the mixing distribution is low, complicated sampling procedures may be needed to distinguish between subtypes, but the value of distinguishing is low. Complicated sampling procedures therefore may be of benefit only in cases of intermediate variance of the mixing distribution.

The problem of patch sampling, particularly discussed by Oaten (1977), is perhaps the most cited example of a case where stochasticity significantly changes the results of a deterministic model (Charnov 1976a) in OFT. However, Oaten certainly does not deal with *the* stochastic case, rather only with a subdivision of the stochastic problem of information: patch sampling (Stephens and Charnov 1982). Furthermore, there will be natural foraging situations in which patch sampling will simply not be important (if subtype recognition is easy) even if there will always be stochasticity of other sorts. In short, it is not fair to conclude from the literature of patch sampling that stochasticity necessarily changes the picture.

Environmental tracking. At the center of the information problem is the data that some characteristic of an animal's foraging environment must change unpredictably. Yet most models assume that the animal "knows," either explicitly or implicitly, certain parameters of its environment. At first, it may seem that a straightforward application of statistical sampling theory and parameter estimation might yield interesting results. There are two problems with this. First, as with Gould's result, the value of knowing a parameter must be defined in terms of a given foraging problem. A statistician wants to know the value of μ, a forager only cares if knowing μ increases its rate of energy gain. Second, statisticians conventionally assume that they are sampling from an unknown but unchanging probability. Such set parameters are probably only rarely obtained in environmental tracking problems.

There have been two promising approaches to this question. One can treat the changing distributions by assuming that a forager faces a novel situation, a simple type of change. Krebs et al. (1978) have considered such a problem experimentally by providing Great Tits with a problem of sampling each of two patches until they find which is best. The experimental paradigm is simple, but the underlying mathematics are usually intractable. Nonetheless, the result shows that at least some theoretically important parameters are also important to animals. A significant and practically uncited (in the OFT literature) family of models asks a more generally applicable question: How frequently should the forager sample in a changing environment (Estabrook and Jespersen 1974; Arnold 1978)? These models assume that the forager faces a series of dichotomous choices between V, a variable system of rewards, and A, a consistently valuable alternative reward. The variable system changes in a simple way (according to a first-order Markov process), alternating between good and bad states. If the forager chooses V and finds a "good," this gives it information about the likelihood of a good at the next choice. Optimal behavior has a simple form, once a "bad" has been encountered on the variable V-system, the forager should exploit the alternative choice A for some fixed number of trials N and then sample to see if the V-system has returned to a good state.

In choosing N, the forager has two problems, if it sets N too small, it samples often and takes many bads when it could have A's with certainty, that is, it makes sampling errors. If it sets N too large, it overruns the run of bads and misses goods, that is, it makes overrun errors. The crucial and appealing result is that the optimal N is determined by the ratio of the costs of sampling errors to overrun errors (Stephens

1982). Although the model is quite specific, the "error" analysis may have general implications for environmental tracking models.

Mechanisms of learning. Many behavioral ecologists feel that the information problem concerns learning. However, it is necessary to distinguish between "learning how," in the sense that a bumblebee learns how to exploit the nectar of a complicated flower (Heinrich 1979b; Hughes, 1979, has considered the implications of this kind of learning for diet theory), and "learning about," which is more or less the information problem.

A number of so-called learning models have simply proposed mechanisms for modifying behavior (Ollason 1980; Pulliam and Dunford 1980; Harley 1981; Killeen 1981; Lester, unpub.). These rules, all derivatives of mathematical psychology learning models (Bush and Mosteller 1955), generally consist of two components. The first component is a rule for combining past and present experience, usually of the general form:

$$A_N = BX_N + (1 - B)A_{N-1}$$

where A_N is the estimated current average, X_N the number of items encountered in the present time interval, and B is a weighting parameter (e.g., $B = 2$ gives exclusive weight to present success). This is, in effect, an exponential decay model of "memory," because after i intervals of nonreward, the value of A_{N+i} has declined by a factor of $(1 - B)^i$. The second component is a response rule for allocating behavior in relation to the values for alternatives of A_N. For example, Lester (unpub.) and Killeen (1981) assume that the animal allocates its time between alternatives according to the equation, $T_1/T_2 = N_1/N_2$. Harley (1981) claims that his rule (the relative payoff sum rule) has some characteristics of the evolutionarily stable learning rule. However, it is relatively easy to find better rules than Harley's for specific cases. For example, Harley's only updates the probability that a behavior will be used, it cannot perform the obviously sensible act of regular sampling. Houston et al. (in press) and Houston (unpub.) have considered several other cases in which Harley's rule is suboptimal.

Another general criticism of the Bush and Mosteller derivative models is referred to by Houston et al.; past experience is incorporated as a single average (perhaps weighted by time), but how the average is achieved is not specified. However, it is known that different patterns of reinforcement (e.g., partial reinforcement vs. continuous reinforcement when both are followed by extinction) have difference consequences for future behavior even if the value of the average over N previous time intervals is the same. Killeen (unpub.) discusses possible modifications to incorporate such effects.

Predator–prey interactions: ecological implications of traditional models

The functional response

The functional response of a predator is the relationship between the number of prey eaten at a given time and the density of prey available (Solomon 1949). One of the most frequently observed forms of this curve is a decelerating rise to a plateau (type 2 functional response of Holling, 1959). The most widely accepted explanation for this shape is that as prey density increases an increasing proportion of the predator's time is spent handling prey, until at the asymptote feeding rate is limited by handling time (Holling 1959). However, where handling time and asymptotic feeding rate have both been measured in the field, handling time has been far too short to account for the plateau. The reciprocal of the asymptote estimated from curve fitting should equal the measured handling time, but the available evidence suggests it does not. In two studies of Redshank (Goss-Custard 1977a,b,c), handling times for worms and *Corophium* were measured directly to be 0.4–6.0 seconds (dependent on size) and 0.5 seconds, respectively, whereas estimates from the functional response curves were 12.0 and 2.2 seconds for the two kinds of prey. (These latter estimates were made by use of unpublished data.) For European Oystercatchers feeding on cockles, the actual values of handling times ranged from 19 to 29 seconds in different sites (Sutherland 1982b), whereas the estimated value was 75 seconds. So what limits feeding rate at the asymptote of the functional response? We suggest, as one possible hypothesis, an explanation based on OFT.

Figure 6.1 illustrates our graphical model of the functional response based on optimal diet theory. The model assumes that prey of a particular species are divided into size classes that are ranked by the predator according to profitability (energy per unit handling time) with the largest sizes being the most profitable. For each size class or set of size classes, attack rate is described by a type 2 functional response because of usual effects of handling time. It is possible to draw a family of functional response curves for different subsets of the total prey distribution. As first the smallest, then successively larger, size classes are

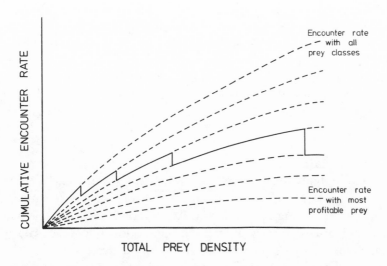

Figure 6.1. A model of functional response. See text for explanation.

excluded from the subset, the curves become shallower and have a lower asymptote, the lowest curve being the functional response for a predator eating only the largest size class. The reasons for this change in the curve are (1) as more size classes are excluded the effective density of prey is lowered producing a shallower curve and (2) the remaining size classes become successively larger, having longer handling times, which lowers the asymptote. Now, and this is the crux of the argument, conventional optimal diet models predict that as total prey density (or more precisely the density of more profitable sizes) increases, the predator should become more selective, excluding less profitable sizes from its diet. This predicted effect of optimal diet theory is shown as a solid curve in Figure 6.2. The predator jumps from one functional response curve to the next one down as it becomes more selective. The model predicts a saw-toothed curve because of all-or-none predictions about exclusion of size classes, but with a continuum of size classes, the expected field result is a smoothed version of the saw tooth with an asymptote below that predicted by the average handling time for all prey in the population.

This is not the only possible explanation for the functional response asymptote. Indeed, the figures given earlier for observed estimated handling times in Redshank and Oystercatchers indicate that the hypothesis does not account completely for these two examples. Two other ideas that seem feasible are (1) a decrease in search speed of the predator as prey density increases, perhaps because the predator allocates time to other activities, such as scanning for predators or fighting rivals (Caraco et al. 1980a), and (2) satiation may limit food intake at high densities of prey. The latter is less likely to apply to the field evidence mentioned earlier than to laboratory experiments where for the convenience of collecting data the feeding rate is usually expressed as items per day regardless of how much of the day is spent searching for prey.

Aggregation
The stability of predator–prey interactions is influenced by aggregation, the tendency of predators to concentrate their searching effort in patches of high prey density. Aggregation was originally modeled by ecologists either by a simple mechanistic model (e.g., Hassell and May's fixed giving-up time model) or by a descriptive equation (Hassell and May 1973). Subsequently, analogous models based on maximizing encounter rate have been found to give results qualitatively similar to those of the earlier models both in theoretical analyses of stability (Comins and Hassell 1979) and in description of field results (Hassell 1980).

Sutherland (unpub.) has shown how patterns of predator aggregation can be predicted from the ideal free distribution (Fretwell 1972). According to this model, the individual predator's time allocation is influenced both by prey density and interference by other predators. If predators maximize their rate of intake and are unconstrained in their choice of patches, they should be distributed according to the equation $\beta_i = c\alpha_i^{1/m}$, where α_i is the proportion of prey in the ith patch and β_i is the corresponding proportion of predators, c is a normalizing constant, and m describes the strength of interference (Sutherland, unpub.).

Switching
A second aspect of predator behavior that can have a stabilizing effect on predator–prey interactions is switching (Murdoch 1969; Murdoch and Oaten 1975). Switching occurs in a two-prey case when the predators prey disproportionally on the more abundant prey. For our discussion, it is important to distinguish two kinds of switching: between habitats and within habitats. Switching between two prey types in different habitats (Royama 1970; Murdoch et al. 1975) can be viewed as a consequence of aggregation and can be predicted by the same ideal free model of aggregation referred to earlier (Figure 6.2).

Several authors have noted that switching within a habitat, in which

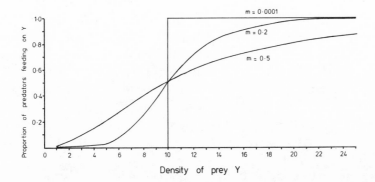

Figure 6.2. How predators should allocate their time between different prey in separate patches. As described in the text, the relationship between the proportion of predators (β_i) in a patch and the proportion of prey (α_i) can be described by

$$\beta_i = c\alpha_i^{1/m}$$

where m is the strength of interference and c is a constant. Now, consider two patches, one containing prey of species X and the other containing prey of species Y. (For convenience, we assume the prey are of equal value to the predator and equally easy to find.) The predators are free to move between patches. The proportion of predators feeding on X, i.e. $\beta_X/(\beta_X + \beta_Y)$, then equals

$$\frac{c_X\alpha_X^{1/m}}{c_X\alpha_X^{1/m} + c_Y\alpha_Y^{1/m}}$$

The switching behavior predicted by this model is shown in the figure. We have assumed a density of 10 individuals of species X per unit area. With negligible interference, and if animals conform to the ideal free distribution, they should all feed on Y when there are more than 10 per unit area, and they should all feed on X when there are less than 10. As the value of m increases, the step function becomes a sigmoid curve.

the predator's diet changes in response to changes in the proportions of the two prey encountered, appears to contradict the prediction of energy maximizing models of prey selection, in which the absolute abundance of the more profitable prey is expected to determine choice (e.g., Hughes 1979). Several resolutions to this apparent conflict have been suggested, the neatest of which is that of Bonser (1981) in his analysis of switching by the freshwater hemipteran *Notonecta*. Bonser points out that the definition of switching depends on measurement of c, the animal's preference when prey are presented in a 50:50 mixture ($c = 1$, means no preference). Because the emphasis in switching studies has been on the proportion of prey presented, little attention has been paid to the absolute abundance at which c is measured. Suppose c is measured at a prey density for which OFT would predict a preference for the more profitable prey. When the mixture in the switching experiment is now changed to, say, 25:75 (profitable to unprofitable), the abundance of the more profitable one may be below the point at which selectivity is predicted by OFT. The experimenter records this as an apparent switch away from the more profitable prey (because the value of c is lower than in the 50:50 treatment). When Bonser measured the response of *Notonecta* to mixtures of two prey at different relative and absolute abundances, he found that the animals fitted the predictions of OFT and not switching. Furthermore, reanalysis of some earlier data showing apparent switching (Lawton et al. 1974) suggested that c might have been measured at too high a density.

Other ways of resolving the apparent conflict between switching theory and OFT have been discussed by Hughes (1979), Visser (1981), and Hubbard et al. (1982). Hughes and Visser favor the idea that the profitability of a prey may increase as a function of its own density, because the predator may learn to handle (Lawton et al. 1974) and detect (Dawkins 1971) it more readily or digest it more efficiently (Al-Jaborae 1979; Partridge 1981). In these cases, switching may be a result of the predator specializing on the currently more profitable prey. Hubbard et al. make a different point: A combination of depletion of the more profitable prey and a greater searching efficiency for the less profitable prey (due, e.g., to greater conspicuousness) could lead to switching.

Conclusions

Success and failure of optimality models

A widely voiced criticism of the use of optimization models in biology is that some aspect of living organisms reflect phylogenetic ontogenetic constraints rather than the honing action of natural selection (Gould and Lewontin 1979; Oster and Alberch 1982; Harper 1982). Does this undermine the use of optimality models to analyze foraging behavior? As we have seen, optimal foraging models have enjoyed a modest success in accounting for observed behavior. We suggest that the constrast between the foraging theorist's enthusiasm for optimality models and the scepticism of authors such as Gould and Lewontin is in part a result of analyzing different types of problems. Those who emphasize phylogenetic and embryological constraints are often concerned with accounting for major morphological trends in evolution: Why do limbs taper toward the distal end? Why have five digits instead of six? Why do bivalve mollusk shells vary in their shape and ribbing? In contrast, the behavioral ecologist accepts as given the morphology and physiology of the species and incorporates them as constraints (e.g., handling time, prey detection time, and digestive bottlenecks) in optimization models. The models of behavioral ecologists ask how behavior might be optimized given these constraints, whereas the morphologist or embryologist asks the much more difficult question of how the constraints arose in the first place. Perhaps, we suggest, it is because of their relatively modest aims that optimization models in behavioral ecology have been successful.

There are, however, a number of successful applications of optimality theory to explain morphological traits (e.g., Wainwright et al. 1976). Most of these appear to be examples of morphological design features related to deforming forces (e.g., the design of trabeculae in bones; G. Oster, pers. comm.). These models are parallel to those of the behavioral ecologist in that they do not try to explain major morphological trends but relatively minor adjustments within the constraints set by major evolutionary trends.

Our point can be further illustrated by referring to an example of differences in the success of optimization models in accounting for behavioral traits. Patterson et al. (1980) experimentally manipulated the cost and benefits of alternative reproductive strategies in two species of North American blackbirds. They found that the birds, in accord with the predictions of a simple optimization model, modified their parental effort but not their mating system. One interpretation of this result is that although the mating system is subject to historical constraints, within these constraints, parental feeding of young is optimized.

Future research in OFT

Our survey of the literature (Table 6.1) shows that there is still a need for quantitative tests of the classical patch, prey, and related models. Too many papers use the catchphrase "optimal foraging" to dress up a study of feeding ecology that has little, or nothing, to do with testing OFT; they should not be confused with proper studies of OFT. The general picture that emerges from proper studies to date is of considerable qualitative success of the models, but slightly less success in quantitative predictions. However, there are still insufficient data to estimate, for example, the average percentage deviation between observed and predicted results and whether the deviations show any systematic pattern. This would be a useful prelude to the construction of models with more constrained strategy sets (ROT) that will have greater accuracy. In the elucidation of ROT, one particularly interesting development is the emerging parallel between OFT and operant psychology (Staddon 1980). This link also is likely to be fruitful in exploring the predictions of stochastic models of risk.

In addition to these links with operant psychology, a promising direction for future research in OFT is the development of more species-oriented models. By using the same basic components and rationale as the classical models but incorporating specific details of the biology of the system under study, it may be possible to generate models of greater predictive accuracy than those that formed the first generation of OFT.

Summary

Optimal foraging models do not test the proposition that animals are optimal. They do test whether or not the particular assumptions about constraints and optimization criteria included in the models give a good account of behavior. The comment that OFT models are too simple to cope with the complexity of the real world is not a criticism of OFT in particular but a criticism of the piecemeal, reductionist approach to studying complex phenomena.

We refer to the early models of prey choice and patch use, and their derivatives, as the "classical" or "traditional" models of OFT. Many papers purporting to test these models do no more than refer to some qualitative agreement between observations and an assumption of the models. There have been only a few more incisive tests. On the whole, the qualitative (and sometimes quantitative) agreement between observations and predictions is quite good. We discuss the implications of some deviations from predicted results.

One extension of traditional models is to incorporate more details of the specific mechanism used by individual species to solve foraging problems. We view these models of "rules of thumb" as being essentially similar to traditional models but including more constraints. Concepts such as "suboptimal" or "imperfectly optimal" in this context are misleading, because optimization can only be defined in relation to the strategy set available to the animal.

Traditional models of OFT are deterministic. We discuss two aspects of stochasticity, risk and information. Models of risk imply sensitivity of animals not just to mean rates of intake but also to variance in rate of intake. There is some experimental evidence for this. The value of information to the foraging animal depends on how it can be used. There are many instances where the value of information is low (e.g., variance in patch quality is low) or information is easy to obtain (e.g., variance in quality is very high). For these reasons, traditional deterministic models may be more robust than is sometimes supposed. Other approaches to the study of information include models of tracking a changing environment and mechanistic models of learning derived from the mathematical psychology literature.

Many of the criticisms leveled at the adaptationist program in general and at optimality models in particular do not apply to OFT, because the constraints in OFT models are more readily identifiable than those in, for example, morphological trends in evolution.

Acknowledgments

We thank Luc-Alain Giraldeau, Alasdair Houston, and Anne Sorensen for their comments on the manuscript. J. R. K. was supported by a Nuffield Foundation Science Fellowship and by NERC, and W. J. S. by a NERC postdoctoral fellowship. We are also grateful to John Goss-Custard for comments and allowing us to use his data.

Literature cited

Al-Jaborae, F. F. 1979. The influence of diet on the gut morphology of the Starling (*Sturnus vulgaris* L. 1758). Ph.D. thesis, Oxford University.

Andersson, M. 1978. Optimal foraging area: size and allocation of search effort. *Theor. Popul. Biol. 13:*397–409.

– 1981. Central place foraging in the Whinchat, *Saxicola rubetra. Ecology 62:*538–544.

Arnold, S. J. 1978. The evolution of a special class of modifiable behaviors in relation to environmental pattern. *Am. Nat. 112:*415–427.

Aronson, R. B., and T. J. Givnish. In press. Optimal central place foragers; a comparison with null hypotheses. *Ecology.*

Barnard, C. J., and C. A. J. Brown. 1981. Prey size selection and competition in the Common Shrew (*Sorex areneus* L.). *Behav. Ecol. Sociobiol. 8:*239–243.

Barnard, C. J., and H. Stephens. 1981. Prey size selection by Lapwings in Lapwing/gull associations. *Behaviour 77:*1–22.

Belovsky, G. E. 1978. Diet optimization in a generalist herbivore, the moose. *Theor. Popul. Biol. 14:*105–134.

Best, L. S., and P. Bierzychudek. 1982. Pollinator foraging on foxglove (*Digitalis pupurea*): a test of a new model. *Evolution 36:*70–79.

Bond, A. B. 1980. Optimal foraging in a uniform habitat: the search mechanism of the Green Lacewing. *Anim. Behav. 28:*10–19.

– 1981. Giving up as a Poisson process: the departure decision of the Green Lacewing. *Anim. Behav. 29:*629–630.

Bonser, R. 1981. Laboratory studies in prey selection by *Notonecta glauca.* Ph.D. thesis, York University.

Breck, J. E. 1978. Suboptimal foraging strategies for a patchy environment. Ph.D. thesis, Michigan State University.

Brooke, M. de L. 1981. How an adult Wheatear (*Oenanthe oenanthe*) uses its territory when feeding nestlings. *J. Anim. Ecol. 50:*683–696.

Bush, R., and F. Mosteller. 1955. *Stochastic models for learning.* New York: Wiley.

Caraco, T. 1980. On foraging time allocation in a stochastic environment. *Ecology 61:*119–128.

– 1981. Energy budgets, risk and foraging preferences in Dark-eyed Juncos (*Junco hymelais*). *Behav. Ecol. Sociobiol. 8:*213–217.

Caraco, T., S. Martindale, and T. S. Whittham. 1980a. An empirical demonstration of risk-sensitive foraging preferences. *Anim. Behav. 28:*820–830.

Caraco, T., S. Martindale, and H. R. Pulliam. 1980b. Flocking: advantages and disadvantages. *Nature (London) 285:*400–401.

Carlson, A., and J. Moreno. 1981. Central place foraging in the Wheatear *Oenanthe oenanthe:* an experimental test. *J. Anim. Ecol. 50:*917–924.

Charnov, E. L. 1976a. Optimal foraging, the marginal value theorem. *Theor. Popul. Biol. 9:*129–136.

– 1976b. Optimal foraging: attack strategy of a mantid. *Am. Nat. 110:*141–151.

Cody, M. L. 1971. Finch flocks in the Mohave desert. *Theor. Popul. Biol. 2:*142–158.

Comins, H. N., and M. P. Hassell. 1979. The dynamics of optimally foraging predators and parasitoids. *J. Anim. Ecol. 48:*335–351.

Cook, R. M., and B. J. Cockrell. 1978. Predator ingestion rate and its bearing on feeding time and the theory of optimal diets. *J. Anim. Ecol. 47:*529–547.

Corbet, S. A., I. Cuthill, M. Fallows, T. Harrison, and G. Hartley. 1981. Why do nectar-foraging bees and wasps work upwards on inflorescences? *Oecologia (Berlin) 51:*79–83.

Cowie, R. J. 1977. Optimal foraging in Great Tits (*Parus major*). *Nature (London) 268:*137–139.

Cowie, R. J., and Krebs, J. R. 1979. Optimal foraging in patchy environments. In R. M. Anderson, B. D. Turner, and L. R. Taylor (eds.). *The British Ecological Society Symposium,* Vol. 20, *Population dynamics,* pp. 183–205. Oxford: Blackwell Scientific Publications

Davidson, D. W. 1978. Experimental tests of the optimal diet in two social insects. *Behav. Ecol. Sociobiol. 4:*35–41.

Davies, N. B. 1977a. Prey selection and social behaviour in wagtails (Aves: Motacillidae). *J. Anim. Ecol. 46:*37–57.

– 1977b. Prey selection and the search strategy of the Spotted Flycatcher (*Muscicapa striata*): a field study on optimal foraging. *Anim. Behav. 25:*1016–1033.

Davies, N. B., and T. R. Halliday. 1979. Competitive mate searching in male Common Toads, *Bufo bufo. Anim. Behav. 27:*1253–1267.

Davies, N. B., and A. I. Houston. 1981. Owners and satellites: the economics of territory defence in the Pied Wagtail, *Motacilla alba. J. Anim. Ecol. 50:*157–180.

Davison, M. 1969. Preference for mixed-interval vs. fixed-interval schedules. *J. Exp. Anal. Behav. 12:*247–252.

Dawkins, M. 1971. Perceptual changes in chicks, another look at the 'search image' concept. *Anim. Behav. 19:*566–574.

DeBenedictis, P. A., F. B. Gill, F. R. Hainsworth, G. H. Pyke, and L. L. Wolf. 1978. Optimal meal size in hummingbirds. *Am. Nat. 112:*301–316.

Doyle, R. W. 1979. Ingestion rate of a selective deposit feeder in a complex mixture of particles: testing the energy-optimization hypothesis. *Limnol. Oceanogr. 24:*867–874.

Dunstone, N., and R. J. O'Connor. 1979. Optimal foraging in an amphibious mammal. I. The aqualung effect. *Anim. Behav. 27:*1182–1194.

Ebersole, J. P., and J. C. Wilson. 1980. Optimal foraging: the response of *Peromyscus leucopus* to experimental changes in processing time and hunger. *Oecologia (Berlin) 46:*80–85.

Elner, R. W., and R. N. Hughes. 1978. Energy maximization in the diet of the shore crab, *Carcinus maenas. J. Anim. Ecol. 47:*103–116.

Emlen, J. M. 1966. The role of time and energy in food preference. *Am. Nat. 100:*611–617.

Erichsen, J. T., J. R. Krebs, and A. I. Houston. 1980. Optimal foraging and cryptic prey. *J. Anim. Ecol. 49:*271–276.

Estabrook, G. F., and D. C. Jespersen. 1974. The strategy for a predator encountering a model-mimic system. *Am. Nat. 108:*443–457.

Evans, R. M. 1982. Efficient use of food patches at different distances from a breeding colony in Black-billed gulls. *Behaviour 79:*28–38.

Fretwell, S. D. 1972. *Populations in a seasonal environment.* Princeton, N.J.: Princeton University Press.

Furnass, T. I. 1979. Laboratory experiments on prey selection by Perch fry (*Perca fluviatilis*). *Freshwater Biol. 9:*33–43.

Gardner, M. B. 1981. Mechanisms of size selectivity by planktivorous fish: a test of hypotheses. *Ecology 62:*571–578.

Gibb, J. A. 1958. Predation by tits and squirrels on the eucosmid *Ernarmonia conicolana* (Heyl.). *J. Anim. Ecol. 27:*375–396.

Gibbon, J. in press. In H. L. Roitblat, T. Bever, and H. Terrace (eds.). *Animal cognition, H. F. Guggenheim Conference 1982.* New York: Columba University Press.

Gibbon, J., and R. M. Church. 1981. Time left: linear vs logarithmic subjective time. *J. Exp. Psychol. 7:*87–107.

Gibson, R. M. 1980. Optimal prey-size selection by Three-spined Sticklebacks (*Gasterosteus aculeatus*): a test of the apparent-size hypothesis. *Z. Tierpsychol. 52:*291–307.

Giller, P. S. 1980. The control of handling time and its effects on the foraging strategy of a heteropteran predator, *Notonecta. J. Anim. Ecol. 49:*699–712.

Gilliam, J. F., R. F. Green, and N. E. Pearson. 1982. The fallacy of the traffic policeman: a response to Templeton and Lawlor. *Am. Nat. 119:*875–878.

Giraldeau, L.-A., and D. L. Kramer. 1982. The marginal value theorem: a quantitative test using load size variation in a central place forager, the Eastern Chipmunk, *Tamias striatus. Anim. Behav. 30:*1036–1042.

Goss-Custard, J. D. 1977a. Optimal foraging and the size selection of worms by Redshank, *Tringa totanus,* in the field. *Anim. Behav. 25:*10–29.

– 1977b. Predator responses and prey mortality in the Redshank, *Tringa totanus* (L.), and a preferred prey, *Corophium volutator* (Pallas). *J. Anim. Ecol. 46:*21–35.

– 1977c. The energetics of prey selection by Redshank, *Tringa totanus* (L.), in relation to prey density. *J. Anim. Ecol. 46:*1–19.

Gould, J. P. 1974. Risk, stochastic preference, and the value of information. *J. Econ. Theory 8:*64–84.

Gould, S. J., and R. C. Lewontin. 1979. The spandrels of San Marco and the Panglossian paradigm: a critique of the adaptationist programme. *Proc. R. Soc. London Ser. B 205:*581–598.

Green, R. F. 1980. Bayesian birds: a simple example of Oaten's stochastic model of optimal foraging. *Theor. Popul. Biol. 18:*244–256.

Greenstone, M. H. 1979. Spider feeding behaviour optimises dietary essential amino acid composition. *Nature (London) 282:*501–503.

Hall, D. J., E. E. Werner, J. F. Gilliam, G. G. Mittelbach, D. Howard, C. G. Doner, J. A. Dickerman, and A. J. Stewart. 1979. Diel foraging behaviour and prey selection in the golden shiner *Notemigonus crysoleucas. J. Fish. Res. Board Can. 36:*1029–1039.

Harley, C. B. 1981. Learning the evolutionarily stable strategy. *J. Theor. Biol. 89:*611–633.

Harper, D. G. C. 1982. Competitive foraging in Mallards: 'ideal free' ducks. *Anim. Behav. 30:*575–584.

Harper, J. L. 1982. After description. In E. I. Newman (ed.). *The plant community as a working mechanism,* pp. 11–25. Oxford: Blackwell Scientific Publications.

Hartling, L. K., and R. C. Plowright. 1979. Foraging by bumblebees on patches of artificial flowers: a laboratory study. *Can. J. Zool. 57:*1866–1870.

Hassell, M. P. 1980. Foraging strategies, population models and biological control: a case study. *J. Anim. Ecol. 49:*603–628.

Hassell, M. P., and R. M. May. 1973. Stability in insect host-parasite models. *J. Anim. Ecol. 42:*693–726.

– 1974. Aggregation of predators and insect parasites and its effect on stability. *J. Anim. Ecol. 43:*567–594.

Hegner, R. E. 1982. Central place foraging in the White-fronted Bee-Eater. *Anim. Behav. 30:*953–963.

Heinrich, B. 1979a. *Bumblebee economics.* Cambridge, Mass.: Harvard University Press.

– 1979b. Resource heterogeneity and patterns of movement in foraging bumblebees. *Oecologia (Berlin) 40:*235–245.

Heithaus, E. R., and T. H. Fleming. 1978. Foraging movements of a frugivorous bat, *Carollia perspicillata* (Phyllostomatidae). *Ecol. Monogr. 48:*127–143.

Heller, R. 1980. On optimal diet in a patchy environment. *Theor. Popul. Biol.* *17*:201–214.

Heller, R., and M. Milinski. 1979. Optimal foraging of sticklebacks on swarming prey. *Anim. Behav. 27*:1127–1141.

Herrnstein, R. S. 1964. Aperiodicity as a factor in choice. *J. Exp. Anal. Behav. 7*:179–182.

Hodges, C. M. 1981. Optimal foraging in bumblebees: hunting by expectation. *Anim. Behav. 29*:1166–1171.

Hodges, C. M., and L. L. Wolf. 1981. Optimal foraging in bumblebees: why is nectar left behind in flowers? *Behav. Ecol. Sociobiol. 9*:41–44.

Holling, C. J. 1959. Some characteristics of simple types of predation and parasitism. *Can. Entomol. 91*:385–398.

Horn, H. S. 1978. Optimal tactics of reproduction and life-history. In J. R. Krebs and N. B. Davies (eds.). *Behavioural ecology, an evolutionary approach,* pp. 411–429. Oxford: Blackwell Scientific Publications.

Houston, A. I. 1980. Godzilla v. the creature of the black lagoon. In M. F. Toates and T. R. Halliday (eds.). *The analysis of motivational systems,* pp. 297–318. New York: Academic Press.

Houston, A. I., A. Kacelnik, and J. M. McNamara. in press. Some learning rules for acquiring information. In D. J. McFarland, (ed.). *Functional ontogeny.* London: Pitman Books.

Houston, A. I., J. R. Krebs, and J. T. Erichsen. 1980. Optimal prey choice and discrimination time in the Great Tit (*Parus major* L.). *Behav. Ecol. Sociobiol. 6*:169–175.

Houston, A. I., and J. M. McNamara. 1982. A sequential approach to risk taking. *Anim. Behav. 30*:1260–1261.

Howell, D. J., and D. L. Hartl. 1980. Optimal foraging in glossophagine bats: when to give up. *Am. Nat. 115*:696–704.

Hubbard, S. F., and R. M. Cook. 1978. Optimal foraging by parasitoid wasps. *J. Anim. Ecol. 47*:593–604.

Hubbard, S. F., R. M. Cook, J. G. Glover, and J. J. D. Greenwood. 1982. Apostatic selection as an optimal foraging strategy. *J. Anim. Ecol. 51*:625–633.

Hughes, R. N. 1979. Optimal diets under the energy maximization premise: the effects of recognition time and learning. *Am. Nat. 113*:209–221.

Hughes, R. N., and R. W. Elner. 1979. Tactics of a predator, *Carcinus maenas,* and morphological responses of the prey, *Nucella lapillus. J. Anim. Ecol. 48*:65–78.

Hughes, R. N., and R. Seed. 1981. Size selection of mussels by the blue crab *Callinectes:* energy maximizer or time minimizer? *Mar. Ecol. Prog. Ser. 6*:83–89.

Hughes, R. N., and C. R. Townsend. 1981. Maximizing net energy returns from foraging. In C. R. Townsend and P. Calow (eds.). *Physiological ecology,* pp. 86–108. Oxford: Blackwell Scientific Publications.

Iwasa, Y., M. Higashi, and N. Yamamura. 1981. Prey distribution as a factor determining the choice of optimal foraging strategy. *Am. Nat. 117*:710–723.

Jaegar, R. G., and D. E. Barnard. 1981. Foraging tactics of a terrestrial salamander: choice of diet in structurally simple environments. *Am. Nat. 117*:639–664.

Janetos, A. C., and B. J. Cole. 1981. Imperfectly optimal animals. *Behav. Ecol. Sociobiol. 9:*203–210.

Jenkins, S. H. 1980. A size-distance relation in food selection by Beavers. *Ecology 61:*740–746.

Kacelnik, A., A. I. Houston, and Krebs, J. R., 1981. Optimal foraging and territorial defence in the Great Tit (*Parus major*). *Behav. Ecol. Sociobiol. 8:*35–40.

Kamil, A. C., and S. J. Yoerg. 1982. Learning and optimal foraging. In P. P. G. Bateson and P. H. Klopfer (eds.). *Perspectives in ethology*, Vol. 5. New York: Plenum.

Kaufman, L. W., and G. Collier. 1981. Economics of seed handling. *Am. Nat. 118:*46–60.

Killeen, P. R. 1981. Averaging theory. In C. M. Bradshaw, E. Szabadi, and C. F. Lowe (eds.). *Quantification of steady state operant behaviour*, pp. 21–34. Amsterdam: Elsevier/North Holland.

Killeen, P. R., J. P. Smith, and S. J. Hanson. 1981. Central place foraging in *Rattus norvegicus. Anim. Behav. 29:*64–70

Kislalioglu, M., and R. N. Gibson, 1976. Some factors governing prey selection by the 15-spined stickleback, *Spinachia spinachia* (L.). *J. Exp. Mar. Biol. Ecol. 25:*159–169.

Kramer, D. L., and W. Nowell. 1980. Central place foraging in the Eastern Chipmunk, *Tamias striatus. Anim. Behav. 28:*772–778.

Krebs, J. R. 1973. Behavioural aspects of predation. In P. P. G. Bateson and P. H. Klopfer (eds.). *Perspectives in ethology*, Vol. I, pp. 73–111. New York: Plenum.

– 1978. Optimal foraging: decision rules for predators. In J. R. Krebs and N. B. Davies (eds.). *Behavioral ecology, an evolutionary approach*. Sunderland, Mass.: Sinauer Associates.

– 1980. Optimal foraging, predation risk and territory defence. *Ardea 68:*83–90.

Krebs, J. R., and M. I. Avery. in press. Test of central place foraging in a single-prey loader. *J. Anim. Ecol.*

Krebs, J. R., and N. B. Davies. 1981. *An introduction to behavioural ecology*. Oxford: Blackwell Scientific Publications.

Krebs, J. R., J. T. Erichsen, M. I. Webber, and E. L. Charnov. 1977. Optimal prey selection in the Great Tit (*Parus major*). *Anim. Behav. 25:*30–38.

Krebs, J. R., A. I. Houston, and E. L. Charnov. 1981. Some recent developments in optimal foraging. In A. C. Kamil and T. D. Sargent (eds.). *Foraging behavior, ecological, ethological and psychological approaches*, pp. 3–18. New York: Garland STPM Press.

Krebs, J. R., A. Kacelnik, and P. Taylor. 1978. Test of optimal sampling by foraging Great Tits. *Nature (London) 275:*27–31.

Krebs, J. R., J. C. Ryan, and E. L. Charnov. 1974. Hunting by expectation or optimal foraging? a study of patch use by chickadees. *Anim. Behav. 22:*953–964.

Larkin, S. B. C. 1981. Time and energy in decision making. Ph.D. thesis, Oxford University.

Lawton, J. H., J. R. Beddington, and R. Bonser. 1974. Switching in invertebrate predators. In M. B. Usher and M. H. Williamson (eds.). *Ecological stability*, pp. 141–158. London: Chapman & Hall.

Lea, S. E. G. 1979. Foraging and reinforcement schedules in the pigeon: optimal and non-optimal aspects of choice. *Anim. Behav. 27*:875–886.

Lessells, C. M., and D. W. Stephens. in press. Central place foraging: single prey loaders again. *Anim. Behav.*

Leventhal, A. M., R. F. Morrell, E. F. Morgan, Jr., and C. C. Perkins, Jr. 1959. The relation between mean reward and mean reinforcement. *J. Exp. Psychol. 57*:284–287.

Lewis, A. R. 1980. Patch use by Gray Squirrels and optimal foraging. *Ecology 61*:1371–1379.

Lobel, P. S., and J. C. Ogden. 1981. Foraging by the herbivorous parrot fish *Sparisoma radians*. *Mar. Biol. 64*:173–183.

MacArthur, R. H., and E. R. Pianka. 1966. On optimal use of a patchy environment. *Am. Nat. 100*:603–609.

Martindale, S. 1982. Nest defense and central place foraging: a model and experiment. *Behav. Ecol. Sociobiol. 10*:85–90.

Mazur, J. E. 1981. Optimization theory fails to predict performance of pigeons in a two-response situation. *Science 214*:823–825.

Maynard Smith, J. 1978. Optimization theory in evolution. *Annu. Rev. Ecol. Syst. 9*:31–56.

McCleery, R. H. 1977. On satiation curves. *Anim. Behav. 25*:1005–1015.

McFarland, D. J. 1977. Decision making in animals. *Nature (London) 269*:15–21.

McNair, J. M. 1982. Optimal giving-up times and the marginal value theorem. *Am. Nat. 119*:511–529.

– in press. Energy reserves and the risk of starvation: some foraging strategies. *Am. Nat.*

McNamara, J. in press. Optimal control of the diffusion coefficient of a simple diffusion process. *Math. Operat. Res.*

McNamara, J., and A. Houston. 1980. The application of statistical decision theory to animal behaviour. *J. Theor. Biol. 85*:673–690.

– in press. Short-term behaviour and lifetime fitness. In D. J. McFarland (ed.). *Functional ontogeny*. London: Pitmans Books.

McNamara, J. M. 1982. Optimal patch use in a stochastic environment. *Theor. Popul. Biol. 21*:269–288.

Milinski, M. 1979. An evolutionary stable feeding strategy in sticklebacks. *Z. Tierpsychol. 51*:36–40.

Milinski, M. and R. Heller. 1978. Influence of a predator on the optimal foraging behaviour of sticklebacks (*Gasterosteus aculeatus* L.). *Nature (London) 275*:642–644.

Milton, K. 1979. Factors influencing leaf choice by howler monkeys: a test of some hypotheses of food selection by generalist herbivores. *Am. Nat. 114*:362–378.

Mittelbach, G. G. 1981. Foraging efficiency and body size: a study of optimal diet and habitat use by bluegills. *Ecology 62*:1370–1386.

Moermond, T., and J. Denslow. in press. Fruit choice in Neotropical birds. *J. Anim. Ecol.*

Morse, D. H. 1980. *Behavioral mechanisms in ecology.* Cambridge, Mass.: Harvard University Press.

Murdoch, W. W. 1969. Switching in general predators: experiments on predator specificity and stability of prey populations. *Ecol. Monogr. 39:*335–354.

Murdoch, W. W., S. Avery, and M. E. B. Smyth. 1975. Switching in predatory fish. *Ecology 56:*1094–1105.

Murdoch, W. W., and A. S. Oaten. 1975. Predation and population stability. *Adv. Ecol. Res. 9:*2–131.

Myers, J. P., P. G. Connors, and F. A. Pitelka. 1981. Optimal territory size and the Sanderling: compromises in a variable environment. In A. C. Kamil and T. D. Sargent (eds.). *Foraging behavior, ecological, ethological and psychological approaches,* pp. 135–158. New York: Garland STPM Press.

Norberg, R. A. 1977. An ecological theory on foraging time and energetics and choice of optimal food-searching method. *J. Anim. Ecol. 46:*511–529.

– 1981. Temporary weight decrease in breeding birds may result in more fledged young. *Am. Nat. 118:*838–850.

Oaten, A. 1977. Optimal foraging in patches: a case for stochasticity. *Theor. Popul. Biol. 12:*263–285.

O'Brien, W. J., N. A. Slade, and G. L. Vinyard. 1976. Apparent size as the determinant of prey selection by bluegill sunfish (*Lepomis macrochirus*). *Ecology 57:*1304–1310.

Ollason, J. G. 1980. Learning to forage-optimally? *Theor. Popul. Biol. 18:*44–56.

Orians, G. H., and N. E. Pearson. 1979. On the theory of central place foraging. In D. J. Horn, G. R. Stairs, and R. Mitchell (eds.). *Analysis of ecological systems,* pp. 155–177. Columbus: Ohio State University Press.

Oster, G. F., and P. Alberch. 1982. Evolution and bifurcation in developmental programs. *Evolution 36:*444–459.

Oster, G. F., and E. O. Wilson. 1978. Caste and ecology in the social insects. Princeton, N.J.: Princeton University Press.

Owen-Smith, N., and P. Novellie. 1982. What should a clever ungulate eat? *Am. Nat. 119:*151–178.

Palmer, A. R. 1979. Prey selection by non-visual invertebrate predators: a field test of optimal foraging. Ph.D. thesis, University of Washington.

Parker, G. A. 1978. Searching for mates. In J. R. Krebs and N. B. Davies (eds.). *Behavioural ecology, an evolutionary approach,* pp. 214–244. Oxford: Blackwell Scientific Publications.

Parker, G. A., and R. A. Stuart. 1976. Animal behavior as a strategy optimizer: evolution of resource assessment strategies and optimal emigration thresholds. *Am. Nat. 110:*1055–1076.

Partridge, L. 1981. Increased preferences for familiar foods in small mammals. *Anim. Behav. 29:*211–216.

Pastorok, R. A. 1980. The effects of predator hunger and food abundance on prey selection by *Chaoborus* larvae. *Limnol. Oceanogr. 25:*910–921.

Patterson, C. B., W. J. Erckmann, and G. H. Orians. 1980. An experimental study of parental investment of polygyny in male blackbirds. *Am. Nat. 116:*757–769.

Pleasants, J. M. 1981. Bumblebee response to variation in nectar availability. *Ecology 62:*1648–1661.

Pubols, B. H., 1962. Constant versus variable delay of reinforcement. *J. Comp. Physiol. Psychol. 55:*52–56.

Pulliam, H. R. 1975. Diet optimization with nutrient constraints. *Am. Nat. 109:*765–768.

– 1976. The principle of optimal behaviour and the theory of communities. In P. P. G. Bateson and P. H. Klopfer (eds.). *Perspectives in ethology,* Vol. 2, pp. 311–332. New York: Plenum.

– 1980. Do Chipping Sparrows forage optimally? *Ardea 68:*75–82.

Pulliam, H. R., and C. Dunford. 1980. *Programmed to learn.* New York: Columbia University Press.

Pyke, G. H. 1978a. Optimal foraging in bumblebees and coevolution with their plants. *Oecologia (Berlin) 36:*281–293.

– 1978b. Optimal foraging in hummingbirds: testing the marginal value theorem. *Am. Zool. 18:*739–752.

– 1978c. Are animals efficient harvesters? *Anim. Behav. 26:*241–250.

– 1979a. Optimal foraging in bumblebees: rule of movement between flowers within inflorescences. *Anim. Behav. 27:*1167–1181.

– 1979b. The economics of territory size and time budget in the Golden-winged Sunbird. *Am. Nat. 114:*131–145.

– 1980. Optimal foraging in bumblebees: calculation of net rate of energy intake and optimal patch choice. *Theor. Popul. Biol. 17:*232–246.

– 1981a. Why hummingbirds hover and honeyeaters perch. *Anim. Behav. 29:*861–867.

– 1981b. Optimal travel speeds of animals. *Am. Nat. 118:*475–487.

Pyke, G. H., H. R. Pulliam, and E. L. Charnov. 1977. Optimal foraging: a selective review of theory and tests. *Q. Rev. Biol. 52:*137–154.

Rapport, D. J. 1980. Optimal foraging for complementary resources. *Am. Nat. 116:*324–346.

Real, L. A. 1980a. Fitness, uncertainty, and the role of diversification in evolution and behavior. *Am. Nat. 115:*623–638.

– 1980b. On uncertainty and the law of diminishing returns in evolution and behaviour. In J. E. R. Staddon (ed.). *Limits to action: the allocation of individual behaviour,* pp. 37–64. New York: Academic Press.

– 1981. Uncertainty and plant-pollinator interactions: the foraging behavior of bees and wasps on artificial flowers. *Ecology 62:*20–26.

Rechten, C., M. Avery, and A. Stevens. in press. Optimal prey selection: why do Great Tits show partial preferences? *Anim. Behav.*

Rechten, C., J. R. Krebs, and A. I. Houston. 1981. Great Tits and conveyor belts: a correction for non-random prey distribution. *Anim. Behav. 29:*1276–1277.

Ringler, N. H. 1979. Selective predation by drift feeding brown trout *Salmo trutta. J. Fish. Res. Board Can. 36:*392–403.

Royama, T. 1970. Factors governing the hunting behaviour and selection of food by the Great Tit (*Parus major* L.). *J. Anim. Ecol. 39:*619–668.

Scheibling, R. E. 1981. Optimal foraging movements of *Oreaster reticulatus* (L.) (Echinodermata: Asteroidea). *J. Exp. Mar. Biol. Ecol. 51:*173–185.

Schluter, D. 1981. Does the theory of optimal diets apply in complex environments? *Am. Nat. 118:*139–147.

Schoener, T. W. 1971. Theory of feeding strategies. *Annu. Rev. Ecol. Syst.* *2*:369–404.

— 1974. Resource partitioning in ecological communities. *Science 185*:27–39.

— 1979. Generality of the size-distance relation in models of optimal foraging. *Am. Nat. 114*:902–914.

Sibly, R., and D. McFarland. 1976. On the fitness of behavior sequences. *Am. Nat. 110*:601–617.

Sih, A. 1980a. Optimal behavior: can foragers balance two conflicting demands? *Science 210*:1041–1043.

— 1980b. Optimal foraging: partial consumption of prey. *Am. Nat. 116*:281–290.

Simon, H. A. 1956. Rational choice and the structure of the environment. *Psychol. Rev. 63*:129–138.

Smith, J. N. M. 1974. The food searching behaviour of two European thrushes. II. The adaptiveness of the search patterns. *Behaviour 49*:1–61.

Smith, J. N. M., P. R. Grant, B. R. Grant, I. J. Abbott, and L. K. Abbott. 1978. Seasonal variation in feeding habits of Darwin's ground finches. *Ecology 59*:1137–1150.

Solomon, M. E. 1949. The natural control of animal populations. *J. Anim. Ecol. 18*:1–35.

Staddon, J. E. R. 1980. Optimality analyses of operant behaviour and their relation to optimal foraging. In J. E. R. Staddon (ed.). *Limits to action: the allocation of individual behaviour,* pp. 101–117. New York: Academic Press.

Stein, R. A. 1977. Selective predation, optimal foraging, and the predator-prey interaction between fish and crayfish. *Ecology 58*:1237–1253.

Stephens, D. W. 1981. The logic of risk-sensitive foraging preferences. *Anim. Behav. 29*:628–629.

— 1982. Stochasticity in foraging theory: risk and information. Ph.D. thesis, Oxford University.

Stephens, D. W., and E. L. Charnov. 1982. Optimal foraging: some simple stochastic models. *Behav. Ecol. Sociobiol. 10*:251–263.

Sutherland, W. J. 1982a. Spatial variation in the predation of cockles by Oystercatchers at Traeth Melynog, Anglesey II. The pattern of mortality. *J. Anim. Ecol. 51*:491–500.

— 1982b. Do Oystercatchers select the most profitable cockles? *Anim. Behav. 30*:857–861.

Templeton, A. R., and L. R. Lawlor. 1981. The fallacy of the averages in ecological optimization theory. *Am. Nat. 117*:390–393.

Tinbergen, J. 1981. Foraging decisions in Starlings (*Sturnus vulgaris* L.). *Ardea 69*:1–67.

Townsend, C. R., and A. G. Hildrew. 1980. Foraging in a patchy environment by a predatory net spinning-caddis larva: a test of optimal foraging theory. *Oecologia (Berlin) 47*:219–221.

Turelli, M., J. H. Gillespie, and T. W. Schoener. 1982. The fallacy of the fallacy of the averages in ecological optimization theory. *Am. Nat. 119*:879–884.

Vadas, R. L. 1977. Preferential feeding: an optimization strategy in sea urchins. *Ecol. Monogr. 47*:337–371.

Visser, M. 1981. Prediction of switching and counter-switching based on optimal foraging. *Z. Tierpsychol. 55*:129–138.

Waage, J. K. 1979. Foraging for patchily-distributed hosts by the parasitoid, *Nemeritis canescens. J. Anim. Ecol. 48:*353–371.

Waddington, K. D., and B. Heinrich. 1979. The foraging movements of bumblebees on vertical "inflorescences": an experimental analysis. *J. Comp. Physiol. 134:*113–117.

Waddington, K. D., and L. R. Holden. 1979. Optimal foraging: on flower selection by bees. *Am. Nat. 114:*179–196.

Wainwright, S. A., W. D. Biggs., J. D. Curry, and J. M. Gosline. 1976. *Mechanical design in organisms.* London: Arnold.

Wehner, R., and M. V. Srinivasan. 1981. Searching behaviour of desert ants, genus *Cataglyphis* (Formicidae, Hymenoptera). *J. Comp. Physiol. 142A:*315–338.

Werner, E. E., and D. J. Hall. 1974. Optimal foraging and the size selection of prey by the Bluegill sunfish (*Lepomis macrochirus*). *Ecology* 55:1042–1052.

Werner, E. E., and G. G. Mittelbach. 1981. Optimal foraging: field tests of diet choice and habitat switching. *Am. Zool. 21:*813–829.

Whitham, T. G. 1977. Coevolution of foraging in *Bombus* and nectar dispensing in *Chilopsis:* a last dreg theory. *Science 197:*593–596.

– 1980. The theory of habitat selection: examined and extended using *Pemphigus* aphids. *Am. Nat. 115:*449–466.

Williams, D. 1982. studies of optimal foraging using operant techniques. Ph.D. thesis, University of Liverpool.

Wilson, E. O. 1980a. Caste and division of labor in leaf-cutter ants (Hymenoptera: Formicidae: *Atta*). I. The overall pattern in *A. sexdens. Behav. Ecol. Sociobiol. 7:*143–156.

– 1980b. Caste and division of labor in leaf-cutter ants (Hymenoptera: Formicidae: *Atta*). II. The ergonomic optimization of leaf cutting. *Behav. Ecol. Sociobiol. 7:*157–165.

Winterhalder, B. 1981. Foraging strategies in the boreal forest: an analysis of Cree hunting and gathering. In B. Winterhalder, and E. A. Smith (eds.). *Hunter-gatherer foraging strategies,* pp. 66–98. Chicago: University of Chicago Press.

Winterhalder, B., and E. A. Smith (eds.). 1981. *Hunter-gatherer foraging strategies.* Chicago: University of Chicago Press.

Ydenberg, R. C. 1982. Studies of foraging and vigilance in the Great Tit. (*Parus major* L.). Ph.D. thesis, Oxford University.

Zach, R. 1978. Selection and dropping of whelks by Northwestern Crows. *Behaviour 67:*134–148.

– 1979. Shell dropping: decision-making and optimal foraging in Northwestern Crows. *Behaviour 68:*106–117.

Zach, R., and J. B. Falls. 1976. Do Ovenbirds (Aves: Parulidae) hunt by expectation? *Can. J. Zool. 54:*1894–1903.

– 1978. Prey selection by captive Ovenbirds (Aves: Parulidae). *J. Anim. Ecol. 47:*929–943.

Zach, R., and J. N. M. Smith. 1981. Optimal foraging in wild birds? In A. L. Kamil and T. D. Sargent (eds.). *Foraging behavior, ecological, ethological and psychological approaches,* pp. 95–109. New York: Garland STPM Press.

Zimmerman, M. 1981. Optimal foraging, plant density and the marginal value theorem. *Oecologia (Berlin) 49:*148–153.

Zwarts, L., and R. H. Drent. 1981. Prey depletion and regulation of predator density: Oystercatchers (*Haematopus ostralegus*) feeding on mussels (*Mytilus edulis*). In N. V. Jones and W. J. Wolf (eds.). *Feeding and survival strategies of estuarine organisms*. New York: Plenum.

Commentary

J. P. MYERS

Optimal foraging theory (OFT), along with its siblings in behavioral ecology – theories on mating systems, inclusive fitness, and the like – has changed the way we look at bird behavior. This is true not only for the conclusions we draw. It also holds for the way we pose questions and conduct studies, even for determining what stand as legitimate and interesting approaches. In no small part, this is due to the success of OFT, its sibling theories, and also its practitioners. It is a well-deserved success, for it has provided a paradigm uniting behavior with evolution through a common theoretical ground, natural selection, and developed a suite of powerful logical and methodological tools.

Chapter 6 conveys the scope of OFT's success. It shows how OFT has developed, it identifies logical errors committed by both opponents and proponents, and it points to new directions.

Krebs and his colleagues and students have been in the thick of this billowing field almost from the beginning. In no small degree, they have given it shape and substance. This imbues their perspective with rich detail and authority. At the same time, it bespeaks a commitment to the theory that ought to wave some cautionary flags.

Is foraging always optimal?

The flags wave most furiously when the chapter dismisses fundamental criticisms of optimal foraging as misunderstandings. Their response to the "apocryphal aardvark connoisseur" – the OFT critic claiming that aardvarks don't forage optimally – handles one level of such criticism neatly. Yet, at the same time it sidesteps a far more basic issue. Optimality models in biology have been criticized intensely when presented as attempts to test whether animals are or are not optimal (e.g., Gould and Lewontin 1979; Lewontin 1979). Krebs et al. maintain that testing optimality per se is not their intent. Rather, they want the test to be whether one particular hypothesis about optimization correctly predicts a particular animal foraging pattern. They leave unstated how we might ultimately derive some general conclusions on the underlying issue, optimality per se.

Presumably, we either will be so convinced by the weight of case after case that the conclusion will be inescapable. Alternatively, we may be so entranced by the neatness of each individual example that the general issue will never resurface. Surely, neither Krebs nor his colleagues will ever be satisfied with the latter, that is, ignoring the general issue, and the critics of OFT would never live with the former (save a rejection, which is unlikely if the ratio of confirming to rejecting papers continues in its current trend).

Whether animals optimize remains a fundamental question. If OFT cannot address that proposition, then it is hardly a robust "attempt to understand the decision rules of foraging animals," the definition with which Krebs et al. began their essay. What can be more general than the issue of optimality versus some pattern of decision making in which optimality is irrelevant?

It strikes me that OFT may have moved too quickly forward, swept ahead by its clear successes and its near successes, and in the process leaped beyond a critical test, comparing optimal versus nonoptimal models. The tests now concentrate on ever more refined versions of optimality, and unless explicitly addressed to the more general questions, may never challenge it again.

At the root of the problem lie four facts, all of which discourage heavy research investments in nonoptimal behaviors:

1. It is easy to develop predictions about behavior based on the assumption of optimality. The theoretical baggage for this has been well developed in economics and linear programming, and perhaps the only surprise is that it took behavioral ecologists so long to discover these tools.

2. It is easy to dismiss quantitative deviations from the predictions of these models if the qualitative trends are in line. The uncertain difficulties of behavioral observation and measurement make this even easier.

3. It is hard to develop quantitative nonoptimal predictions and especially to make them testable alternatives to predictions from optimal models.

4. If one accepts the logical consequences of natural selection, it is hard to imagine why animals might not forage optimally.

Support for the first three points comes from the literature: the plethora of optimality models, the easy reinterpretations of data, the dearth of nonoptimality models. Serious attention is now being given to the last, for example, Janetos and Cole's (1981) analysis of imperfectly optimal animals, and the undercurrent of discussion about satisficing (Simon 1956). These are particularly important developments because ultimately they may help fill in holes left by OFT's leapfrog gait, if they don't erode its underpinnings altogether.

Reasons for nonoptimal foraging

Why shouldn't animals forage optimally? There are several possible reasons, some of which are not necessarily inconsistent with the OFT logic developed by Krebs et al.

Constraints

Phylogenetic constraints will prevent some solutions that would otherwise be optimal (Janetos and Cole 1981). As Krebs et al. point out, this may not be suboptimality so much as a mistaken set of assumptions by the investigator. Once real constraints are incorporated within the predictions, optimality could yet still prevail. These constraints could result from physical limits or merely from inadequate time for natural selection to act.

Trade-offs

Animals must solve many problems at once, and as a result their actual behavior may result from a compromise between simultaneous, conflicting demands. Again, this does not have to be inconsistent with optimality theory. What Krebs et al. call second-generation OFT models regularly incorporate trade-offs between foraging and other ecological requirements, such as predator avoidance or territorial defense.

There are two ways, however, to make those trade-offs: One entails simultaneous optimization of multiple parameters, in the classic linear programming sense. The other is to let the less important slide while focusing on the more critical, as might occur when trading off foraging (where lessened efficiency may not result in death) versus predator avoidance (where mistakes are fatal). Studies of trade-offs frequently show a change in foraging behavior when predation risk is increased, but they have yet to show a true dynamic optimization (e.g., Milinski and Heller 1978; Sih 1980).

Moreover, as the problem changes from simultaneous optimization of two parameters to one of many parameters, the challenge to the animal compounds. Janetos and Cole (1981) suggest that at some point the marginal benefits for machinery capable of solving complex optimizations may not justify the costs.

The EMP hypothesis or adaptive roadblocks to optimization

Differences in the timing and scale of importance of conflicting demands might obviate trade-offs. This is best illustrated by analogy with a current defense policy issue. Much recent attention has focused on one of the physical effects on semiconductor-based information technology of exploding a single nuclear device high above the earth's surface (Broad 1981). The explosion causes an electromagnetic pulse (EMP) to radiate through the atmosphere. The EMP wreaks havoc in any semiconductor-based equipment that is not highly protected against such an event. Protection is very difficult, if not impossible, for installations and equipment that cannot be totally insulated from the radiation. The consequences of EMP's disruptive effects for communication and computer functioning have only slowly dawned on the defense establishment. A single explosion may eliminate most of the hardware on which current defense strategy rests. The irony is that much of the tube-based equipment that semiconductors replaced was immune to EMP. Semiconductors offered incomparable speed and capacity, but they also carried a potentially disastrous Achilles heel.

The relevance to optimality in biology should be clear. EMP represents an unpredictable selective event of devastating magnitude, one of low frequency relative to other selective factors, or one yet to occur. An animal, as an individual, that proceeds willy-nilly toward some optimal solution may expose itself to such effects and be at a long-term disadvantage relative to individuals slower to optimize. The EMP analogy suggests that short-term optimal solutions can be detrimental in the long run. Hence, we might expect inherent restrictions to slow the approach to optimality.

Several counterarguments can be raised immediately. This could be interpreted as simply another time scale of optimization. Perhaps it is, in which case the critics of optimality in biology are not going to be any happier with it than with current reasoning. What it does, on the other hand, is offer a scenario where we might expect less than optimal tactics within the time scale of concern to OFT. Under appropriate conditions, an organism should do enough to get by, and it should satisfice and not risk some unpredictable, low-frequency, perhaps even unspecified event. A more serious criticism is the implausibility of sub-

220 J. P. Myers

optimal tactics persisting through periods between catastrophic events. Imagine two selective factors X (the EMP analog) and Y (strong and incessant). For Y, there are two solutions, y (optimal) and y' (suboptimal). These solutions do not handle X equally well: If X occurs, animals employing y are eliminated, whereas those using y' persist. The question is whether the differential between y and y' is so strong under Y that no y' will remain when X finally occurs. Clearly, the plausibility of the EMP hypothesis will depend upon the relative frequencies and magnitudes of different selective events. The conditions may be quite restrictive. On the other hand, even if restrictive in equations, they may be common in nature. Logic alone will not resolve this issue.

Krebs et al. dismiss satisificing as simply a minor variation on OFT. I disagree. Its basic premise is fundamentally different from optimization, because it entails minimum threshold criteria for success or failure of particular behavioral tactics, rather than maximization of net benefit. Satisficing predicts that animals should be riddled with behavioral traditions, drifting through routines that meet each requirement but optimize none. They will switch from their traditions only once they no longer work – jerked out of complacency by the sudden appearance of a predator or cumulative deterioration in feeding conditions.

Summary

In summary, there are several reasons why animals might not forage optimally. Some are more artifacts than anything else – a discordance between behavior and the extrapolation of theory rather than between behavior and the theory itself. In the final analysis, they will be consistent with a world view based on optimization. Others, however, are not artifactual. They do not sit well with traditional models of natural selection and fitness, for which optimization is the natural currency. Whether they offer a better set of premises for predicting foraging behavior remains to be seen.

The challenge is to develop empirical tests that discriminate among optimal, suboptimal, and nonoptimal models. In the process, we will broaden OFT to FT, and thereby strengthen our science.

Literature cited

Broad, W. J. 1981. Nuclear pulse. 1. Awakening to the chaos factor. *Science* *212*:1009–1012.

Gould, S. J., and R. C. Lewontin. 1979. The spandrels of San Marco and the Panglossian paradigm: a critique of the adaptationist programme. *Proc. R. Soc. London Ser. B 205:*581–598.

Janetos, A. C., and B. J. Cole. 1981. Imperfectly optimal animals. *Behav. Ecol. Sociobiol. 9:*203–210.

Lewontin, R. C. 1979. Fitness, survival and optimality. In D. J. Horn, G. R. Stair, and R. Mitchell (eds.). *Analysis of ecological systems,* pp. 3–21. Columbus: Ohio State University Press.

Milinski, M., and R. Heller. 1978. Influence of a preditor on the optimal foraging behavior of sticklebacks *Gasterosteus aculeatus. Nature (London) 275:*642–644.

Sih, A. 1980. Optimal foraging: partial consumption of prey. *Am. Nat. 116:*281–290.

Simon, H. A. 1956. Rational choice and the structure of the environment. *Psychol. Rev. 63:*129–130.

7   Biochemical studies of
microevolutionary processes

GEORGE F. BARROWCLOUGH

In any scientific endeavor, it is the understanding of process, rather than the description of pattern that is ultimately of interest. Only when we know why a pattern exists, that is, when we understand its cause, can we make generalizations with any real degree of confidence. This is as true of systematics and evolutionary biology as it is of any other field (Eldredge and Cracraft 1980).

The detailed investigation of microevolutionary processes has lagged behind the description of microevolutionary patterns just as a deep understanding of evolutionary forces has been retarded in other areas, such as behavioral ecology, biogeography, and macroevolution. For example, we realize that a cline may be the result of gene flow, natural selection, or even random drift, but, as with other areas of evolutionary biology, it is common merely to assert that selection was responsible for the pattern. Consequently, anything more than a subjective understanding of microevolutionary mechanisms has been slow to develop. What is needed to proceed beyond this illusion of knowledge are analytical models and statistical tests to examine the concordance of patterns to the expectations of given processes. Consistency with one's world view is not understanding; alternatives must be eliminated through quantification.

What we know about microevolutionary processes today has come mainly through quantitative studies of morphological variation. These have involved observations of changes associated with potent selective agents and analyses of correlations of morphological variation with environmental factors. For example, James (1970) investigated the relationship between geographical variation in the size of 12 species of birds and climate in eastern North America. She confirmed the existence of a pattern consistent with Bergmann's ecogeographic rule by establishing concordance of patterns across taxa that were in turn correlated with temperature and humidity. Similarly, selection due to climatic factors has been identified as the probable causal agent for some

223

of the geographic variation of morphology in North American populations of the House Sparrow, *Passer domesticus*. In this case also, the pattern could be related to a process through correlation with environmental variables (Johnston and Selander 1971) and through the results of natural experiments showing the selective effects of winter storms and cold (O'Donald 1973; Rising 1973). Additional studies have found similar concordance to ecomorphological rules (Power 1970). Equally important is that others have either not found concordance, found results opposite to those expected, or found association with other factors (Niles 1973; Baker and Moeed 1979; Baker 1980). Nevertheless, it seems clear that detailed, quantitative investigations of the extent, pattern, and processes responsible for geographic variation of morphology in birds are now underway. This program of research, quantitative and statistical in orientation, has certainly been one of the most productive areas of avian systematics during the last 15 years.

Despite this progress, the universality of the results of these morphological studies it not clear. Size and shape may or may not respond to the same forces and in the same fashion as other aspects of the genotype and phenotype. Only by examining several categories of characters can we learn about the relative magnitudes and efficacy of various microevolutionary mechanisms. Also, little has been discovered about mechanisms of speciation. A few patterns are known that suggest extrinsic processes of undeniable importance in the origin of isolation (Mengel 1964; Hubbard 1973). However, the process of speciation is a phenomenon comprised of several levels of causation. Even given knowledge of a dispersal or vicariant event, there still remain questions about the intrinsic mechanism leading to reproductive isolation – genetic revolutions, karyotypic changes, gradual adaptive divergence, and so forth. Thus, the study of microevolutionary processes is still in its infancy, and new techniques and approaches are necessary to foster progress in this undertaking.

Biochemical techniques and analysis

Most ornithologists are aware of the use of various biochemical techniques to address higher taxonomic problems. In recent years, however, there also have been attempts to use some of these same methods to analyze systematic problems at the species and within-species levels. Thus, the geographic pattern of genetic variation has been investigated for a few avian species, as has the nature of genic differentiation among closely related species. In addition to this systematic activity by ornithologists, there have been, for the last 15 to 20 years, extensive theoretical and empirical investigations of the nature of this variation by population geneticists. This research has attempted to evaluate the relative roles of such microevolutionary processes as natural selection, random drift, mutation, gene flow, time and isolation, and demographic processes. A natural extension of this work is an attempt to explore the nature of microevolutionary processes in birds and to identify salient differences in these processes between birds and other vertebrate taxa. This field of research is the subject of this review. The process of speciation will be included with the discussion of microevolutionary processes. This seems reasonable because the processes leading to geographic variation may, in many cases, also lead to speciation (White 1978; but see Eldredge and Cracraft 1980), and speciation is one of the evolutionary processes of greatest interest to avian systematists.

There are several advantages of molecular techniques over more traditional types of characters. First, biochemical methods allow us to examine characters with a relatively simple and well-understood, developmental origin. In most cases, the direct translated products of a gene or the genes themselves, are analyzed. Problems of pleiotropy or complex genetic–environmental interactions during ontogeny are reduced greatly. In this sense, biochemical characters may be "cleaner" than are phenotypic ones. Hence, any biochemical confirmation of microevolutionary patterns based on other characters would be reassuring. Frelin and Vuilleumier (1979) have discussed this aspect of biochemical methods and reasoning at some length.

A second advantage of molecular methods was pointed out by Lewontin (1974). In order to investigate details of evolutionary processes, it is necessary to know how much genetically based variation there is on which selection can act. Accordingly, biochemical methods yield information concerning the magnitude and distribution of genic variation, at least for structural genes. To obtain equivalent data for morphological or behavioral traits would necessitate difficult breeding experiments and heritability studies.

Finally, some molecular methods, because they yield information concerning the state of single genes, can be analyzed in terms of a well-developed population genetic theory of process. Comparable theory for polygenic and quantitative traits is not yet very well developed, although the work of Lande (1976, 1980, 1981) is beginning to rectify that situation. If biochemical studies are to make a major contribution to our understanding of microevolution in birds, then it will be through access to this theory of process, equations of gene mechanics that relate genetic patterns to evolutionary mechanisms. Some relatively simple biochemical methods may be useful for this purpose.

Electrophoresis. The separation of proteins that differ in charge, molecular weight, or conformation by placing them in an electric field on some supporting medium has become known to most ornithologists. Reviews of the basic techniques are available elsewhere, and methods have improved enormously since the early seventies (Sibley 1970; Brewer 1970). Instead of general protein stains, it is now customary to rely principally on specific isozyme staining methods that identify bands through their specific activity. With these techniques, there are usually only one or a few bands on a gel, and the determination of homology of protein morphs does not present a serious problem. The most commonly used gel medium has been refined starch. Other media, such as polyacrylamide, agar, and cellulose acetate, are used in special situations and have varying resolving power depending on the nature of the molecules being separated. The resolution of different alleles may depend on pH, buffer type, and gel concentration; consequently, it may be productive to try several systems (Brush 1979). Isoelectric focusing, usually on polyacrylamide, separates proteins by the pH at which they are intrinsically neutral rather than by size or conformation (see, e.g., Sibley and Frelin 1972).

The analysis of electrophoretic data is straightforward. Bands on gels (electromorphs) are usually readily understood in terms of modern molecular genetics. The patterns obtained can be related to frequencies of various alleles at single loci, and each individual can be assigned a genotype. Heterozygosity, the extent of genic variation within individuals can be calculated once genotypes are known. Several algorithms exist to convert allelic frequency data into a measure of the genetic distance between populations (Hedrick 1971; Rogers 1972; Nei 1978). Nei's measure, D, is most widely used, and a correction for error due to small sizes can be computed, an advantage not available with the other two measures. Wright (1978) provides a method for performing what amounts to an analysis of variance on allelic frequencies, enabling the calculation of an among-population component of genetic variance, F_{st}. This statistic is important in studies of geographic variation. Several other ways are available to compute F_{st}, but Wright's method contains a correction for sampling error due to small numbers of individuals.

Chromosome preparation and staining. Chromosomes, large aggregates of DNA and proteins, have been useful in systematic studies of *Drosophila,* mammals, and other groups. Karyotypic changes may be related to speciation in some groups (White 1978) and can be related to quantitative models of genetics (Lande 1979; Hedrick 1981). A review of the methods and results of karyotypic analysis in birds is available (Shields 1982).

DNA restriction mapping. This is a relatively new technique (Lansman et al. 1981), and only preliminary studies have been published using this technique on birds (Glaus et al. 1980). Special enzymes are used to cleave DNA at particular sites. The DNA fragments can then be separated by weight on a gel. Mitochondrial DNA frequently is used so that there will not be an overwhelming number of pieces of DNA after the enzymatic digestion. Restriction mapping has been used in other vertebrate groups especially with mitochondrial DNA, and it is probably only a question of time before some avian systematists investigate the approach.

The enzymes used to dissect the DNA are extremely site specific; hence, if two individuals have different patterns of DNA fragments on a gel after this digestion, then the DNAs of the two must differ. The presence or absence of the various fragments can be used as characters. Nei and Li (1979) provide a method to convert these data into a DNA distance measure. These techniques have been used at several taxonomic levels including studies of the phylogeny of hominids (Brown et al. 1979) and the relationships among populations of gophers (Avise et al. 1979a). It is certainly a promising method for future studies of microevolution.

Other methods. Many other biochemical techniques are currently available with potential application to systematic work. However, most of them will not be useful for studies of microevolution for several reasons. For studies of taxonomy at higher levels, it may often be sufficient to examine one or a few exemplars of each taxon of interest. Thus, expensive or labor-intensive methodologies may be feasible. Microevolutionary studies, however, are based on populations, not one or two individuals. For studies of microevolution, it will usually be essential to deal with sufficiently large samples to get estimates of within-as well as among-population variances. For techniques such as protein sequencing (Ibrihimi et al. 1979), peptide mapping (Corbin 1968), microcomplement fixation (Prager et al. 1974), and DNA sequencing or hybridization (Shields and Straus 1975; Sibley and Ahlquist 1981), obtaining complete matrixes of variances among individuals, as well as among populations, will be neither quick nor inexpensive.

Review of avian studies

The recent reviews of the use of biochemical methods in avian systematics have concentrated on the use of these techniques in taxonomy at higher levels (Selander 1971; Sibley et al. 1974). Macrotaxonomy is not the subject of this review. It is unfortunate, though, that only since the time of those reviews, have useful quantitative methods of analysis of biochemical data become available. Now that the tools of analysis are available, the earlier techniques, such as peptide mapping and electrophoresis, that were abandoned by most avian systematists working at higher levels require fresh inquiry to determine their actual usefulness. Such renewed investigation is just beginning (Gutiérrez et al., in press).

Molecular data, before being used to address questions of process, must be organized to yield descriptions of two important and distinct variables: the extent of variation within individuals and that among populations. The first is measured by heterozygosity, and the latter by either genetic distances or F statistics.

Heterozygosity

Selander and Johnson (1973) first reviewed the patterns of electrophoretic variation in vertebrate species. At that time, only one species of bird, *Passer domesticus,* had been assayed for genetic variation. In the following decade, much work was done in birds at the species level and below. A basic statistic calculated from these data is heterozygosity, which is an estimate of the fraction of genes at which an individual is polymorphic. Nei (1978) provides a method to compute this measure and correct for sampling error due to small numbers of individuals; Hardy–Weinberg equilibrium is assumed. Table 7.1 presents a summary of estimated heterozygosities using data reported for *breeding populations* of birds for which both the number of loci and the number of individuals were relatively large (i.e., total number of genes examined approximately 1,000 or more). The mean of these heterozygosities is 0.053. Nevo (1978) reviewed estimates of genic heterozygosity among vertebrates, and this average for the largest avian studies is approximately the same as the one reported for other vertebrate

Table 7.1 *Estimates of genic heterozygosity* ($\hat{H}$) *in birds: results based on electrophoretically detected variation from single breeding populations*

Species	No. of loci	No. of individuals	$\hat{H}$	Reference
Ardea herodias	28	46	0.007	Guttman et al. (1980)
Cygnus buccinator	19	43	0.013	Barrett and Vyse (1982)
Lagopus lagopus	23	269	0.082	Gyllensten et al. (1979)
Lagopus mutus	23	45	0.044	Gyllensten et al. (1979)
Callipepla squamata	27	20	0.037	Gutiérrez et al. (in press)
Lophortyx gambelii	27	19	0.025	Gutiérrez et al. (in press)
Cyrtonyx montezumae	27	23	0.024	Gutiérrez et al. (in press)
Coturnix pectoralis	36	47	0.041	Baker and Manwell (1975)
Sphyrapicus varius	37	15	0.039	N. K. Johnson (unpub.)
Empidonax difficilis	45	28	0.053	N. K. Johnson (unpub.)
Eremophila alpestris	35	17	0.102	N. K. Johnson (unpub.)
Hirundo tahitica	15	31	0.078	Manwell and Baker (1975)
Petrochelidon ariel	15	33	0.065	Manwell and Baker (1975)
Pomatostomus temporalis	20	80	0.094	Johnson and Brown (1980)
Passerella iliaca	38	51	0.039	Zink (unpub.)
Zonotrichia capensis	14	41	0.054	Zink (1982)
Zonotrichia leucophrys	44	21	0.057	Corbin (1981)
Junco hyemalis	37	25	0.061	Barrowclough (unpub.)
Amphispiza belli	43	20	0.047	N. K. Johnson (unpub.)
Geospiza fortis	27	53	0.057	Yang and Patton (1981)
Geospiza fuliginosa	27	80	0.062	Yang and Patton (1981)
Geospiza scandens	27	26	0.058	Yang and Patton (1981)
Geospiza conirostris	27	21	0.056	Yang and Patton (1981)
Camarhynchus parvulus	27	21	0.030	Yang and Patton (1981)
Dendroica coronata	32	23	0.036	Barrowclough (1980b)
Vireo solitarius	42	20	0.045	N. K. Johnson (unpub.)
Icterus galbula	19	28	0.060	Corbin et al. (1979)
Passer domesticus	33	57	0.147	Cole and Parkin (1981)
Aplonis cantoroides	18	32	0.026	Corbin et al. (1974)
Aplonis metallica	18	42	0.055	Corbin et al. (1974)

classes, 0.049. Neither the avian studies nor those reviewed by Nevo involved exhaustive searches for the class of "hidden" alleles that has been the subject of much recent work (e.g., Coyne et al. 1979). Thus, although the absolute levels of genic heterozygosity in birds may ultimately prove higher than those reported here, there is no reason to believe it is any greater or smaller than in other vertebrates.

Several additional caveats concerning estimates of genic heterozygosity must be pointed out. Not only are absolute levels of genic variability in some doubt, but, depending upon the comparison, relative values are as well. Not all loci examined electrophoretically are equally variable. Therefore, unless the same, or very nearly the same, loci are examined for a pair of species, it is risky to make claims about the relative levels either of variability or of other parameters, such as population size, that covary with it. Thus, the general pattern of average genic variability in birds resembles that of other vertebrate taxa (Table 7.1). However, specific comparisons of any two arbitrary avian taxa may not be meaningful because of large standard errors due to numbers of individuals, numbers of loci, and choice of loci. Similar cautions apply to statements about the evolutionary potential of taxa, management goals and decisions, history of inbreeding of the taxa, and so forth. A large number of assumptions underlie all such conclusions.

A major issue precipitated by the initial surveys of the extent of genetic variation in natural populations, and subsequently argued throughout the 1970s, was the relative efficacy of natural selection versus random drift (neutrality) in the maintenance of this variability. The issue has not been resolved, but there has been steady attrition from the selectionist side (Barrowclough 1981). Initially, the neutral camp was the province of a few theoreticians. They were opposed by most experimental workers, scientists who had come of age in the selectionist climate of the 1950s and 1960s. This has now changed somewhat, and, sensitive to other possibilities, a pluralism is spreading among evolutionary biologists (Gould and Lewontin 1979). Many ecological and population geneticists now consider much of the observed variation to be neutral or near neutral. In part, this change may be due to a better understanding of what neutrality actually implies. It does not mean, for example, that the protein products of two alleles are not "seen" by natural selection, or have no function but only that they perform approximately equally as well at their function. That is, the selective difference between alternates is small enough that they are approximately equivalent given the effective population size (i.e., $N_{ce}s < 1$). In addition, differences among loci in variability, related to size or function (O'Brien et al. 1980), may still be consistent with the mutation-drift theory. For instance, the observed pattern of general monomorphism of supernatant malate dehydrogenase in birds (Kitto and Wilson 1966) suggests that there may be severe constraints on this molecule, but this does not mean that variation, when it is observed, is not effectively neutral (Karig and Wilson 1971). It is just that a small fraction of the mutations at the locus are neutral and not quickly eliminated. Such phenomena are encompassed by the variable mutation rate formulation of the neutral models.

The neutral and selection hypotheses are statistical and have to be addressed using distributions of allelic frequencies over many loci. It is not acceptable, for example, to correlate the frequencies of each allele repeatedly until something is significant and then move on to the next locus. This cannot fail to be successful given sufficient independent variables (Schnell and Selander 1981). The critical tests will involve ensemble distributions across loci and few independent variables. Indeed, the strong point of the neutral theorists is the fact that they are able to predict overall patterns of allelic frequencies with few free parameters (Chakraborty et al. 1980). The selectionist hypothesis, when it has been possible to generate a priori predictions, has not been as successful as the neutralist theory when compared to actual data (Li 1978; Chakraborty et al. 1980). A recent analysis of the most thorough studies of genic variation within avian populations found general quantitative agreement with several predictions of the mutation-drift theory (Barrowclough, Johnson, and Zink, unpub.). An example is given in Figure 7.1. The observed pattern of allelic frequencies is quite similar to the predicted pattern for a sample from a breeding population of Yellow-rumped Warblers. Here 27 individuals were examined for 32 loci (Barrowclough 1980b). The predicted pattern of allelic frequencies for the neutral model is based solely on one statistic, the overall heterozygosity. The specific neutral theory used was the infinite allele–constant mutation rate model. The slight excess of rare alleles (alleles with a frequency less than 0.05) observed may be due to a variable mutation rate or to a bottleneck of population numbers in the last 100,000 years. However, it is not consistent with such selective models as overdominance, which produce excess numbers of alleles at higher frequencies. It would seem useful to perform such tests routinely and before invoking selective hypotheses to explain genic variation.

If most of the genic variability present in natural populations is neutral or near neutral, then it could provide a useful tool for the study of both micro- and macroevolutionary processes because it could then be used as an indicator of population structure and relative divergence times.

Geographic variation

Perhaps the most important use of electrophoretic data so far, in terms of the elucidation of microevolutionary patterns, has been its role in examining the genetic structure of natural populations. This latter term is meant to encompass several related aspects of the organization of populations, including their effective size, the extent of inbreeding, and the degree of connectedness of local demes through

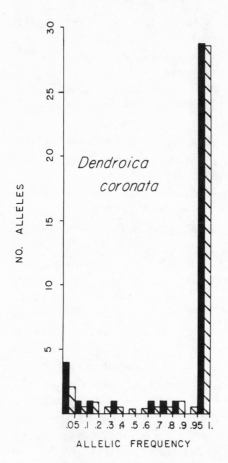

Figure 7.1. Observed (solid) and expected (hatched) numbers of electrophoretically detected alleles in a sample from a breeding population of *Dendroica coronata*. Expected distribution calculated using the infinite allele–constant mutation rate model for neutral-drift genetic variation. Heterozygosity was 0.034.

gene flow. The genetical structure is an attribute of populations that is of crucial importance for the quantitative understanding of the mechanisms and dynamics of microevolution, including speciation. The relative efficacy of various modes of speciation depend upon the size and connectedness of demes (Templeton 1980b). The response of populations to selection and stochastic influences depends upon the same factors (Wright 1970). In spite of this importance, however, little was known about this aspect of natural populations of most animals, including birds, until quite recently. This is a consequence of the fact that genetic structure is the product of demographic processes. Thus, to deduce the genetic structure of populations, it is necessary to document carefully their demographics and especially the extent of dispersal. For birds, such studies are difficult, labor-intensive projects. Until the recent rebirth of interest in the subject, only initial, exploratory work was possible (Miller 1947).

The development of efficient electrophoretic techniques has allowed the widespread sampling of relatively large numbers of individuals from moderate numbers of populations across the range of a species. The computation of allelic frequencies from such data has enabled direct quantitative estimates of the genetic structure of populations. These results depend in part on the validity of the assumption of the approximate neutrality of electrophoretic variation, but each variable protein locus can be used to estimate independently such parameters as Wright's F_{st}, and the variance among these estimates can be used to get some sense of the adequacy of the neutrality assumption (Lewontin and Krakauer 1973). In addition, it is encouraging that estimates of the same genetic parameters based on demographic modeling yield conclusions similar to those based on the avian electrophoretic data (Barrowclough 1980a,b).

Table 7.2 is a compilation of estimates of F_{st}, the among-population component of genetic variance, for birds based on electrophoresis of various structural proteins and enzymes. This statistic varies from zero to one. A value of zero indicates panmixia or a lack of differentiation among populations, and a value of one indicates the fixation of alternate alleles in different populations and, hence, a lack of effective gene flow between them. The values reported in Table 7.2 were calculated from data in the literature and a few studies in progress. Wright's (1978) formula was used to estimate F_{st} and results in a correction for error associated with small sample sizes of individuals and probably represents the algorithm of choice. The among-locus portion of the sampling error is not easily removed. Thus, estimates based on one or a few loci should be viewed with some caution. Nevertheless, it seems clear that most of the estimates are relatively small, especially those from studies with larger samples of both populations and loci. For the most part, they are less than 0.06. This implies, for example, that 94% of the total genetic variance is found within populations, and only 6% is distributed among populations. Some of the cases involve island populations, a particularly good test of the effects of isolation. It appears that the endemic finches of the Galapagos Islands have slightly larger values of F_{st} than average. This is definitely true of the relatively sedentary Warbler Finch, *Certhidea olivacea* (Yang and Patton 1981). Also, the

Table 7.2. *Geographic component ($\hat{F}_{st}$) of genetic variation in birds: estimates based on electrophoretic variation using the method of Wright (1978)*

Taxon	No. of loci	No. of populations	$\hat{F}_{st}$	Region	Reference
Somateria mollissima	1	6	0.057	Northern Europe	Milne and Robertson (1965)
Dendragapus obscurus	1	9	0.000	British Columbia	Redfield et al. (1972)
Phasianus colchicus	1	4	0.006	Iowa	Vohs and Carr (1969)
Columba livia	1	2	0.030	Ireland	Ferguson (1971)
Columba palumbus	1	2	0.051	Ireland	Ferguson (1971)
Passerella iliaca	14	31	0.016	Western U.S.	Zink (unpub.)
Zonotrichia capensis	6	5	0.015	Northwest Argentina	Handford and Nottebohm (1976)
Zonotrichia leucophrys	12	8	0.032	West coast of U.S.	Corbin (1981)
Zonotrichia leucophrys	3	9	0.047	Western U.S.	Baker (1975)
Junco hyemalis	9	6	0.008	Western U.S.	Barrowclough (unpub.)
Junco spp. complex	9	9	0.084	Western U.S.	Barrowclough (unpub.)
Geospiza magnirostris	7	4	0.046	Galapagos Islands	Yang and Patton (1981)
Geospiza fortis	12	8	0.065	Galapagos Islands	Yang and Patton (1981)
Geospiza fuliginosa	12	10	0.054	Galapagos Islands	Yang and Patton (1981)
Geospiza scandens	11	3	0.020	Galapagos Island	Yang and Patton (1981)
Camarhynchus parvulus	8	4	0.057	Galapagos Island	Yang and Patton (1981)
Certhidea olivacea	8	4	0.125	Galapagos Islands	Yang and Patton (1981)
Dendroica coronata	8	5	0.029	Western North America	Barrowclough (1980b)
Pipilo erythrophthalmus	1	6	0.229	U.S.	Sibley and Corbin (1970)
Icterus g. galbula	2	5	0.012	Eastern U.S.	Corbin et al. (1979)
Icterus galbula (incl. *bullockii*)	2	8	0.018	U.S.	Corbin et al. (1979)
Agelaius phoeniceus	1	3	0.037	U.S.	Brush (1968, 1970)
Aplonis cantoroides	2	4	0.127	New Guinea and Bismarck Archipelago	Corbin et al. (1974)
Aplonis metallica	2	6	0.029	New Guinea	Corbin et al. (1974)
Aplonis metallica	2	4	0.040	Bismarck Archipelago	Corbin et al. (1974)

six populations of Shining Starlings (*Aplonis metallica*) from the large island of New Guinea show less genetic differentiation than do the populations from the four smaller islands of the neighboring Bismarck Archipelago. These data are consistent with the results we would expect based on a presumption of reduced gene flow among islands (Corbin et al. 1974).

The results in Table 7.2 indicate a general lack of extensive genetic differentiation among conspecific bird populations. The relevance of this observation becomes apparent when the results of similar surveys for other vertebrate classes are viewed. The appendix table provides a summary of estimates of F_{st} for several classes of vertebrates. All values were recomputed using Wright's (1978) method. The sample was restricted to include data only from studies involving two or more loci from a moderately large part of the range of continuously distributed species. That is, island and other isolated populations were excluded from this survey in order to obtain a reflection of the extent of genetic differentiation strictly due to demography and vagility differences and not due to the effects of fragmented ranges. The survey is not exhaustive but probably includes data from most studies meeting these qualifications published before 1982.

The distributions of the estimates of F_{st} for the various classes are shown in Figure 7.2; summary statistics for these distributions appear in Table 7.3. The sample sizes for the various classes of vertebrates are small, and consequently the standard errors for the summary statistics are relatively large. However, it does appear that there are some differences among classes, with birds showing a reduced range of values of F_{st}. Such results are consistent with a presumed generally greater vagility in birds than in such animals as salamanders and rodents, but this result might also reflect differences in the ages of populations.

An important, but poorly understood, factor is the time it takes for parameters that reflect the genetic structure of populations to reach equilibrium. For example, if populations of many species of salamanders are very old (Highton and Webster 1976; Wake 1981), and many of the other organisms surveyed have only come to occupy their ranges in the last 100,000 years, and if it takes on the order of 10^5 years or more for genetic differentiation to approach equilibrium (Nei et al. 1975), then some of the differences among classes may only reflect differences in the ages of the species. This subject will require some extensive investigation.

If there are differences in the distributions of F_{st} among vertebrate classes and if, in particular, birds do have a small among-population

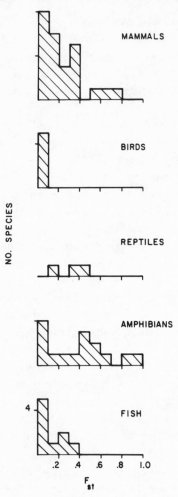

Figure 7.2. Distributions of estimated values of F_{st} among populations across the breeding ranges of species of several classes of vertebrates.

component of genetic variance due to vagility or age, then the work of Templeton (1980a,b) suggests that the predominant modes of speciation in birds may differ from those of such groups as urodeles and relatively sedentary rodents. This will be explored in the following section.

A second index of the extent of differentiation among populations is genetic distance. As mentioned earlier, several measures have been suggested. Nei's D and Roger's S are most frequently used. Unlike F_{st}, these estimates are useful at transpecific levels. D has been of wide-

Table 7.3. *Genetic differentiation among populations of vertebrates: empirical estimates of measures of the distribution of* $\hat{F}_{st}$ *for each taxon based on electrophoretic data summarized in appendix table*

Class	No. of species examined	$\hat{F}_{st}$	SD	SE
Fish	9	0.114	0.137	0.046
Amphibians	15	0.383	0.283	0.073
Reptiles	3	0.304	0.179	0.103
Birds	5	0.022	0.011	0.005
Mammals	25	0.230	0.183	0.037

spread interest because of its theoretical relationship to divergence time (see the section "Time").

Patterns of variation of avian genetic distances have been reviewed recently (Avise et al. 1980; Barrowclough 1980b; Barrowclough et al. 1981). The results are similar to the F_{st} data. The extent of avian genetic differentiation, as measured by D or S, is smaller than is that of other vertebrates and invertebrates of the same taxonomic level. Possible causes for this phenomenon are discussed in aforementioned papers and in the following commentary by J. C. Avise.

Beyond pattern: inferences of process and history

Inferences about evolutionary processes may be facilitated by the examination of biochemical data in two ways. First, the data may directly suggest the dominance of some processes over others or, through the use of genetic modeling, lead to conclusions about the history of taxa and aspects of their biology. Second, the data may lead to inferences in an indirect fashion. For instance, the absence of concordance of biochemical variation with variation of other characters may lead to insights about the relative importance of different processes operating on the various suites of characters.

Among the topics about which direct inferences might be made are two subjects already discussed: the problem of the adaptive nature of genic variation and the details of the genetic structure of populations. A third topic, the estimation of times of evolutionary divergence, is also of particular interest. Secondary inferences, such as details of the mechanisms of speciation in birds, may also flow from these primary inferences. All the observations are enhanced when comparisons can be made with other classes of vertebrates.

Time

If the variation detected by molecular methods is neutral and if population sizes have been in equilibrium, then the genetic distance between two species or isolated populations ought to be a linear function of the time since they became isolated (Nei 1975; Sarich 1977). Of course, population sizes may not have been constant forever, and this may have a decided effect on the substitution rate of amino acids (Korey 1981). Also, selection will undoubtedly be important for the spread or maintenance of some alleles. Nevertheless, these problems do not necessarily rule out a molecular clock (Wilson et al. 1977). If, for instance, over the long run, population bottlenecks and selective processes are approximately uniformly distributed among lineages, then something like a clock would still obtain. This would mean, though, that the clock would only be useful for estimating relatively older divergence times, specifically, for those events sufficiently far in the past that the central limit theorem might apply. This limitation might well apply anyway as the accumulation of substitutions is a stochastic process. Because 30 to 40 proteins typically are used in electrophoretic surveys, there is an among-locus sampling error. This error may be large enough that small values of genetic distance may not be reliable for estimating either divergence times of less than hundreds of thousands of years or the relative order of branching events that are close to each other in time. These problems greatly reduce the utility of genetic distances for the investigation of microtaxonomy in birds. This is because estimates of genetic distances among closely related species of birds have been found to be quite small, as discussed earlier, and often not significantly different from zero. Therefore, the use of a molecular clock to estimate ages and divergence times in birds has been mainly at the intrageneric and higher levels (Prager and Wilson 1975; Smith and Zimmerman 1976; Avise et al. 1980; Sibley and Ahlquist 1981; Zink 1982). In other groups, such as salamanders and rodents, genic distances between conspecific populations can be quite large (Patton and Yang 1977; Wake 1981), and molecular techniques are proving useful in studies of the evolutionary systematics of these organisms.

It is apparent from the aforementioned caveats that the molecular clock issue, especially when based on a single locus or a few proteins, is a complex one (Wilson et al. 1977). Consequently, the eventual acceptance of the validity of the estimation of times and dates will depend on empirical support. If a reasonably accurate clock does emerge, it is possible that it will have to be calibrated independently for different groups and in particular for birds (Barrowclough and Corbin 1978; Avise et al. 1980; Gutiérrez et al., in press). As things currently stand, the factor for converting genetic distance to time is generally agreed to vary by at least a factor of 5, depending upon the genic loci examined (Sarich 1977; see Commentary to Chapter 7). Thus, it definitely seems that, at least for birds, an "electrophoretic clock" is not going to be sufficiently accurate to be useful for precise dating of microevolutionary events.

It should be noted, however, that even if molecular distances cannot be treated as a linear function of time, there is nevertheless an empirically observed increase of these distances with taxonomic rank (Barrowclough et al. 1981), and it seems likely that these distances will be monotonically increasing functions of time since divergence. Thus, in the absence of a good fossil record, molecular distances still may be better than anything else for the estimation of relative divergence of higher level taxa.

Genetics of speciation

One of the persistent criticisms of population genetics has been its lack of relevance to "real" problems and data and its failure to address such obviously important phenomena as species formation (Lewontin 1974; Templeton 1981). Recently, however, what amounts to a minor revolution has occurred in the field. Through the theoretical work of a few investigators, most notably Alan Templeton and Russell Lande, speciation is now being investigated, and the domain of evolutionary population genetics is being enlarged to include quantitative genetics and, hence, encompass morphological variation. These results are of interest because they enable us to explore such questions as how particular populational processes will affect different types of characters (Lande 1976, 1980, 1981) and how population structure, speciation mechanisms, and genetic variation are interrelated (Templeton 1980a,b). This body of work is too extensive and intricate to summarize in detail but deals with the unification of quantitative measures of populational variation with population genetics theory. Because it is potentially one of the most important areas of future growth of microevolutionary studies, an example will be given.

Templeton has been able to associate probabilities to particular combinations of speciation mechanisms and genetic population structure. This is a significant advance over just acknowledging the existence of several modes of speciation. Table 7.4 summarizes part of Templeton's (1980a,b) quantitative investigation of the genetics of speciation. Furthermore, electrophoretic "signatures" of the various modes also have been predicted. The electrophoretic results are most informative when reduced to estimates of the genetic distance between the parent population and the geminate species. In general, speciation in populations with a genetic structure consisting of small demes (large F_{st}) will be associated with large genetic distances. Genetic distances will be small when speciation occurs in populations characterized by large population sizes and significant gene flow (small F_{st}).

The data currently available on geographic variation of allelic frequencies (Tables 7.2 and 7.3) indicate that birds, or at least temperate-zone passerines, are characterized by relatively large, panmictic populations. Consequently, Table 7.4 reveals that cases of avian speciation will be characterized predominantly by those processes listed in the bottom half of the table. This leads to the conclusion that not all possible modes of speciation are likely for birds. In particular, we can predict that chromosomal speciation, whether stasipatric (White 1978) or allopatric, is not apt to be common in birds. Table 7.4 would suggest, however, that for rodents it may be appreciable. Furthermore, when isolation of populations is due to geological processes resulting in the rifting of a species' distribution into two or more fragments, speciation can be expected to proceed slowly in birds and as a by-product of gradual adaptive change. Dispersal events leading to a few individuals colonizing a new piece of habitat are most likely to lead to speciation through a genetic transilience (a concept related to Mayr's "genetic revolution"; see Templeton 1980a).

These results do not offend our preconceptions. However, it must be realized that they are not based on ornithological lore but are predictions independently derived from population genetics theory and electrophoretic data on geographic variation in birds. Furthermore, they enable us to understand why evolutionary systematists working on other classes and phyla have such different views of the process of speciation than do ornithologists. There are differences in modes of speciation as an inevitable consequence of differences in population structure and vagility.

One way to explore the usefulness of these techniques and theory is to reexamine some well-studied group and consider the reasonableness of the results. The Galapagos finches are a suitable taxon. Their taxonomy (Swarth 1931), ecology (Abbott et al. 1977; Grant and Grant 1982), morphology (Bowman 1961), and evolutionary history (Lack

Table 7.4. *Probability of speciation by several different mechanisms as a function of the genetic structure of the population and the process of fissioning.*

Population structure	Isolating event	Probability of speciation by various modes			
		Adaptive divergence	Genetic "transilience"	Chromosomal rearrangement	Clinal divergence
Small, semiisolated demes	Founder (small propagule)	Moderate	Negligible	Large	–
	Vicariance (large fragments)	Large	Negligible	Moderate	–
	None[a]	–	Negligible	Moderate	Large
Large, integrated populations	Founder	Small	Large	Very small	–
	Vicariance	Moderate	Negligible	Negligible	–
	None	–	Negligible	Negligible	Small

Note: For simplicity only two population structures are given; however, a continuum exists.
[a]Sympatric or parapatric divergence.
Source: Based on Templeton (1980b).

1961; Grant and Grant 1979; Yang and Patton 1981) have been examined in detail.

Electrophoretic data on the Galapagos finches describe the patterns of among–species differentiation and the variation within species among islands (Yang and Patton 1981). As is apparently typical of birds, the genetic distances among 11 species of these finches were found to be small by comparison with nonavian taxa. The among-island differentiation within species, as measured by F_{st}, was also small, with the exception of *Certhidea olivacea*. Thus, in spite of the insular nature of their habitat, the genetic structure of these populations is characteristic of species with moderately large populations united by gene flow. Because it takes only an individual immigrant or two per generation to maintain genetic coherence and because interisland migrants have been observed (Lack 1961), this seems reasonable. Again, this indicates that we should consider the lower half of Table 7.4 in our attempt to categorize the likely mechanisms of speciation in these finches.

Vicariance events are very unlikely to have been responsible for geographical isolation on this archipelago. The islands are volcanic, and the channels between islands are deep. Thus, presuming the origin of incipient species to have been the immigration of a propagule consisting of a small number of individuals to some especially isolated island, we observe that the most probable reason for the development of reproductive incompatibility between the new species and the old parental population is a genetic transilience (Table 7.4). The other two possibilities are much less likely to occur. First, a gradual accumulation of adaptive substitutions leading to genetic incompatibility is only likely if the isolation is prolonged and the environments on the various islands differ to a moderate or large degree. Neither of these situations seem probable. Second, the case of a chromosomal change facilitating speciation only becomes probable if the new population remains very small for a considerable length of time.

The latter alternative can be ruled out in two additional ways. First, and directly, are the findings (Jo, in press) that the karyotypes of all these finches are apparently identical (however, no special chromosomal banding studies were performed). Second, if the initial population had remained small for a sufficiently long time to allow a real chance for chromosomal mutants to arise, then the results of Nei et al. (1975) indicate that we would indirectly observe this reduction in N_e through a heterozygosity depression. For example, if a population goes through a bottleneck of size 5 to 10 breeding individuals for 25 or so generations, that is, long enough for there to be some chance of karyotypic change arising, then the level of genic heterozygosity would be reduced rapidly and would take on the order of 10^5 to 10^6 years to recover. These populations do not show such a reduced level of $\overline{H}$ compared to other passerines (Table 7.1).

Grant (1981) and Grant and Grant (1979) have suggested that sympatric speciation may have played a role in the diversification of these finches. However, Table 7.4 indicates that clinal speciation is not very likely given a genetic structure consisting of large, nearly panmictic populations. Clinal divergence is only likely in situations where the populations are organized into semi-isolated demes. Moreover, in panmictic situations, selection has to be large to produce this kind of speciation (Endler 1977). For sympatric speciation, selection has to be considerably greater, and the ultimate probability is less because isolating mechanisms must arise that are linked to the morphological structures responding to the ecological driving force (Smith 1966). Thus, although possible, this does not appear to have been one of the more prevalent processes of speciation in these birds.

These results imply that the only process of speciation quantitatively consistent with what is known about Galapagos finches is that of dispersal of a small group of founders followed by a genetic transilience and rapid increase in population size. Again, this seems reasonable and leaves our world view inviolate, but it is important to acknowledge that, prior to these qualitative and quantitative arguments based on biochemical results and population genetics theory, our preconceptions about speciation processes in these birds were only that, unsupported notions not readily distinguished from other, if less popular, possibilities.

Character sets and modes of evolution

This chapter began with a brief review of another field that has contributed to the beginnings of an understanding of microevolutionary patterns and mechanisms – morphometrics. An obvious question will then arise: Might it be fruitful to examine both biochemical and morphometric traits for the same population?

One reasonable way to proceed in addressing this question is to contrast the patterns of variation among multiple suites of characters. This could be done using ordination or clustering techniques. Until very recently, this approach had not been attempted. This may simply be because the biochemical methods were relatively new. Investigators using them were not familiar with multivariate morphometrics and vice versa. Zink (1982, unpub.), however, has pursued this line of inquiry at both the within- and among-species levels. He contrasted the pattern of electrophoretic variation for large samples of loci, individuals, and populations to that of multivariate morphological variation. Among populations and subspecies of the Fox Sparrow (*Passerella iliaca*), Zink found the morphometric variation to conform more closely to geography and traditional ideas of taxonomy than did the electrophoretic variation. However, the genetic distances among these populations were small and not significantly different from each other. Among species of emberizine finches, he found clustering of electrophoretic distances to better express taxonomic relationships than did the phenetic data, which seemed to reflect size.

This comparison of microevolutionary patterns of different suites of characters conveys some information about evolutionary processes, but this information is qualitative. The various character sets may yield identical patterns or different patterns. If identical, then it may be parsimonious to presume the characters are responding to the same evolutionary forces. If different, then we must assume that the same history and environments have had different effects on the various characters of the same birds. Thus, Zink's results cast doubt on the validity of the nonspecificity hypothesis. That is, the idea that different sets of characters generally will yield similar patterns of relationship when analyzed phenetically. Still, this approach has not taken us as far as we would like to proceed in terms of identifying the relative efficacy of various, potentially important, evolutionary mechanisms. Nevertheless, this may be the right direction.

There is a general consensus, in part based on the results of progress in multivariate morphometrics, that patterns of geographical variation in morphology are the adaptive products of selection operating through ecological conditions. Alternately, there is also the presumption, again to some extent based on both theory and empirical results, that the patterns of much of the biochemical variation are the products of time, distance, and gene flow – demographic events and stochastic processes. If these presumptions are in fact true, then there is no a priori reason to expect concordance of these patterns. In fact, biochemical variation may offer a possible null hypothesis – a pattern against which patterns in other character sets can be tested. Lack of concordance would suggest that the second suite of characters is responding to something other than distance or time since separation. One probable alternative is selection. This line of investigation requires further exploration, but if reasonable statistical tests can be performed, then the contrasting of patterns of morphology and plumage with those of biochemical variation could be an extremely useful approach to the problem of inferring process given pattern. Research along these lines has only just begun. In the absence of examples from the literature, I will again turn to the Galapagos finches, as a model for analysis of this type.

I have computed measures of distance (taxonomic distance) among 10 species of Galapagos finches for both plumage coloration and skeletal variation (Table 7.5). The plumage distances are based on X, Y, and Z coordinates from spectrophotometric curves (Hardy 1936) from the back, nape, throat, and breast of six adult males from populations from single islands of each species. The skeletal distances are based on 25 measurements from several bones of adult males from single islands. These skeletal measurements were log-transformed before distances were computed. The matrix of Rogers' distances published by Yang and Patton (1981) was used as a measure of genetic distance among the same species.

The best statistical method for comparing these data sets is not immediately clear. Correlation coefficients among the entries of the plumage and genetic matrixes (0.686), the plumage and skeletal matrixes (0.111), and the skeletal and genetic matrixes (0.299) are all relatively small. This is probably not surprising. Bowman (1961) and Boag and Grant (1981) have shown that at least the skull and beak morphology of these finches probably are associated with ecological variables. Yang and Patton (1981) have interpreted the genetic variation to be the product of time and demographic processes, that is, to be essentially neutral.

Phylogenetic trees may be inferred from each of the three distance matrixes. When either the distance Wagner algorithm (Farris 1972) or the UPGMA method (Sneath and Sokal 1973) was used, three topologically different trees were obtained. This suggests that the processes of evolution of the various characters have been independent. These correlations and trees impart some information about the processes involved in the evolution of the Galapagos finches. It should be possible to extract more detailed knowledge of the relative evolutionary history of the three sets of traits and the nature of the apparent differences in evolutionary rates of these plumage, skeletal, and structural genic characters.

I attempted to get information on relative rates of evolution of the various sets of characters by forcing all three distance matrixes onto the same tree. To obtain a tree on which to make the comparisons, I used a computer program that finds a maximum likelihood tree given arc sine

Table 7.5. *Plumage and morphometric distances among Galapagos finches*

Taxon	Taxon 1	2	3	4	5	6	7	8	9	10
1. *G. magnirostris*	–	0.23	0.21	0.23	0.10	0.19	1.01	1.26	2.71	2.16
2. *G. fortis*	1.29	–	0.11	0.03	0.15	0.21	0.86	1.09	2.51	1.96
3. *G. fuliginosa*	2.07	0.79	–	0.11	0.12	0.11	0.89	1.16	2.55	2.00
4. *G. difficilis*	2.01	0.76	0.26	–	0.15	0.20	0.85	1.10	2.51	1.96
5. *G. scandens*	1.34	0.35	0.86	0.80	–	0.13	0.94	1.20	2.62	2.07
6. *G. conirostris*	0.66	0.73	1.49	1.42	0.71	–	0.95	1.24	2.61	2.06
7. *C. crassirostris*	0.83	1.09	1.74	1.64	1.04	0.64	–	0.43	1.95	1.34
8. *C. parvulus*	2.32	1.07	0.36	0.45	1.14	1.75	1.97	–	1.86	1.27
9. *C. pallidus*	1.26	.50	1.00	0.91	0.37	0.67	0.85	1.22	–	0.71
10. *C. olivacea*	3.13	1.91	1.20	1.26	1.92	2.57	2.75	0.92	1.99	–

Note: Spectrophotometric distance computed from three tristimulus values each from back, crown, throat, and breast of six adult male specimens for each species. Morphometric distances computed from log-transformed measurements of five or six adult male specimens from single islands for each species. Twenty-five measurements used were based on Robins and Schnell (1971) (their measurement Nos. 1, 3, 5, 6, 8, 11, 12, 14, 16, 18, 21, 22, 24, 29, 30, 33, 34, 36, 37, 39, 40, 42, 43, 45, and 47).

square-root-transformed allelic frequencies (Felsenstein 1981). The branching structure obtained represents the most probable evolutionary history of the group given alleles are approximately neutral and lineages evolve independently of each other. There is no way of finding most probable evolutionary trees with the other two data sets, because reasonable models of evolutionary dynamics currently do not exist for such quantitative data (Felsenstein 1981). I used the allelic frequency data reported by Yang and Patton (1981) to obtain the tree. Once the branching pattern of the tree was known, I forced each distance matrix onto the tree using linear programming. That is, for each distance matrix, I sought a solution to the problem of obtaining an overall mimimum length tree subject to the constraint that the sum of the distances along the branches on the tree connecting a pair of taxa be greater than or equal to the pairwise distance in the distance matrix. For each character set, this procedure yields a single objective function (the total tree length) and a series of constraints (one for each pairwise distance). Linear programming yields an optimal solution to problems of this kind (Au and Stelson 1969). I followed this procedure for each character set and then normalized each set of distances so that they would be comparable. This was done by rescaling the distances to make the total tree length for each character set equal to 1.0. This rescaling makes it possible to find instances of heterogeneous amounts of evolution by searching for branches that differ in length for the three suites of characters. The results are shown in Figure 7.3. The particular root chosen for the tree is arbitrary. The relative amounts of evolution for the three traits vary widely on some branches. For example, there has been much evolution of plumage color (0.198 unit) along the branch separating the *Geospiza* and *Camarhynchus* clades, but little genic (0.050) and almost no skeletal change (0.002). Thus, this procedure facilitates the identification of periods in the evolutionary history of a group of taxa during which interesting evolutionary events occurred. If we take the electrophoretic distances along the branches to represent, approximately, the expectations of Brownian motion (neutral drift), then several episodes of strong selection for plumage and skeletal characters can be identified in the history of these finches.

Conclusions and new directions

Two general patterns have emerged from studies of biochemical variation within species of birds. First, the amount of genetic variation within species examined thus far is of the same order of magnitude as is that of other vertebrates. Thus, most species of birds are not lacking in genetic variation. Second, the distribution of the degree of genetic differentiation among conspecific populations of birds is shifted to small values in comparison with the distributions for such other vertebrates as mammals and amphibians.

These facts, the patterns of genetic variation in birds, are straightforward, but there are many questions and uncertainties associated with them. The reduced levels of genetic differentiation among populations is interpreted to be a consequence of increased gene flow and effective population sizes due to avian vagility. Consequently, species in which effective gene flow or population sizes might be expected to be reduced should be investigated. For example, nonpasserines, tropical species, very small and isolated populations, and relict and fugitive species all need to be examined. Species with distinctive mating and social systems, for instance lekking species, also deserve inquiry.

These two generalizations, by themselves, are of some interest. For example, before the advent of widespread electrophoretic surveys, it was not obvious that there might be general patterns of differences in the population structures among vertebrate classes. The inferences about evolutionary processes, which may be facilitated by these bio-

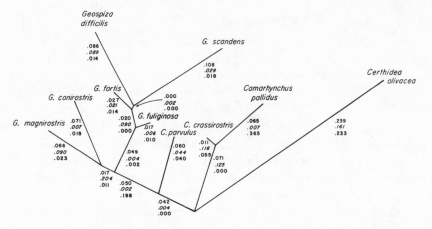

Figure 7.3. Evolution of three sets of characters in the Galapagos finches. Electrophoretic (top number), skeletal (middle number), and plumage (lower number) distances (each normalized) were fitted, using linear programming, to the maximum likelihood branching structure (based on allelic frequencies). Differences between the three numbers indicate different evolutionary rates among the character sets along a given branch.

chemical results, are also of great interest. Three lines of inquiry into this area are already being pursued.

The distributions of genic variation within populations appear to be consistent with the patterns expected if such variation were approximately neutral and predominately influenced by random drift. This conclusion should be examined further, and the methods for doing so are available. If this variation is governed principally by stochastic factors, then genetic distances can be used to estimate divergence times, *if* demographic events along various lineages were not very different. Considerable effort needs to be put into developing methods for testing these assumptions, for calibrating this "clock," and for establishing empirically the range of genetic distances over which the clock is reliable. That is, the point at which the signal-to-noise ratio of the clock is sufficiently large to yield confident estimates must be determined.

The data on the genetic structure of populations, in conjunction with recent advances in population genetics, have opened the possibility of dissecting the process of speciation at a level not previously available. It has become possible to identify the relative probabilities with which different evolutionary mechanisms might be responsible for the origin of genetic incompatibilities between sister species. However, there are several levels to the process of speciation. Other problems, such as the nature of the event initially leading to geographic isolation and the nature of premating isolating mechanisms, remain. Nevertheless, the advance here is an important one, and several case histories, involving disparate taxa, need to be undertaken.

Finally, now that methods have become available for obtaining and analyzing quantitative data on geographic variation of populations for several different traits it will be worthwhile to compare these measures of differentiation in different characters on the same set of populations. Thus, for example, the extent of genetic and morphological differentiation can be contrasted. Again, considerable effort is needed in developing techniques: in this case for the efficacious analysis of differences in patterns among the different characters. For instance, if the biochemical results can be interpreted as reflecting time, then the genetic distances could be treated as a null hypothesis to be used in the computation of how much selection has occurred in the evolution of plumage or skeletal variation.

Time, unfortunately, remains the outstanding issue in all attempts to interpret molecular data in terms of evolutionary processes. The problem is that patterns of genic variation do not simply reflect either the current situation or the events at some known time in the past. Rather the observed patterns are the results of a convolution of the current genetic state of the population with the cumulative effects of changes that have taken place over the last 10^5 to 10^6 years (Nei et al. 1975). Thus, many apparently straightforward comparisons may not have simple interpretations. For example, heterozygosity, as discussed earlier, theoretically bears a simple relationship to population size. This could be used to investigate the relative sizes of populations and, consequently, the effects of mating systems and so forth. However, because of its complex relationship with the time course of population size, heterozygosity may strongly reflect a bottleneck of population numbers 50,000 years ago and not current size. A similar argument, not yet quantitatively investigated, applies to comparisons of interpopulational statistics. Major theoretical and empirical studies are needed in this area.

There are other areas of microevolutionary research in which biochemical data may take a contribution. Many will probably only become apparent after much more work is done and comparative data are reported. The availability of techniques for the quantitative analysis of geographic variation of allelic frequencies suggests some areas of taxonomic interest.

Corbin's (1981) suggestion that secondary contact can, in some cases, be detected with electrophoresis is complementary to Schueler and Rising's (1976) similar suggestion of increased variability of morphological characters in hybrid zones. This concept requires additional theoretical and empirical analysis. It would superficially seem, however, that the longer the period of isolation, the greater the possibility of a significant shift in allelic frequencies. Because the increased variability is a function of the differences in allelic frequencies, subspecies based on this criterion would only be those with a significant history of isolation. This seems highly desirable. For such cases, there would be good reason to suspect that the two taxa actually represent different gene pools (Barrowclough 1982). "Nonsense subspecies," those based on small, often clinal, adaptive changes of one or two characters, would not qualify under this criterion.

The analysis of isolates of uncertain species status and of possible sibling species would also be facilitated with biochemical data. Large differences in allelic frequencies among populations, or the fixation of alternate alleles, indicate a long history of isolation and lack of gene flow. Hence, such data can be construed as support for species status of taxa for which other information is lacking or marginal. Again, more empirical studies are necessary. There can be no standards for comparison until there is a wide empirical and theoretical data base for background.

It is necessary to remember that many of these techniques are new and not well established. What we have really learned so far is merely that there is potential in the approach. It may take another decade of work to explore adequately the limits of this potential. Nevertheless, almost anything learned about microevolutionary processes during this exploration will be new.

Appendix. *Estimates of among-population normalized genetic variance* ($\hat{F}_{st}$) *for some approximately continuously distributed species of vertebrates: estimates computed using the method of Wright (1978)*[a]

Species[a]	No. of loci	No. of populations	$\hat{F}_{st}$	Reference
Fish				
Chanos chanos	9	14	0.041	Winans (1980)
Gadua morhua	4	5	0.018	Jamieson (1975)
Menidia menidia	5	5	0.013	Johnson (1975)
Menidia peninsula	8	4	0.104	Johnson (1975)
Oncorynchus gorbuscha	3	32	0.014	Aspinwall (1974)
Platichthys stellatus	2	4	0.006	Johnson and Beardsley (1975)
Salmo clarki bouvieri	6	4	0.208	Loudenslager and Gall (1980)
Salmo clarki henshawi	10	8	0.406	Loudenslager and Gall (1980)
Zoarces viviparous	5	46	0.213	Christiansen et al. (1976), Hjorth and Simonsen (1975)
Amphibians				
Ambystoma macrodactylum	10	8	0.341	Howard and Wallace (1981)
Aneides flavipunctatus	19	15	0.280	Larson (1980)
Desmognathus ochrophaeus	12	29	0.467	Tilley et al. (1978)
Plethodon cinereus	15	15	0.645	Highton and Webster (1976)
Plethodon dorsalis	23	17	0.906	Larson and Highton (1978)
Acris crepitans	3	15	0.837	Dessauer and Nevo (1969), Salthe and Nevo (1969)
Bufo americanus	5	25	0.078	Guttman (1975)
Bufo arenarum	9	15	0.181	Matthews (1975)
Bufo cognatus	5	5	0.022	Rogers (1973)

Appendix (cont.)

Species[a]	No. of loci	No. of populations	$\hat{F}_{st}$	Reference
Bufo punctatus	10	4	0.533	Feder (1979)
Bufo speciosus	5	6	0.019	Rogers (1973)
Hyla chrysoscelis	8	9	0.435	Ralin and Selander (1979)
Hyla regilla	11	17	0.501	Case et al. (1975)
Rana pipiens	4	16	0.457	Dunlap (1981)
Ranidella insignifera	4	10	0.048	Blackwell (1978)
Reptiles				
Anolis carolinensis	12	3	0.122	Webster et al. (1972)
Sceloporus undulatus	7	8	0.310	Spohn and Guttman (1976)
Uta stansburiana	12	14	0.479	McKinney et al. (1972)
Birds				
Zonotrichia leucophrys	12	8	0.032	Corbin (unpub.)
Junco hyemalis	9	6	0.008	Barrowclough (unpub.)
Dendroica coronata	8	5	0.029	Barrowclough (1980b)
Icterus galbula	2	5	0.012	Corbin et al. (1979)
Aplonis metallica	2	6	0.029	Corbin et al. (1974)
Mammals				
Didelphis virginiana	13	3	0.097	Kovacic and Guttman (1979)
Desmodus rotundus	3	4	0.025	Honeycutt et al. (1981)
Macrotus californicus	7	4	0.156	Greenbaum and Baker (1976)
Macrotus waterhousii	6	3	0.096	Greenbaum and Baker (1976)
Macaca fuscata fuscata	8	9	0.051	Nozawa et al. (1975)
Macaca mulatta	4	3	0.064	Darga et al. (1975)
Procyon lotor	4	5	0.016	Dew and Kennedy (1980)
Spermophilus mexicanus	11	10	0.294	Cothran et al. (1977)
Spermophilus spilosoma	11	12	0.310	Cothran et al. (1977)

S. tridecemlineatus	10	10	0.349	Cothran et al. (1977)
Thomomys bottae	10	5	0.169	Patton et al. (1972)
Thomomys umbrinus (76S)	13	3	0.203	Patton and Feder (1978)
Thomomys umbrinus (76N)	11	4	0.340	Patton and Feder (1978)
Thomomys umbrinus (78)	14	6	0.508	Patton and Feder (1978)
Dipodomys merriami	9	7	0.089	Johnson and Selander (1971)
Dipodomys ordii	9	9	0.720	Johnson and Selander (1971)
Peromyscus attwateri	2	8	0.211	Kilpatrick and Zimmerman (1975)
Peromyscus boylii rowleyi	6	8	0.150	Kilpatrick and Zimmerman (1975)
Peromyscus californicus	16	13	0.355	Smith (1979)
Peromyscus maniculatus	6	18	0.161	Avise et al. (1979b)
Peromyscus pectoralis	9	19	0.636	Kilpatrick and Zimmerman (1975)
Peromyscus polionotus	15	25	0.349	Selander et al. (1971)
Sigmodon hispidus	12	8	0.121	Johnson et al. (1972)
Mus musculus	5	9	0.198	Selander et al. (1969)
Alces alces	5	18	0.089	Ryman et al. (1980)

[a]Survey restricted to studies reporting two or more variable protein loci and encompassing a large geographical area.

Literature cited

Abbott, I. J., L. K. Abbott, and P. R. Grant. 1977. Comparative ecology of Galapagos ground finches (*Geospiza* Gould). Evaluation of the importance of floristic diversity and interspecific competition. *Ecol. Monogr.* 47:151–184.

Aspinwall, N. 1974. Genetic analysis of North American populations of the pink salmon, *Oncorhynchus gorbuscha*, possible evidence for the neutral mutation–random drift hypothesis. *Evolution* 28:295–305.

Au, T., and T. E. Stelson. 1969. *Introduction to systems engineering, deterministic models.* Reading, Mass.: Addison-Wesley.

Avise, J. C., C. Giblin-Davidson, J. Laerm, J. C. Patton, and R. A. Lansman. 1979a. Mitochondrial DNA clones and matriarchal phylogeny within and among geographic populations of the pocket gopher, *Geomys pinetis. Proc. Natl. Acad. Sci. USA.* 76:6694–6698.

Avise, J. C., M. H. Smith, and R. K. Selander. 1979b. Biochemical polymorphism and systematics in the genus *Peromyscus.* VII. Geographic differentiation in members of the *truei* and *maniculatus* species groups. *J. Mammal.* 60:177–192.

Avise, J. C., J. C. Patton, and C. F. Aquadro. 1980. Evolutionary genetics of birds. I. Relationships among North American thrushes and allies. *Auk* 97:135–147.

Baker, A. J. 1980. Morphometric differentiation in New Zealand populations of the House Sparrow (*Passer domesticus*). *Evolution* 34:638–653.

Baker, A. J., and A. Moeed. 1979. Evolution in the introduced New Zealand populations of the Common Myna, *Acridothers tristis* (Aves, Sturnidae). *Can. J. Zool.* 57:570–584.

Baker, C. M. A., and C. Manwell. 1975. Molecular biology of avian proteins. XII. Protein polymorphism in the Stubble Quail *Coturnix pectoralis* – and a brief note on the induction of egg white protein synthesis in wild birds by hormones. *Comp. Biochem. Physiol.* 50B:471–477.

Baker, M. C. 1975. Song dialects and genetic differences in White-crowned Sparrows (*Zonotrichia leucophrys*). *Evolution* 29:226–241.

Barrett, V. A., and E. R. Vyse. 1982. Comparative genetics of three Trumpeter Swan populations. *Auk* 99:103–108.

Barrowclough, G. F. 1980a. Gene flow, effective population sizes, and genetic variance components in birds. *Evolution* 34:789–798.

– 1980b. Genetic and phenotypic differentiation in a wood warbler (genus *Dendroica*) hybrid zone. *Auk* 97:655–668.

– 1981. Mammalian population genetics: progress report on a world view in transition. *Evolution* 35:1255–1256.

– 1982. Geographic variation, predictiveness, and subspecies. *Auk* 99:601–603.

Barrowclough, G. F., and K. W. Corbin. 1978. Genetic variation and differentiation in the Parulidae. *Auk* 95:691–702.

Barrowclough, G. F., and K. W. Corbin, and R. M. Zink. 1981. Genetic diffentiation in the Procellariiformes. *Comp. Biochem. Physiol.* 69B:629–632.

Blackwell, J. M. 1978. Intra-specific divergence in the western Australian frog *Ranidella insignifera.* I. The evidence from gene frequencies and genetic distance. *Heredity* 40:339–348.

Boag, P. T., and P. R. Grant 1981. Intense natural selection in a population of Darwin's Finches (Geospizinae) in the Galapagos. *Science 214:*82–85.

Bowman, R. I. 1961. Morphological differentiation and adaptation in the Galapagos finches. *Univ. Calif. Publ. Zool. 58:*1–302.

Brewer, G. J. 1970. *An introduction to isozyme techniques.* New York: Academic Press.

Brown, W. M., M. George, Jr., and A. C. Wilson. 1979. Rapid evolution of animal mitochondrial DNA. *Proc. Natl. Acad. Sci. USA 76:*1967–1971.

Brush, A. H. 1968. Conalbumin variation in populations of the Redwinged Blackbird, *Agelaius phoeniceus. Comp. Biochem. Physiol. 25:*159–168.

– 1970. An electrophoretic study of eggwhites from three blackbird species. *Univ. Conn. Occas. Pap. 1:*243–264.

– 1979. Comparison of egg-white proteins: effect of electrophoretic conditions. *Biochem. Syst. Ecol. 7:*155–165.

Case, S. M., P. C. Haneline, and M. F. Smith. 1975. Protein variation in several species of *Hyla. Syst. Zool. 24:*281–295.

Chakraborty, R., P. A. Fuerst, and M. Nei. 1980. Statistical studies on protein polymorphism in natural populations. III. Distribution of allele frequencies and the number of alleles per locus. *Genetics 94:*1039–1063.

Christiansen, F. B., O. Frydenberg, J. P. Hjorth, and V. Simonsen. 1976. Genetics of *Zoarces* populations. IX. Geographic variation at the three phosphoglucomutase loci. *Hereditas 83:*245–256.

Cole, S. R., and D. T. Parkin. 1981. Enzyme polymorphisms in the House Sparrow, *Passer domesticus. Biol. J. Linn. Soc. 15:*13–22.

Corbin, K. W. 1968. Taxonomic relationships of some *Columba* species. *Condor 70:*1–13.

– 1981. Genic heterozygosity in the White-crowned Sparrow: a potential index to boundaries between subspecies. *Auk 98:*669–680.

Corbin, K. W., C. G. Sibley, and A. Ferguson. 1979. Genic changes associated with the establishment of sympatry in orioles of the genus *Icterus. Evolution 33:*624–633.

Corbin, K. W., C. G. Sibley, A. Ferguson, A. C. Wilson, A. H. Brush, and J. E. Ahlquist. 1974. Genetic polymorphism in New Guinea starlings of the genus *Aplonis. Condor 76:*307–318.

Cothran, E. G., E. G. Zimmerman, and C. F. Nadler. 1977. Genic differentiation and evolution in the ground squirrel subgenus *Ictidomys* (genus *Spermophilus). J. Mammal. 58:*610–622.

Coyne, J. A., W. F. Eanes, J. A. M. Ramshaw, and R. K. Koehn. 1979. Electrophoretic heterogeneity of α-glycerophosphate dehydrogenase among many species of *Drosophila. Syst. Zool. 28:*164–175.

Darga, L. L., M. Goodman, M. L. Weiss, G. W. Moore, W. Pryychodko, H. Dene, R. Tashian, and A. Koen. 1975. Molecular systematics and clinal variation in macaques. In C. L. Markert (ed.). *Isozymes,* Vol. IV, *Genetics and evolution,* pp. 797–812. New York: Academic Press.

Dessauer, H. C., and E. Nevo. 1969. Geographic variation of blood and liver proteins in cricket frogs. *Biochem. Genet. 3:*171–188.

Dew, R. D., and M. L. Kennedy. 1980. Genic variation in raccoons, *Procyon lotor. J. Mammal. 61:*697–702.

Dunlap, D. G. 1981. Geographic variation of proteins and call in *Rana pipiens* from the northcentral United States. *Copeia 1981:*976–879.

Eldredge, N., and J. Cracraft. 1980. *Phylogenetic patterns and the evolutionary process.* New York: Columbia University Press.

Endler, J. A. 1977. *Geographic variation, speciation, and clines.* Princeton, N.J.: Princeton University Press.

Farris, J. S. 1972. Estimating phylogenetic trees from distance matrices. *Am. Nat. 106:*645–668.

Feder, J. H. 1979. Natural hybridization and genetic divergence between the toads *Bufo boreas* and *Bufo punctatus. Evolution 33:*1089–1097.

Felsenstein, J. 1981. Evolutionary trees from gene frequencies and quantitative characters: finding maximum likelihood estimates. *Evolution 35:*1229–1242.

Ferguson, A. 1971. Geographic and species variation in transferrin and ovotransferrin polymorphism in the Columbidae. *Comp. Biochem. Physiol. 38B:*477–486.

Frelin, C., and F. Vuilleumier. 1979. Biochemical methods and reasoning in systematics. *Z. Zool. Syst. Evolutionsforsch. 17:*1–10.

Glaus, K. R., H. P. Zassenhaus, N. S. Fechheimer, and P. S. Perlman. 1980. Avian mtDNA: structure, organization and evolution. In A. M. Kroon and C. Saccone (eds.). *The organization and expression of the mitochondrial genome,* pp. 131–135. Amsterdam: North Holland.

Gould, S. J., and R. C. Lewontin. 1979. The spandrels of San Marco and the Panglossian paradigm: a critique of the adaptationist programme. *Proc. R. Soc. London 205 B:*581–598.

Grant, B. R., and P. R. Grant. 1979. Darwin's finches: population variation and sympatric speciation. *Proc. Natl. Acad. Sci. USA 76:*2359–2363.

– 1982. Niche shifts and competition in Darwin's finches: *Geospiza conirostris* and congeners. *Evolution 36:*637–657.

Grant, P. R. 1981. Speciation and the adaptive radiation of Darwin's finches. *Am. Sci. 69:*653–663.

Greenbaum, I. F., and R. J. Baker. 1976. Evolutionary relationships in *Macrotus* (Mammalia:Chiroptera): biochemical variation and karyology. *Syst. Zool. 25:*15–25.

Gutiérrez, R. J., R. M. Zink, and S. Y. Yang. in press. Genic variation and systematic relationships of some galliform birds. *Auk.*

Guttman, S. I. 1975. Genetic variation in the genus *Bufo.* II. Isozymes in northern allopatric populations of the American toad, *Bufo americanus.* In C. L. Markert (ed.). *Isozymes,* Vol. IV, *Genetics and evolution,* pp. 679–697. New York: Academic Press.

Guttman, S. I., G. A. Grau, and A. A. Karlin. 1980. Genetic variation in Lake Erie Great Blue Herons (*Ardea herodias*). *Comp. Biochem. Physiol. 66B:*167–169.

Gyllensten, U., C. Reuterwall, and N. Ryman. 1979. Genetic variability in Scandinavian populations of willow grouse (*Lagopus lagopus* L.) and rock ptarmigan (*Lagopus mutus* L.). *Hereditas 91:*301.

Handford, P., and F. Nottebohm. 1976. Allozyme and morphological variation in population samples of Rufous-collared Sparrow, *Zonotrichia capensis,* in relation to vocal dialects. *Evolution 30:*802–817.

Hardy, A. C. 1936. *Handbook of colorimetry.* Cambridge, Mass.: MIT Press.

Hedrick, P. W. 1971. A new approach to measuring genetic similarity. *Evolution 25:*276–280.

— 1981. The establishment of chromosomal variants. *Evolution 35:*322–332.

Highton, R., and T. P. Webster. 1976. Geographic protein variation and divergence in populations of the salamander *Plethodon cinereus. Evolution 30:*33–45.

Hjorth, J. P., and V. Simonsen. 1975. Genetics of *Zoarces* populations. VIII. Geographic variation common to the polymorphic loci HbI and Est III. *Hereditas 81:*173–184.

Honeycutt, R. L., I. F. Greenbaum, R. J. Baker, and V. M. Sarich. 1981. Molecular evolution of vampire bats. *J. Mammal. 62:*805–811.

Howard, J. H., and R. L. Wallace. 1981. Microgeographical variation of electrophoretic loci in populations of *Ambystoma macrodactylum columbianum* (Caudata: Ambystomatidae). *Copeia 1981:*466–471.

Hubbard, J. P. 1973. Avian evolution in the aridlands of North America. *Living Bird 12:*155–196.

Ibrahimi, I. M., E. M. Prager, T. J. White, and A. C. Wilson. 1979. Amino acid sequence of California Quail lysozyme. Effect of evolutionary substitutions on the antigenic structure of lysozyme. *Biochemistry 13:*2736–2744.

James, F. C. 1970. Geographic size variation in birds and its relationship to climate. *Ecology 51:*365–390.

Jamieson, A. 1975. Enzyme types of Atlantic cod stocks on the North American banks. In C. L. Markert (ed.). *Isozymes,* Vol. IV, *Genetics and evolution,* pp. 491–515. New York: Academic Press.

Jo, N. in press. Karyotypic analysis of Darwin's finches. In R. I. Bowman and A. E. Leviton (eds.). *Recent advances in Galapagos science.* San Francisco: American Association for the Advancement Science.

Johnson, A. G., and A. J. Beardsley. 1975. Biochemical polymorphism of starry flounders, *Platichthys stellatus,* from the northwestern and northeastern Pacific Ocean. *Anim. Blood Groups Biochem. Genet 6:*9–18.

Johnson, M. S. 1975. Biochemical systematics of the atherinid genus *Menidia. Copeia 1975:*662–691.

Johnson, M. S., and J. L. Brown. 1980. Genetic variation among trait groups and apparent absence of close inbreeding in Grey-crowned Babblers. *Behav. Ecol. Sociobiol. 7:*93–98.

Johnson, W. E., and R. K. Selander. 1971. Protein variation and systematics in kangaroo rats (genus *Dipodomys*). *Syst. Zool. 20:*277–405.

Johnson, W. E., R. K. Selander, M. H. Smith, and Y. J. Kim. 1972. Biochemical genetics of sibling species of the cotton rat (*Sigmodon*). *Stud. Genet. 7:*297–305.

Johnston, R. F., and R. K. Selander. 1971. Evolution in the House Sparrow. II. Adaptive differentiation in North American populations. *Evolution 25:*1–28.

Karig, L. M., and A. C. Wilson. 1971. Genetic variation in supernatant malate dehydrogenase of birds and reptiles. *Biochem. Genet. 5:*211–221.

Kilpatrick, C. W., and E. G. Zimmerman,. 1975. Genetic variation and systematics of four species of mice of the *Peromyscus boylii* species group. *Syst. Zool. 24:*143–162.

258   G. F. BARROWCLOUGH

Kitto, G. B., and A. C. Wilson. 1966. Evolution of malate dehydrogenase in birds. *Science 153:*1408–1410.

Korey, K. A. 1981. Species number, generation length, and the molecular clock. *Evolution 35:*139–147.

Kovacic, D. A., and S. I. Guttman. 1979. An electrophoretic comparison of genetic variability between eastern and western populations of the opossum (*Didelphis virginiana*). *Am. Midl. Nat. 101:*269–277.

Lack, D. 1961. *Darwin's finches.* New York: Harper.

Lande, R. 1976. Natural selection and random genetic drift in phenotypic evolution. *Evolution 30:*314–334.

– 1979. Effective deme sizes during long-term evolution estimated from rates of chromosomal rearrangement. *Evolution 33:*234–251.

– 1980. Genetic variation and phenotypic evolution during allopatric speciation. *Am. Nat. 116:*463–479.

– 1981. Models of speciation by sexual selection on polygenic traits. *Proc. Natl. Acad. Sci. USA 78:*3721–3725.

Lansman, R. A., R. O. Shade, J. F. Shapira, and J. C. Avise. 1981. The use of restriction endonucleases to measure mitochondrial DNA sequence relatedness in natural populations. *J. Mol. Evol. 17:*214–226.

Larson, A. 1980. Paedomorphosis in relation to rates of morphological and molecular evolution in the salamander *Aneides flavipunctatus* (Amphibia, Plethodontidae). *Evolution 34:*1–17.

Larson, A., and R. Highton. 1978. Geographic protein variation and divergence in the salamanders of the *Plethodon welleri* group (Amphibia, Plethodontidae). *Syst. Zool. 27:*431–448.

Lewontin, R. C. 1974. *The genetic basis of evolutionary change.* New York: Columbia University Press.

Lewontin, R. C., and J. Krakauer. 1973. Distribution of gene frequency as a test of the theory of the selective neutrality of polymorphisms. *Genetics 74:*174–195.

Li, W.-H. 1978. Maintenance of genetic variability under the joint effect of mutation, selection and random drift. *Genetics 90:*349–382.

Loudenslager, E. J., and G. A. E. Gall. 1980. Geographic patterns of protein variation and subspeciation in cutthroat trout, *Salmo clarki. Syst. Zool. 29:*27–42.

Manwell, C., and C. M. A. Baker. 1975. Molecular genetics of avian proteins. XIII. Protein polymorphism in three species of Australian passerines. *Aust. J. Biol. Sci. 28:*545–557.

Matthews, T. C. 1975. Biochemical polymorphism in populations of the Argentine toad, *Bufo arenarum. Copeia 1975:*454–465.

McKinney, C. O., R. K. Selander, W. E. Johnson, and S. Y. Yang. 1972. Genetic variation in the side-blotched lizard (*Uta stansburiana*). *Stud. Genet. 7:*305–318.

Mengel, R. M. 1964. The probable history of species formation in some northern wood warblers (Parulidae). *Living Bird 3:*9–43.

Miller, A. H. 1947. Panmixia and population size with reference to birds. *Evolution 1:*186–190.

Milne, H., and F. W. Robertson. 1965. Polymorphisms in egg albumen pro-tein and behavior in the eider duck. *Nature (London) 205:*367–369.

Nei, M. 1975. *Molecular population genetics and evolution.* Amsterdam: North Holland.

– 1978. Estimation of average heterozygosity and genetic distance from a small number of individuals. *Genetics 89:*583–590.

Nei, M., and W.-H. Li. 1979. Mathematical model for studying genetic varia-tion in terms of restriction endonucleases. *Proc. Natl. Acad. Sci. USA 76:* 5269–5273.

Nei, M., T. Maruyama, and R. Chakraborty. 1975. The bottleneck effect and genetic variability in populations. *Evolution 29:*1–10.

Nevo, E. 1978. Genetic variation in natural populations. *Theor. Popul. Biol. 13:*121–177.

Niles, D. M. 1973. Adaptive variation in body size and skeletal proportions of Horned Larks of the southwestern United States. *Evolution 27:*405–426.

Nozawa, K., T. Shotake, and Y. Okura. 1975. Blood protein polymorphisms and population structure of the Japanese macaque, *Macaca fuscata fuscata.* In C. L. Markert (ed.). *Isozymes,* Vol. IV, *Genetics and evolution,* pp. 225–241. New York: Academic Press.

O'Brien, S. J., M. H. Gail, and D. L. Levin. 1980. Correlative genetic variation in natural populations of cats, mice and men. *Nature (London) 288:*580–583.

O'Donald, P. 1973. A further analysis of Bumpus' data: the intensity of nat-ural selection. *Evolution 27:*398–404.

Patton, J. L., and J. H. Feder. 1978. Genetic divergence between populations of the pocket gopher, *Thomomys umbrinus* (Richardson). *Z. Säugetierkd. 43:*17–30.

Patton, J. L., R. K. Selander, and M. H. Smith. 1972. Genic variation in hy-bridizing populations of gophers (genus *Thomomys*). *Syst. Zool. 21:*263–270.

Patton, J. L., and S. Y. Yang. 1977. Genetic variation in *Thomomys bottae* pocket gophers: macrogeographic patterns. *Evolution 31:*697–720.

Power, D. M. 1970. Geographic variation of Red-winged Blackbirds in central North America. *Univ. Kans. Publ. Mus. Nat. Hist. 19:*1–83.

Prager, E. M., A. H. Brush, R. A. Nolan, M. Nakanishi, and A. C. Wilson. 1974. Slow evolution of transferrin and albumin in birds according to micro-complement fixation analysis. *J. Mol. Evol. 3:*243–262.

Prager, E. M., and A. C. Wilson. 1975. Slow evolutionary loss of the potential for interspecific hybridization in birds: a manifestation of slow regulatory evolution. *Proc. Natl. Acad. Sci. USA 72:*200–204.

Ralin, D. B., and R. K. Selander. 1979. Evolutionary genetics of diploid-tri-ploid species of treefrogs of the genus *Hyla. Evolution 33:*595–608.

Redfield, J. A., F. C. Zwickel, J. F. Bendell, and A. T. Bergerud. 1972. Tem-poral and spatial patterns of allele and genotype frequencies at the *Ng* locus in blue grouse (*Dendragapus obscurus*). *Can. J. Zool. 50:*1657–1662.

Rising, J. D. 1973. Age and seasonal variation in dimensions of House Spar-rows, *Passer domesticus* (L.), from a single population in Kansas. In S. C. Kendeigh and J. Pinowski (eds.). *Productivity, population dynamics, and syste-matics of granivorous birds,* pp. 327–336. Warsaw: Polish Science Publications.

Robins, J. D., and G. D. Schnell. 1971. Skeletal analysis of the *Ammodramus-Ammospiza* grassland sparrow complex: a numerical taxonomic study. *Auk* 88:567–590.

Rogers, J. S. 1972. Measures of genetic similarity and genetic distance. *Stud. Genet.* 7:145–153.

— 1973. Protein polymorphism, genic heterozygosity and divergence in the toads *Bufo cognatus* and *B. speciosus*. *Copeia 1973*:322–330.

Ryman, N., C. Reuterwall, K. Nygren, and T. Nygren. 1980. Genetic variation and differentiation in Scandinavian moose (*Alces alces*): are large mammals monomorphic? *Evolution 34*:1037–1049.

Salthe, S. N., and E. Nevo. 1969. Geographic variation of lactate dehydrogenase in the cricket frog, *Acris crepitans*. *Biochem. Genet. 3*:335–341.

Sarich, V. M. 1977. Rates, sample sizes, and the neutrality hypothesis for electrophoresis in evolutionary studies. *Nature (London) 265*:24–28.

Schnell, G. D., and R. K. Selander. 1981. Environmental and morphological correlates of genetic variation in mammals. In M. H. Smith and J. Joule (eds.). *Mammalian population genetics, pp. 60–99. Athens: University of Georgia Press.*

Schueler, F. W., and J. D. Rising. 1976. Phenetic evidence of natural hybridization. *Syst. Zool. 25*:283–289.

Selander, R. K. 1971. Systematics and speciation in birds. In D. S. Farner and J. R. King (eds.). *Avian biology*, Vol. I, pp. 57–147. New York: Academic Press.

Selander, R. K., and W. E. Johnson. 1973. Genetic variation among vertebrate species. *Annu. Rev. Ecol. Syst. 4*:75–91.

Selander, R. K., M. H. Smith, S. Y. Yang, W. E. Johnson, and J. B. Gentry. 1971. Biochemical polymorphism and systematics in the genus *Peromyscus*. I. Variation in the old-field mouse (*Peromyscus polionotus*). *Stud. Genet. 6*:49–90.

Selander, R. K., S. Y. Yang, and W. G. Hunt. 1969. Polymorphism in esterases and hemoglobin in wild populations of the house mouse (*Mus musculus*). *Stud. Genet. 5*:271–338.

Shields, G. F. 1982. Comparative avian cytogenetics: a review. *Condor 84*:45–58.

Shields, G. F., and N. A. Straus. 1975. DNA-DNA hybridization studies of birds. *Evolution 29*:159–166.

Sibley, C. G. 1970. A comparative study of the egg-white proteins of passerine birds. *Peabody Mus. Nat. Hist. Bull. 32*:1–131.

Sibley, C. G., and J. E. Ahlquist. 1981. The phylogeny and relationships of the ratite birds as indicated by DNA-DNA hybridization. In G. G. E. Scudder and J. L. Reveal (eds.). *Evolution today, proceedings of the Second International Congress of Systematic and Evolutionary Biology,* pp. 301–335. Pittsburgh, Pa.: Carnegie-Mellon University Press.

Sibley, C. G., and K. W. Corbin. 1970. Ornithological field studies in the Great Plains and Nova Scotia. *Discovery 6*:3–6.

Sibley, C. G., K. W. Corbin, J. E. Ahlquist, and A. Ferguson. 1974. Birds. In C. A. Wright (ed.). *Biochemical and immunological taxonomy of animals,* pp. 89–176. New York: Academic Press.

Sibley, C. G., and C. Frelin. 1972. The egg white protein evidence for ratite affinities. *Ibis 114:*377–387.

Smith, J. K., and E. G. Zimmerman. 1976. Biochemical genetics and evolution of North American blackbirds, family Icteridae. *Comp. Biochem. Physiol. 53B:* 319–324.

Smith, J. M. 1966. Sympatric speciation. *Am. Nat. 100:*637–650.

Smith, M. F. 1979. Geographic variation of genic and morphological characters in *Peromyscus californicus. J. Mammal. 60:*705–722.

Sneath, P. H. A., and R. R. Sokal. 1973. *Numerical taxonomy.* San Francisco: Freeman.

Spohn, R. T., and S. I. Guttman. 1976. An electrophoretic study of inter- and intrapopulation genetic variation in the northern fence swift, *Sceloporus undulatus hyacinthinus. Comp. Biochem. Physiol. 55B:*471–474.

Swarth, H. S. 1931. The avifauna of the Galapagos Islands. *Occas. Pap. Calif. Acad. Sci. 18:*1–299.

Templeton, A. R. 1980a. The theory of speciation via the founder principle. *Genetics 94:*1011–1038.

– 1980b. Modes of speciation and inferences based on genetic distances. *Evolution 34:*719–729.

– 1981. Mechanisms of speciation–a population genetic approach. *Annu. Rev. Ecol. Syst. 12:*23–48.

Tilley, S. G., R. B. Merritt, B. Wu, and R. Highton. 1978. Genetic differentiation in salamanders of the *Desmognathus ochrophaeus* complex (Plethodontidae). *Evolution 32:*93–115.

Vohs, P. A., Jr., and L. R. Carr. 1969. Genetic and population studies of transferrin polymorphisms in Ring-necked Pheasants. *Condor 71:*413–417.

Wake, D. B. 1981. The application of allozyme evidence to problems in the evolution of morphology. In G. G. E. Scudder and J. L. Reveal (eds.). *Evolution today, Proceedings of the Second International Congress of Systematic Evolutionary Biology,* pp. 257–270. Pittsburgh, Pa: Carnegie-Mellon University Press.

Webster, T. P., R. K. Selander, and S. Y. Yang. 1972. Genetic variability and similarity in the *Anolis* lizards of Bimini. *Evolution 26:*523–535.

White, M. J. D. 1978. *Modes of speciation.* San Francisco: Freeman.

Wilson, A. C., S. S. Carlson, and T. J. White. 1977. Biochemical evolution. *Annu. Rev. Biochem. 46:*573–639.

Winans, G. A. 1980. Geographic variation in the milkfish *Chanos chanos.* I. Biochemical evidence. *Evolution 34:*558–574.

Wright, S. 1970. Random drift and the shifting balance theory of evolution. In K. Kojima (ed.). *Mathematical topics in population genetics,* pp. 1–31. New York: Springer-Verlag.

– 1978. *Evolution and the genetics of populations,* Vol. IV, Chicago. University of Chicago Press.

Yang, S. Y., and J. L. Patton. 1981. Genic variability and differentiation in the Galapagos finches. *Auk 98:*230–242.

Zink, R. M. 1982. Patterns of genic and morphological variation among sparrows in the genera *Zonotrichia, Melospiza, Junco,* and *Passerella. Auk 99:*632–649.

Commentary

JOHN C. AVISE

It is appropriate that this American Ornithologists' Union centennial volume *Perspectives in Ornithology* should include summaries of recent work on molecular genetics of birds. A molecular approach to the study of avian systematics and evolution provides at least two new perspectives not available in the more traditional ornithological literature dealing with morphology, behavior, ecology, and biogeography. First, molecular techniques frequently yield information concerning specific arrays of genes or their products; hence, there is little ambiguity about the particular genetic bases of the characters studied or the amount of genetic information assayed. Second, because analogous, and often homologous, sets of genes and proteins can often be assayed, molecular techniques can provide "common yardsticks" for quantitatively contrasting genetic characteristics of phylogenetically distinct arrays of organisms. For example, in Chapter 7, Barrowclough summarized the allozyme-based estimates of population differentiation in birds and other vertebrates. Compared to most nonavian vertebrates, the magnitude of genetic differentiation among continuously distributed conspecific avian populations is remarkably small ($\hat{F}_{st}$ values in Table 7.3 and Figure 7.2).

Barrowclough also alluded briefly to the small genetic distances commonly observed between closely related avian species. Here I will explicitly summarize this literature and compare it to similar data gathered for the nonavian vertebrates. I will initially confine this summary to estimates of genetic divergence based on multilocus protein electrophoresis, by far the most commonly employed molecular technique to date. This brief commentary is based on a more comprehensive presentation by Avise and Aquadro (1982), which should be consulted by interested readers.

Genetic distances, derived from conventional electrophoretic analyses of proteins encoded by 19 or more loci, are now available for at least 12 avian genera. Comparable data, obtained from electrophoretic analyses of similar arrays of proteins, are also available for more than 30 nonavian vertebrate genera. Results are presented in Figure 7.4. Some strong trends are apparent. Avian congeners are very "conservative" in level of protein divergence relative to most nonavian vertebrate congeners. The contrast with amphibians and most reptiles is particularly

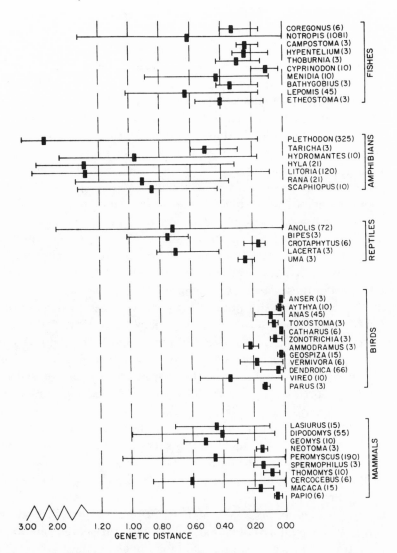

Figure 7.4. Means (and ranges) of genetic distance between species within assayed vertebrate genera, plotted along a common scale of Nei's (1972) distance
statistic. Numbers of pairwise comparisons of species are given in parentheses.

striking. Amphibian congeners are roughly 10–20 times more divergent in protein composition (electrophoretically assayed) than are
species of birds placed within a taxonomic genus. Overall, across the
five vertebrate classes, there is a fairly strong correlation between magnitude of protein differentiation among conspecific populations and

magnitude of protein differentiation among congeneric species (Figure 7.5). I doubt that the remarkably conservative pattern of genetic divergence in birds would have been predicted or anticipated 10 years ago.

There are at present few other molecular data germane to the corroboration (or refutation) of this conservative pattern of divergence among closely related avian species. A recent search for "hidden" protein variation in the thrush genus *Catharus*, using a variety of electrophoretic running conditions, failed to distinguish any new electromorphs that had not already been detected under the standard electrophoretic assay (Aquadro and Avise 1982). Immunological techniques such as microcomplement fixation have been employed to quantitate protein relationships among avian families and orders, and, at these higher taxonomic levels as well, protein distances have proved to be unexpectedly small (Prager et al 1974; Wilson et al. 1977; Prager and Wilson 1980). Charles Sibley has initiated an ambitious survey of avian single-copy DNAs, using DNA hybridization techniques. At the time of this writing, only part of these data is as yet published (e.g., Sibley and Ahlquist 1980, 1981), but within a short time a comparative review of nuclear DNAs in birds versus other vertebrate classes may be attempted. Newly introduced restriction enzyme assays of mitochondrial DNA (mtDNA) hold great promise for comparing closely related avian species (Lansman et al. 1981), but to date only a single study has been completed. Glaus et al. (1980) report estimates of mtDNA nucleotide sequence divergence (p) among genera and families of galliform birds to be in the range of $p=0.09-0.18$. Although these values are no longer than the p's between the mtDNA's of two congeneric rat species (*Rattus rattus* and *R. norvegicus*; Brown and Simpson 1981), it remains to be seen whether avian mtDNA will generally prove to be conservative relative to mtDNA's of other vertebrates.

Documentation of a conservative pattern of protein evolution in birds will come more easily than will a satisfactory understanding of the evolutionary processes responsible for the conservatism. One of many conceptual frameworks for dealing with some of the issues is presented in Table 7.6. Fundamentally, two alternative hypotheses might explain protein conservatism in birds: either avian congeners are, on the average, much younger (i.e., have shared a common ancestor more recently) than most nonavian vertebrate congeners or else rate of protein evolution is decelerated in birds relative to other vertebrates. A variety of corollaries, ramifications, and experimental tests flow from these alternatives (Table 7.6).

The case for a uniform "electrophoretic protein clock" is, in my

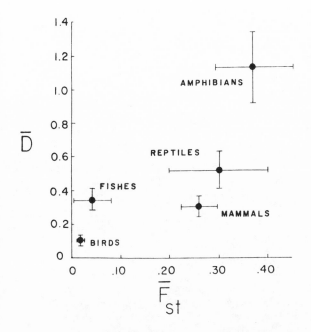

Figure 7.5. Observed mean levels of genetic differentiation among conspecific vertebrate populations ($\overline{F}_{st}$) and among congeneric species per vertebrate genus ($\overline{D}$). $\overline{F}_{st}$ values are from Chapter 7, and $\overline{D}$ values are from Figure 7.4 (see also Avise and Aquadro 1982). Horizontal and vertical bars indicate ± 1 SE.

opinion, overstated in the literature. Actually, a number of widely different clock calibrations have been employed by different authors (Avise and Aquadro 1982) although almost every study that has addressed the issue has concluded that protein-based estimates of times of species divergence correspond very well to independent estimates of divergence times obtained from fossil or zoogeographic evidence. Figure 7.6 plots two distinct but commonly employed electrophoretic protein clocks against genetic distance. From these two clocks that might be employed, one could conclude that an average pair of avian congeners last shared a common ancestor anywhere from about 80,000 to 2 million years ago. Some avian paleontologists argue that most living species of birds arose in the Pleistocene (Brodkorb 1960; Moreau 1963), but others suggest that many congeneric species diverged in the Miocene or Pliocene (Wetmore 1951; Mengel 1964). According to these same electrophoretic clocks, an average pair of assayed amphibian species would have separated between 1 and 20 million years ago.

If it is true that avian congeners are younger than most nonavian vertebrate congeners, this could in part be an artifact of the taxonomic

Table 7.6. *A conceptual framework for dealing with the empirical observation of a conservative pattern of protein differentiation among closely related avian species*

Fundamental alternative possibilities
A. Avian congeners are younger than most nonavian vertebrate congeners
B. Protein evolution is decelerated in birds
Possible corollaries and ramifications of A
1. Avian genera have been taxonomically "oversplit" relative to other vertebrate genera
2. Avian speciations occur rapidly
3. Rates of morphological (i.e., plumage) and behavioral divergence can be different than rates of protein evolution
4. Morphological and behavioral differences may represent only rather superficial genetic changes
5. Should be reflected in small genetic distances for entire avian genome
Possible corollaries and ramifications of B
1. Uniform electrophoretic clock in vertebrates does not exist
2. May not be reflected in small genetic distances for entire avian genome
3. Rate of protein evolution influenced by intrinsic properties of birds
    a. Ecological (i.e., mating system, generation length)
    b. Evolutionary (i.e., mode of speciation)
    c. Genetic (i.e., heterozygosity, mutation rate)
    d. Physiological (i.e., body temperature)
Possible empirical tests of A versus B
1. Examine other portions of avian genome (i.e., by restriction enzymes, sequencing, DNA hybridization)
2. More careful study of fossil/biogeographic evidence of ages of avian speciations

---

procedures employed by avian and other systematists. If avian taxonomic procedure has involved excessive "splitting" at the generic level, genetic distances among avian congeners would be truncated at low values for this reason alone. By the criterion of protein distances as electrophoretically assayed, avian genera *are* taxonomically oversplit relative to amphibian genera, for example. By what other criteria could we ascertain whether avian genera are oversplit? As already noted, molecular techniques provide one of the few common yardsticks for comparing groups of organisms as different as birds and amphibians. By the criterion of census numbers of living species per genus, birds are oversplit relative to reptiles and amphibians but are not oversplit relative to fishes and mammals (Avise and Aquadro 1982).

Many avian congeners, such as *Dendroica* warblers or *Geospiza* finches, are strikingly different in plumage coloration, behavior, bill size and shape, and other specializations associated with niche use and sexual isolation. Yet those arrays of species are very similar in protein composition. If the small protein distances do indeed reflect recency

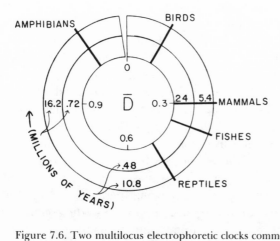

Figure 7.6. Two multilocus electrophoretic clocks commonly employed in the literature. Inner circle; genetic distance (Nei's, 1972, D); outer circles; times of species divergence according to the two extreme clocks. Overall mean genetic distances between congeneric vertebrates species (see Figure 7.5) are indicated by solid bars.

of speciation, then the entire generic diversity in morphology and behavior must have arisen rapidly, certainly within 2 million years. The process of speciation itself must then have been rapid. Perhaps the ecological differences among these bird species stem from rather superficial genetic changes, involving relatively few genes. In any event, morphological divergence and protein divergence can apparently occur at independent rates. This theme has long been argued by Wilson and colleagues (e.g., Wilson et al 1977).

On the other hand, if the protein similarities do not reflect unusually recent speciations among birds, avian protein evolution must be decelerated. Avian species must then differ from most other vertebrates in some intrinsic characteristic that is related to and influences protein divergence. Such characteristics are not readily apparent; among the vertebrates, counterexamples to most proposed ecological, evolutionary, or genetic properties that might be unique to or prevalent among birds can readily be advanced. However, I find one uniquely avian characteristic appealing as a candidate for possibly inhibiting rate of amino acid substitution: body temperature. Perhaps the relatively high and stable body temperatures of most birds provide an extreme internal physiological environment to which only a small number of alternative forms of an enzyme or other protein are functionally well adapted. In such a stringent environment, most amino

acid substitutions would be selected against. This hypothesis is motivated by the important effect that the increased thermal energy of
high-temperature regimes is known to have on weak chemical bonds
of protein structure, rates of chemical reactions, and the ordering of
enzyme–substrate systems (Hochachka and Somero 1973; Somero
1978). Mammals are also homeothermic, but have body temperatures
lower than most birds by about 3–10°; more mammals also exhibit
various forms of torpor or hibernation and hence experience fluctuating intracellular temperature regimes during their lifetimes.

The temperature-regime hypothesis has a corollary that may prove
especially relevant to new empirical studies of total avian DNA. If high
temperatures primarily influence enzyme operation and structure, the
conservative pattern of divergence in protein-coding genes may be
atypical of the entire avian genome, only a small fraction of which is
involved in protein-coding function. Alternatively, if the conservative
pattern of protein divergence in birds reflects recent common ancestry
rather than a decelerated rate of divergence, the conservative magnitude of genetic divergence should be genome-wide. For these reasons,
it will be most interesting to learn results of new genetic work on
restriction enzyme and hybridization assays of avian DNAs.

The hypotheses of recent speciation versus protein deceleration are
extreme alternatives to account for the empirical pattern of conservatism in protein divergence within and among closely related avian
species. In reality, the final answer may involve elements of both possibilities. Thus, avian congeners are probably much younger than, for
example, amphibian congeners (in part, because of taxonomic conven-
tions of generic limits in the groups), but it is also likely that rate of
protein evolution in birds is somewhat slower. In the future, I see a
critical need for more empirical data along two fronts. The new techniques of DNA assay must be applied to birds and other organisms,
with care taken to assure that methods of data acquisition and analysis
in different groups are sufficiently similar that meaningful direct comparisons can be made. There is a critical need also for better non–
molecular-based estimates of ages of species separation. The major
impediment to resolution of the possibility of a uniformly calibrated
molecular clock is the lack of critical fossil-based or biogeographic
knowledge of avian (or other vertebrate) speciation times. I am less
optomistic about progress in this second area. The fossil record for
birds is notoriously poor, and the fossil remains are often inadequate to
provide much more than generic or subfamilial assignments. Nonetheless, increased communication between molecular evolutionists and

avian paleontologists is desirable. The recent, unanticipated findings about molecular divergence in birds should provide a strong stimulus for greater interaction.

Literature cited

Aquadro, C. F., and J. C. Avise. 1982. Evolutionary genetics of birds VI. A reexamination of protein divergence using varied electrophoretic conditions. *Evolution 36:*1003–1019.

Avise, J. C., and C. F. Aquadro. 1982. A comparative summary of genetic distances in the vertebrates: patterns and correlations. In M. K. Hecht, B. Wallace, and G. T. Prance (eds.). *Evolutionary biology*, Vol. 15, pp. 151–185. New York: Plenum.

Brodkorb, P. 1960. How many species of birds have existed? *Fl. St. Mus. Bull.* 5:41–53.

Brown, G. G., and M. V. Simpson. 1981. Intra- and interspecific variation of the mitochondrial genome in *Rattus norvegicus* and *Rattus rattus:* restriction enzyme analysis of variant mitochondrial DNA molecules and their evolutionary relationships. *Genetics 97:*125–143.

Glaus, K. R., H. P. Zassenhaus, N. S. Fechheimer, and P. S. Perlman. 1980. Avian mtDNA: structure, organization and evolution. In A. M. Kroon and C. Saccone (eds.). *The organization and expression of the mitochondrial genome*, pp. 131–135. Amsterdam: Elsevier/North Holland.

Hochachka, P. W., and G. N. Somero. 1973. *Strategies of biochemical adapation.* Philadelphia: Saunders.

Lansman, R. A., R. O. Shade, J. F. Shapira, and J. C. Avise. 1981. The use of restriction endonucleases to measure mitochondrial DNA sequence relatedness in natural populations. III. Techniques and potential applications. *J. Mol. Evol. 17:*214–226.

Mengel, R. M. 1964. The probable history of species formation in some northern wood warblers (Parulidae). *Living Bird 3:*9–43.

Moreau, R. E. 1963. The distribution of tropical African birds as an indicator of past climatic changes. In F. C. Howell and F. Bourliere (eds.). *African ecology and human evolution*, pp. 28–42. Chicago: Aldine.

Nei, M. 1972. Genetic distance between populations. *Am. Nat. 106:*283–292.

Prager, E. M., A. H. Brush, R. A. Nolan, M. Nakanishi, and A. C. Wilson. 1974. Slow evolution of transferrin and albumin in birds according to microcomplement fixation analysis. *J. Mol. Evol. 3:*263–278.

Prager, E. M., and A. C. Wilson. 1980. Phylogenetic relationships and rates of evolution in birds. In *Acta XVII Congr. Int. Ornithol.*, pp. 1209–1214.

Sibley, C. G., and J. E. Ahlquist. 1980. The relationships of the "primitive insect eaters" (Aves: Passeriformes) as indicated by DNA × DNA hybridization. In *Acta XVII Congr. Int. Ornithol.*, pp. 1215–1219.

Sibley, C. G. and J. E. Ahlquist. 1981. The phylogeny and relationships of the ratite birds as indicated by DNA-DNA hybridization. In G. Scudder and J. Reveal (eds.). *Evolution today, proceedings of the 2nd International Congress of*

Systematic Evolutionary Biology, pp. 303–337. Pittsburgh, Pa.: Carnegie-Mellon University Press.

Somero, G. N. 1978. Temperature adaptation of enzymes. *Annu. Rev. Ecol. Syst. 9:*1–29.

Wetmore, A. 1951. Recent additions to our knowledge of prehistoric birds, 1933–1949. In *Proceedings of the X International Ornithological Congress, Uppsala*, pp. 51–74.

Wilson, A. C., S. S. Carlson, and T. J. White. 1977. Biochemical evolution. *Annu. Rev. Biochem. 46:*573–639.

8   Organization of the avian genome

GERALD F. SHIELDS

The term "genome" has been used in its most restrictive context by eukaryotic geneticists to refer to the genetic component of a complete, or haploid, set of chromosomes of an individual. Alternatively, discussions of the genome might involve considerations of the amount of DNA per haploid nucleus or the morphology and number of an entire (diploid) complement of chromosomes. Furthermore, the genome might be discussed in terms of how DNA sequences are arranged in relation to one another and how individual genes or groups of genes are structured and how they function. The broader use of the term "genome" will be emphasized here, because it is the purpose of this review to interrelate the several aspects of the avian genome as we now understand them in terms of their structure and apparent function.

With the exception of a number of domesticated species of economic or special interest, studies directed at an understanding of the formal genetics of birds have not been widespread. A tremendous amount of information exists, however, concerning the distribution, morphology, reproductive biology, and behavior of wild birds. Thus, it seems that birds would be excellent models for study at genetic and molecular levels. Recent innovations in technique, particularly those directed at an understanding of the organization of the genome, have resulted in significant progress.

Genome size

Amounts of DNA in nuclei of cells have traditionally been determined by measuring the uptake of the DNA-specific Feulgen stain through microdensitometry. Results have been reported in two forms: either in picograms (1.0×10^{-12} g) of DNA per cell (haploid) or as the number of nucleotides of DNA per haploid cell. Genome size in the nearly 50 species of birds studied appears constant regardless of

271

chromosome number. Modal haploid DNA amounts for birds are typically 1.8 pg, which is about half that of reptiles and only one-third that of mammals (Atkin et al. 1965; Bachman et al. 1972a,b; Sparrow et al. 1972). Although the ratio between the largest and smallest genome sizes is small (1.3 to 1), species at the extremes of the range of variation in diploid chromosome number have not been studied in this way. Several authors have associated the apparent decrease in genome size of birds and mammals relative to other vertebrates with increased organismal complexity (Ohno 1970; Hinegardner 1976) although the evidence is indirect. The relevance of genome size variability to the organization of DNA is discussed in a later section.

Chromosomes

Our knowledge of the chromosomes of birds is limited, since only about 4% of the nine thousand extant species have been studied and the large majority of these in only a cursory way. Knowledge of the extent of variability in chromosome morphology and number in birds has been summarized recently (Shields 1982).

Birds are unique as a vertebrate class because all species thus far studied possess both ordinary macrochromosomes and at least several pairs of minute microchromosomes (<1 μm long). Microchromosomes also occur in reptiles as well as some holocephalian, chondrostean, and holostean fishes. Accumulated evidence for birds suggests that their microchromosomes are stable A (essential)-type chromosomes and should not be confused with the more variable B-type chromosomes, some of which are minute, unstable, and variable and occur in a variety of other types of organisms. Nonetheless, due to the difficulty in counting and identifying the small microchromosomes, no studies have been directed at determining whether or not B chromosomes exist in birds.

Shoffner (1974) suggested that retention of so many microchromosomes in birds is inefficient because more spindle fibers and centromeres must be maintained than if there were fewer and larger chromosomes. Since the presence of microchromosomes in birds seems universal, they might possibly be maintained because they preserve selectively favorable linkage groups whose function might be lost by coalescence of chromosomes. Some species possess few microchromosomes and the reduction in microchromosome number appears not to have been accompanied by increase in number and size of macrochromosomes. The Stone Curlew (*Burhinus oedicnemus*) possesses the lowest diploid chromosome number ($2n=40$) of any bird studied. The karyotype includes only four sets of microchromosomes, yet the remaining 12 sets of macrochromosomes appear identical in morphology and number to other species of shore birds that possess large numbers of microchromosomes (Bulatova et al. 1971; Bulatova, pers. comm.). On the other hand, species with high diploid chromosome numbers have many microchromosomes and few macrochromosomes. The Hoopoe (*Upupa epops*) possesses 60 pairs of microchromosomes and only 3 of macrochromosomes (Misra and Srivastava 1975). Comparative studies of differentially banded chromosomes of closely related species should reveal the extent to which chromosomes are fused or fragmented.

Clearly, birds have conserved their total chromosome number. Nearly 70% of all birds studied possess diploid chromosome numbers in the range 76–84 (Shields 1980). This conservation of total chromosome number may not, however, be a direct consequence of the considerable range in size of chromosomes in birds. Lizard chromosomes are similar in size to those of birds; yet lizards are far more variable in chromosome numbers (White 1973).

Some evidence suggests differences in function for macro- and microchromosomes. Chicken (*Gallus gallus*) microchromosomes appear to replicate their DNA earlier in the synthesis (S) period of the cell cycle than do macrochromosomes (Schmid 1962; Donnelly and Newcomer 1963). Alternatively, Galton and Bredbury (1966) observed late replication of DNA of microchromosomes in the Rock Dove (*Columba livia*). Variation in labeling time between macro- and microchromosomes also occurred in six species of Passeriformes (Takagi 1972).

Stefos and Arrighi (1974) used DNA reassociation, density gradient centrifugation, and in situ hybridization of chromosomes to reveal that in the chicken repeated DNA is localized primarily in the centromeric regions of microchromosomes and not the macrochromosomes. This situation is in marked contrast to that in mammals where centromeric regions of all chromosomes almost always contain considerable amounts of highly repeated DNA. The repetitious DNA fraction of the chicken comprised 17% of the genome and formed two separate peaks upon CsCl density gradient analysis, a heavy subfraction (1.7152 g cm^{-3}) localized on the W sex chromosome and a lighter subfraction (1.6985 g cm^{-3}) localized at the centromeric regions of the microchromosomes.

Such localization studies directed at specifically determining the chromosomal localization of DNA types are sorely needed in birds. It is known that the W sex chromosome in some female birds contains significant amounts of C band heterochromatin (Stephos and Arrighi 1971; Shields et al. 1982). In such studies, the normally cryptic W chromosome is nearly totally C banded, whereas other chromosomes of comparable size are banded only at their centromeres or not at all. A comparative study of bird sex chromosomes using current differential banding studies and in situ DNA hybridization techniques would be informative, because sex chromosomes in birds have been unequivocally identified in only a few of the nine thousand extant species.

DNA reassociation

Much of what we know about the organization of DNA sequences in the eukaryotic genome has been a direct result of the application of nucleic acid reassociation techniques (Britten and Kohne 1968). Accordingly, DNA and or RNA molecules are isolated, purified, and subsequently reassociated in vitro in an attempt to better understand their structural and ultimately functional features as they exist in the native state.

Although our knowledge of genome sizes, chromosome morphologies, and numbers in birds appears meager due to the relatively small numbers of species analyzed, our knowledge of the molecular organization of DNA sequences in birds parallels that obtained for mammals. Most of the detail of this work is based on investigation of the chicken genome, and although some features of the molecular organization of DNA of this species differ considerably from representatives of other vertebrate groups, there seems little reason to doubt that what we learn from analysis of chicken DNA will be true for most other birds as well.

The first analysis of DNA reassociation in solution was completed as part of a study to estimate the DNA sequence relatedness of several problematic taxa in the genus *Junco* (Shields and Straus 1975). In such studies, which have now been done for representatives of nearly every major phylogenetic lineage of eukaryotes, nuclear DNA is isolated in pure form, fractionated into appropriate piece size (usually 450–500 nucleotides, NT) for reassociation, melted briefly at 100°, and then allowed to reassociate to completion in an appropriate incubation regime. This is the methodological basis behind the construction of the so-called $C_o t$ plot. One of the most significant observations to result from such studies in eukaryotes was that a major portion of DNA reassociated far more rapidly than expected (Britten and Kohne 1968). It was hypothesized and later proven by a variety of methods that this rapidly reassociating fraction was repeated DNA, identical or similar sequences of DNA that were present many thousands of times and in some cases even a million times in the genome. Subsequently, it became clear that this repeated fraction was a heterogenous assemblage of sequences varying not only in number but in length as well. For convenience, the repeated DNAs have been divided into the following categories: simple sequence, highly repeated DNA (including satellite DNA), and intermediately repeated DNA, which are separate from the unique, single-copy components, which are represented only once in the genome.

A feature that appears to be common to birds is their possession of relatively small amounts of repeated DNA (Shields and Straus 1975; Eden and Hendrick 1978; Arthur and Straus 1978). This feature may in part be associated with the reduced genome size of birds compared to other vertebrates; amphibians and fish possessing large genomes correspondingly possess large amounts of repeated DNA.

The very rapidly reassociating fraction of DNA often consists of sequences reiterated many times, in some mammals 10^6 times per genome. These sequences are often represented in the satellite DNA components, which differ in buoyant density gradient centrifugation from main band DNA. These components have intrigued molecular biologists, because there is no evidence that they act as DNA coding sequences for proteins. Indeed, their function is unknown and the subject of considerable controversy (Doolittle and Sapienzo 1980; Orgel and Crick 1980).

Highly repeated DNAs have been hypothesized to be involved in chromosome recognition events, rearrangement events, and, more broadly, in speciation events. However, detailed manipulation of satellite DNA sequences in species of rats (Miklos et al. 1980) indicate only that they have an important influence on the meiotic recombination system (reviewed in John and Miklos 1979; Miklos and John 1979; Yamamoto 1979). Very rapidly reassociating sequences often contain foldback (palindromic) sequences. Self-complementary sequences are often located at the 5' terminal region of eukaryotic and prokaryotic genes. Those DNA sequences, reiterated from 10^1 to 10^5 times, are often very heterogeneous in composition, and it is believed that closely related species whose genome sizes differ probably differ because of variations in the kinds and amounts of middle repetitive DNA (Schmidtke et al. 1979). Few comparative analyses have been conducted on avian repeated DNAs. However, Eden and Hendrick (1978) found that one-third and two-thirds of the chicken repeated DNA sequences are repeated 15- and 1500-fold, respectively. Further, Eden et al. (1978) found that sequences present in repeated DNA of chickens were represented by a factor of 1/100 in Ostrich repeated DNA. Thus, they conclude that there has been a substantial change in repetition frequency of many repeated sequences during the course of avian evolution. The fact that the genome of birds is small by vertebrate standards and that birds seem to possess unusually small amounts of repeated DNA argues for the idea that bird genomes might provide biologists with a unique model for the study of control of gene expression.

DNA sequence organization

The fact that eukaryotes possessed repeated and unique sequences automatically led to the question of how these different sequences were arranged in relation to one another in the genome. Were all of the repeated sequences clustered in one part of the genome while the uniques resided elsewhere, or was there a linear alternation of the sequence types from one end of the DNA molecule to another?

The answer to this question came when moderately fragmented DNA was reassociated so that only repeated DNA could react. Large "network" structures were formed that suggested that each strand that reassociated contained several different repeated sequences. These networks contained much of the DNA and thus must have contained repeated sequences interspersed with unique ones (Bolton et al. 1966).

DNA reassociation experiments were later designed (Davidson et al. 1973) to determine specifically the spacial relationships of unique and repeated DNA sequences. Labeled long fragments were reassociated with a great excess of short fragments. These mixtures were incubated to driver DNA $C_o t$'s so that only repetitive sequences could reassociate. The concentration of long fragments was so low that they did not measurably reassociate with one another. After incubation, the fraction of labeled fragments that reassociated was measured. Long reassociated fragments contained repeated sequence elements bound to short DNA fragments. When this procedure was repeated for a variety of labeled fragment lengths, the results suggested that as the fragment size of the labeled DNA increased the amount of nonrepeated sequences increased. The conclusion drawn for a clawed frog, Xenopus laevis, the first animal studied in this way, was that about three-fourths of the nonrepeated DNA is interspersed with three-fourths of the repeated DNA. For X. laevis, the length of single-copy sequence elements in short-period interspersed DNA averages about 700 to 800 nucleotides. The average length of the short interspersed repeated sequences is about 300 nucleotides. From 50 to 55% of the *Xenopus* DNA is organized in this short-period interspersion pattern. Some 25 to 40% of the rest of the genome is arranged in a long-period interspersion pattern. Repeated sequences longer than 1,000 base pairs (bp) occur in less than 10% of the genome.

This *Xenopus*-like interspersion pattern for DNA was subsequently described for a variety of eukaryotic organisms (Davidson et al. 1975a), and it seemed possible that this short-period alternation of repeated and unique sequences might be involved with the manner in which structural genes were expressed (Britten and Davidson 1969). Such observations led to the hypothesis that both repeated and unique DNA sequences in an interspersed pattern might function in control of gene expression through the rearrangement of control elements. Support for this came from the fact that the length of short-period interspersed unique sequences (700–800 NT) fit reasonably well with the presumed average length of a structural gene and that in 2 sea urchin (*Strongylocentrocus purpuratus*) structural genes were shown to be adjacent to intermediately repeated DNA (Davidson et al. 1975b). These observations led to the formulation of a model that predicted that two different types of RNA transcripts of interspersed DNA templates, one including the structural gene transcript and the other lacking it, might interact to form RNA–RNA hybrids. This nuclear hybridization of RNA types functioned in the manner through which structural transcripts were processed (Davidson and Britten 1979). Independent evidence that evolutionary change in structural genes (genes coding for protein), as analyzed by electrophoresis and microcomplement fixation of proteins, seems not to be sufficient enough to explain morphological diversity lent support to the model. Thus, rearrangement of control elements rather than change in structural genes might be at the basis of gross evolutionary change (Wilson 1976). It was hypothesized further that change in control elements of DNA sequences might be reflected in diversity in chromosome number and morphology within a lineage (Bush et al. 1977).

It is now clear, however, that not all organisms have their DNA ordered into a *Xenopus*-like short-period interspersion pattern. The majority of *Drosophila* DNA is organized in a long-period interspersion pattern wherein repeated sequences of at least 5,600 bp are interspersed with unique sequences which average 13,000 bp (Manning et al. 1975; Crain et al. 1976b). Long-period *Drosophila*-like interspersion patterns have now been observed in a variety of organisms and seemingly long-period interspersion correlates with small genome size (Crain et al. 1976b).

Long-period interspersion patterns have been observed in the chicken (Arthur and Straus 1978; Eden and Hendrick 1978; Epplen et al. 1978); domestic Muscovy duck, *Cairina moschata* (Epplen et al. 1978); and the insects *Chrionomus tentans* (Wells et al. 1976), *Apis mellifera* (Crain et al. 1976b), and *Sarcophaga bullata* (Samols and Swift 1979). In the chicken, repeated sequences average from 2 to 4 kilobases (kb), whereas single-copy sequences are about 4.5 kb long in more than 40% of the genome. Longer stretches of unique sequences occur in the remainder of the genome. Genomes of all these animals, however, contain long-period interspersed DNA, which differs significantly in overall length. Their only common feature is that they lack a 0.2- to 0.4-kb-long interspersed repeated segment. It thus seems questionable to assign a firm regulatory role to the short-period *Xenopus*-like pattern. Even more surprising is the recent observation by Ginatulin and Ginatulina (1979) that the Royal Albatross (*Diomedea epomophora*) and the Common Murre (*Uria aalge*) both possess small genomes but very short period interspersed patterns (0.3-kb repeats alternating with 0.15-kb uniques).

Furthermore, some lower eukaryotes have been shown to lack interspersion of DNA sequences all together. Hudspeth et al. (1977) found no interspersion of mold (*Achlya bisexualis*) DNA tested to a fragment length of 135,000 bp. Beauchamp et al. (1979) found no interspersion of nematode (*Panagrellus silusiae*) DNA, which was tested to 2,000 bp in length. It thus seems unlikely that a short-period pattern of DNA interspersion is an absolute prerequisite to gene processing.

Only 5–10% of unique sequence DNA actually codes for protein, yet in many organisms, 40–50% of the unique sequence DNA exists in an interspersed pattern. Thus, it seems likely that too large an amount of DNA is interspersed to represent alternation of control elements and structural genes. Evidence from DNA sequence analysis indicates that most transcriptional units in eukaryotes are too large to fit the short-period interspersion pattern. Analysis of nucleotide sequences of coding and noncoding gene regions should provide definitive answers as to the possibility of an ascribed regulatory role for interspersed sequences of DNA.

Some insects and birds have both small genomes and long-period interspersion patterns. Moreover, many of these species possess small amounts of repeated DNA. It has been suggested (Crain et al. 1976a; Epplen et al. 1978) that these long-period patterns have been created from short-period ones by excision of interspersed repeated and unique DNA. Furthermore, it has been suggested that organisms possessing short-period interspersion patterns retain the potential for specialization; species with long-period patterns have lost this potential. If long-period DNA sequence patterns have, in fact, evolved from short-period ones in the course of specialization, then ecotypically specialized species of birds should be expected to possess long-period patterns of DNA interspersion. The albatross and murre data cited earlier argue against this hypothesis. Much of the difficulty regarding genomic evolution was reviewed recently by Doolittle (1981).

Species that are ecotypic generalists or at least not believed to be specialized might still posses short-period patterns. Interestingly, our studies (Shields and Straus 1975) of passerines revealed a relatively large amount of repeated DNA (35% of the genome) as compared to ranges of 10–20% for other birds. It would be informative to investigate the interspersion patterns of DNA in a group of unspecialized, radiating, species-rich birds for evidence of short-period interspersion. To my knowledge, no such group of birds has been studied in this way. Birds should prove to be especially valuable in such studies, because species complexes can be readily identified at extremes of the specialization continuum.

Specific genes of birds

The recent discovery of the utility of restriction endonuclease enzymes of prokaryotic origin that cleave DNA at specific base pair sequences together with advances in recombinant DNA technology and molecular cloning have finally enabled us to become intimately familiar with the actual sequence of nucleotides in the native DNA molecule. No longer do we have to infer the structure of the gene in terms of RNA or proteins that it produces; sequence structure of the gene in its entirety can be determined outright. Entire genomes of viruses (Sanger et al. 1977) as well as an increasing number of eukaryotic genes have now been sequenced in total. Many significant discoveries have resulted from this research, and there is little doubt that the major structural components of many eukaryotic genes will soon be well understood.

Direct DNA sequence studies have revealed that some genes of prokaryotes are overlapping. That is, the same sequence of DNA can produce portions of different messenger RNA molecules that code for portions of different proteins. Moreover, many gene sequences of eukaryotes are not colinear with the mature messenger RNA molecules that they produce. Sequences of RNA located between eventual coding regions of RNA are cleaved out of the molecule before it matures and leaves the nucleus. Thus, interpretations of genic evolution based on the structure of mature messenger RNA molecules or the proteins they produce are now known to have been misleading, and it seems likely that overlapping as well as split genes would never have been discovered without the development of DNA sequence technology.

These techniques involve the isolation of DNA from the study organism and subsequent digestion using specific restriction endonucleases that cleave the DNA at prescribed regions. A number of restriction enzymes can be used to divide the DNA molecule to workable size, and partial digests using lesser amounts of enzymes and shorter digest incubations result in variable fragment sizes. Double-digests allow for the linear ordering of DNA sequences. This then results in a restriction map of the genome, and the individual fragments are separated by size upon agarose gel electrophoresis. A major problem in studying many eukaryotic genes has been their infrequent occurrence in the genome. Many are believed to be single-copy sequences. However, once the specific sequence is obtained by electrophoresis and blotting, it can be inserted into a bacterial plasmid or phage that can be made to replicate the sequence in quantity. Restriction enzymes are again used to insert the sequence into the vector and to retrieve it in quantity after replication. The sequence can then be characterized as to the order of its nucleotides through a variety of sequencing methods (Sanger and Coulson 1975; Maxam and Gilbert 1977).

As a result of these recent studies, discrete structural components of a number of eukaryotic genes are being described. Thus, promotor regions are being described at the 5′ end of the transcription unit, which seemingly function as a starting point for RNA polymerase activity and the beginning of transcriptional events.

Promotor regions are followed in a 5′ to 3′ direction by so-called leader DNA sequences, which are untranslated sequences that code for the 5′ end of the mature messenger RNA molecule preceding the adenosine–uridine–guanosine (AUG) initiation codon. Next follows a variable number of coding and noncoding sequences, which in total define the functional gene. In most eukaryotic genes thus far sequenced, these regions are split. That is, although the entire region is transcribed, portions in the interior of the initial transcript are removed by cleavage and do not form part of the mature messenger RNA. Thus, this region of the DNA is separated into sequences that give rise to complementary RNA sequences that will eventually code for protein and those that are removed during RNA processing.

The former have been termed "exons" or coding sequences, whereas the latter have been termed "introns" or intervening sequences. The numbers of exons and introns are specific for particular genes, but there is great variety in their number between different genes. For example, rat insulin I gene is composed of two coding regions interrupted by only one intervening sequence (Cordell et al. 1979), whereas the chick conalbumin gene is composed of 17 coding regions separated by 16 intervening sequences (Cochet et al. 1979).

The last coding sequence is followed by a trailing sequence. Homologies between trailing sequences of some genes have been observed. This suggests potential for secondary structure formation and a possible regulatory role. It is not clear, however, if these sequences are transcribed. Polyadenylation of the 3' end of the messenger RNAs of structural genes is a post-transcriptional process.

It has been shown that tubular gland cells in the oviduct of young chickens can be stimulated by estrogen injection to synthesize and secrete large amounts of the major egg white proteins: ovalbumin, ovotransferrin, lysozyme, and ovomucoid (McKnight and Palmiter 1979). Because detailed sequence analysis of any gene is aided by our ability to obtain it or its transcripts in quantity, systems such as oviduct-associated genes have been used to great advantage. The fact that genes for oviduct secretory proteins can be directly stimulated in young hens by injection of hormones has made this system a model for gene sequence study. The majority of this work has been carried out in the laboratories of P. Chambon (Strasbourg), B. W. O'Malley (Houston), and R. Palmiter (Seattle). Spacial restrictions do not allow me to discuss each oviduct gene in detail here. Rather, I will discuss structural components of the genes and attempt to identify similarities and differences as they seem to exist. A discussion of the evolution of these components and their hypothesized role in gene processing is included. Readers are referred to the excellent accounts of this topic in Lewin (1980), Schmidtke and Epplen (1980), and Breathnach and Chambon (1981).

The DNA sequence of the "promotor" region of the chick ovalbumin gene contains a heptanucleotide pattern (TATAAAA) (Gannon et al. 1979), identical to that originally described for the presumed promotor region of the histone gene cluster in *Drosophila* by D. Hogness. Thus, this region has become known as the "Hogness box." An identical sequence has been observed in the same region of the chick ovomucoid gene (Lai et al. 1979). That this sequence has been conserved in evolution and that it may function as a promotor of RNA transcription is supported by the fact that it occurs in mouse β major globin (Konkel et al. 1978); mouse immunoglobin light chain (Bernard et al. 1978); rat insulin I and II (Cordell et al. 1979); adenovirus late (Ziff and Evans 1978), and *Bombyx mori* silk fibroin (Tsuijimoto and Suzuki 1979). These presumed promotor sequences occur about 30 nucleotides upstream from the start of the initiation codons. Interestingly, they are similar in structure to the Pribnow box (TAT$_G^A$ATG) of prokaryotes; however, the latter resides closer (10 bp) to the first coding sequence. The promotor role of the Hogness box has been hypothesized only on the basis of its structural similarity to the Pribnow box. However, elimination of the Hogness box in chick ovalbumin in vitro systems reduces and even halts initiation of transcription. The same effect is observed in β globin and adenovirus late region genes. Mutation of the third base of the Hogness box from T to G reduces transcription by more than 95% in the chick conalbumin gene (Minty and Newmark 1980). Several viral genes do not have Hogness boxes, and some that do (e.g., SV40 virus) seem not to experience inhibited transcription upon deletion of the box. Initiation of transcription is probably a precise interplay between the promotor region and regions both up- and downstream from it.

It seems likely that Hogness boxes are separated from the initiation codon by a leader sequence of variable length. Gannon et al. (1979) have described an untranslated leader sequence of 45 nucleotides in the chick ovalbumin gene. However, Royal et al. (1979) describe a 47-NT leader sequence in ovalbumin. Introns have been observed in some 5′ leader sequences (Lewin 1980). In contrast to ovalbumin, and like other eukaryotic messenger RNAs, conalbumin does not have an untranslated leader sequence encoded in a separate exon at its 5′ end (Cochet et al. 1979). Current investigations are being directed at determining the importance of such leader sequences in transcription by monitoring results of transcription in systems from which they have been removed. Most leader sequences are dispensable under in vivo conditions in simian virus 40 genes (Chambon 1977). Similarly, the first 23–28 nucleotides of the 5′ noncoding region of the β globin messenger RNA are not essential in the wheat germ cell-free translation system. More work is needed to determine the importance of the interaction of leader sequences with other elements of the genome. Complementarity between the 18 S ribosomal RNA sequence, 3′-UAGGAAGGCGU-5′ and sequences of the 5′ noncoding regions of a number of eukaryotic messenger RNAs including ovalbumin RNA has been observed (Cochet et al. 1979).

Increasing numbers of split genes in eukaryotes and their viruses are being reported. They are, however, not universal in occurrence in eukaryotic systems. None has been observed in yeast cytochrome *c*, sea urchin histones, *Drosophila* heat shock genes, or small ribosomal RNA genes of yeast (*Saccharomyces* and *Dictyostelium*) (Lewin 1980). Although introns have been observed in large ribosomal RNA genes of chloroplasts and mitochondria of some species, none has been observed in maize chloroplasts or bacteria.

All of the avian oviduct genes that have been sequenced possess introns. Ovalbumin and ovomucoid genes of chick each possess 8 exons separated by 7 introns, although the lengths of each differ greatly. Lysozyme of chick possesses 4 exons separated by 5 introns. Chick ovotransferrin contains 17 exons and 16 introns. To my knowledge, none of these genes has been sequenced in any other bird species.

Although only a limited number of genes have been sequenced in their entirety, some trends seem to be emerging regarding the structural features of their components. The total length of intervening sequences in most split genes far outnumbers the total length of coding sequences. Ratios of minimal transcript length to length of mature messenger RNAs of chick are 4.0 for ovalbumin, 6.3 for ovomucoid, and 6.0 for lysozyme (Lewin 1980).

Intervening sequences appear to be evolving more rapidly than their associated coding sequences within the same gene. This holds whether one compares sequences of the same gene in different species or multiple copies of related genes in one species. For example, van Ooyen et al. (1979) observed little sequence homology (40–53%) between introns of β globin genes of rabbit and mouse. Homology of coding sequences within these genes was significantly higher (81%). Similar values were reported by Heilig et al. (1980) for components of chick ovalbumin and the linked and related X and Y genes; these data are in general agreement with those from other mammals.

It is difficult to speculate on the relationships of introns. Earlier studies (Jeffreys and Flavell 1977; Mandel et al. 1978; Roop et al. 1978; Woo et al. 1978) suggest that individual introns in globin and ovalbumin genes appear unrelated to one another or to sequences found elsewhere in the genome. However, Cochet et al. (1979) have observed high- and low-frequency repeats both upstream of the chick ovotransferrin gene and within one intron. Moreover, Heilig et al. (1980) have observed what appears to be a tandem repeat of the ovalbumin gene in chicken. They described X, Y, and ovalbumin genes through sequence analysis within a 40-kb region. Four repeat regions were observed (two in X, a third in the 3' region of the poly(A) site of X, and a fourth upstream from the Y leader sequence). Thus, introns may not always be unique as was earlier thought. Hybridization studies show that repeated sequences within the X gene have no apparent relationship with those found between X and Y genes. The four repeats described do not cross-hybridize with apparent repeated sequences found in chick conalbumin. Thus, there seems to be no evidence, at least in this system, that some of these repeated sequences are common regulatory elements for steroid-induced genes. In all globin genes of birds thus far described, the position and occurrence of introns appear to be conserved regardless of the cellular source of DNA (Lewin 1980).

Ends of coding sequences and beginnings of intervening sequences have been compared within the chick ovalbumin gene (Breathnach et al. 1978). The canonical sequences AG↓GTA and TXCAG↓ (X, Xanthine) result for the left and right junctions, respectively. Very similar intron–exon junction sequences have been identified in globin, insulin, and immunoglobin genes (Konkel et al. 1978; Tonegawa et al. 1978).

Avvedimento et al. (1980) have identified three possible internal splice sites on an intron of the chick α2-collagen gene. This observation argues against a hypothesis that introns are removed in one piece. The details of RNA splicing are incomplete (Ohno 1980; Lewin 1980) and no splicing enzymes have yet been identified.

The question of why intervening sequences occur in some eukaryotic genes is difficult to answer. Their absence in some genes argues against their necessity. Moreover, in *Drosophila*, ribosomal genes with and without introns exist. In this system, genes with introns seem not to be transcribed, whereas uninterrupted genes are transcribed (Glätzer 1979; Long and Dawid 1979).

The fact that stop codons have been observed in significant numbers in some introns argues against the fact that they might readily become functional coding sequences through rearrangement. An amazing feature of these sequences is that although divergence has occurred between corresponding introns of the same genes in different species, their lengths and positions in relation to exons seem to be conserved.

Several hypotheses have been proposed for functions of introns (Gilbert 1978; Crick 1979; Lewin 1980). It is possible that because introns are generally larger than the exons that they separate recombination between them is more likely. If this is so, they may in fact function to bring formerly separate coding sequences closer together. This rearrangement of coding sequences might function to test various associations for advantageous effects. Because point mutation within exons is not implemented, the original functions of the coding sequences may not be altered. Thus, introns might function as a "sink" for mutations. Introns might expedite the shuffling of exons and new combinations of coding sequences could thus be brought together more rapidly than would otherwise be the case. Sequencing of genes that code for identical or similar proteins in related organisms should provide answers to these and other questions.

Some sequence homologies have been observed in trailing regions beyond the termination codon. Some of these have been of the palindromic or inverted repeat types described by Busslinger et al. (1979) in sea urchin histones and in chicken globin genes (Dodgson et al. 1979). The hexanucleotide, 5'-AAUAAA-3', has been described in the 3' noncoding region of a number of messenger RNAs (Proudfoot and Brownlee 1976). However, not all messengers possess this sequence (Robertson 1979), and, therefore, the proposed role of this sequence as a recognition site is unclear.

Most eukaryotic messenger RNAs bear a polyadenylic acid tail of about 200 nucleotides, which is added to the 3' region after transcription. Experimental deletion of poly(A) tails in *Xenopus* oocyte messengers significantly decreased their half-life (Huez et al. 1974). Readdition of poly(A) tails to frog oocyte messenger RNAs stabilized their expression rate (Huez et al. 1975).

This is truly an exciting time to study molecular biology. The advances made in DNA sequence technology are providing not only clues to the structural aspects of eukaryotic genes, the knowledge of which is a prerequisite to an understanding of function, but also DNA sequence descriptions can now be used in a comparative context to estimate degrees of nucleotide divergence and rates of evolutionary change. Continued study of the structure and organization of the avian genome seems warranted if for no other reason than the fact that such an abundance of potential corroborative data on morphology, distribution, and behavior of birds already exists.

Literature cited

Arthur, R. R., and N. A. Straus. 1978. DNA sequence organization in the genome of the domestic chicken (*Gallus*). *Can. J. Biochem. 56*:257–263.
Atkin, N. B., G. Mattinson, W. Becak, and S. Ohno. 1965. The comparative DNA content of nineteen species of placental mammals, reptiles, and birds. *Chromosoma 17*:1–10.

Avvedimento, V. E., G. Vogeli, Y. Yamada, J. V. Maizel, I. Pastan, and B. de Crombrugghe. 1980. Correlation between splicing sites within an intron and their sequence complementarity with U1 RNA. *Cell 21*:689–696.

Bachman, K., O. B. Goin, and C. J. Goin. 1972a. Nuclear DNA amounts in vertebrates. In H. H. Smith (ed.). *Evolution of genetic systems*, pp. 419–450. New York: Gordon & Breach.

Bachman, K., B. A. Harrington, and J. P. Craig. 1972b. Genome size in birds. *Chromosoma 37*:405–416.

Beauchamp, R. S., J. Pasternak, and N. A. Straus. 1979. Characterization of the free living nematode *Panagrellus silusiae:* absence of short period interspersion. *Biochemistry 18*:245–250.

Bernard, O., N. Hozumi, and S. Tonegawa. 1978. Sequences of mouse immunoglobin light chain genes before and after somatic changes. *Cell 15*:1133–1144.

Bolton, E. T., R. J. Britten, D. B. Cowie, R. B. Roberts, P. Szafranski, and M. J. Waring. 1966. Biophysics. *Carnegie Inst. Washington Yearbk. 64*:313–348.

Breathnach, R., C. Benoist, K. O'Hare, F. Gannon, and P. Chambon. 1978. Ovalbumin gene: evidence for a leader sequence in mRNA and DNA sequences at the exon-intron boundaries. *Proc. Natl. Acad. Sci. USA 75*:4853–4857.

Breathnach, R., and P. Chambon. 1981. Organisation and expression of eukaryotic split genes coding for proteins. *Ann. Rev. Biochem.* 50:349–383.

Britten, R., and P. Kohne. 1968. Repeated sequences in DNA. *Science 161*: 529–540.

Britten, R. J., and E. H. Davidson. 1969. Gene regulation for higher cells: a theory. *Science 165*:349–357.

Bulatova, N., E. N. Panov, and S. I. Radzhabli. 1971. Description of the karyotypes of certain species of birds of the U. S. S. R. *Dokl. Biol. Sci. 191*:479–483.

Bush, G. L., S. M. Case, A. C. Wilson, and J. L. Patton. 1977. Rapid speciation and chromosomal evolution in mammals. *Proc. Natl. Acad. Sci. USA 74*:3942–3946.

Busslinger, M., R. Portmann, and M. Birnstiel. 1979. A regulatory sequence near the 3' end of sea urchin histone genes. *Nucleic Acids Res. 6*:2997–3008.

Chambon, P. 1977. The molecular biology of the eukaryotic genome is coming of age. *Cold Spring Harbor Symp. Quant. Biol. 42*:1209–1234.

Cochet, M., F. Gannon, R. Hen, L. Maroteaux, F. Perrin, and P. Chambon. 1979. Organisation and sequence studies of the seventeen-piece chicken conalbumin gene. *Nature (London) 282*:567–574.

Cordell, B., G. Bell, E. Tischer, F. DeNoto, A. Ullrich, R. Pictet, W. J. Rutter, and H. M. Goodman. 1979. Isolation and characterization of a cloned rat insulin gene. *Cell 18*:533–543.

Crain, W. R., E. H. Davidson, and R. J. Britten. 1976a. Contrasting patterns of DNA sequence arrangement in *Apis mellifera* (honeybee) and *Musca domestica* (housefly). *Chromosoma 59*:1–12.

Crain, W. R., F. C. Eden, W. R. Pearsob, E. H. Davidson, and R. J. Britten. 1976b. Absence of short period interspersion of repetitive and nonrepetitive sequences in the DNA of *D. melanogaster*. *Chromosoma 56*:309–326.

Crick, F. 1979. Split genes and RNA splicing. *Science 204:*264–271.

Davidson, E. H., and R. J. Britten. 1979. Regulation of gene expression: possible role of repetitive sequences. *Science 204:*1052–1059.

Davidson, E. H., G. A. Galau, R. C. Angere, and R. J. Britten. 1975a. Comparative aspects of DNA organization in metazoa. *Chromosoma 51:*253–259.

Davidson, E. H., B. R. Hough, W. H. Klein, and R. J. Britten. 1975b. Structural genes adjacent to interspersed repetitive DNA sequences. *Cell 4:*217–238.

Davidson, E. H., B. R. Hough, C. S. Amenson, and R. J. Britten. 1973. General interspersion of repetitive with nonrepetitive sequence elements in the DNA of *Xenopus. J. Mol. Biol. 77:*1–24.

Dodgson, J. B., J. Strommer, and J. D. Engel. 1979. Isolation of the chicken B-globin gene and a linked embryonic B-like globin gene from a chicken DNA recombination library. *Cell 17:*879–887.

Donnelly, G. M., and E. H. Newcomer. 1963. Autoradiographic patterns in cultured leucocytes of the domestic fowl. *Exp. Cell. Res. 30:*363–368.

Doolittle, W. F. 1981. Prejudices and preconceptions about genomic evolution. *in Evolution today*, G. G. E. Scudder and J. L. Reveal (eds.). *Proceedings of the Second International Congress Systematic and Evolutionary Biology*, pp. 197–205. Pittsburgh, Pa.: Carnegie-Mellon University Press.

Doolittle, W. F. and C. Sapienza. 1980. Selfish genes, the phenotype paradigm and genome evolution. *Nature (London) 284:*601–603.

Eden, F. C., and J. P. Hendrick. 1978. Unusual organization of DNA sequences in the chicken. *Biochemistry 17:*5838–5844.

Eden, F. C., J. P. Hendrick, and S. S. Gottlieb. 1978. Homology of single copy and repeated sequences in chicken, duck, Japanese Quail, and Ostrich DNA. *Biochemistry 17:*5113–5121.

Epplen, J. T., M. Leipoldt, W. Engel, and J. Schmidtke. 1978. DNA sequence organisation in avian genomes. *Chromosoma 69:*307–321.

Galton, M., and P. Bredbury. 1966. DNA replication pattern of the sex chromosomes of the pigeon (*Columba livia domestica*). *Cytogenetics 5:*295–306.

Gannon, F., K. O'Hare, F. Perrin, J. P. LePennic, C. Benoist, M. Cochet, R. Breathnach, A. Royal, A. Garapin, B. Cami, and P. Chambon. 1979. Organisation and sequences at the 5′ end of a cloned complete ovalbumin gene. *Nature (London) 278:*428–434.

Gilbert, W. 1978. Why genes in pieces. *Nature (London) 271:*501.

Ginatulin, A. A., and L. K. Ginatulina. 1979. Molecular structure of genome of vertebrates. In V. A. Krassilov (ed.). *Evolutionary studies: parallelism and convergence*. Vladivostok.

Glätzer, K. H. 1979. Lengths of transcribed rDNA repeating units in spermatocytes of *Drosophila hydei*: only genes without an intervening sequence are expressed. *Chromosoma 75:*161–175.

Heilig, R., F. Perrin, F. Gannon, J. L. Mandel, and P. Chambon. 1980. Ovalbumin gene family: structure of the X gene and evolution of duplicated split genes. *Cell 20:*625–637.

Hinegardner, R. 1976. Evolution of genome size. In F. J. Ayala (ed.). *Molecular evolution*, pp. 179–199. Sunderland, Mass: Sinauer Associates.

Hudspeth, M. E. S., W. E. Timberlake, and R. B. Goldberg. 1977. DNA Sequence organization in the Water Mold *Achlya. Proc. Natl. Acad. Sci. USA 74:*4332–4336.

Huez, G., G. Marbaix, E. Hubert, Y. Cleuter, M. Leclerc, H. Chantrenne, R. Devos, H. Soreq, U. Nudel, and U. Z. Littauer. 1975. Readenylation of polyadenylate-free globin messenger RNA restores its stability *in vivo. Eur. J. Biochem. 59:*589–592.

Huez, G., G. Marbaix, E. Hubert, M. Leclercq, U. Nudel, H. Soreq, R. Solomon, B. Lebleu, M. Revel, and U. Z. Littauer. 1974. Role of the polyadenylate segment in the translation of globin messenger RNA is *Xenopus* oocytes. *Proc. Natl. Acad. Sci. USA 71:*3143–3146.

Jeffreys, A. J., and R. A. Flavell. 1977. The rabbit B globin gene contains a large insert in the coding sequence. *Cell 12:*1097–1108.

John, B., and G. L. G. Miklos. 1979. Functional aspects of satellite DNA and heterochromatin. *Int. Rev. Cytol. 58:*1–114.

Konkel, D. A., S. M. Tilghman and P. Leder. 1978. The sequence of the chromosomal mouse B globin major gene: homologies in capping, splicing and poly(A) sites. *Cell 15:*1125–1132.

Lai, E. C., J. P. Stein, J. F. Catterall, S. L. C. Woo, M. L. Mace, A. R. Means, and B. W. O'Malley. 1979. Molecular structure and flanking nucleotide sequences of the natural chicken ovomucoid gene. *Cell 18:*829–842.

Lewin, B. 1980. *Gene expression.* New York: Wiley.

Long, E. O., and I. B. Dawid. 1979. Expression of ribosomal DNA insertion in *Drosophila melanogaster. Cell 18:*1185–1196.

Mandel, J. L., R. Breathnach, P. Gerlinger, M. Le Muir, F. Gannon, and P. Chambon. 1978. Organisation of coding and intervening sequences in the chicken ovalbumin split gene. *Cell 14:*641–653.

Manning, J. E., C. W. Schmid, and N. Davidson. 1975. Interspersion of repetitive and nonrepetitive DNA sequences in the *D. melanogaster* genome. *Cell 4:*141–156.

Maxam, A. M., and W. Gilbert. 1977. A new method for sequencing DNA. *Proc. Natl. Acad. Sci. USA 74:*560–564.

McKnight, G. S., and R. D. Palmiter. 1979. Transcriptional regulation of the ovalbumin and conalbumin genes by steroid hormones in chick oviduct. *J. Biol. Chem. 252:*2060–2068.

Miklos, G. L. G., and B. John. 1979. Heterochromatin and satellite DNA in man: properties and prospects. *Am. J. Hum. Gen. 31:*264–280.

Miklos, G. L. G., D. A. Willcocks, and P. R. Baverstock. 1980. Restriction endonuclease and molecular analysis of three rat genomes with special reference to chromosome rearrangement and speciation problems. *Chromosoma 76:*339–363.

Minty, A., and P. Newmark. 1980. Gene regulation: new, old and remote controls. *Nature (London) 288:*210–211.

Misra, M., and M. D. L. Srivastava. 1975. Chromosomes of two species of Coraciiformes. *Nucleus 18:*89–92.

Ohno, S. 1970. *Evolution by gene duplication.* New York: Springer-Verlag.

– 1980. Origin of intervening sequences within mammalian genes and the universal signal for their removal. *Differentiation 17:*1–15.

Orgel, L. E., and F. H. C. Crick. 1980. Selfish DNA: the ultimate parasite. *Nature (London) 284:*604–607.

Proudfoot, N. J., and G. G. Brownlee. 1976. 3' non-coding region sequences in eukaryotic messenger RNA. *Nature (London) 263:*211–214.

Robertson, J. S. 1979. 5' and 3' terminal nucleotide sequences of the RNA genome segments of influenza virus. *Nucleic Acids Res. 6:*3745–3757.

Roop, D. R., J. L. Nordstrom, S. Y. Tsai, M. J. Tsai, and B. W. O'Malley. 1978. Transcription of structural and intervening sequences in the ovalbumin gene and identification of potential ovalbumin mRNA precursors. *Cell 15:*671–685.

Royal, A., A. Garapin, B. Cami, F. Perrin, J. L. Mandel, M. Le Muir, F. Brégégére, F. Gannon, J. P. Le Pennec, P. Chambon, and P. Hourilsky. 1979. The ovalbumin gene region: common features in the organization of three genes expressed in chicken oviduct under hormonal control. *Nature (London) 279:*125–132.

Samols, D., and H. Swift. 1979. Genomic organization in the Flesh Fly, *Sarcophaga bullata. Chromosoma 75:*129–143.

Sanger, F., G. M. Air, B. G. Barrell, N. L. Brown, A. R. Coulson, J. C. Fiddes, C. A. Hutchison III, P. M. Slocombe, and M. Smith. 1977. Nucleotide sequence of bacteriophage φX174 DNA. *Nature (London) 265:*687–695.

Sanger, F. and A. R. Coulson. 1975. A rapid method for determining sequences in DNA by primed synthesis with DNA polymerase. *J. Mol. Biol. 94:*441–448.

Schmid, W. 1962. DNA replication patterns of the hetero-chromosomes in *Gallus domesticus. Cytogenetica 1:*344–352.

Schmidtke, J. and J. T. Epplen. 1980. Sequence organization of animal nuclear DNA. *Hum. Genet. 55:*1–18.

Schmidtke, J., E. Schmitt, M. Leipoldt, and W. Engel. 1979. Amount of repeated and non-repeated DNA in the genomes of closely related fish species with varying genome sizes. *Comp. Biochem. Physiol. B 64:*117–120.

Shoffner, R. N. 1974. Chromosomes of birds. In H. Busch (ed.). *Cell nucleus,* pp. 223–261. New York: Academic Press.

Shields, G. F. 1980. Avian cytogenetics: new methods and comparative results. In R. Nöhring (ed.). *Acta XVII Congressus Internationalis Ornithologici,* pp. 1226–1231.

– 1982. Comparative avian cytogenetics: a review. *Condor 84:*45–58.

Shields, G. F., G. H. Jarrell, and E. Redrupp. 1982. Enlarged sex chromosomes of woodpeckers (Piciformes). *Auk 99:*767–771.

Shields, G. F., and N. A. Straus. 1975. DNA-DNA hybridization studies of birds. *Evolution 29:*159–166.

Sparrow, A. H., H. J. Price, and A. G. Underbrink. 1972. A survey of DNA content per cell and per chromosome of prokaryotic and eukaryotic organisms: some evolutionary considerations. In H. H. Smith (ed.). *Evolution of genetic systems,* pp. 451–494. New York: Gordon & Breach.

Stefos, K., and F. E. Arrighi. 1971. Heterochromatic nature of W chromosome in birds. *Exp. Cell. Res. 68:*228–231.

– 1974. Repetitive DNA of *Gallus domesticus* and its cytological locations. *Exp. Cell. Res. 83:*9–14.

Takagi, N. 1972. A comparative study of the chromosome replication in six species of birds. *Jpn. J. Genet. 47:*115–123.

Tonegawa, S., A. M. Maxam, R. Tizard, O. Bernard, and W. Gilbert. 1978.

Sequence of a mouse germ line gene for a variable region of an immuno-globulin light chain. *Proc. Natl. Acad. Sci. USA 75:*1485–1489.

Tsuijimoto, Y., and Y. Suzuki. 1979. Structural analysis of the fibroin gene at the 5' end and its surrounding regions. *Cell 16:*425–436.

van Ooyen, A., J. van den Berg, N. Mantel, and C. Weissmann. 1979. Comparison of total sequence of a cloned rabbit B-globin gene and its flanking regions with a homologous mouse sequence. *Science 206:*337–344.

Wells, R., H. D. Royer, and C. P. Hollenberg. 1976. Non-Xenopus-like DNA sequence organization in the *Chrionomus tentans* genome. *Mol. Gen. Genet. 147:*45–51.

White, M. J. D. 1973. *Animal cytology and evolution.* Cambridge: Cambridge University Press.

Wilson, A. C. 1976. Gene regulation in evolution. In F. J. Ayala (ed.). *Molecular evolution,* pp. 225–231. Sunderland, Mass. : Sinauer Associates.

Woo, S. L. C., A. Dugaiczyk, M. J. Tsai, E. C. Lai, J. F. Catterall, and B. W. O'Malley. 1978. The ovalbumin gene: cloning of the natural gene. *Proc. Natl. Acad. Sci. USA 75:*3688–3692.

Yamamoto, M. 1979. Cytological studies of heterochromatin function in the *Drosophila melanogaster* male: autosomal meiotic pairing. *Chromosoma 72:*293–328.

Ziff, E. B., and R. M. Evans. 1978. Concidence of the promotor and capped 5' terminus of RNA from the adenovirus 2 major late transcription unit. *Cell 15:*1463–1475.

9   The origin and early radiation of birds

LARRY D. MARTIN

Historical perspective

According to Fisher (1967), the first described avian paleospecies was *Larus toliapicus* (Konig 1825). Thus, the story of fossil birds barely exceeds 150 years, and the 100 years of the American Ornithologists' Union include nearly two-thirds of this. During most of this time, a major burden for paleornithologists has been a lack of comparative skeletons of recent birds. We are now beginning to solve this problem, and modern collections exist with thousands of skeletons covering about two-thirds of the world's species.

The other major problem is the incompleteness of most avian fossils. Fragmentary materials rarely contribute to improving avian systematics and thus limit the impact of paleornithology. The few studies that have used excellent specimens (Marsh 1880; Wellnhofer 1974; Harrison and Walker 1976a; Martin and Tate 1976; Olson 1977), illustrate how much more convincing conclusions are when based on more than one skeletal element. Such studies are possible for less than 10% of the total fossil record of birds, and studies of distribution, both geographic and temporal, must rely on less complete specimens.

Fisher (1967) summarized the distribution of fossil birds and estimated the early avian diversification. Shufeldt and Lambrecht were among the first to estimate the diversity of past avifaunas. Shufeldt realized knowing the osteology of Recent birds was fundamental to understanding fossil ones, and he made numerous osteological contributions (Shufeldt 1909). Many of the outstanding publications in avian paleontology have been catalogs (Milne-Edwards 1867–1871; Lydekker 1891; Lambrecht 1933; Wetmore 1956; Brodkorb 1963b, 1964, 1967, 1971a, 1978). Modern paleornithology is largely due to three American ornithologists: Alexander Wetmore, Hildegarde Howard, and Pierce Brodkorb. Their efforts gave North America the best-known fossil avifaunas and have inspired the present generation of workers. Paleornithology is no longer the province of a few geographically restricted workers and is undergoing a renaissance. This is reflected by the two recent Festschrift volumes in honor of Wetmore and Howard that contain more new information on fossil birds than was published in the previous decade and include contributions from 30 authors.

Accordingly, our concepts of the origin of birds and their radiation of the Mesozoic through early Tertiary have been completely revised. This change includes the spectacular arguments on bird origins presented by dinosaur specialists like John Ostrom and Robert Bakker. I consider this most provocative area to emphasize in this chapter.

The founding of the American Ornithologists' Union occurred only slightly later than another momentous ornithological event – the publication of the *Odontornithes* in 1880 by O. C. Marsh. This may still be the single most influential publication on a group of fossil birds, as it fixed in the public mind most of the attributes which we ascribe to birds from the age of dinosaurs. Ultimately, it had a strange and unexpected effect on paleornithology. Instead of inspiring new research, it gave such a sense of completeness that the Cretaceous toothed birds were set aside as a thoroughly known group.

During the next 100 years, new discoveries of Cretaceous toothed birds were made and ignored, with a few interesting exceptions. The first was an effort to fit these birds into the scheme of modern orders. Marsh (1880) noted many similarities between *Hesperornis* and loons and compared *Ichthyornis* with terns and gulls. He concluded that these similarities were not due to special reationship and maintained his toothed birds in a separate subclass (Marsh 1873, 1875). Later workers also notice the similarity of *Hesperornis* to loons and grebes (Thompson 1890; Heilmann 1926; Howard 1950) and regarded the hesperornithiform birds as either the progenitors or sister group of loons and grebes. Such thinking probably led Shufeldt (1915) to produce a loonlike palatal restoration for *Hesperornis*. Now that the structure is known (Gingerich 1976), we realize that his restoration was entirely imaginary. Similar thinking probably underlay other misinterpretations of the Mesozoic birds including the well-known idea of Gregory (1951, 1952) that the toothed jaws found with *Ichthyornis* were really those of a baby mosasaur. This observation, since proven to be incorrect, struck a sensitive chord among ornithologists. There were several attempts to disassociate other parts from the toothed birds or what Brodkorb (1971b) called the "Fable of the Toothed Birds." This period constituted an aberration in thinking about Mesozoic birds and reached its extreme when Swinton (1958) questioned the assignment of toothed jaws to members of the Hesperornithiformes and Bock (1969, 1977) questioned the association of amphicoelous vertebrae with *Ichthyornis*. Neither of these suspicions are correct (Gingerich 1972; Martin and Stewart 1977, 1982), and Marsh's original descriptions are once again accepted. In fact, Swinton (1958) alone had the misfortune to illustrate a mosasaur jaw as that of a bird, as the specimen labeled in Swinton (1958:Fig. 16) as *Hesperornis regalis* really is that of a small mosasaur.

The origin of birds

The dinosaur hypothesis

Many people would be surprised to learn that a dinosaurian origin for birds is an old idea that was suggested almost as soon as the first whole dinosaur skeleton was described (Cope 1867). Huxley (1868), impressed with the similarity of *Archaeopteryx* to the contemporary coelurosaurian dinosaur *Compsognathus*, suggested that birds must have come from a similar root. This was well received among dinosaur specialists, and papers supporting a dinosaurian origin of birds followed (Marsh 1877; Williston 1879; see Ostrom 1976a for a full listing). The early suggestion and acceptance of this idea are easy to understand. Many characteristic features of birds other than feathers are related to bipedal locomotion. For most of the nineteenth century, the only truly bipedal reptilian models available to paleontologists were dinosaurs. Recognition of bipedality in dinosaurs was still a novelty when *Archaeopteryx* was described in 1861. There was also controversy about which dinosaurs were avian progenitors. Most, including Huxley (1868) and Marsh (1877), favored theropods because of the easily made comparisons with *Compsognathus*. Galton (1970) supported ornithischian dinosaurs because, like birds, they have an opisthopubic pelvis. This viewpoint garnered little additional support because many specializations of ornithischians (form of the teeth, predentary bone, etc.) are not found in birds. Perhaps Mudge (1879) had the clearest vision when he wrote "dinosaurs vary so much from each other that it is difficult to give a single trait that runs through the whole. But no single genus or set of genera have many features in common with the birds, or a single persistent, typical element or structure which is found in both." This generalization is probably as true now as it was then.

Dinosaurs held a secure position as avian ancestors until the early 1900s, by which time a wide variety of bipedal reptiles were known from the Triassic of South Africa. Most of these were put into an odd assortment of primitive archosaurs called the Pseudosuchia. One branch or another of the Pseudosuchia was thought to have given rise to the more specialized archosaurs like crocodilians and dinosaurs and it is understandable that this group was proposed by Broom (1913) for the origin of birds. This idea almost completely replaced the dinosaur hypothesis, because of Heilmann's book *The Origin of Birds* (1926). Lowe (1933, 1944) attempted to show a polyphyletic origin of birds from dinosaurs, but the pseudosuchian theory of bird origins prevailed almost unchanged from Heilmann's exposition of it until Galton (1970) attempted to revive the ornithischian hypothesis, Walker's (1972) attempt to relate birds and crocodilians, and Ostrom's (1973) revival of the theropod hypothesis. Since Ostrom (1973, 1974, 1975a,b, 1976a,b, 1978) began work on the subject, the idea of a theropod dinosaur origin for birds has become firmly established in the public and scientific mind. Walker's (1972) attempt to relate birds and crocodilians was never generally accepted, probably in part because most people are familiar with modern crocodilians and realize how wholly unlike birds they are. Dinosaurs, on the other hand, are not really familiar to most biologists, and it is easy to imagine some unknown form evolving feathers or to conceive of feathers on the known forms. If birds are most closely related to dinosaurs, we might reasonably suppose that characters birds and crocodilians share must have occurred in dinosaurs and that dinosaurs also had some distinctly avian features such as feathers and endothermy. However, if birds and crocodilians are more closely related to each other than either is to dinosaurs, we can make few confident extrapolations about dinosaur behavior and soft anatomy beyond that indicated by their skeletons and tracks.

The theropod argument has been argued thoroughly by Ostrom. Its less obvious implications should be considered. Most dinosaurs lack the derived features believed to unite dinosaurs and birds, and thus the origin of birds should be sought within a resticted group of dinosaurs. A sister group relationship should not be established with all saurischian dinosaurs or even with Theropoda as a whole. The "advanced" coelurosaurs (*Velociraptor, Saurornithoides,* etc.) most like birds are all Cretaceous in age. It is unclear whether many of the derived characters shared with birds have great antiquity in coelurosaurian phylogeny (many are absent from the Triassic coelurosaur *Coelophysis*); hence, a very late origin for birds is usually argued by supporters of a coelurosaurian ancestry. In some works (Ostrom 1978), *Archaeopteryx* is considered as a coelurosaur, and *Enaliornis,* from the early Cretaceous, as the earliest bird. Most derived characters relating coelurosaurs to birds are also related to the type of advanced bipedality found in coeluro-saurs. The earliest birds, if derived from coelurosaurs, should also have this kind of bipedality, but an obligate biped with a rigid tail, such as the Lower Cretaceous theropod *Deinonychus*, would be almost incapable of climbing trees. Thus, it is not surprising that Ostrom (1979) is the chief advocate of a cursorial origin for avian flight.

If a cursorial origin for avian flight is accepted, we must develop a model by which feathers achieved their modern size and form other than as an airfoil. Ostrom (1979) suggests that feathers evolved first as an insulating cover for small, highly active endotherms, second, as traps for insects, and third, for lift in powered flight. If one accepts an arboreal origin for avian flight, feathers can be derived directly for their aerodynamic qualities (Parkes 1966).

The most important features thought by Ostrom (1973, 1975a,b, 1976a) to support a close relationship between coelurosaurian dino-saurs and birds are as follows: (1) large orbit in a triangular skull with an antorbital fenestra; (2) large deltoid crest on the humerus; (3) nar-row elongate scapula; (4) "biceps tubercle" on the coracoid; (5) acces-sory fenestrae in the antorbital fenestra; (6) semilunate carpal bone in the carpus; (7) hand reduced to the first three digits; (8) penultimate digits of the manus elongate; (9) pubic expansion on the pubic symphy-sis; (10) high astragular prominence on the astragalus; (11) loss of the outer digit on the pes; and (12) distally located articulation for the first metatarsal. Of these characters, 1 through 4 all occur on reptiles that are not dinosaurs but that are thought by some to be related to croco-dilians (for instance, *Sphenosuchus* and *Scleromochlus*). The quality of character 4, the biceps tubercle, seems doubtful, as the supposed ho-mologous structure on *Archaeopteryx* is probably for the attachment of the furcula rather than a muscle attachment as was suggested for the structure labeled "biceps tubercle" in *Sphenosuchus* and *Deinonychus* (Os-trom 1976a). Character 5 (the accessory fenestrae) is visible only in the Eichstatt specimen, and its interpretation may be doubtful (Wellnhofer 1974). The semilunate carpal (character 6) is the most striking feature in Ostrom's argument. In *Deinonychus*, Ostrom (1969) considered it to be the radiale and described its movement in articulation with the radius. He also described the ulna and argued "that these two bones alone composed the functional wrist joints" (Ostrom 1969). If we accept this interpretation of the wrist of *Deinonychus*, we cannot accept the homology of the semilunate bone in *Deinonychus* with the semilunate bone in the carpus of birds (which is a series of fused distal carpals articulating proximally with the cuneiform and scapholunar). The homology of the digits present in the manus of birds and theropods has been attacked by Tarsitano and Hecht (1980), so that although I might accept character 8, other workers strongly object. Character 9, the pubic expansion, is not so similar in birds and dinosaurs as generally thought. It is laterally compressed in theropods and apparently dorsoventrally flattened in *Archaeopteryx*. Character 10, the high astragular process in dinosaurs, is apparently nonhomologous to the similar structure in birds (pretibial bone), which serves the same function (Martin et al. 1980). This is a basic feature necessary to stabilize the mesotarsal joint for advanced bipedal locomotion. If the "pretibial bone" is unique in birds, as I believe, the common ancestor of birds and dinosaurs was not an advanced biped, and shared features related to that locomotion are probably convergent. The reduction or loss of the outer digit of the pes (character 11) occurs in some advanced bipedal crocodilian relatives (Walker 1970), and the interpretation of the first metatarsal as distally situated in theropods has been attacked by Tarsitano and Hecht (1980). In summary, the argument for a theropod origin of birds is unequivocally supported only by the elongate penultimate digits (a feature that also occurs in pterosaurs). Alternatively, we must weigh the evidence supporting a crocodilian origin for birds (Walker, 1972).

The crocodilian hypothesis

Walker (1972, 1974, 1977) centered his case for crocodilian affinities of birds around a Triassic reptile, *Sphenosuchus*, which he thought to be a crocodilian. Crompton and Smith (1980) reject *Sphenosuchus* as a crocodilian and after examination of the specimen, I agree. However, both Triassic and Recent true crocodilians are more like birds than is *Sphenosuchus*, which must now be peripheral to the question of bird origins. Although modern crocodilians and birds are totally unlike in habitat and locomotion, the differences from birds are considerably less in Triassic and Jurassic crocodilians, some of which were bipedal (*Hallopus*). There is also an extensive but little known radiation of Triassic archosaurs that share crocodilian features (Walker 1964). The most interesting of these little archosaurs is *Cosesaurus aviceps* from the late Triassic of Spain. This small animal has been described as a "protobird" (Ellenberger 1977), and associated feather impression have been claimed. Through the kindness of P. Ellenberger, I examined the original specimen in detail, and although I could not confirm the presence of feathers, I agree that impressions of some sort of integumentary structures are preserved.

Ostrom (1976a) suggested that *Cosesaurus* might be related to a similarly small late Triassic archosaur from Scotland, *Scleromochlus*. I examined the original material of that genus and generally agree with Ostrom's conclusion. *Scleromochlus* and *Cosesaurus* share a skull with large orbits, a pointed snout, triangular antorbital fenestra, long slender lower jaw lacking a mandibular fenestra and having an elongate retroarticular process, and a narrow elongate scapula. The *Cosesaurus* specimen is evidently from a young individual, and I estimate that the adult would have been nearly the size of *Scleromochlus*. Walker (1964) argued that *Scleromochlus* was an aberrant aetosaur, a group he related to the crocodilomorphs. *Scleromochlus* and *Cosesaurus* share with *Longisquama*, from the early Triassic of Russia, similarities of the shoulder girdle including long narrow scapulae. *Longisquama* and *Cosesaurus* also have structures of similar shape and position, which resemble closely the furcula of *Archaeopteryx*. They possess enlarged integumentary structures, which were interpreted as scales in *Longisquama* (Sharov 1970) and as feathers in *Cosesaurus* (Ellenberger 1977). The presence or absence of these features cannot be determined for *Scleromochlus*. Unfortunately, the cranial characters used to relate birds to crocodilians (Whetstone and Martin 1979, 1981; Martin et al. 1980) cannot be ascertained from this material. However, the possibility of avian affinities for these small Triassic forms should be taken seriously.

Tarsitano and Hecht (1980:171) provide the following list of Walker's (1972, 1974, 1977) characters thought to unite crocodilians and birds: (1) the possession of laterosphenoids, an external mandibular foramen, and an antorbital fenestra; (2) forward position of the quadrate head, articulating with the squamosal and prootic; (3) a kinetic skull with a streptosylic quadrate; (4) crescentic shape of the occipital surface; (5) a short paraoccipital process projecting behind the quadrate and forming the posterior wall of the tympanic cavity; (6) inferred possession of paired salt glands in *Sphenosuchus;* (7) similar carotid circulation based on paired grooves and a pneumatic basisphenoid in *Sphenosuchus;* (8) an elongate cochlear duct; (9) similarity of the manus, carpus, and elbow joint; (10) similar pattern of digit reduction; (11) posteriorly directed pubis of the "Jurassic crocodilian *Hallopus*"; (12) the scapulocoracoid in *Archaeopteryx* can be derived from that of *Sphenosuchus;* (13) similar morphology of the palatines; (14) crocodilians were originally arboreal as indicated by the climbing ability of juvenile crocodilians, the morphology of the tarsus, long humerus, pneumatization of the skull and limbs of fossil crocodilomorphs, marked inward and forward curvature of the lower half of the tibia and the reduction of the first metatarsal; (15) eustachian tube system.

Tarsitano and Hecht (1980) present an extended critique of these characters, some of which seems to be justified, and some of which appears to be a misunderstanding of the criteria for determining primitive versus derived character states. They apparently think that the presence of a similar character state in widely disparate phylogenetic groups suffices to show plesiomorphy irrespective of the distribution of the character within these groups. I would agree with them that characters 1, 4, 5, 12, and 13 are probably primitive and that 2, 3, 6, 11, 14, and 15 probably contribute nothing to the argument as we now understand it. Contrary to Tarsitano and Hecht (1980), I think that characters 7 and 8 may have value and should not be rejected simply because they are shared by certain other reptiles besides crocodiles. Again, however, it really doesn't matter how we view Walker's original characters, because his arguments were mostly based on *Sphenosuchus,* which is no longer crucial to ideas on avian ancestry.

Features now supporting a crocodile/bird relationship (Whetstone and Martin 1979, 1981; Martin et al. 1980) include: (1) unserrated teeth (theropod dinosaurs normally have serrated teeth); Osborne (1903) reported absence of serrations in the dinosaur *Ornitholestes,* but Whetstone (pers. comm.) observed serrations on the posterior teeth of the same material studied by Osborn; (2) a distinct constriction between the crown and root of the teeth; (3) expanded bony root covered with cementum and connected to the jaw by peridontal ligaments; (4) circular or oval resorption pits formed by the tilting of the developing tooth labially and its main development within the pulp cavity of its predecessor; (5) implantation of the teeth in a groove in the young individuals at least; (6) formation of the lingual walls of the major tooth-bearing bones by extensions of dense bone, rather than by attachment bone, forming interdental plates as in the Theropoda (Martin and Stewart, unpubl.); (7) pneumatic quadrate; (8) foramen aerosum in the lower jaw; (9) periotic pneumatic cavities in the dorsal, central, and rostral positions; (10) a quadrate cotylus at the anterior base of the parocciput; (11) a bipartite quadrate articulation with dermal and endochondral bones – anteriorly with the prootic, squamosal, and laterosphenoid, posteriorly with the prootic and otoccipital; (12) a squamosal shelf over the ear region; (13) anteriomedial origin of the temporal musculature; (14) two pneumatic cavities surrounding the cerebral carotid arteries.

These characters occur in both modern and Mesozoic birds and crocodilians and are apparently primitive for each group, suggesting that a common ancestor could have split very early from the main crocodilian stem. Conversely, they are very complex derived structures in Archosauria and provide a stronger argument for relationship than any uniting the Theropoda as a whole. If they occurred in one "theropod dinosaur" as they do in almost any crocodilian, the strongest argument at present would be to remove that "dinosaur" from the Theropoda and place it in the avian/crocodilian group. The implications of an avian/crocodilian sister group relationship are quite different from those of a theropod origin of birds. Typical crocodilians occur in the late Triassic and the avian/crocodilian divergence would have to be middle Triassic or earlier. The earliest members of the avian line might not have been obligate bipeds and were likely to have been arboreal. Feathers might have evolved with powered flight rather than before it.

Origin of feathers and flight

Feathers of modern birds serve various functions but primarily they provide insulation and the lift for powered flight. Whether one of these two functions came first is presently uncertain. Ostrom (1974) argued for a preflight development of endothermy and the evolution of feathers as insulative structure analogous to mammalian hair. Ostrom's proavis is a small, fast-running, bipedal theropod dinosaur, covered by contour feathers, which evolved on the forelimbs into a sort of "insect net" (see Ostrom 1979). This insect trap was then further modified as an airfoil for short glides and finally for powered flight. This model is a more sophisticated version of the cursorial origin of bird flight proposed by Williston (1879) and Nopcsa (1907). The cursorial model has always seemed flawed because it does not provide much explanation for stages between leaping and true powered flight. Any gliding would probably be brief, at an altitude not much higher than the initial leap, and at a speed less than the initial ground speed. None of these features seems likely to enhance either the capture of prey or escape from predators. The use of the feathers as an insect trap requires that air pass through the feathers easily, a condition incompatible with flight. The rotations of the limbs required to capture prey and convey it to the mouth are quite different from those required for flight, and the two sets of motions cannot be easily reconciled. Furthermore, the position of the wings in Heilmann's (1926) restoration of *Archaeopteryx* is unlike that in modern birds. This position is not supported by additional study of the available material of *Archaeopteryx*, which shows that the wing is avian in orientation (see Figure 9.1).

An arboreal origin of avian flight was proposed by Marsh (1880) and has been the most widely accepted model. Bock (1965) thoroughly

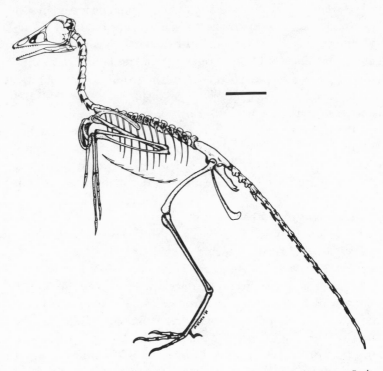

Figure 9.1 Restoration of the left side of the skeleton of *Archaeopteryx*. Scale =
4 cm.

reviewed this model and the various implicit intermediate stages. Its
obvious advantages are that the air speed for early phases of flight may
be provided by gravity, and escape from predators seems to be an
adequate behavioral motivation. The strongest objection to the arboreal
model is the difficulty that an obligate biped would have climbing trees.
Any dinosaurian ancestor would have been such a biped.

Archaeopteryx was apparently an adequate if uninspired flyer (Olson
and Feduccia 1979a; Feduccia 1980). This viewpoint contradicts the
interpretation that the primitive shoulder girdle of *Archaeopteryx* coupled with the lack of an ossified sternum would preclude true powered
flight (Ostrom 1974, 1979). However, the sternal flight muscles also
originate from the coracoids and furcula, which were the likely primary
sites for the origins of these muscles. Longer coracoids and sternal
ossification in later birds may be related to the lengthening of the flight
muscles by moving their origins posteriorly. The furcula itself extends
ventrally and may have functioned much as the sternal keel does in

more advanced birds. I am convinced that *Archaeopteryx* was capable of some powered flight and was basically arboreal.

Neither the pseudosuchian nor crocodilian models of avian origins require that the early proavian lineage include obligate bipeds, and both these models are compatible with an arboreal origin of avian flight. Possibly, endothermy, feathers, and the unique avian respiratory system are all results of the development of flight and may have developed either at the same time or after powered flight was achieved.

The Mesozoic radiation

Fossils that are not birds

Many features separating birds from modern reptiles (large brain and eyes, bipedal locomotion, loss of teeth, etc.) occurred among a wide variety of Mesozoic reptiles. Thus, isolated elements may be misallocated to the Class Aves. Brodkorb (1978) recognized this and published a long list of *Nomina Non Avium*. I agree with him on most Mesozoic forms listed and have examined the original specimens of nearly all of these to check his assignments. *Bradycneme draculae* and *Heptasteornis andrewsi* (Harrison and Walker 1975) from the Upper Cretaceous of Transylvania are theropod dinosaurs as suggested by Brodkorb (1978) and appear to be struthiomimids in view of the extreme enlargement of the astragalus and reduction of the calcaneum. Probably *Elopteryx nopcsai* Andrews from the same site is also a dinosaur, although Brodkorb (1963b) retains it in the Pelecaniformes. I also examined the excellent Mongolian material of *Oviraptor* (Osmolska 1976), and undoubtedly *Caenagnathus* from the Upper Cretaceous of Canada is a coelurosaurian dinosaur near *Oviraptor* as suggested by Brodkorb (1978). The brain cavity with evidence of an advanced flocculus and reptilian ear region also confirms Brodkorb's (1978) assignment of *Laopteryx priscus* Marsh from the Upper Jurassic of Wyoming to the Pterosauria.

The status of the Upper Jurassic bones collected by James Jensen and thought by some to be birds (Madsen 1978; Ostrom 1978) is uncertain. I have only examined casts of the original two fragments that were too incomplete to assign confidently to class. We have to await the description of additional material collected by Jensen to evaluate his discovery. The genus *Wyleyia* (Harrison and Walker 1973) is listed by Brodkorb (1978) as based on "almost certainly a reptilian humerus," but I regard present knowledge of *Wyleyia* too incomplete to make a judgment. *Gobipteryx minuta* (Elzanowski 1976) was listed by Brodkorb

(1978) as a small dinosaur. This is now demonstratively incorrect, as it is avian (see the discussion of the Sauriurae in the next section).

All other supposed birds from the Mesozoic listed as reptilian by Brodkorb (1978), including the famous Triassic or Lower Jurassic "bird tracks" (Hitchock 1836), seem to be obviously reptilian, and *Archaeopteryx* stands alone as the only known Jurassic bird.

The Sauriurae: Archaeopteryx and the terrestrial birds of the Mesozoic

Only 30 species of fossil birds antedate the description of *Archaeopteryx* (Brodkorb 1971b), which has held a central position since the start of the study of fossil birds. It has been the subject of numerous publications (Owen 1864; Heilmann 1926; de Beer 1954; Wellnhofer 1974; Ostrom 1976b; Tarsitano and Hecht 1980). It is thus surprising that so little of its basic anatomy can be gleaned from the literature.

Heilmann's (1926) restoration of the skull of the Berlin specimen seems to be based more on his understanding of the then current restorations of the skulls of *Aetosaurus* and *Euparkeria* than on anything visible on the Berlin example. In fact, the Berlin specimen only shows the size and shape of the outline of the skull and lower jaws, the external nares, and antorbital fenestra, and the orbit. Much of the remainder of Heilmann's restoration can now be shown to be wrong.

Knowledge of the skull and lower jaws were enhanced greatly by the discovery of the Eichstatt specimen and the preparation of the cranium on the London specimen (Figure 9.7). These new specimens and preparations (Wellnhofer 1974; Martin, unpubl.) show that the restorations of the skull by Heilmann (1926), the palate by Kleinschmidt (1951), and the brain by Jerison (1973) were all somewhat in error. The skull is triangular with a prominent antorbital fenestra apparently divided into three parts by vertical bars (Wellnhofer 1974). The premaxilla is short, and the nasals meet in the midline. Apparently, the skull could be neither mesokinetic nor rhynchokinetic and streptostyly would be limited by the extension of a part of the quadratojugal up the lateral side of the quadrate. There is no evidence for an attachment for the postorbital on the London cranium, and the impression thought to relate to the postorbital on the Eichstatt specimen (Wellnhofer 1974) is part of the supraorbital shelf. The bone identified as the squamosal by Wellnhofer (1974) is the paraoccipital process, and the squamosal is either very reduced or absent. The quadrate is unlike that of modern birds and has a single head that appears to be the articulation with the prootic and otoccipital rather than the head that articulates with the prootic, squamosal, and laterosphenoid. This would be a derived condition for *Archaeopteryx,* as it differs from known reptiles as well as from ornithurine birds. As in modern birds, the skull of *Archaeopteryx* lacked the middle bar separating the diapsid openings.

The Eichstatt and Berlin skulls show clearly that *Archaeopteryx* had a narrow pointed snout with short premaxillaries bearing four teeth with flattened, unserrated crowns and expanded bony roots on each side. The premaxillaries bear numerous small foramina on their external surfaces and might have had a horny covering. The premaxillaries form the anteroventral, anterior, and anterodorsal border of the external nares but do not extend dorsally posterior to the posterior margin of the external nares, and the nasal bones form the posterior margin of the external nares. The medial margins of the two nasals abut along the midline except for their anteriormost extension, where they are separated by the premaxillaries. They form only a tiny portion of the dorsal margin of the external nares. The ventral borders of the external nares are formed primarily by the maxillaries. The external nares are large and oval in shape. At the anteriormost margin of the external nares, the premaxillaries bear a large posteriorly directed foramen. The maxillary has a high dorsal process, which joins the nasals to form the anterior margin of the antorbital fenestra. The maxillary is slender posterior to this process and forms almost all the lower border of the antorbital fenestra. On the Eichstatt specimen, it bears nine teeth and on the Berlin specimen at least eight. It meets the jugal at a posteriorly slanting suture almost at the posterior margin of the antorbital fenestra, but the lacrimal has its ventral contact entirely with the jugal. The nasals meet the frontals above the lacrimal. The palate can be seen on only the Eichstatt specimen, and it can be properly interpreted on that specimen only with preparation of the other side of the skull. There appear to be long slender palatines separated from each other that terminate on the maxillopalatines. It is unclear whether the bones identified as pterygoids by Wellnhofer (1974) are extensions of the palatines or not. If they are not pterygoids, then none are visible on the slab unless the bone identified as an ectopterygoid is one. *Archaeopteryx* is the only bird described as having an ectopterygoid; the pterygoids of the hesperornithiforms are short triangular bones somewhat similar to this supposed ectoptergyoid. However we interpret the various elements, the palate of *Archaeopteryx* does not resemble that of any modern bird.

The brain is small (about 0.9 ml in the London specimen) but basically avian and appears sufficient for powered flight. The lower jaw lacks a mandibular fenestra and has a long well-developed retroarticular process. The symphysis was not fused, and there does not appear to be an intermandibular joint.

Wellnhofer (1974) shows nine cervical vertebrae for *Archaeopteryx*. This is quite a short neck and is three vertebrae less than in Heilmann's much copied restoration (Heilmann 1926). The three posterior vertebrae are fused in the Berlin specimen, and it seems certain the Wellnhofer's count is correct for that specimen too. The vertebrae of *Archaeopteryx* are amphicoelous, and they have small lateral pleurocoels in the lumbar region. All of the ribs are double-headed, and there is a full complement of well-ossified stomach ribs (gastralia), which begin just behind the coracoids so that there is little room for a cartilaginous sternum and probably no ossified one occurred. The absence of an ossified sternum and ossified uncinate processes suggests that breathing with parabronchi and an air sac system as found in modern birds did not occur in *Archaeopteryx*.

The tail of *Archaeopteryx* includes 22 unfused vertebrae, most of which are elongate. There was no pygostyle. The postzygapophyses are especially elongate from about the fourth through the eleventh caudal vertebra, and the prezygapophyses are elongate from the twenty-first through the seventeenth caudal vertebra. There are large flat intracentra from about the sixth to the eighteenth caudal vertebra. Each caudal vertebra bore a single pair of feathers. The tail provides an airfoil with about the same surface area of the wings. The shoulder girdle of *Archaeopteryx* as in all known birds was rotated 90° from the plane found in dinosaurs and almost all other reptiles. In birds, the coracoids are oriented anteroventrally, and the scapula lies alongside and parallel to the vertebral column. Several of the anterodorsal vertebrae of *Archaeopteryx* are fused together, and a facet on the scapula articulates with these vertebrae. This helps to separate and stabilize the shoulder girdle. The other stabilizing factor is the furcula, which is relatively larger than in any other known bird. The furcula and the coracoids would have been the major areas of origin of the *pectoralis* musculature, and this probably explains their relatively large size.

The humerus of *Archaeopteryx* is unlike that of any other known bird or reptile having the deltoid crest centered on the palmar surface of the proximal end so that the proximal end appears teardrop shaped. The ulna is of fairly avian form and slightly shorter than the humerus.

The carpus had a least three carpal bones. The most important of these is a large semilunate carpal. In modern birds, this carpal bone forms the *trochlea carpalis*. It articulates proximally with the radiale and the ulnate. The three metacarpal bones, which fuse to form the carpometacarpus in modern birds, are unfused in *Archaeopteryx*.

On the manus of the Berlin *Archaeopteryx*, the outer digit has three phalanges and a claw (fourth phalanx). The first two phalanges are short, whereas the third is elongate. Phalanges 2–4 show postmortem rotation and are seen from the side. The proximal ends are expanded as is normal for phalanges. The Marburg specimen (Heller 1959) confirms the presence of four phalanges on the outer digit as does the Eichstatt specimen.

The argument of Tarsitano and Hecht (1980) that the phalangeal formula for the manus of *Archaeopteryx* is 2,3,3 rather than 2,3,4 was expressed previously by others including Heinroth (1923), who was refuted by Heilmann (1926:26). Heilmann stated that Heinroth "often mentions 'den Bruch des 3. Fingers,' but the statement is founded merely upon inaccurate observation. It is a dislocated joint, for the thickened articular surfaces of the phalanges are plainly visible." Both Heinroth (1923) and Tarsitano and Hecht (1980) based their interpretation largely on the Berlin specimen. My study of that and the Eichstatt specimen leads me to agree with Heilmann and, in fact, Tarsitano and Hecht's (1980) excellent photographs also support this. I could not confirm any significant differences in the lengths of the first and second phalanges from the right to the left side, contrary to Tarsitano and Hecht (1980).

It is difficult to understand Tarsitano and Hecht's (1980) insistence that the outer digit of *Archaeopteryx* had only three phalanges except to make the fossil evidence agree with embryological argument. Embryologists commonly identify the outer digit on the bird manus as the fourth digit, but primitively this digit has only three phalanges, whereas the third digit has four. Because *Archaeopteryx* clearly has four phalanges on its outer digit, I am forced to agree that the digital formula is 2,3,4. Because of the close similarity in the parts of the hand between modern birds and *Archaeopteryx*, I agree with Heilmann (1926) and Ostrom (1976b) that the digits of modern birds are 1–3.

The pelvis has a large rounded preacetabular ilium and a short pointed postacetabular ilium (Figure 9.2). The ilium forms about two-thirds of the acetabulum and the ischium most of the remaining one-third; the pubis was practically excluded from the acetabulum. As in other Mesozoic birds, the acetabulum is partly closed. The pubis is opisthopubic, as shown by Tarsitano and Hecht (1980). The pubes are much longer than the ischia and are united distally for much of their length. They form a dorsoventrally flattened pubic apron expanded slightly at its distal end.

The femur is long and slender with a shallow anterior curve. A distinct lesser trochanter is separated from the greater trochanter by a deep groove. The distal end of the femur has a rounded ridge just above the external condyle on the posterior side, which acts as a stop when the tibiotarsus is flexed.

The tibiotarsus has two distinct distal condyles, a medial one composed of the astragalus and a lateral one composed of the calcaneum. There is a triangular pretibial bone just above the calcaneum. This distal end is like that in modern birds (Martin et al. 1980) except that it is less fused. As in most other Mesozoic birds, there is no supratendinal bridge on the tarsometatarsus. The fibula is long but does not reach the calcaneum distally. The tarsometatarsus is fused proximally but not distally; the distal tarsals are fused individually to the metatarsals or are lost. The hallux is reflected posteriorly, and the distal end is arched suggesting arboreal rather than terrestrial habits.

Brodkorb (1971b:34) argued that *Archaeopteryx* was too close in time to *Gallornis* to be directly ancestral to that genus or any other known Cretaceous bird. The new preparation and study of the *Archaeopteryx* specimens strongly confirms Brodkorb's insight. The reduction or loss of the squamosal and reduction of the proximal articulation of the quadrate along with the special articulation of the scapula to the vertebral column and the proximal fusion of the tarsometatarsus are all derived features that exclude *Archaeopteryx* from direct ancestry to modern birds. However, these same features unite *Archaeopteryx* with a remarkable radiation of Cretaceous birds, part of which has been reported recently by C. A. Walker (1981). Walker's work is based on an extensive collection of bird bones made by Jose Bonaparte from the late Cretaceous (Maestrichian) Lecho Formation of northwestern Argentina. This collection includes about 60 individual bones, a few of which were found associated. They are three dimensional and prepared completely. Almost all major elements of the skeleton are represented (but not for any one taxon) except for the skull, dentary, pubis, and tail. Four or five genera and species are present. Walker erected a new subclass, the Enantiornithes, for these birds and characterized it with a combination of features. The most notable of these is a scapular "boss" on the coracoid, which fitted into a coracoidal "facet" on the scapula, the reverse of the normal condition in modern birds. The Enantiornithes also share some of the peculiar derived characters (Figures 9.2, 9.3) that place *Archaeopteryx* off of the main

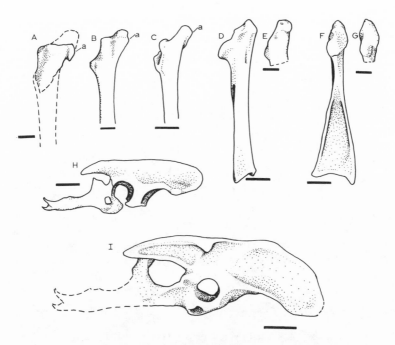

Figure 9.2. (A–C). Dorsal views of the left scapulae (a = articulation with the vertebral column): (A) *Alexornis*, (B) *Gobipteryx*, (C) enantiornithine, (D–G). Right coracoids (lateral views, D,E; dorsal views, F,G): (D) enantiornithine, (E) *Alexornis*, (F) enantiornithine, (G) *Alexornis*. (H,I). Lateral views of right ilia and ischia: (H) *Archaeopteryx*, (I) enantiornithine. Dashed lines, restored. scales: A, E, G = 0.1 cm; B, C, D, F, H, I = 1 cm.

line to modern birds. In other words, I think that there is a major dichotomy near the base of known avian evolution and that the Enanti-ornithes, but not modern birds, belong on the branch with *Archaeop-teryx*. This can be reflected in a taxonomic arrangement by placing the Enantiornithes into the Subclass Sauriurae with *Archaeopteryx*. The advances related to the improvement of powered flight in the Enantior-nithes including an ossified sternum, elongate coracoids (Figure 9.2), and a fused carpometacarpus serve readily to separate *Archaeopteryx* from the Enanthiornithes. The Sauriurae may be divided into two in-fraclasses, which have formally been used at the subclass level, namely, the Infraclass Archaeornithes for *Archaeopteryx* and the Infraclass Enan-tiornithes (Figure 9.4). It is difficult to determine whether *Archaeopteryx* was directly ancestral to the Enantiornithes, because we do not know if all of the features by which they differ are less derived in *Archaeopteryx*. On the whole, *Archaeopteryx* appears to be more primitive than is any enantiornithine bird.

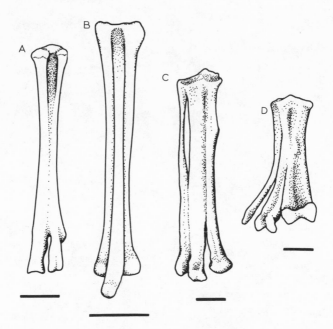

Figure 9.3. Anterior views of right tarsometatarsi: (A) Immature ornithurine bird (*Baptornis*) showing large triangular tarsal cap, the posterior position of the proximal end of the middle metatarsal, and the early fusion of the distal portions of the metatarsals. (B–D) Sauriurine birds: (B) *Archaeopteryx*, (C) enantiornithine bird (reversed), (D) enantiornithine bird. Scales = 1 cm.

Walker described one order within the Enantiornithes, the Enantiornithiformes, but that order varies considerably. All are advanced flying birds with well-developed wings and carpometacarpi, but the sterna were apparently poorly developed, and the coracoids may have been important as muscle origins. Unfortunately, the furculum is unknown. The vertebrae are primarily amphicoelous, but some cervicals were beginning to evolve a heterocoelous state. The pelvis, as far as known, closely resembles that of *Archaeopteryx* with a short, pointed postacetabular ilium and a broad ischium with a dorsal process buttressing against the ilium (Figure 9.2). The femur has the same kind of "stop" as *Archaeopteryx* just above the posterior ridge of the external condyle. The tibiotarsus and tarsometatarsus are especially interesting. There is no supratendinal bridge, and in one genus, the distal end of the tibiotarsus has the outer trochlea expanded into a hemispherical structure that fits a concavity on the proximal end of the tarsometatarsus. This permits the whole foot to rotate sideways with the reduction of the outer metatarsal to a "splint" (Figure 9.3D). These birds

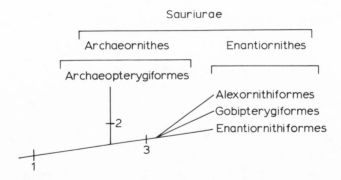

Figure 9.4. Proposed relationships of the Sauriurae. Characters utilized: (1) characters 1–5, p. 310 of text; (2) loss or reduction of the squamosal; (3) elongation of the coracid, fusion of the carpometacarpus, ossification of the sternum, and rotation of the deltoid crest on the humerus.

were apparently predators, as at least one kind has highly developed raptorial claws. They range from about the size of a Common Crow (*Corvus brachyrhynchos*) to that of a Great Blue Heron (*Ardea herodias*). The long narrow coracoids with their peculiar articulations and the scapula with a facet to buttress against the fused anterior dorsals unite the enantiornithine birds of Argentina with *Gobipteryx minuta* (Figure 9.2).

G. *minuta* was described from a fragmentary skull and lower jaws from the late Cretaceous (Campanian) Barun Goyot Formation in Mongolia. Originally put in the Palaeognathae by Elzanowski (1976), the specimen is so distorted and broken that no confidence can be placed in the palatal restoration that led to that taxonomic assignment. Enough of the quadrate is preserved to show a general similarity with that of *Archaeopteryx*. Brodkorb (1976, 1978) suggested that *Gobipteryx* was more likely a reptile than a bird, but new material reported recently by Elzanowski (1981) decisively settles the matter. Besides the two skulls of *Gobipteryx*, the Polish Mongolian Expedition also collected fossil eggs in the same area and sediments. Some of these eggs are avian and contain well-preserved skeletons with skulls very similar to the type of G. *minuta* (Elzanowski 1981). Unlike all other known Mesozoic bird jaws, teeth are absent. They resemble *Archaeopteryx* but not *Hesperornis* and *Ichthyornis* in lacking an intermandibular joint. The jaw articulation and retroarticular process are unusual but very similar to the single known enantiornithine jaw. Many features can be seen in the unhatched skeletons, and it is likely that the chicks were well formed and precocious at hatching. The anterior dorsal vertebrae are fused, and the scapula is attached to them by

the characteristic sauriurine facet. The coracoid is elongate and exca-
vated dorsally as in *Enantiornis* and bears a convex scapular articulation
as in *Enantiornis*. Clearly, *Gobipteryx* is an enantiornithine bird, and eventually the characteristic proximally fused tarsometatarsus of the Sauriurae should be found associated with it.

Alexornis antecedens is another sauriurine bird. It was described by
Brodkorb (1976) as the type of a new order, the Alexornithiformes. All
elements of the known skeleton of *Alexornis* closely resemble those of
the enantiornithine birds, and probably its affinities are with the Sauriurae. Perhaps the most convincing elements are the coracoid and the
scapula, which are so unlike the coracoid and scapula in neornithine
birds that Brodkorb (1976) reversed them so that his "coracoid" is in
fact the scapula and the "scapula" is the coracoid (Figure 9.2E, G). At
this time, the only other sauriurine bird from North America is represented by the proximal end of a humerus from the Upper Cretaceous
(Campanian) of New Mexico. This specimen is presently being studied
by Cyril Walker. *Alexornis* is a sparrow-sized bird, whereas *Gobipteryx* is
about chicken size.

Apparently the Sauriurae all share the following characters: (1) scapula with an anteromedial facet abutting against a series of fused anterior dorsal vertebrae; (2) fusion of the anteriormost thoracic vertebrae
with a special process on the scapula abutting them; (3) tarsometatarsus
fused proximally but not distally; (4) metatarsal bones fused in a
straight line; and (5) distal tarsal bones either absent or fused as small
individual bones (not forming a large tarsal cap). *Archaeopteryx* shows a
suite of characters that may or may not be present in the Enantiornithes including: (6) reduction or loss of the squamosal; and (7) an
unusually robust furcula. The Enantiornithes are characterized by (8)
the shape of the lower jaw articulation (unknown in *Alexornis*); (9) elongate narrow coracoids with a boss that fits into a "coracoidal facet" on
the scapula (present in all known forms); and (10) humerus with a
large flat deltoid crest oriented laterally.

The hypothesis of relationship that I derive from these data is outlined in Figure 9.4. In this figure, I recognize the Subclass Sauriurae
Haeckel 1886, for *Archaeopteryx, Gobipteryx, Alexornis*, and the Enantiornithes. It seems clear that the Sauriurae radiated considerably during
the Cretaceous and occupied many terrestrial habitats.

The Ornithurae: aquatic birds of the Mesozoic

The other side of the great Mesozoic avian dichotomy is the
ornithurines, which include all presently living birds. They must have

existed as a separate lineage at least as far back as the earliest known sauriurine bird, *Archaeopteryx*, but even a late Jurassic origin does not allow much time for their subsequent diversification into two fundamental adaptive types by the early Cretaceous. The earliest well-known members of this subclass are the foot-propelled diving birds of the Infraclass Odontoholcae. These birds (Hesperornithiformes) were already highly specialized divers in the process of loss of flight, and (although we do not presently have fossils of them) other ornithurine birds must have diversified and developed flight while the Hesperornithiformes were beginning to radiate (Martin 1980b). Perhaps *Gallornis* belongs to this other radiation, which led eventually to all modern birds. It is from a Lower Cretaceous (Neocomian) marine deposit in France and thus is slightly older than the earliest known hesperornithiform bird and only slightly younger than *Archaeopteryx*. Unfortunately, we will probably never know its affinities. The type and referred specimen were only a proximal end of femur and a fragment of a humerus and they both have been lost for decades. Lambrecht's (1931) figures may be inadequate to assign it to subclass, and it certainly does not belong to any modern order. It may have been a wading bird as suggested by Brodkorb's (1963b) assignment to the Ardeiformes, Suborder Phoenilopteri, but Aves *incertae sedis* seems to be the most reasonable placement. All Mesozoic Ornithurae are aquatic in one way or another. The basic adaptive zone of the group seems to have been wading along the shoreline. This is still an important avian adaptive zone and is one for which few other Mesozoic vertebrates would have been well suited. The relatively short tarsometatarsus of the Upper Cretaceous bird *Ichthyornis* suggests a bird that swam on the water, and this is a functional stage through which the predecessor of the hesperornithiform birds must have passed.

The early Cretaceous avian record (Figure 9.5A) is tantalizing but not informative. Certainly birds had a worldwide distribution achieved during the Jurassic or earlier. The earliest Western Hemisphere records of birds appear to be tracks from the Lower Cretaceous Dakota Sandstone of Kansas and Colorado (Brodkorb 1978) and from the Lower Cretaceous of Alberta, Canada (Currie 1981). If identified correctly, they indicate the presence of nonhesperornithiform birds larger than *Ichthyornis*, which occupied beaches and freshwater margins. The African and Australian records of this time are isolated feathers (Waldmann 1970; Schlee 1973); The African record (Schlee 1973) is from the Lebanon amber; Lebanon was then connected with Africa (Figure 9.5).

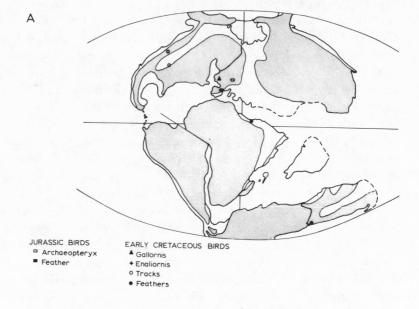

A

JURASSIC BIRDS
  □ Archaeopteryx
  ■ Feather

EARLY CRETACEOUS BIRDS
  ▲ Gallornis
  ✦ Enaliornis
  ○ Tracks
  ● Feathers

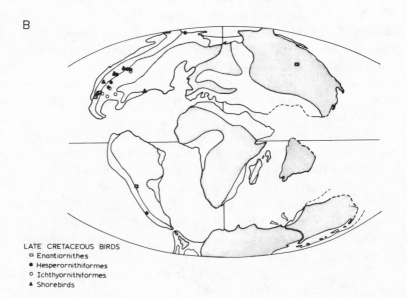

B

LATE CRETACEOUS BIRDS
  □ Enantiornithes
  ● Hesperornithiformes
  ○ Ichthyornithiformes
  ▲ Shorebirds

Figure 9.5. (A) Distribution of known Jurassic and early Cretaceous birds. (B) Distribution of known late Cretaceous birds. Continental positions and epicontinental seas are from Tedford (1974).

312

The oldest known hesperornithiform bird is *Enaliornis* from the upper part of the early Cretaceous (Wealden) in England. The remains of *Enaliornis* are uncrushed but are often badly abraded and broken. There is no known association of skeletal parts, but much of the skeleton can be reassembled. The cranium is known from several specimens, all of which are separated from the rest of the skull at the juncture between the parietals and the frontals. This contact was a movable (mesokinetic) joint in *Enaliornis* and in the much later *Parahesperornis* (Whetstone and Galton, unpub.). The ear region resembles that of *Parahesperornis* and *Hesperornis* and is the primitive condition in ornithurine birds (Whetstone, pers. comm.). The cervicals are becoming fully heterocoelous, but the posterior dorsals are amphicoelous. Nothing is known about the wing and shoulder girdle, but the extremely thickened pachyostosic bones suggest to me that flight had already been lost. The femur was short and broad much as in *Baptornis*. The cnemial crest was elongate and triangular, and the tibiotarsus was also elongate. It had no supratendinal bridge on the distal end. The tarsometatarsus was compressed and, as in other Hesperornithiformes, had a high anterior ridge on its lateral margin. It was marine as were virtually all other hesperornithiform birds.

Enaliornis resembles the loon level of adaptation for foot-propelled diving, and Brodkorb (1963b) actually assigned it to the Gaviiformes. There is no support for such an assignment beyond the general similarities of all foot-propelled diving birds (Storer 1960a). The characters of the skull (mesokinetic joint), tibiotarsus (broad triangular cnemial crest), and tarsometatarsus (external anterior ridge and enlarged outer trochlea) clearly support affinity with the other hesperornithiforms.

Enaliornis is unrecorded outside of England, and a hiatus exists in fossils of hesperornithiform birds until the late Cretaceous Niobrara Chalk of Kansas. This hiatus is only slightly broken by an *Enaliornis*-like tarsometatarsus from the Cenomanian, Greenhorn Formation, Kansas, in the Sternberg Museum, Fort Hays, Kansas.

The Niobrara Chalk reveals much about the hesperornithiforms (Figure 9.5B, 9.6). They were predators of small fish as shown by coprolites found with both *Baptornis* (Martin and Tate 1976) and *Hesperornis*. The body and neck were elongate and the bones nonpneumatic. The walls of the bones were pachyostosic, and the birds would have had a relatively high specific gravity to facilitate diving. Feathers are preserved with two of the *Parahesperornis* specimens. The feathers seem to have been largely plumulaceous, and the birds would have looked "furry." Impressions of the foot show that it was scutellate reticulate in *Parahesperornis* and probably also in *Baptornis* (Martin and Tate 1976).

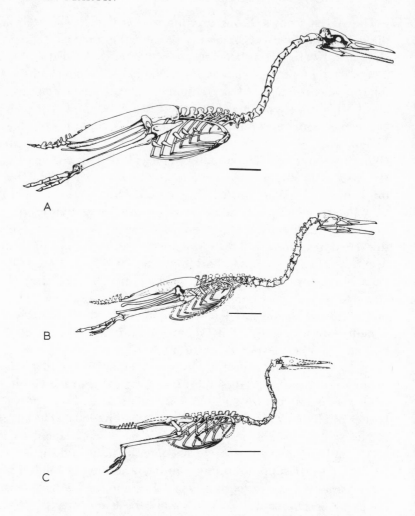

Figure 9.6. Restorations of the skeletons of three hesperornithiform birds found in the same levels of the Late Cretaceous Niobrara Chalk in Kansas: (A) *Hesperornis regalis*, (B) *Parahesperornis alexi*, (C) *Baptornis advenus*. Scale = 10 cm.

Toe rotation on the recovery stroke was well developed in the Hesperornithidae and was beginning to evolve in the Baptornithidae. Probably the feet in all Hesperornithiformes were lobed. As in foot-propelled divers in general (Storer 1960a,b), the feet are moved posteriorly on the body. The femur and tibiotarsus were locked in place and could no longer rotate ventrally so that the feet could not lie under the center of gravity. This meant that on land the hesperornithiforms pushed themselves along on their stomachs probably with an undulating motion as seals do today.

We know virtually nothing of the nesting of the hesperornithiform birds. *Baptornis* is the only genus for which young are known (Martin and Bonner 1977). The narrow pelvis would require an elongate egg, but little else can be said. The adults entered estuaries on occasion as Fox (1974) recorded from Alberta.

The hesperornithiform birds are diverse with the two Niobrara Chalk families (*Baptornithidae* and *Hesperornithidae*) containing four genera and six species (Figure 9.6). The later (Campanian) Pierre Shale contained at least two genera and five species. All of these were marine and range from central Kansas to above the Arctic Circle in Canada (Russell 1967; Martin and Stewart 1982). They ranged in size from that of a small grebe to birds nearly 1.8 m long (Figure 9.7) and were the small, endothermic seal-like pursuit predators of the late Cretaceous epicontinental seas (Figure 9.5)

Neogaeornis is a small hesperornithiform with foot-propelled diving adaptation like *Baptornis*. It is from the late Cretaceous of Chile (Lambrecht 1929; Brodkorb 1963b) and, until the description of the Enantiornithes, was the only Mesozoic bird known from South America. It is still the only known Southern Hemisphere hesperornithiform bird.

Hesperornithiform birds closely resemble other foot-propelled divers. Birds that swim primarily on the surface usually hold the feet under the body in a position that not only drives the bird forward but also provides lift and is not radically different from the posture used in terrestrial locomotion. The lift by the feet and the natural buoyancy of the body become a problem for underwater swimming. The solution for hesperornithiform birds was in part pachyostosis. This increased the specific gravity and was developed further in the hesperornithiforms than in any other known birds. Increases in specific gravity also produce increased wing loading, and an upper limit is set for pachyostosis if flight is to be maintained. Commonly, surface swimmers lack pachyostosis and must expend energy to stay under water, much like other animals expend energy to remain at the surface. The normal position of the feet under the body also provides lift that must be countered by having the legs out to the sides of the body. This lateral placement of legs and feet is permanently fixed in the hesperornithiforms and makes possible swimming directly forward without climbing or descending. Slight changes of the orientation of the feet make possible a full range of motion in three dimensions. The femur cannot be swung under the body (center of gravity), and, in fact, no articular

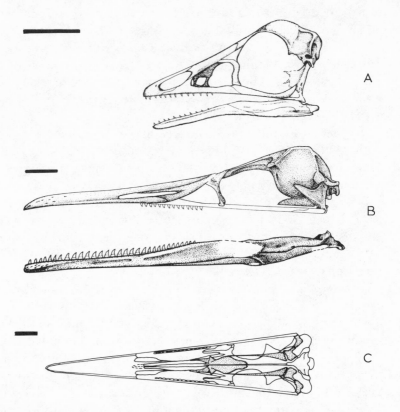

Figure 9.7. Skulls of Mesozoic birds. (A) *Archaeopteryx*, (B) *Parahesperornis alexi*, (C) hesperornithiform palate. (Modified from Gingerich, 1976). Quadrates, pterygoids, palatines, and vomers stippled. Scale = 2 cm.

facets exist for that position of the femur in the acetabulum. Foot-propelled divers usually do not use their wings for propulsion, and the forelimb may produce drag. With loss of flight, the forelimb is usually reduced. The femur no longer serves directly in propulsion but helps to position the foot. This orientation reduces drag and enhances the effectiveness of the feet for propulsion. Accompanying this, the tibia is affixed along the elongate postacetabular ilium (Fig. 9.8).

The hesperornithiform birds have large, well-ossified uncinate processes and sterna that suggest a typical avian respiratory system with parabronchi and sternal breathing. This implies the presence of an air sac system, although no direct evidence is available in the form of pneumatopores in the bones. The function of the large pleurocoels is uncertain. The shoulder girdle is remarkably primitive with *Archa-*

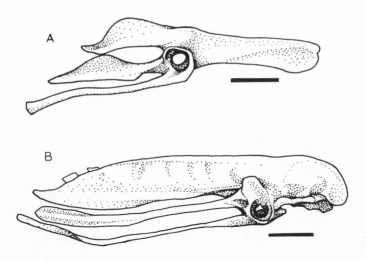

Figure 9.8. (A) *Apatornis* pelvis, (B) *Hesperornis regalis* pelvis. Scales: A = 1 cm; B = 5 cm. (From Marsh, 1880.)

eopteryx-like coracoids. The clavicles are not fused to form a furcula, and the wings are extremely reduced. The radius and ulna are known only in *Baptornis* and if present in *Hesperornis* must have been very small.

The palate of *Parahesperornis* and *Hesperornis* is now known in detail from three specimens. Gingerich (1973) restored the skull and palate of *Hesperornis* from *Hesperornis regalis*, YPM (Yale Peabody Museum) 1206, and the University of Kansas specimen, KUVP 2287. The two specimens are not congeneric, the latter being the holotype of *Parahesperornis alexi*. Consequently, Gigerich's restoration (see Figure 9.7) is a composite and does not totally represent either genus. The features of the palate are similar in the two genera, and Gingerich's restoration of the palate seems basically correct. McDowell (1978) questioned Gingerich's identification of some palatal elements, but examination of the actual specimens leaves little doubt. The quadrate, pterygoid, and palatine were essentially articulated on KUVP 2287. Although Gingerich (1976) called this a palaeognathous palate, it is so only in the broadest sense of the term, and the hesperornithiform palate with its very short pterygoids and long slender palatines appears to be a unique complex, characteristic of this group.

The hesperornithiform birds lack teeth on the premaxilla (Figure 9.7). However, teeth are retained in the maxilla and the dentary. The premaxilla is covered with small foramina in the region of the bill

indicating that it was highly vascularized, and a horny bill (rhampho-
theca) was probably present. The skulls of the hesperornithiform
birds were loosely sutured throughout life, and both *Hesperornis* skulls
are considerably disassociated. *Enaliornis* and *Parahesperornis* are fully
mesokinetic with a joint at the juncture of the frontals and the parietals. This joint is fused in *Hesperornis* apparently because the origins
of the temporalis muscles have moved anteriorly over it. Gingerich
(1976) points out that rhynchokinesis as in modern birds would be
impossible and proposes instead a unique kinesis (maxillokinesis)
where the maxillae can slide anteroposteriorly on the nasal–premaxil-
lary subnarial bars (Gingerich 1973).

The skull of *Ichthyornis* has not been described adequately, the bill
and the maxilla are essentially unknown, but certainly the dentaries
carried typical avian teeth (Gingerich 1972; Martin and Stewart 1977;
Martin et al. 1980). These are fully socketed in older adults, but in
younger specimens they are in a groove as in *Hesperornis*. The lower jaw
has a well-developed intermandibular joint between the splenial and
the angular precisely as do *Hesperornis* and the contemporary marine
lizards (mosasaurs). This joint permits some lateral flexion of the jaws
(Gregory 1951, 1952). It appears that a coronoid bone was present.
Thus, *Ichthyornis* is presently the only bird where this bone seems to be
clearly definable, although I suspect that it occurred in *Archaeopteryx*
and *Hesperornis*.

The cervical vertebrae of *Ichthyornis* have large pleurocoels and are
amphicoelous with a tendency toward heterocoely. Evolutionarily heterocoely begins at the front of the vertebral column and moves posteriorly. Many living birds have nonheterocoelous posterior vertebrae,
and this is presumably primitive. The synsacrum contains 10 fused
vertebrae. Marsh restored *Ichthyornis victor* with five free caudal vertebrae and three fused into the pygostyle. However, the tail was possibly
longer as no articulated tail has been found. The wing is typically
carinate with a totally fused carpometacarpus. The distal segments are
short and robust, and the three segments of the wing (primary, secondary, humeral) are all of similar length. The distal end of the
humerus has a deep depression for *brachialis anticus* and generally
resembles the distal end in charadriiform humeri. The proximal end
has the deltoid crest rotated nearly 90° from the position in *Archaeopteryx*. This position was considered derived for *Ichthyornis* by Harrison
(1973), who concluded that the Ichthyornithiformes are not directly
ancestral to any known Tertiary birds. I agree.

The pelvises of *Apatornis* (Figure 9.8) and *Ichthyornis* have long preacetabular and relatively short postacetabular ilia. The acetabulum is partially closed. The ischium is short, broad, and separate from both the ilium and pubis except in the acetabular area. The femur has a modern appearance. The tibiotarsus has a low cnemial crest and lacks a supratendinal bridge on the distal end in both *Ichthyornis* and *Apatornis*. The tarsometatarsus of *Ichthyornis* is short and broad. There is a small projection in the region of the hypotarsus but no well-defined hypotarsus. The distal end is not highly arched and the inner condyle is situated proximally, a derived condition. There is a distinct enclosed distal foramen.

The sternum is keeled and the coracoid, scapula, and furcula all join to form a typical triosseal canal. This is the earliest evidence for the typical carinate flight apparatus. The coracoid has a deep round scapular facet, which seems to be characteristic of the early carinate birds. In spite of this close similarity to carinate birds in general and to charadiiform birds in particular, the Ichthyornithiformes are clearly a side branch, and our main insight into the origins of modern birds must lie with other Mesozoic birds. The fauna of the Niobrara epicontinental sea (Figure 9.9) is thus composed of evolutionary "dead-ends."

The only known birds that might elucidate the origin of the Tertiary avifauna are from the New Jersey Green Sands and from the Upper Cretaceous of Wyoming, Montana, and Alberta (Figure 9.5). The most important papers on these specimens are Brodkorb (1963a) and Cracraft (1972, 1973a), who thought they recognized loons (Gaviiformes), flamingos (Phoenicopteri), shorebirds (Charadriiformes), and a gruiform. Olson and Feduccia (1980b) contend that the supposed loon *Lonchodytes* and the supposed flamingo *Torotix* are incorrectly assigned and resemble charadriiforms. Cracraft concluded that *Telmatornis* belonged in the Charadriiformes. The affinities of *Laornis* are uncertain. Cracraft (1973b) placed it in the gruiformes within its own superfamily, the Laornithoidea. It has a broad supratendinal bridge and is one of the oldest birds clearly showing this character. The distal margin of the internal condyle is notched as it often is in Gruiformes but there is no tubercle on the supratendinal bridge. Cracraft (1973a) considered *Laornis* close to the Rallidae. Brodkorb (1978) considered it *incertae sedis* within the Aves, and this correctly reflects present knowledge. This is not because we lack sufficient evidence to assign it to a group but because we lack sufficient evidence to fully characterize the group to which it belongs. Indeed, it may well be that the Mesozoic birds now assigned to the Charadriiformes, if they were fully known, would all be assigned to extinct orders. Probably they and their close relatives con-

Figure 9.9. Reconstructions (from left to right) of *Hesperornis regalis, Baptornis advenus,* and *Ichthyornis victor.* (Modified from restorations in Spinar and Burian, 1972; Martin and Tate, 1976.)

tain the progenitors of the entire Tertiary radiation. The earliest Paleocene birds that I have seen represent a continuance of basic patterns in these late Mesozoic "shorebirds." It would not be surprising if they occupied an evolutionary position analogous to that of the Mesozoic "insectivores" at the base of the Tertiary mammalian radiation. It is also not surprising that the earliest representatives of many bird orders like the Anseriformes, Phoenicopteri, and Galliformes show similarities with shorebirds and waders (Olson and Feduccia 1980a, b).

The relationships of the Mesozoic Ornithurae are summarized in Figure 9.10. The Hesperornithiformes are grouped (Figure 9.10) with modern birds in the Ornithurae on the basis of the following: (1) ossified uncinate processes; (2) fusion of the metatarsals beginning distally; (3) middle metatarsal (metatarsal III) posterior to the medial and lateral metatarsals on the proximal end; and (4) large triangular tarsal cap on the tarsometatarsus. The Hesperornithiformes are united by (5) elongated postacetabular pelvis; (6) shortened femur; (7) reduced wing; (8) palatal structure (the combination of short broad pterygoids with long narrow palatines described by Gingerich, 1976, is unique for the Class Aves); (9) maxillokinesis (Gingerich 1976); (10) enlarged triangular cnemial crest on the tibiotarsus; (11) large trihedral patella with a foramen for the *ambiens* tendon; (12) compressed tarsometatarsus with the outer metatarsal emphasized; and (13) distinct outer metatarsal ridge. All carinate birds including *Ichthyornis* may be united by the (14) carinate sternum; (15) semilunate articular surface on the dis-

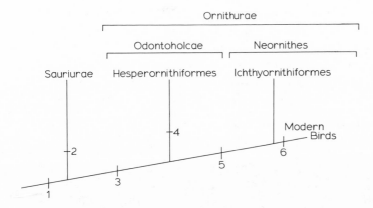

Figure 9.10. Phylogenetic relationships of the Sauriurae and the Cretaceous Ornithurae. Characters utilized: (1) feathers and pretibial bone; (2) characters 1–5, p. 310 of text; (3) characters 1–4, p. 320 of text; (4) characters 5–13, p. 320 of text; (5) characters 14–16, pp. 320–321 of text; (6) characters 17–28, p. 321 of text.

tal end of the ulna (absent on the hesperornithiform *Baptornis*, but the ulna is unknown in *Hesperornis*); and (16) coracoid with scapular facet significantly below the humeral end (primitively terminal, but it is unclear whether this condition is primitive or derived in ratites). All modern carinates are united by the following: (17) loss of teeth from the dentary and maxillary bones; (18) loss of the coroid bone from the lower jaw; (19) reduction of the parietals to a very posterior portion of the skull roof; (20) rhychokinesis; (21) neognathous palate; (22) elongated pterygoid; (23) fused uncinate processes; (24) pneumatic bones; (25) pelvis with a completely open acetabulum; (26) ilium and ischium united posteriorly; (27) supratendinal bridge on the tibiotarsus (absence of the bridge is primitive, but it is unclear whether its absence is a reversal in owls and parrots); and (28) a distinct hypotarsus on the tarsometatarsus.

The palaeognathous birds are excluded in this analysis because of difficulties in determining polarities. I have divided the Ornithurae Haeckel, 1866 into two infraclasses: the Odontoholcae Stejneger 1884 (new rank) for the hesperornithiform birds and the Neornithes Gadow, 1893 (new rank), for *Ichthyornis* and all later birds.

Ossification of the sternum appears to have evolved independently in the later Sauriurae and in the Ornithurae. Heterocoelous vertebrae probably evolved independently three times: (1) in the Sauriurae, where they

may never have become fully developed; (2) in the Hesperornithiformes; and (3) in the carinates in a stage later than *Ichthyornis*. A tendency toward this type of vertebral articulation occurs in anterior cervical vertebrae in *Ichthyornis*. *Hesperornis* and *Baptornis* are more advanced
than many living carinates (as, for instance, most Charadriiformes) in
having all the presacral vertebrae (except the atlas) heterocoelous.

Early Tertiary radiation

Far less is known about the Paleocene avian radiation than
about that of the Cretaceous. This situation may improve with a revival
of interest in Paleocene fossil mammals. With rare exceptions, the finding of Cenozoic fossil birds has been a by-product of collecting the
more common and extensively studied fossil mammals. Like the mammals, the earliest Paleocene birds seem to be mostly a continuance of
the latest Cretaceous kinds. These belong to a basic shorebird adaptive
type rather than any modern taxonomic entity. Probably many of these
birds, if fully known, would be members of extinct groups or would
show trends culminating in very different taxonomic groups later in
the Tertiary.

Presently, the most important Paleocene avifauna comes from the
late Paleocene of France. Russell (1964) recently collected from these
deposits, which were also important sources of fossil mammals and
birds in the last century. The most common bird is a gigantic flightless
form, *Gastornis*. Knowledge of *Gastornis* comes largely from work by
Lemoine (1878–1881), who published a restoration of *Gastornis edwardsi*
that was widely copied and compared to the contemporaneous North
American bird *Diatryma*. Although some similarity is evident, Lemoine's
restoration of *Gastornis* differs considerably from Matthew and
Granger's (1917) restoration of *Diatryma*. Most of these differences disappeared when I examined Lemoine's original material, which proves
to be a composite of a variety of animals. The restored skull contains
some pieces that are probably avian, but the entire lower jaw is composed of various fish bones and the quadrate of a turtle. The shoulder
girdle and wing are composed of turtle and champosaur, a primitive
crocodilelike reptile, elements. The elongate tarsometatarsus is a composite of the distal end of a tarsometatarsus along with fragments of a
femur and a tibiotarsus. When all extraneous parts are removed, only a
little remains of the original skeletal restoration, and this closely resembles *Diatryma*. There seems to be no adequate evidence to keep the
Diatrymidae separate from the Gastornithidae, although the genera
Diatryma and *Gastornis* should be kept as separate on the basis of details

of the tarsometatarsus. Very little evidence links the Gastornithes with the cranes as did Matthew and Granger (1917). The skeleton, including the palate, has some features that are usually associated with ratites. The Gastornithes have generally been considered predaceous birds, although objections have been raised because they do not seem to have an adequate "tearing bill." They have a very large head and, because most Paleocene mammals were small, *Diatryma* and *Gastornis* might well have swallowed their prey whole.

The Paleocene fauna of France also contains a sizable and diverse fauna of fossil owls. Diurnal birds of prey (Falconiformes) are presently not known before late Eocene (Cracraft 1969), and the first Paleocene owl was only described a few years ago (Rich and Bohaska 1976). If many late Cretaceous and early Tertiary mammals were small and nocturnal, an early radiation of owls seems plausible. From the middle Eocene until the Oligocene, the strigiform radiations in North America and Europe seem to have been separate as the American Eocene owls belong to an archaic family, the Protostrigidae, which is unknown from Europe. A purported protostrigid, *Eostrix*, from Britain (Harrison 1980) does not resemble other protostrigids. The modern strigiform radiation seems to be derived from the Old World owls.

Along with *Gastornis*, the French fauna also contains *Remiornis heberti*, which seems best interpreted as a ratite. It was large and, judging from its thick-walled limb bones, flightless. *Remiornis* shares many features with the fossil ratites *Mullerornis* (Pleistocene of Madagascar), *Eremopezus* (Eocene of Africa), and *Stromeria* (Oligocene of Africa). These three genera are often placed in the Aepyornithidae with the elephant birds. The relationships of the ratites have been extremely controversial. I presently think that they come from a very early phase of modern avian radiation, and some of the traits thought to unite them, including many aspects of the palate, may be retentions of primitive characters that occurred in many Paleocene birds.

This idea is supported by the description (Houde and Olson 1981) of volant early Tertiary palaeognaths from the Paleocene of Montana and the Lower Eocene of Wyoming. These birds are apparently like the hesperornithiforms *Enaliornis* and *Parahesperornis* in having a fully functional joint between the frontals and the parietals (Houde and Olson 1981). The palate is quite different from that in the *Hesperornithiformes*, as the pterygoids are elongate and fused to the palatines. In the hesperornithids, the pterygoids are short, broad, and articulated by a movable joint to the palatines. The very slender palatines of the hesperornithids also do not fuse with the vomer as described for these early Tertiary birds. I suspect that the loss of flight occurred independently in several different lineages. In contrast, Cracraft (1974) attempted to identify primitive and derived character states in the osteology of ratites and concluded that palaeognathous birds are strictly monophyletic (holophyletic). The following features occurred in their ancestors: tibiotarsus with a supratendinal bridge, narrow intercondylar fossa, rounded and relatively small condyles, internal condyle with no ligamental pit or palatelike ligamental prominence, tarsometatarsus with a simple hypotarsus of one low ridge and with a shallow anterior groove. If these assumptions are accepted, then we may extend Cracraft's phylogeny to include *Remiornis* between the Apteryges (moas and kiwis) and the Struthiones, placing them close to the basal stock of the elephant birds and the ostriches, emus, and cassowaries. This position is not unreasonable for the oldest known ratite.

The earliest African ratites are Eocene (*Eremopezus* and *Psammornis*) and Oligocene (*Stromeria*). *Eremopezus* and *Stromeria* are not well known but appear similar to *Remiornis*. Ratites occurred in Europe from the late Paleocene through the middle Eocene (*Eleutherornis*) and early Oligocene (*Proceriavis*). South American ratites appear in the Miocene with *Opisthodactylus*. If the argument (Cracraft 1974) that the closest relative of the rhea is the ostrich is valid, then the biogeographic situation is analogous to that of both caviomorph rodents and platyrhine primates. Both mammalian groups first appeared in the South American Oligocene and are thought to have their closest relatives in Africa. They are unknown in North America. Whatever route explains the distribution of these animals should also work for birds, and it is unnecessary to explain the present distribution of rheas and ostriches by the rifting of South America and Africa. Such an explanation also requires an earlier separation of the rhea–ostrich common lineage from the rest of the avian radiation than seems justified by the fossil record.

It is not surprising that the Gastornithidae occurred in both North America and Europe, because these land masses were connected across the North Atlantic (McKenna 1975) through the Paleocene and early Eocene. The mammalian faunas of the two regions were similar then, and the avifauna should have been also, and we might expect to find birds like *Remiornis* in North America as well as Europe. The recently discovered fauna on Ellesmere Island is critical (West and Dawson 1978), because it confirms the presence of *Diatryma* on a remnant of the old North Atlantic connection and shows that paratropical vegetation and fauna, which included large testudinid tortoises and alligators, extended practically to the North Pole. Hence, no high-latitudinal climactic barriers affected distribution. In other words, the latitudinal zonation as it exists today developed progressively during the Tertiary. Tropical forms probably composed essentially the entire avifauna during the late Paleocene and early Eocene. We would expect also that European and North American Paleocene or early Eocene avifaunas would contain about the same taxa. These faunas would diverge throughout the Eocene and tropical faunas would also become more isolated as high-latitude climatic barriers became effective.

As far as is known, all toothed birds became extinct at the end of the Cretaceous. It is likely that the early shorebirds found with the mammalian faunas of the late Cretaceous were not toothed. When birds reradiated, teeth did not reappear, and functional substitutes for teeth developed in the horny bill and occasionally as the bone of the jaws themselves. The most remarkable of these were the bony-toothed birds Odontopterygiformes (Harrison and Walker 1976b). They were important members of the avifauna from the lower Eocene through the early Pliocene when they occurred in Europe, North America, Africa, and New Zealand, indicating a worldwide distribution. Some had wingspreads approaching 5.5 m, making them nearly the largest flying birds known.

Knowledge of Eocene avifaunas is based largely on the Lower Eocene, Green River Shales, and associated sediments in North America (Figure 9.11). Most remains in these deposits are of waders or other water birds. The Green River lakes covered thousands of square miles in the present states of Colorado, Wyoming, and Utah. Through the early Eocene the lakes fluctuated in size, shape, and salinity, and they provided enormous lengths of shoreline and marginal marshy habitat. The surrounding vegetation was tropical or subtropical forest with a wide variety of plants including palms. The lake fauna included crocodilians, and the marginal vegetation was used by extinct lemuroid primates and tapirs. The best analog of this environment may be lakes in the rift valleys of Africa today (Feduccia 1980).

The most important fossil bird found in the Green River Lakes was *Presbyornis*, a wading anseriform whose disarticulated remains have been described as a recurvirostrid (Wetmore 1926), a flamingo (Howard 1955), and finally as an anseriform (Harrison and Walker 1976c; Olson and Feduccia 1980a). This confusion is to be expected with a basal member of a group using fragmentary remains far back in the geologic record. This difficulty reflects more our ignorance of what these early birds looked like than lack of diagnostic characters. *Presbyornis* is significant in demonstrating the existence in early Eocene of an

Figure 9.11. Reconstructions (from left to right) of *Diatryma*, *Presbyornis*, and *Primobucco*. (Modified from restorations in Spinar and Burian, 1972; Olson, 1977; Feduccia, 1980.)

important radiation of anseriforms like flamingos occupying shallow saline lakes in vast numbers. The fact that they combine many primitive features with long wading legs suggests that swimming might be a secondary adaptation in anseriforms. True flamingos are contemporary with *Presbyornis* and had the bill adaptations of modern flamingos (Olson and Feduccia 1980b). Olson and Feduccia (1980a,b) derive both the Anseriformes and the Phoenicopteri from shorebirds (Charadriiformes). Olson (1977) also described a frigatebird, *Limnofregata azygosternon*, in the Green River lacustrine sediments. Feduccia (1980) suggests that it preyed on the chicks of *Presbyornis*.

The forest around the Green River lakes held a diversity of birds and included many forms restricted to the tropics and subtropics today. The dominant perching birds of this time appear to be an extinct family of early Piciformes, the Primobucconidae (Feduccia and Martin 1976), which are most closely related to the puffbirds (bucconidae) today found primarily in South America. The Primobucconidae may link the Piciformes with the Coraciiformes. The slightly younger (middle Eocene) European Messel Fauna lacks primobucconids but does contain a variety of coraciiforms. This seems to be a general pattern in Europe, and probably *Primoscens minutus* and *Parvicuculus minor* from the English Eocene are coraciiforms rather than the oldest known passeriform and cuculid, respectively. *Parvicuculus* was considered a primobucconid by Olson and Feduccia (1979b), but my examination of the specimen convinces me that their assignment is wrong. The Primobucconidae seem to be restricted to North America.

The family Neocathartidae is represented only by *Neocathartes*, a long-legged cursorial bird with a curved raptorial beak, which might be related to New World vultures. It was apparently an analog of the African Secretarybird. Like many early ground predators, it had many gruiform characters. New World vultures (Cathartidae) occurred in Europe in the middle Eocene of Germany and apparently radiated substantially in the European Oligocene (Cracraft and Rich 1972). Two middle Eocene localities of Germany, the Messel Local Fauna and the Geisaltal locality, contain remarkable fossil avifauna (Hoch 1980). In both places, the fossils were preserved in oil shale and are generally articulated specimens with some preserved feathers. This avifauna is mostly unstudied, but over 100 specimens, comprising 10 to 20 genera, are involved. Many of these genera show tropical African affinities, and the presence of a ground roller (*Brachypteraciidae*) shows affinities with Madagascar. Probably *Primoscens minutus* from the Upper Eocene of England (Harrison and Walker 1977) was also a ground roller. The African affinities of the European avifauna remained even in the late Eocene and Oligocene of France.

The Bathornithidae in North America and the Ergilornithidae in Asia are large gruiform members of Oligocene faunas. The Ergilornithidae became cursorially adapted only to the extent that the outer toe was reduced. I regard this reduction as convergent with that in ostriches, although some believe it indicates relationship (Feduccia 1980). Probably the ergilornithids and at least some bathornithids were moderately large ground predators. This theme was extended in South America in the Oligocene where the first phororhachids became very large and represent an important component of the South American predators during the rest of the Tertiary.

Most of the earlier assignments of fossil birds older than late Eocene have proven to be wrong, and later Eocene to Oligocene records are still the oldest for many families, such as the Anatidae, Ciconiidae, Gaviidae, and Cuculidae. Surprisingly, the Eocene through Oligocene faunas lack a significant early record of the Passeriformes (Olson 1976; Feduccia 1977, 1980). This order might have diverged before its fossil record indicates, but certainly the bulk of the radiation was Miocene or later. Clearly, some avifaunas of truly modern aspect date back to the Oligocene, and many birds presently confined to the Old and New World tropics were widely distributed in middle latitudes then. In Europe, the faunas show distinctly African affinities (Rich 1974), and North America contained birds that are presently considered South American (Feduccia 1980). Tertiary avifaunas have recently been summarized for Australia (Rich 1975a), Asia (Kurochkin 1976), and South America (Tonni 1980).

Basic patterns

Because most fossils are fragmentary, the classification of fossil birds is conservative. Many fossil birds have been referred for convenience to a modern group that they resemble most closely, although clear relationships may not be established. It is unlikely that many avian orders are much older than comparable mammalian orders. Bird evolution has probably been affected by factors similar to those guiding mammalian evolution, and we might expect the two to parallel each other. Only a few mammalian orders go back to the Cretaceous, and many Paleocene mammals belong to primitive orders, which became extinct by Oligocene times. Most modern avian orders date back to Paleocene and Eocene with the modern families beginning in the late Eocene or Oligocene.

The importance of convergence is established among fossil mammals, and we know that certain lineages acquired great morphological similarity independently. This can be seen easily in saber-toothed mammals, where the highly specialized dirk-toothed morphology evolved at least three times in placentals and once in marsupials (Martin 1980a). Similarly in birds, the highly specialized Hesperornithiformes resemble modern grebes (Storer 1960a,b) and large, wing-propelled divers evolved in the Sphenisciformes, Charadriiformes (Great Auk), and Pelecaniformes (Plotopteridae) (Storer 1960a, b; Olson 1980). The existence of detailed convergence in the fossil record suggests caution when we use such similarities to unite modern birds.

Summary and conclusions

The origin of birds remain controversial, and three viable models exist. Two of these, the crocodilian sister group relationship (Walker 1972; Whetstone and Martin 1979) and the pseudosuchian origin (Heilmann 1926; Tarsitano and Hecht 1980) are rather similar, as the common ancestor of birds and crocodiles would probably be placed in the paraphyletic grade Pseudosuchia in most current classifications. Both models require a separation of the avian lineage in the middle Triassic or earlier, do not require that the ancestral form be an advanced biped, and are consistent with an arboreal origin of avian flight. Endothermy would probably be a result of the evolution of flight as would feathers, and it seems unlikely that either would precede it. The third model, a theropod dinosaur origin of birds (Marsh 1877; Ostrom 1973), is based on characters only found in very late theropods; hence, the separation of the avian lineage is more recent. the extreme version treats *Archaeopteryx* as a theropod and derives birds in the early Cretaceous (Ostrom 1978). The model demands that the proavis be an obligate biped and is most consistent with a cursorial origin of avian flight. Endothermy and feathers presumably would have evolved in a preflight running phase.

The pseudosuchian model was based primarily on primitive archosaurian characters and is thus difficult to evaluate. The crocodilian model is supported primarily by skull and dental characters, whereas the strongest evidence for a theropod origin of birds seems to be in the wrist and manus. I feel that the weight of evidence has shifted in favor of a crocodilian relationship for birds, but many authors still support Ostrom's theropod model.

Archaeopteryx remains the only known Jurassic bird and the oldest known bird. However, it can no longer be considered the progenitor of all later birds but is, instead, the oldest known member of the Subclass Sauriurae. This implies that birds of more modern types (Subclass Ornithurae) were present in the late Jurassic and that the common ancestor of the Sauriurae and the Ornithurae must be older. This also puts an upper limit on the time of bird origins as it must be earlier than the late Jurassic. The Sauriurae had an essentially worldwide distribution in the late Cretaceous (Figure 9.5). They were the only known Mesozoic terrestrial birds and varied greatly in size and adaptation. They are not related to any living group of birds but, like the dinosaurs, became extinct at the end of the Cretaceous. The Cretaceous forms seem to have been strong flyers. I accept the arguments (Feduccia and Tordoff 1979; Olson and Feduccia 1979a) that the related *Archaeopteryx* had limited powers of flight. It seems clear that it was a small, arboreal predator with essentially modern feathers.

The early Ornithurae appear to be either shoreline dwellers or were fully aquatic. The oldest member of the subclass with a recognizable fossil record is the foot-propelled diving bird *Enaliornis*. *Enaliornis* was a member of the extinct order Hesperornithiformes, which was highly adapted for foot-propelled diving. These birds shared with *Archaeopteryx* a number of features that are undoubtedly primitive for birds. Heterocoelous vertebrae probably evolved independently twice in the Ornithurae and perhaps also in the Sauriurae. Some features of the Hesperornithiformes may be confined to that group or are primitive ornithurine characters lost in the modern birds. These include a mesokinetic skull, intermandibular joint, a palate with broad short pterygoids, and long slender palatines. Some aspects of the *Hesperornis* palate were considered palaeognathous by Gingerich (1976), and possibly these aspects occurred widely in late Cretaceous and early Tertiary ornithurine birds. The presence of an ossified sternum and ossified uncinate processes in *Hesperornis* suggests that the modern avian mode of breathing had developed by the early Cretaceous. It seems unlikely that *Archaeopteryx* breathed this way as it lacked an ossified sternum. The Hesperornithiformes were diverse and differed greatly in size. They were small endothermic predators of the epicontinental seas. They represent an important fraction of the known Mesozoic avian radiation but became extinct by the end of the Cretaceous (Martin 1980b).

The oldest known carinate (*Ichthyornis*) was a small, toothed, ternlike bird with amphicoelous vertebrae. The Ichthyornithiformes became extinct at the end of the Cretaceous, but similar birds like *Cimolopteryx* probably survived. These birds are usually put in the Charadriiformes but may contain the progenitors of the entire Tertiary radiation. The Paleocene radiation of birds is still largely unknown, and we are just beginning to explore the Eocene record. Most probably the pattern of the avian radiation followed closely that of the mammals. By extension, the modern orders of birds should generally date back to the Paleocene and early Eocene, whereas the modern families appeared in the late Eocene and Oligocene. Most modern genera should be even younger.

Many Paleocene and Eocene forms probably belong to extinct orders and families, and efforts to force them into modern groups has resulted in confusion. During the early and middle Eocene, paratropical faunas and floras occurred at high latitudes, eliminating any opportunity for latitudinal zonation. The earliest members of modern orders may have all evolved under tropical conditions although not necessarily near the equator. Hence, members of groups that still have large suites of primitive characters would logically be tropical today. The development of latitudinal zonation appeared gradually with more seasonal climates developing progressively as the tropics contracted toward the equator. The effects of plate tectonics should also be considered (Cracraft 1973b; Rich 1975b), but many plate movements occurred before the origin of the bird groups that have been related to them. The contraction of the tropics and the development of high-latitude environmental filters may be more important in explaining the distribution of modern birds than is plate tectonics.

One of the most striking aspects of the fossil record of mammals was the nearly simultaneous appearance of most of the modern families in Europe and North America in the early Oligocene. Apparently birds show a similar pattern. If tropical conditions prevailed at high latitudes in the early Eocene, then the present latitudinal zonation of avifaunas is more recent. I contend that modern temperate avifaunas originated in high latitudes and then shifted into middle latitudes as the tropics contracted. Thus, we might expect to find the earliest fossil members of modern groups in northern Canada or Asia. Unfortunately, these areas have produced practically no Tertiary faunas, and the seemingly sudden appearance in the Oligocene of avifaunas of modern aspect in Europe and North America might represent the shift of these faunas into the middle latitudes where paleontologists have been active. Oligocene faunas in North America and Europe retained a subtropical aspect, and many forms presently confined to Africa occurred in Europe, whereas North America showed affinities with South America.

The history of birds resembles mammals in showing the repeated evolution of distinctive adaptive types separated by either geography or time as seen in the close similarity of the foot-propelled divers or in the wing-propelled divers. As the fossil record becomes better understood, patterns will appear that will make additional contributions to our understanding of avian evolution. To date, the chief contributions of the avian fossil record have been biogeographic and the strong evidence for evolution provided by the Mesozoic toothed birds whose importance should not be underestimated. Barbour (1902), who described the reaction of individuals in the last century who were opposed to evolution when confronted with the display of Cretaceous toothed birds, wrote:

The writer hopes that it may be germane to the subject to mention parenthetically that as a student, and subsequently as an assistant, he repeatedly saw parties of men and women, far more religiously zealous than wise, urging Professor Marsh to consider the advisability of concealing this specimen because it savored too much of evolution. They admitted its genuineness, seeing it was before their eyes in the cabinet, but denied that the facts should be made known by allowing it to stand so publicly in exhibition in the cases, and proposed as a remedy that an opaque curtain be so arranged as to be drawn over the specimen to conceal it. The case containing the polydactyl horses, as well as that containing *Hesperornis* and *Ichthyornis*, seemed to trouble them especially.

The fossil record will continue to trouble such people, but it may also trouble the ornithological community. Many of us are involved in the reconstruction of evolutionary processes. Fossils provide tests of the assumptions which we use to create these reconstructions. As the fossil record improves, we must be prepared to abandon some preconceptions, but we may be repaid for this with a more complete and stable picture of avian evolution.

Acknowledgments

For allowing me to examine specimens, I thank J. P. Lehman, D. Goujet, F. Poplin, and D. E. Russell (Museum National d'Histoire Naturelle, Paris); A. J. Charig, A. Milner, and C. A. Walker (British Museum (Natural History), London); G. S. Cowles and C. J. O. Harrison (British Museum, Ornithological subdepartment, Tring); A. D. Walker (University of Newcastle upon Tyne); G. Viohl (Jura Museum, Eichstatt); H. Jaeger and H. Fischer (Humboldt Museum für Naturhunde, Berlin); P. Wellnhofer (Bayerische Staatssammlung, Munich); D. S. Peters (Forschungsinstitut Senchenberg); O. Feist (Muhltal); W. v. Koenigswald (Hesschian Landesmuseum, Darmstadt); Z. Kielan-Jaworowska, A. Elzanowski, and H. Osmolska (Polska Akademia Nauk, Warsaw); P. Ellenberger (Laboratoire de Paleontologie des Vertebres, Montpellier); J. H. Ostrom and M. Turner (Yale Peabody Museum, New Haven); R. J. Zakrewski (Sternberg Memorial Museum, Hays); C. B. Schultz, L. G. Tanner, and M. Voorhies (University of Nebraska State Museum); S. L. Olson and C. Ray (United States National Museum, Washington, D.C.); and T. Ferrusquia (Institute de Geologia, Mexico City).

I have benefited from stimulating conversations with S. Olsen, J. Cracraft, P. Rich, A. Fedducia, R. Mengel, P. Brodkorb, C. Harrison, C. A. Walker, A. Milner, J. D. Stewart, and K. N. Whetstone. M. A. Klotz prepared the figures. M. Jenkinson and J. D. Stewart critically read the manuscript. Funding was provided by the University of Kansas (sabbatical leave) and University General Research Grant 3251–5038, NSF DEV 7821432, and National Geographic Grant 2228-80.

Literature cited

Barbour, E. H. 1902. President's address – the progenitors of birds. In *Proceedings of the Nebraska Ornithological Union, Third Annual Meeting*, pp. 8–39.
Bock, W. J. 1965. The role of adaptive mechanisms in the origin of higher levels of organization. *Syst. Zool. 14*:272–287.

– 1969. The origin and radiation of birds. *Ann. N. Y. Acad. Sci. 167*:147–155.
– 1977. Ichthyornithiformes. In *McGraw-Hill encyclopedia of science and technology*, Vol. 7; p. 7. New York: McGraw-Hill.
Brodkorb, P. 1963a. Birds from the Upper Cretaceous of Wyoming. In *Proceedings of the 13th International Ornithological Congress*, pp. 55–70.
– 1963b. Catalogue of fossil birds, Part 1. *Bull. Fl. State Mus. Biol. Sci. 7*:179–293.
– 1964. Catalogue of fossil birds, Part 2. *Bull. Fl. State Mus. Biol. Sci. 8*:195–335.
– 1967. Catalogue of fossil birds, Part 3. *Bull. Fl. State Mus. Biol. Sci. 11*:99–220.
– 1971a. Catalogue of fossil birds, Part 4. *Bull. Fl. State Mus. Biol. Sci. 15*:163–266.
– 1971b. Origin and evolution of birds. In D. S. Farner and J. R. King (eds.). *Avian biology*, Vol. 1, pp. 19–55. New York: Academic Press.
– 1976. Discovery of a Cretaceous bird, apparently ancestral to the Orders Coraciiformes and Piciformes (Aves: Carinatae). *Smithson. Contrib. Paleobiol. 27*:67–73.
– 1978. Catalogue of fossil birds, Part 5. *Bull. Fl. State Mus. Biol. Sci. 23*:139–228.
Broom, R. 1913. On the South African pseudosuchian *Euparkeria* and allied genera. *Proceedings of the Zoological Society of London*, pp. 619–633.
Cope, E. D. 1867. An account of the extinct reptiles which approached the birds. *Proc. Acad. Natl. Sci. Philadelphia, 19*:234–235.
Cracraft, J. 1969. Notes on fossil hawks (Accipitridae). *Auk 86*:353–354.
– 1972. A new Cretaceous charadriiform family. *Auk 89*:36–46.
– 1973a. Systematics and evolution of the Gruiformes (Class Aves) 3. Phylogeny of the suborder Grues. *Bull. Am. Mus. Nat. Hist. 151*:1–127.
– 1973b. Continental drift, paleoclimatology, and the evolution and biogeography of birds. *J. Zool. London 169*:445–546.
– 1974. Phylogeny and evolution of the ratite birds. *Ibis 115*:494–521.
Cracraft, J., and P. V. Rich. 1972. The systematics and evolution of the Cathartidae in the Old World Tertiary. *Auk 74*:272–283.
Crompton, A. W., and K. K. Smith. 1980. A new genus and species of crocodilian from the Kayenta Formation (late Triassic?) of northern Arizona. In L. Jacobs (ed.) *Aspects of vertebrate history*, pp. 193–217. Flagstaff: Museum of Northern Arizona Press.
Currie, P. J. 1981. Bird footprints from the Gething Formation (Aptian, Lower Cretaceous) of northeastern British Columbia, Canada. *J. Vertebr. Paleontol. 1*:257–264.
de Beer, G. 1954. *Archaeopteryx lithographica. A study based upon the British Museum specimen.* London: British Museum (Natural History).
Ellenberger, P. P. 1977. Quelques precisions sur l'anatomie et la place systematique tres speciale de *Cosesaurus aviceps*. (Ladinien superieur de Montral, Catalogne). *Caudernos Geol. Iberica 4*:169–188.
Elzanowski, A. 1976. Palaeognathous bird from the Cretaceous of Central Asia. *Nature (London) 264*:51–53.
– 1981. Embryonic bird skeletons from the late Cretaceous of Mongolia. *Palaeontol. Polon. 42*:147–179.
Feduccia, A. 1977. A model for the evolution of perching birds. *Syst. Zool. 26*:19–31.
– 1980. *The age of birds.* Cambridge, Mass.: Harvard University Press.

Feduccia, A., and L. D. Martin. 1976. The Eocene zygodactyl birds of North America (Aves: Piciformes). *Smithson. Contribut. Paleobiol. 27:*101–110.

Feduccia, A., and H. B. Tordoff. 1979. Feathers of *Archaeopteryx:* asymmetric vanes indicate aerodynamic function. *Science 203:*1021.

Fisher, J. 1967. Fossil birds and their adaptive radiation. In W. B. Harland (ed.). *The fossil record,* pp. 133–154. London: Geological Society of London.

Fox, R. C. 1974. A middle Campanian, nonmarine occurrence of the Cretaceous toothed bird *Hesperornis* (Marsh). *Can. J. Earth Sci. 11:*1335–1338.

Galton, P. M. 1970. Ornithischian dinosaurs and the origin of birds. *Evolution 24:*448–462.

Gingerich, P. D. 1972. A new partial mandible of *Ichthyornis. Condor 74:*471–473.

– 1973. Skull of *Hesperornis* and early evolution of birds. *Nature (London) 243:*70–73.

– 1976. Evolutionary significance of the Mesozoic toothed birds. *Smithson. Contribut. Paleobiol. 27:*23–33.

Gregory, J. T. 1951. Convergent evolution: the jaws of *Hesperornis* and the mosasaurs. *Evolution 5:*345–354.

– 1952. The jaws of the Cretaceous toothed birds, *Ichthyornis* and *Hesperornis. Condor 54:*73–88.

Harrison, C. J. O. 1973. The humerus of *Ichthyornis* as a taxonomically isolating character. *Bull. Br. Ornithol. Club 93:*123–126.

– 1980. A small owl from the lower Eocene of Britain. *Tert. Res. 3:*83–87.

Harrison, C. J. O., and C. A. Walker. 1973. *Wyleyia:* a new bird humerus from the Lower Cretaceous of England. *Palaeontology 16:*721–728.

– 1975. The Bradycnemidae, a new family of owls from the Upper Cretaceous of Romania. *Palaeontology 18:*563–570.

– 1976a. A reappraisal of *Prophaethon shrubsolei* Andrews (Aves). *Bull. Br. Mus. (Nat. Hist.) Geol. 27:*30.

– 1976b. A review of the bony-toothed birds (Odontopterygiformes): with descriptions of some new species. *Tert. Res. Spec. Pap. 2:*1–61.

– 1976c. Birds of the British Upper Eocene. *J. Linn. Soc. London (Zool.) 59:*323–351.

– 1977. Birds of the British Lower Eocene. *Tert. Res. Spec. Pap. 3:*1–52.

Heilmann, G. 1926. *The origin of birds.* London: Witherby.

Heinroth, O. 1923. Die Flügel von *Archaeopteryx. J. Ornithol. 71:*277–283.

Heller, F. 1959. Ein dritter *Archaeopteryx*–Fund aus den Solnhofener Plattenkalken von Langenaltheim/Mfg. *Erlanger Geol. Abh. 31:*3–25.

Hitchcock, E. 1836. Ornithichnology–description of the feet marks of birds (Ornithichnites) on New Red Sandstone in Massachusetts. *Am. J. Sci. 29:*307–340.

Hoch, E. 1980. A new Middle Eocene shorebird (Aves: Charadriiformes, Charadrii) with columboid features. *Contrib. Sci. Nat. Hist. Mus. Los Angeles County 330:*33–49.

Houde, P., and S. L. Olson. 1981. Paleognathous carinate birds from the early Tertiary of North America. *Science 214:*1236–1237.

Howard, H. 1950. Fossil evidence of avian evolution. *Ibis 92:*1–21.

– 1955. A new wading bird from the Eocene of Patagonia. *Am. Mus. Novit. 1710:*1–25.

Huxley, T. H. 1868. On the animals which are most nearly intermediate between the birds and reptiles. *Ann. Mag. Nat. Hist. 4*:66–75.

Jerison, H. J. 1973. *Evolution of the brain and intelligence.* New York: Academic Press.

Kleinschmidt, A. 1951. Über eine Rekonstruktion des Schadels von *Archaeornis siemensi* Dames 1884 im Naturhist. Museum, Braunschweig. *Proceedings of Xth International Ornithological Congress, Uppsala, June 1950,* pp. 631–635.

Kurochkin, E. N. 1976. A survey of the Paleocene birds of Asia. *Smithson. Contrib. Paleobiol. 27*:75–86.

Lambrecht, K. 1929. *Neogaeornis wetzeli* n. g. n. sp. der erste Kreidevogel der sudlichen Hemisphere. *Paleontol. Z. 11*:121–128.

– 1931. *Gallornis straeleni* ng. n. sp., ein Kreidevogel aus Frankreich. *Bull. Mus. R. Hist. Nat. Belg. 7*:1–6.

– 1933. *Handbuch der Palaeornithologie.* Berlin: Gebruder Borntraeger.

Lemoine, P. 1878–1881. *Recherches sur les oiseaux fossiles des terrains tertiares inferieures des environs de Reims.* Reims: Keller.

Lowe, P. R. 1935. On the relationships of the Struthiones to the dinosaurs and to the rest of the avian class, with special reference to the position of *Archaeopteryx. Ibis 13th Ser., 5*:398–432.

– 1944. Some additional remarks on the phylogeny of the Struthiones. *Ibis 86*:37–43.

Lydekker, R. 1891. *Catalogue of the fossil birds in the British Museum. (Natural History).* London: British Museum (Natural History).

Marsh, O. C. 1873. On a new subclass of fossil birds (Odontornithes). *Am. J. Sci., 3rd Ser. 5*:161–162.

– 1875. On the Odontornithes, or birds with teeth. *Am. J. Sci., 3rd Ser. 10*:402–408.

– 1877. Introduction and succession of vertebrate life in America. *Proc. Am. Assoc. Advmt. Sci. 1877*:211–258.

– 1880. *Odontornithes: a monograph on the extinct toothed birds of North America,* Vol. 7, *Report of the geological exploration of the fortieth parallel,* Professional Papers of the Engineer, No. 18. Washington, D.C.: Department of U.S. Army.

Martin, L. D. 1980a. Functional morphology and the evolution of cats. *Trans. Neb. Acad. Sci. 8*:141–154.

– 1980b. The foot-propelled diving birds of the Mesozoic. In R. Nöhring (ed.). *Acta XVII Congressus Internationalis Ornithologici,* pp. 1237–1242. Berlin: Verlag der Deutsche Ornithologen-Gesellschaft.

Martin, L. D., and O. Bonner. 1977. An immature specimen of *Baptornis advenus* from the Cretaceous of Kansas. *Auk 94*:787–789.

Martin, L. D., and J. D. Stewart. 1977. Teeth in *Ichthyornis* (Class: Aves). *Science 195*:1331–1332.

– 1982. An ichthyornithiform bird from the Campanian of Canada. *Can. J. Earth Sci. 19*:324–327.

Martin, L. D., J. D. Stewart, and K. Whetstone. 1980. The origin of birds: structure of the tarsus and teeth. *Auk 97*:86–93.

Martin, L. D., and J. Tate, Jr. 1976. The skeleton of *Baptornis advenus* from the Cretaceous of Kansas. *Smithson. Contrib. Paleobiol. 27*:35–66.

Masden, J. L. 1978. The oldest fossil bird: a rival for *Archaeopteryx? Science* *199*:284.

Matthew, W. D., and W. Granger. 1917. The skeleton of *Diatryma*, a gigantic bird from the Lower Eocene of Wyoming. *Bull. Am. Mus. Nat. Hist. 37*:307–326.

McDowell, S. 1978. Homology mapping of the primitive archosaurian reptile palate on the palate of birds. *Evol. Theor. 4*:81–94.

McKenna, M. C. 1975. Fossil mammals and early Eocene North Atlantic land continuity. *Ann. Mo. Bot. Gard. 62*:335–353.

Milne-Edwards, A. 1867–1871. *Recherches anatomiques et paleontologiques pour servir a l'histoire des oiseaux fossiles de la France*, Vols. 1 and 2. Paris: Massou.

Mudge, B. F. 1879. Are birds derived from dinosaurs? *Kansas City Rev. Sci. 3*:224–226.

Nopcsa, F. 1907. Ideas on the origin of flight. *Proc. Zool. Soc. London*, 223–236.

Olson, S. L. 1976. Oligocene fossils bearing on the origin of the Todidae and the Momotidae (Aves: Coraciiformes). *Smithson. Contrib. Paleobiol. 27*:111–119.

– 1977. A lower Eocene frigatebird from the Green River Formation of Wyoming (Pelecaniformes: Fregatidae). *Smithson. Contrib. Paleobiol. 35*:1–33.

– 1980. A new genus of penguin-like pelecaniform bird from the Oligocene of Washington (Pelecaniformes: Plotopteridae). *Contrib. Sci. Nat. Hist. Mus. Los Angeles County 330*:51–57.

Olson, S. L., and A. Feduccia. 1979a. Flight capability and the pectoral girdle of *Archaeopteryx. Nature (London) 278*:247–248.

– 1979b. An Old World occurrence of the Eocene avian family, Primobucconidae. *Proc. Biol. Soc. Washington 92*:494–497.

– 1980a. *Presbyornis* and the origin of the Anseriformes (Aves: Charadriomorphae). *Smithson. Contrib. Zool. 323*:1–24.

– 1980b. Relationships and evolution of flamingos (Aves: Phoenicopteridae). *Smithson. Contrib. Zool. 316*:1–73.

Osborn, H. F. 1903. *Ornitholestes hermanni*, a new compsognathid dinosaur from the Upper Jurassic. *Bull. Am. Mus. Nat. Hist. 19*:459–464.

Osmolska, H. 1976. New light on the skull anatomy and systematic position of Oviraptor. *Nature (London) 262*:683–684.

Ostrom, J. H. 1969. Osteology of *Deinonychus antirrhopus*, an unusual theropod from the Lower Cretaceous of Montana. *Bull. Peabody Mus. Nat. Hist. 30*:1–165.

– 1973. The ancestry of birds. *Nature (London) 242*:136.

– 1974. *Archaeopteryx* and the origin of flight. *Q. Rev. Biol. 49*:27–47.

– 1975a. The origin of birds. *Annu. Rev. Earth Planet. Sci. 3*:55–77.

– 1975b. On the origin of *Archaeopteryx* and the ancestry of birds. *Proc. Centre Nat. Res. Sci. 218*:519–532.

– 1976a. *Archaeopteryx* and the origin of birds. *Biol. J. Linn. Soc. 8*:91–182.

– 1976b. Some hypothetical anatomical stages in the evolution of avian flight. *Smithson. Contrib. Paleobiol. 27*:1–21.

– 1978. A new look at dinosaurs. *Natl. Geogr. Mag. 154*:152–185.

– 1979. Bird flight: how did it begin? *Am. Sci. 67*:46–56.

Owen, R. 1864. On the *Archaeopteryx* of von Meyer, with a description of the fossil remains of a long-tailed species, from the lithographic stone of Solnhofen. *Phil. Trans. R. Soc. 153*:33–47.

Parkes, K. C. 1966. Speculations on the origin of feathers. *Living Bird 5:*77–86.

Rich, P. V. 1974. Significance of the Tertiary avifaunas from Africa (with) emphasis on a mid to late Miocene avifauna from Southern Tunisia. *Ann. Geol. Surv. Egypt 4:*167–210.

– 1975a. Changing continental arrangements and the origin of Australia's non-passeriform continental avifauna. *Emu 75:*97–112.

– 1975b. Antarctic dispersal routes, wandering continents, and the origin of Australia's non-passeriform avifauna. *Mem. Nat. Mus. Victoria, Melbourne, 36:*63–126.

Rich, P. V., and D. J. Bohaska. 1976. The world's oldest owl: A new strigiform from the Paleocene of southwestern Colorado. *Smithson. Contrib. Paleobiol. 27:*87–93.

Russell, D. A. 1967. Cretaceous vertebrates from the Anderson River, N.W.T. *Can. J. of Earth Sci. 4:*21–38.

Russell, D. E. 1964. Les Mammiferes Paleocenes d'Europe. *Mem. Mus. Natl. Hist. Nat. Paris Ser. C. 13:*1–324.

Schlee, D. 1973. Harzkonservierte fossile Vogelfedern aus der untersten Kreide. *J. Ornithol. 114:*207–219.

Sharov, A. G. 1970. An unusual reptile from the Lower Triassic of Fergana. *Paleontol. J. 1:*112–116.

Shufeldt, R. W. 1909. Osteology of birds. *Bull. N.Y. State Mus. 130:*5–381.

– 1915. On a restoration of the base of the cranium of *Hesperornis regalis*. *Bull. Am. Paleonol. 5:*73–85.

Spinar, Z. V., and Z. Burian. 1972. *Life before man.* New York: American Heritage Press.

Storer, R. W. 1960a. Evolution in diving birds. *Proceedings of the 12th International Ornithological Congress,* pp. 694–707.

– 1960b. Adaptive radiation in birds. In A. J. Marshall (ed.), *Biology and comparative physiology of birds,* Vol. 1, pp. 15–55. New York: Academic Press.

Swinton, W. E. 1958. *Fossil Birds.* London: British Museum (Natural History).

Tarsitano, S., and M. K. Hecht. 1980. A reconsideration of the reptilian relationships of *Archaeopteryx*. *Zool. J. Linn. Soc. 69:*149–182.

Tonni, E. P. 1980. The present state of knowledge of the Cenozoic birds of Argentina. *Contrib. Sci. Nat. Hist. Mus. Los Angeles County 330:*105–114.

Tedford, R. H. 1974. Marsupials and the new paleogeography. *Soc. Econ. Paleontol. Mineral. Spec. Publ. 21:*109–125.

Thompson, D'A. W. 1890. On the systematic position of *Hesperornis*. *Univ. College Dundee Stud. Zool. 10:*1–15.

Waldmann, M. 1970. A third specimen of a Lower Cretaceous feather from Victoria, Australia. *Condor 72:*377.

Walker, A. D. 1964. Triassic reptiles from the Elgin area: *Stagonolepis, Dasygnathus* and their allies. *Philos. Trans. R. Soc. London Ser. (B) 244:*103–204.

– 1970. A revision of the Jurassic reptile *Hallopus Victor* (Marsh), with remarks on the classification of crocodiles. *Phil. Trans. R. Soc. London Ser. B 257:*323–372.

– 1972. New light on the origin of birds and crocodiles. *Nature (London) 237:*257–263.

– 1974. Evolution, organic. In D. N. Lapedes (ed.), *Yearbook of science and technology, 1974,* pp. 177–179. New York: McGraw-Hill.

– 1977. Evolution of the pelvis in birds and dinosaurs. In S. M. Andrews, R. S. Miles, and A. D. Walker (eds.) *Problems in vertebrate evolution,* pp. 319–358. New York: Academic Press.

Walker, C. A. 1981. New subclass of birds from the Cretaceous of South America. *Nature (London) 292:*51–53.

West, R. M., and M. R. Dawson. 1978. Vertebrate paleontology and the Cenozoic history of the North Atlantic region. *Polar Forschung 38:*103–119.

Wetmore, A. 1926. Fossil birds from the Green River deposits of eastern Utah. *Ann. Carnegie Mus. 16:*391–402.

– 1956. A check-list of the fossil and prehistoric birds of North America and the West Indies. *Smithson. Misc. Collect. 131:*1–105.

Wellnhofer, P. 1974. Das funfte Skelettexamplar von *Archaeopteryx. Palaeontographica (Abt. A) 147:*169–216.

Whetstone, K. N., and L. D. Martin. 1979. New look at the origin of birds and crocodiles. *Nature (London) 279:*234–236.

–1981.Common ancestry of birds and crocodiles: a reply. *Nature (London) 289:*98.

Williston, S. W. 1879. "Are birds derived from dinosaurs?" *Kansas City Rev. Sci. 3:*347–360.

Commentary

DAVID W. STEADMAN

Chapter 9 presents a current summary of knowledge of avian systematics and evolution in the Mesozoic and early Cenozoic eras. Many of Martin's ideas are new and untested; to treat each one in detail would exceed the limits of this commentary. Instead, I will confine my remarks to two of Martin's major topics – The origin of birds and The Mesozoic radiation.

The origin of birds

The origin of birds remains highly controversial. Most workers favor one of three different groups of archosaurs as the ancestors of birds – the thecodonts, the crocodilians, or the theropods (for the most current comprehensive review of this topic, see Feduccia 1980). As in several earlier papers, Martin continues to support a sister group relationship between birds and crocodilians. The history of the bird–crocodilian theory has been a little complicated for the past decade but may be very briefly summarized as follows. Tarsitano and Hecht (1980) analyzed the 15 characters used by Walker (1972, 1974, 1977) to propose monophyly of birds and crocodilians, finding that all are either primitive for archosaurs or are based on erroneous data and conclusions. Tarsitano and Hecht (1980) and McGowan and Baker (1981) also refuted the inner ear characters that Whetstone and Martin (1979) used to ally birds and crocodilians. Whetstone and Martin (1981) then replied by listing 11 shared derived features of birds and crocodilians. These characters are repeated in Chapter 9. Martin et al. (1980) pointed out that avian and crocodilian teeth are similar to each other and fundamentally very different from the laterally compressed, serrated teeth of theropods. The similarity of the teeth in Mesozoic birds and in modern crocodilians seems to be genuine. How to interpret this similarity phylogenetically awaits knowledge of dental morphology in a variety of Triassic and Jurassic thecodonts, crocodilians, theropods, and birds.

Tarsitano and Hecht (1980) have made the most thorough presentation of the growing evidence against the well-publicized theory of a theropod origin for birds, from which they support a thecodont ancestry for Aves. Considering the available information, I also subscribe, albeit cautiously, to a thecodont origin for birds, that is, that birds evolved directly from thecodonts without passing through a crocodilian or theropod grade of morphology. Yet I feel that much of the debate over avian origins is premature until more specimens are discovered, described fully, and compared. McDowell (1978), who pointed out crucial inadequacies in our understanding of homologies in the avian versus reptilian palate, aptly stated, "unfortunately, in a time when some students claim knowledge of the physiology and metabolism of dinosaurs, it is still impossible to get information on dinosaurian osteology." We certainly need a much larger osteological data base for archosaurs, including birds themselves, before the ancestry of birds can be stated without high levels of speculation. For example, we await anxiously thorough descriptions and comparisons of two tiny, at least superficially birdlike, Triassic thecodonts. These are *Cosesaurus* from Spain and *Longisquama* from Russia. (See Ellenberger and de Villalta, 1974; Melendez, 1977; and Sharov, 1970, for brief descriptions of these taxa.) Both of these apparent thecodonts have elongated scales that are reminiscent of feathers. Their ages and sizes are also reasonable for being somewhere near the ancestry of birds.

The Mesozoic record of birds is extremely incomplete. *Archaeopteryx* is the only known Jurassic bird, and contrary to Martin's proposals discussed later, I believe that the relationships of *Archaeopteryx* to any other known birds are remote at best. There is little reason to believe that *Archaeopteryx* was the most advanced bird of its time. In fact, the occurrence of two rather advanced early Cretaceous birds makes it seem probable that *Archaeopteryx* coexisted with other birds whose morphologies were more typically avian and less like that of reptiles (Olson, in press). These two early Cretaceous birds are *Ambiortus* (Ambiortiformes, new order) from deposits of Neocomian age in Mongolia (Kurochkin 1982) and *Enaliornis* from the Albian of England. *Enaliornis* is a specialized diving bird of the Order Hesperornithiformes, whereas *Ambiortus*, based on a partial associated skeleton in which the coracoid, scapula, sternum, and cervical vertebrae are especially well preserved, is the earliest record of a volant, carinate bird of modern appearance. *Enaliornis* and *Ambiortus* occurred only approximately 10–30 and 40 million years later than *Archaeopteryx*, respectively. Together they provide strong evidence that *Archaeopteryx* lagged behind some of its contemporaries in its grade of birdlike morphology. I do not imply here that *Archaeopteryx* is an unfortunate choice for being our only example of a Jurassic bird; indeed, we are fortunate to have the specimens of *Archaeopteryx* available for study. We have, however, very little idea of the morphological diversity of Jurassic birds.

Except for *Ambiortus* and *Enaliornis*, all well-diagnosed taxa of Cretaceous birds are confined to the late Cretaceous. This leaves a huge gap of approximately 65–70 million years between *Archaeopteryx* and the next youngest bird known from a complete, associated specimen. Isolated fossil elements are frequently misleading and often are referred to various higher taxa with little justification. The origins and relationships of Mesozoic birds will be resolved only by study of associated specimens.

To conclude, I cautiously agree with, and can do little to improve upon, these passages from Tarsitano and Hecht (1980:177, 179):

It is in the group [of thecodonts] characterized by the lack of dermal armor and the presence of the mesotarsus that the closest relatives to *Archaeopteryx* should be sought. Similarly the relatives of *Longisquama* and the dinosaurian groups must be associated with such a hypothetical ancestor. To state more at this time is to assume a data base that does not exist.

In conclusion, *Archaeopteryx* is a bird and the origin of birds is at the earliest levels of thecodont evolution. It is impossible to state whether *Archaeopteryx* represents the ancestral bird, although it cannot be distinguished from any hypothetical ancestor derived from cladistic methods. In order to determine the sister group relations or the possible ancestral relationships sought by many paleontologists, new and more reliable data are required. Biochemical systematics may measure genetic similarity in living forms to produce such data, but the analysis will be hindered because crucial groups are extinct. Thus the contribution of this biochemical method will be limited. The relationship of birds to thecodonts remains a paleontological problem.

The subsequent discovery of the early Cretaceous *Ambiortus* permits a modification of Tarsitano and Hecht's statement on the ancestral potential of *Archaeopteryx*, for it would now appear improbable that *Archaeopteryx* is the "ancestral bird."

The Mesozoic radiation

Martin's main thesis is that Mesozoic birds can be divided into two subclasses – the Sauriurae and the Ornithurae. Within the Sauriurae, he proposes two infraclasses: the Archaeornithes for *Archaeopteryx* and the Enantiornithes for three late Cretaceous taxa – the Enantiornithiformes of Argentina, *Gobipteryx* of Mongolia, and *Alexornis* of Baja California. The Ornithurae is divided into two subclasses – the Odontoholcae for the Cretaceous Hesperornithiformes and the Neornithes for the late Cretaceous *Ichthyornis* along with all modern birds. Although some workers may disagree with Martin's groupings within the Neornithes, any quibbling here should be postponed pending the appearance of a detailed comparative osteology of *Ichthyornis*. The Sauriurae as defined by Martin, however, is a controversial taxon with little evidence to justify it.

Before analyzing the characters used by Martin to define the Sauriurae, several pertinent facts about certain "sauriurines" merit discussion. First, it seems likely that at least some of the specimens of Enantiornithiformes are reptilian. Their only description is the understandably brief one of C. A. Walker (1981), who named the new Subclass Enantiornithes on the basis of approximately 60 fossils from the late Cretaceous of Argentina. Walker's Enantiornithes included only the Argentinian fossils; the inclusion of *Gobipteryx* and *Alexornis* in this subclass is Martin's idea. Strangely, C. A. Walker (1981) did not propose any new familial or ordinal names in his vague description (in the caption of an illustration) of the new genus and species *Enantiornis leali*, based upon an associated coracoid, scapula, and humerus among the Argentinian fossils. Most specimens of the Enantiornithiformes are still unillustrated and essentially undescribed, and thus it is very difficult at present to evaluate their systematic position among birds or other vertebrates. Three "distinct types" of enantiornithiform humeri were reported by Walker, although the morphology of the forelimb does suggest possible monophyly of the group, at least at the level of subclass. However, no forelimb elements of enantiornithiforms were found in association with hindlimb elements, and the three extremely different types of enantiornithiform tarsometatarsi illustrated by Walker seem, among themselves, not to belong to closely related animals. In fact, the lack of a hypotarsus derived from tarsal bones in two of these tarsometatarsi means that they are really only metatarsi. This condition is known in birds only in *Archaeopteryx;* in late Cretaceous fossils, it is probably evidence that the specimens represent reptiles, not birds.

Second, *Gobipteryx* was known originally from two partial skulls from the Barun Goyot Formation (?middle Campanian in age) of Mongolia (Elzanowski 1974, 1976, 1977). Elzanowski (1981) referred fossils of several embryonic skeletons to *Gobipteryx*. These remarkable fossils are from the red beds of Khermeen Tsav I, an apparent stratigraphic equivalent of the Barun Goyot Formation. Although it is possible that the embryos belong to the same taxon as the two skulls, the poor preservation of the skulls in the embryos, combined with the damaged nature of the two original skulls, prevents a conclusive assignment of all specimens to the same genus. It should be recognized that the embryos and the skulls may represent different taxa. This is significant because the characters that Martin uses to assign *Gobipteryx* to the Sauriurae and to the Enantiornithes pertain to postcranial as well as cranial elements. For now it is enough to say that both the skulls and the embryos do seem to be avian.

In Chapter 9, Martin uses the following five characters to define the Sauriurae: "(1) scapula with an anteromedial facet abutting against a series of fused anterior dorsal vertebrae; (2) fusion of the anteriormost thoracic vertebrae with a special process on the scapula abutting against them; (3) tarsometatarsus fused proximally but not distally; (4) metatarsal bones fused in a straight line; and (5) distal tarsal bones either absent or fused as small individual bones (not forming a large tarsal cap)." Characters 1 and 2, which appear to be equivalent, are not apparent in any of the taxa involved nor in any birds that I have ever seen. Regardless, thoracic vertebrae are not known for either *Alexornis* or the Enantiornithiformes. Characters 3–5 are not demonstrable in either *Gobipteryx* or *Alexornis*, because no tarsometatarsus or metatarsus is known for these taxa. Thus, we are left with not one character that will diagnose all of the Sauriurae. Characters 3–5 possibly unite only *Archaeopteryx* and the Enantiornithiformes. However, even these characters are questionable, particularly in light of the extreme diversity in the metatarsus of the Enantiornithiformes, as mentioned earlier.

Martin goes on to characterize the Enantiornithes as follows: "(8) the shape of the lower jaw articulation (unknown in *Alexornis*); (9) elongate narrow coracoids with a boss that fits into a 'coracoidal facet' on the scapula (present in all known forms); and (10) humerus with a large flat deltoid crest oriented laterally." The lower jaw articulation is undescribed and unillustrated for the Enantiornithiformes and therefore cannot be evaluated. It is poorly preserved in *Gobipteryx* and not known in *Alexornis*. Character 9 is incorrect for *Alexornis* and probably *Gobipteryx* as well. Martin wrongly accuses Brodkorb (1976) of mistaking the scapula of *Alexornis* for the coracoid and vice versa. The proximal end of the humerus is unknown in *Alexornis*. We are left with a single character (character 8) to diagnose the Enantiornithes, yet this character, if valid, possibly unites the Enantiornithiformes only with *Gobipteryx*. None of Martin's characters supports the inclusion of *Alexornis* in his expanded Enantiornithes.

Thus, both the Subclass Sauriurae and the Infraclass Enantiornithes of Martin are not defined by his characters. There is no reason to regard these taxa as monophyletic. Character states cannot be based on nonexistent specimens, even in a field as plagued by a paucity of specimens as Mesozoic birds.

Conclusions

Avian paleontology has progressed rapidly in the past decade and will undoubtedly continue to do so. Several recent studies on Tertiary birds, for example, have produced new data that bear directly on the systematics, evolution, and zoogeography of living birds. The last 10 years have also seen a rejuvenation of interest in Mesozoic birds, but there has been little agreement among researchers in the systematic and evolutionary interpretations of Mesozoic avian fossils. This is perhaps because of the large morphological and chronological gaps that separate known Mesozoic birds from their reptilian ancestors, as well as the equally large gaps that separate them from living birds. Martin's paper is a rather thorough statement of his own views on avian systematics and evolution in the Mesozoic, but it will do little to resolve the controversies surrounding Mesozoic birds. Although many of Martin's ideas cannot be faulted for lack of originality, they often can be questioned for their lack of supportive evidence. Martin's controversial statements should stimulate new research by his colleagues and thereby contribute significantly to our understanding of avian evolution.

Literature cited

Brodkorb, P. 1976. Discovery of a Cretaceous bird, apparently ancestral to the orders Coraciiformes and Piciformes (Aves: Carinatae). *Smithsonian Contrib. Paleobiol. 27:*67–73.

Ellenberger, P., and J. F. de Villalta. 1974. Sur la presence d'un ancetre probable des oiseaux dans le Muschelkalk superior de Catalogne (Espagne). Nota preliminar. *Acta Geol. Hisp. 9:*162–168.

Elzanowski, A. 1974. Preliminary note on the palaeognathous bird from the Upper Cretaceous of Mongolia. *Palaeontol. Polon. 30:*103–109.

– 1976. Palaeognathous bird from the Cretaceous of central Asia. *Nature (London)* 264:51–53.

– 1977. Skulls of Gobipteryx (Aves) from the upper Cretaceous of Mongolia. *Palaeontol. Polon. 37:*153–165.

– 1981. Embryonic bird skeletons from the late Cretaceous of Mongolia. *Palaeontol. Polon. 42:*147–179.

Feduccia, A. 1980. *The age of birds.* Cambridge, Mass.: Harvard University Press.

Kurochkin, E. N. 1982. Novyy otjad ptits iz Nizhnyelo Mela Mongolii. *Dok. Akad. Nauk SSR 262:*452–455.

Martin, L. D., J. D. Stewart, and K. N. Whetstone. 1980. The origin of birds: structure of the tarsus and teeth. *Auk 97:*86–93.

McDowell, S. 1978. Homology mapping of the primitive archosaurian reptile palate on the palate of birds. *Evol. Theory 4:*81–94.

McGowan, C., and A. J. Baker. 1981. Common ancestry for birds and crocodiles? *Nature (London) 289:*97–98.

Melendez, B. 1977. Descubrimiento de un reptil "proaviano" en el triassico de la Sierra de Prades (Tarragona). *Coloquios Catedra Paleontol. 32:*10–11.

Olson, S. L. in press. A selective synopsis of the fossil record of birds. In D. S. Farner, J. R. King, and K. C. Parkes (Eds.). *Avian biology,* Vol. 7. New York: Academic Press.

Sharov, A. G. 1970. An unusual reptile from the lower Triassic of Fergana. *Paleontol. Zh. 1970:*127–130.

Tarsitano, S., and M. K. Hecht. 1980. A reconsideration of the reptilian relationships of *Archaeopteryx. Zool. J. Linn. Soc. 69:*149–182.

Walker, A. D. 1972. New light on the origin of birds and crocodiles. *Nature (London) 237:257–263.*

– 1974. Evolution, organic. In D. N. Lapedes (ed.). *McGraw-Hill yearbook of science and technology;* pp. 177–179. New York: McGraw-Hill.

– 1977. Evolution of the pelvis in birds and dinosaurs. In S. M. Andrews, R. S. Miles, and A. D. Walker (eds.). *Problems in vertebrate evolution,* pp. 319–357. New York: Academic Press.

Walker, C. A. 1981. New subclass of birds from the cretaceous of South America. *Nature (London) 292:*51–53.

Whetstone, K. N., and L. D. Martin. 1979. New look at the origin of birds and crocodiles. *Nature (London) 279:234–236.*

– 1981. [Reply to McGowan and Baker.] *Nature (London) 289:*98.

Commentary

PAT V. RICH

Paleornithology is one branch of vertebrate paleontology that is in a phase of renaissance, as Chapter 9 indicates. Today paleornithology has expanded far beyond the bounds of its immediate past. Before 1965, only a handful of workers dealt with fossil birds, which ranged from ostriches to crows collected from all parts of the world. Now continents have indigenous investigators, many in areas outside of the classic avian fossil research centers of North America and Europe. Examples such as Edwardo Tonni in Argentina, Keichi Ono and Yoshikazu Hasegawa in Japan, Phil Millener, Ewan Fordyce, and Ron Scarlett in New Zealand, and Lian-hai Hou and Xiangk'wei Yeh in the Peoples' Republic of China emphasize just how expansive the field is now. Although much study is still at the alpha taxonomic level, where even more effort is certainly needed to broaden our data base, a number of studies involve detailed phylogenetic and methodological analyses, consider biases that might effect interpretation, and often propose multiple, rather than single, working hypotheses to explain the often complex data being generated.

With such an enlarged effort in paleornithology, collections of both recent and fossil specimens have expanded greatly, especially those from previously unknown, or poorly known, geographic areas. Likewise, there is more cooperative work, in part due to the greater number of workers. Such efforts result in imaginative, thought-provoking studies that have often been thoroughly argued and defended even before publication, a way of enhancing rapid scientific advancement (e.g., Feduccia and Tordoff 1979; Martin and Tate 1976; Olson and Feduccia 1980a,b). There are interactions between paleornithologists and workers in other disciplines, especially in the case of "border disputes," for example, *Archaeopteryx* (Ostrom 1975, 1976), which has led to healthy debate about the relative value of certain characters used in taxonomic studies and led to evaluations and reevaluations of osteological morphology in reptiles, *Archaeopteryx,* and birds. Such disputes have forced investigators to refine their arguments and reexamine their interpretive theory. Together with this have come better field and preparation techniques, improved molding and casting techniques and materials, more exchange of specimens between scientists, and increased monetary support for collection, processing, and study of fossil birds.

All of this has increased the understanding of composition and development of fossil avifaunas in the late Mesozoic and particularly in the Cenozoic.

Paleornithologists are still quite interested in pursuing the phylogenetic relationships of fossil birds, but in the past decade there has been a marked effort to outline carefully the reasons why a character or character suite is "advanced" (apomorphic) or "primitive" (plesiomorphic). These attempts make understanding and evaluating the logic and reasoning in such papers much easier. Larger collections have enabled detailed analyses of the biases that affect samples, thus making estimates of (past) avifaunal composition more realistic. There has also been a renewed interest in biogeography in the light of the proposal of continental drift and plate tectonic theories in the early 1960s.

Martin has summarized the current state of systematic studies on late Mesozoic and early Cenozoic birds. More sampling is needed, however, to determine the timing of the modern radiation of avian families relative to that of the mammals. Certainly many birds in the early Tertiary were mosaics that combined characteristics of many modern families, and, like the mammal faunas, the modern avifaunas were not in evidence until Oligo–Miocene times. The Paleocene, one of the two major gaps in the avian fossil record (Figure 9.12), needs to be filled, however, before the early development of modern families is well understood. This early record, as it becomes better known, not only gladdens the diehard fossil bird enthusiast but will allow a better understanding of the relationships of the higher taxonomic levels of birds (e.g., Olson and Feduccia 1980a, b). The early Tertiary avian mosaics demonstrate how living groups so far apart morphologically were so similar to one another in the past (e.g. Charadriiformes, Phoenicopteridae, and Anseriformes). The early Tertiary birds demonstrate further how the ecological niches of certain avian groups have changed and how certain groups were replaced by others. For example, in the Paleocene faunas the large, flightless Diatryma was a prime terrestrial carnivore, and the Primobucconidae occupied many of the niches that today are used by the Passeriformes. Granivorous birds are represented in low percentages in the early Tertiary faunas, which may be explained by biased sampling. Perhaps they really were absent, but a much larger sample is needed to properly evaluate this apparent paucity.

The study of taphonomy, what happens to a carcass between the time an animal dies and the time it is deposited in a sediment (or perhaps is found by a paleontologist), is a relatively new field of investigation in

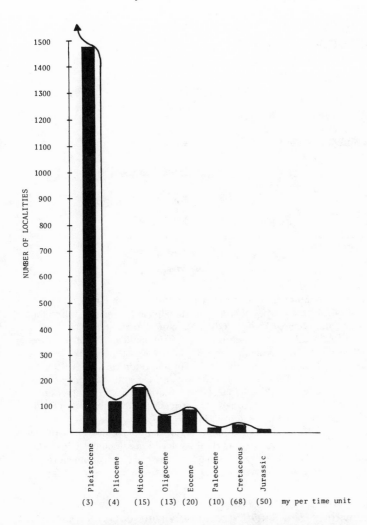

Figure 9.12. An estimate of the number of localities producing fossil birds, on a worldwide scale, during certain time periods (periods and epochs) from Jurassic to Pleistocene.

paleornithology. When large samples of fossil birds are examined from a taphonomic viewpoint (e.g., the Pliocene Langebaanweg avifauna of South Africa; Rich 1980), it is clear that most samples are biased (see Figure 9.13). Bones, such as the humerus, tibiotarsus, and tarsometatarsus, are much more frequently represented in the fossil assemblage than are other elements that are just as numerous or even more so (e.g., phalanges and vertebrae) in the original skeleton. Tumbler ex-

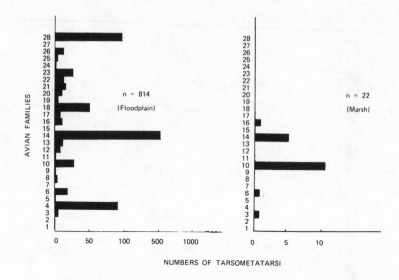

NUMBERS OF TARSOMETATARSI

Figure 9.13. Total numbers of tarsometatarsi of each avian family represented in two different types of sediments (floodplain and marsh) within the
Quartzose Sand Member Units I and II, Varswater Formation, Langebaanweg, South Africa. Birds represented are (1) Struthionidae, (2) Spheniscidae,
(3) Podicipedidae, (4) Procellariiformes, (5) Pelecaniformes, (6) Phalacroco-
racidae, (7) cf. Sulidae, (8) Ciconiidae, (9) Threskiornithidae, (10) Anatidae,
(11) cf. Gypaetinae, (12) Falconidae, (13) Accipitridae, (14) Phasianidae, (15)
Gruidae, (16) Rallidae, (17) Otididae, (18) Charadriiformes, (19) Pteroclidae,
(20) Columbidae, (21) Psittaciformes, (22) Strigidae, (23) Coliidae, (24) Coraciiformes, (25) cf. Alcedinidae, (26) Piciformes, (27) Apodidae, (28) Passeriformes. Floodplain sediments are dominated by more terrestrial birds, primarily francolins, whereas the marsh-derived sediments are dominated by ducks.
It must be admitted that a larger sample from the marsh environment is
needed to confirm this observation. (Corrected; after Rich, 1980).

periments providing simulated stream environments (Napawongse
1981; Napawongse, Berra, and Rich, unpub.) have shown that of all
bones in the avian skeleton these are the most resistant to erosion and
reinforce the idea that small bones, bones with very thin walls, and
bones with large surface-to-volume ratios are not often preserved in
fossil samples derived from fluviatile environments (see Figure 9.14).
Other studies show clearly that birds associated with water, and not
those truly terrestrial forms, are much more often represented in fossil
collections (Figure 9.15). This must be considered when asking how
close an approximation the fossil sample is to the avifauna that provided the raw materials. A taphonomic analysis *must* be done in order
to clarify the biases.

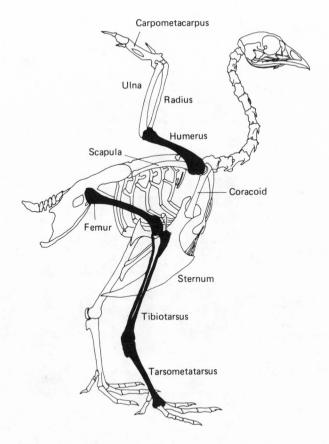

Figure 9.14. Bird skeleton illustrating bones (black) that are the most resistant to wear in simulated fluviatile environments, due in part to shape, size, and degree of ossification of individual bones. Least durable bones were the cranium, vertebrae, synsacrum and pelvis, ribs and sternum, and phalanges, whereas the remaining had an intermediate degree of wear. In the fossil record, it is these same resistant bones that are most frequently preserved (see Napawongse 1981; Napawongse, Berra, and Rich, unpub.).

A third major area of recent paleornithological interest is that of paleozoogeography. The development of plate tectonic theory in the early 1960s encouraged biogeographers to rethink many current theories on the origin and dispersal of birds. Sometimes this meant replacing a theory with a single set of definite answers, such as that of Mayr's on the origin of the Australian avifauna (Mayr 1944), with a set of multiple hypotheses (Cracraft 1973; Rich 1975a,b; Olson and Steadman, 1981; Rich et al., in press). Such exchange of certainty of avifaunal origin for uncertainty (or multiple hypotheses) is not the most

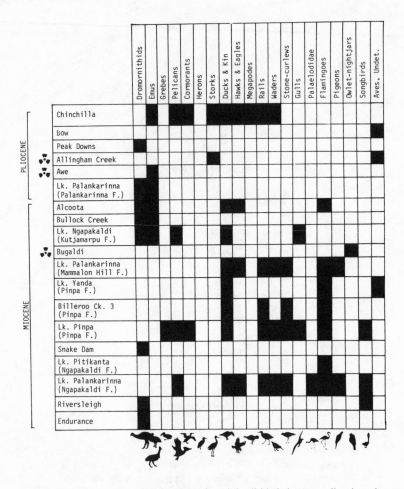

Figure 9.15. Localities that have produced fossil birds in Australia plotted against their avifaunas. Black symbols on left indicate what few sites have radiometric age control. Note that wetland birds are most frequently represented and that such terrestrial birds as the parrots and pigeons, now very diverse in Australia, are noticeably missing, probably caused by a preservation bias.

palatable medicine, but a necessity if the purpose of science is to seek the best estimate of reality based on the data available. In any analysis, the records of different continents are not of equal quality (see Figure 9.16), and thus if a group is absent on a continent with a poor record, such as Asia, this factor should be considered in evaluating the significance of such an absence. Similarly different periods of time have

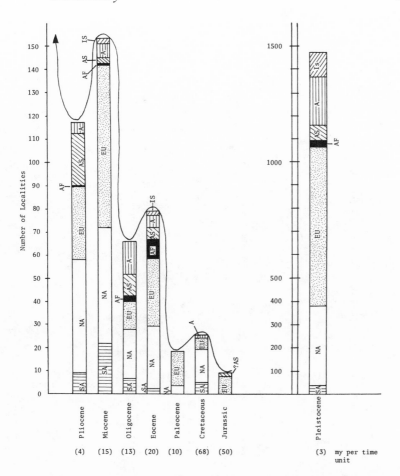

Figure 9.16. Comparison of the number of localities producing fossil birds in different parts of the world during different time periods. Abbreviations include: A, Australia, AF, Africa, AS, Asia, EU, Europe, IS, Islands, NA, North America, SA, South America. For much of the existence of birds, only the records of Europe and North America are very good.

better or poorer quality records, and this too should be recognized (see Figure 9.12, 9.16) when evaluating the diversity of avifaunas (Figure 9.17).

To sum up, more work is in progress in paleornithology today than at any time before. Looking back from this vantage point, we should all greatly value the firm and broad foundation prepared by Wetmore, Brodkorb, and Howard and their predecessors.

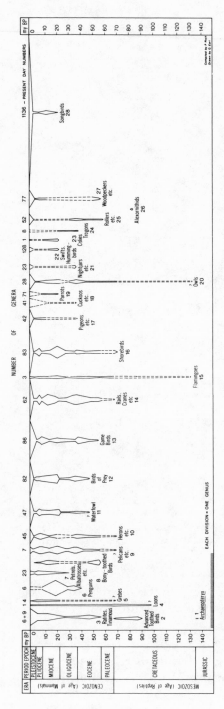

Figure 9.17. An estimate of the diversity (at the generic level) of avifaunas through time for each of the known orders of birds in pre-Holocene deposit. The record of parrots should now be extended in to the Eocene; and the early Cretaceous record of owls is questionable. Enantiornithiformes, *Gobipteryx*, and *Ambiortus* have not been included.

Literature cited

Cracraft, J. 1973. Continental drift, paleoclimatology, and the evolution and biogeography of birds. *J. Zool. 169:*455–545.

Feduccia, A., and H. B. Tordoff. 1979. Feathers of *Archaeopteryx:* asymmetric vanes indicate aerodynamic function. *Science 203:*1021–1022.

Martin, L. D., and J. Tate, Jr. 1976. The skeleton of *Baptornis advenus* (Aves: Hesperornithiformes). *Smithson. Contrib. Paleobiol. 27:*35–66.

Mayr, E. 1944. The birds of Timor and Sumba. *Bull. Am. Mus. Nat. Hist. 83*(2):127–194.

Napawongse, P. 1981. A taphonomic study of bird bone degredation in fluviatile environment. B.S.(Hons.) thesis, Monash University, Clayton, Victoria.

Olson, S. L., and A. Feduccia. 1979. Flight capability and the pectoral girdle of *Archaeopteryx. Nature (London) 278:*247–248.

— 1980a. Relationships and evolution of Flamingos (Aves: Phoenicopteridae). *Smithson, Contrib. Zool. 316:*1–73.

— 1980b. *Presbyornis* and the orgin of the Anseriformes (Aves: Charadriomorphae). *Smithson. Contrib. Zool. 323:*1–24.

Olson, S. L., and D. W. Steadman. 1981. The relationships of the Pedionomidae (Aves: Charadriiformes). *Smithson. Contrib. Zool. 337:*1–25.

Ostrom, J. H. 1975. The origin of birds. *Annu. Rev. Earth Planet. Sci. 3:*55–77.

— 1976. Some hypothetical anatomical stages in the evolution of avian flight. *Smithson. Contrib. Paleobiol. 27:*1–21.

Rich, P. V. 1975a. Antarctic dispersal routes, wandering continents, and the origin of Australia's non-passeriform avifauna. *Mem. Nat. Mus. Victoria, Melbourne 36:*63–126.

— 1975b. Changing continental arrangements and the origin of Australia's non-passeriform continental avifauna. *Emu 75:*97–112.

— 1980. Preliminary report on the fossil avian remains from late Tertiary sediments at Langebaanweg (Cape Province), South Africa. *S. Afr. J. Sci. 76:*166–170.

Rich, P. V., G. F. van Tets, and C. Balouet. in press. The birds of Western Australasia. *Proceedings of the XVIII International Ornithological Congress, Moscow.*

10 Avian community ecology: an iconoclastic view

JOHN A. WIENS

Community ecology is concerned with the patterns that characterize natural assemblages of species and understanding how these patterns came to be and how general they are in nature. The historical roots of community ecology lie in botanical studies of more than a century ago, but perspectives gained from investigations of birds have made major contributions to the development of the conceptual foundation that has guided and dominated activity in the discipline over the past quarter century. Avian studies have also contributed much of the evidence that supports community theory; in Schoener's (1974) review, for example, 34% of the 50 studies of resource partitioning in terrestrial animal communities cited were of birds. There is thus justification for considering the recent development, current problems, and future prospects of community ecology in the context of studies of bird communities.

In order to understand contemporary community ecology and its problems, it is necessary to consider its development over the past two or three decades. During the 1950s, two views of community structuring, each originally developed earlier in the century, were actively discussed, and each was expressed in studies of bird communities. On the one hand, ecologists following the view of Gleason (1917; see McIntosh 1975) and Curtis (1959) suggested that the species within communities were arrayed along environmental gradients more or less independently of one another, each according to its own ecological requirements. Communities were thus viewed as assemblages of largely noninteracting species, and they exhibited rather little repeatability in structure or organization from place to place. Bond (1957) and Beals (1960) conducted their studies of breeding bird communities in Wisconsin woodlands within this framework and rather clearly documented that the bird species were distributed along gradients of forest vegetational composition more or less smoothly but largely independently of one another. Bond could distinguish no repeatable, discrete bird communities in these forests and concluded that the independence in the geographic and ecological ranges of the different bird species and the absence of any clearly distinguishable or highly structured community assemblages of bird species supported Gleason's particulate view of communities.

During the late 1950s and early 1960s, an alternative view gained force. Communities were considered to be tightly integrated entities containing suites of interacting species and exhibiting clearly defined and repeatable structure under similar environmental conditions. This view received its primary impetus from the work of Robert MacArthur and his colleagues, but it gained support also from the more population-oriented studies of David Lack; both, of course, worked primarily with birds. Perhaps because it offered exciting new ideas and posed new questions, perhaps because of the apparent mathematical elegance of its theories or the seeming neatness of the explanations of nature that it provided, perhaps because it seemed easy and fun to do, perhaps because community ecology during the 1950s was somewhat stagnant—for whatever reasons, the MacArthur approach to community studies took hold and rapidly assumed a position as the guiding focus of the discipline. It developed the clear attributes of a Kuhnian paradigm (Kuhn 1970; but see Lakatos and Musgrave 1970; Gutting 1980). It presented a coherent, integrated tradition of scientific research that defined which problems were worthy of study and which not, that formed the intellectual and conceptual framework for the education and training of practitioners in the discipline, and that discouraged or actively suppressed alternative or opposing views of communities.

One of the characteristics of a scientific discipline that is dominated by a single paradigm, according to Kuhn's thesis, is that it enters into a phase of "normal science," in which most research activity is directed to a restricted set of questions or problems that seem most relevant to the overall conceptual framework of the paradigm. Moreover, the research tradition established by the paradigm may dictate rather strongly the perspectives and procedures to be followed in investigating these problems. The MacArthur community paradigm established by example a number of characteristics of the appropriate approach to community questions. With the benefit of hindsight, these features can readily be discerned from MacArthur's own writings (e.g., MacArthur and Connell 1966; MacArthur 1971, 1972) and those of his students and colleagues (e.g., Cody 1974; Cody and Diamond 1975; Fretwell 1975; Diamond 1978; Pianka 1981; see also Ricklefs 1975; Wiens 1976, 1977a). Thus, within this paradigm of community ecology:

1. The emphasis is clearly upon the detection and explanation of the *patterns* of nature. These patterns are presumed to be caused by natural

selection. At times, however, MacArthur seemed less interested in the biological reality of the object of study than in the patterns that were expressed. For example, "the question is not whether such communities exist but whether they exhibit interesting patterns about which we can make generalizations" (MacArthur 1971).

2. The approach stresses *generality* – the more general a pattern and its explanation, the greater its interest and neatness. Cody (1974) clearly stated this philosophy: "any and all insights are at a premium; the detailed infrastructure of broad relationships can be illuminated later."

3. The primary process producing the patterns of communities is *competition* among species. Such interactions lead to the exclusion of marginally adapted species from communities, producing a community structure that is well ordered and highly integrated. As Cody and Diamond (1975) put it, "it is natural selection, operating through competition, that makes the strategic decisions on how sets of species allocate their time and energy; the outcome of this process is the segregation of species along resource-utilization axes."

4. The assemblages of species are presumed to be at or close to *equilibrium*. If environmental conditions vary through time, the species populations respond to this variation in such a way that the optimal community structuring and composition "tracks" resource variation. This assumption of equilibrium leads, in turn, to the conclusion that suitable habitat is fully occupied by individuals – that the resource opportunities, and the community structure, are "saturated."

5. Investigations are guided by *theory*, by the predictions of algebraic or graphical models that attempt to portray how natural systems might be structured. The theory thus determines which measures or observations are warranted; "data are used as ornaments for ideas" (Schoener 1972). Fretwell (1975) praised this approach for gaining respectability for hypotheticodeductive procedures in ecology, although much of its application to community ecology has been only marginally hypotheticodeductive, at least in the Popperian sense (Popper 1968a,b).

6. Patterns are discerned, or theories tested, through broad *comparisons* of different states of nature – so-called natural experiments, which capitalize on the variation that occurs naturally in the environment. This approach is held to facilitate "a rapid progression through hypothesis and theory to acceptable and accepted explanations for natural phenomena with economy of time and effort" (Cody 1974).

7. Explanations of the patterns are usually derived from the consistency of the match between *selected examples* or case studies and the

predictions of theory, which are often expressed in the form of *scenarios* (or even metaphors; see Mertz and McCauley 1980). There is generally little effort made to analyze relationships statistically, and, despite the algebra of the theory, the approach is much more qualitative than quantitative.

8. Perhaps largely as a consequence of the assumption of community equilibrium and saturation, only a *few features* of habitat or resource dimensions, those that "are already identified with environmental variables for which competition takes place, as indicated by resultant displacement pattern" (Cody 1974), need be observed or measured.

9. Experimental manipulations of field situations are generally not considered to have much potential for contributing to the resolution of community questions. Moreover, the observations that are used to address the predictions of theory or to confirm the scenarios are usually gathered with little or no consideration of the adequacy or requirements of observational or sampling *methodology or design*. Observations are often obtained in an opportunistic, nonsystematic fashion, and field data are often treated with rather little respect. It is this attitude that presumably enabled Cody to report on community patterns that included in the analyses species that were "expected" to occur in a census but did not do so (Cody 1975) or to base an analysis of resource tracking in a bird community (Cody 1981) upon data that were lost but then regenerated (qualitatively) from memory. MacArthur (1971) expressed a somewhat similar view when he wrote that "there is no such thing as a 'correct' or 'incorrect' measure of diversity or stability or anything else. The virtue of a measure resides solely in the neatness of the relations constructed from it."

These perspectives and procedures established the acceptable manner of studying natural communities, and they are much in evidence in most research conducted on avian communities during the past two decades. The dominant feature of this conceptual framework, the view that competition is the process that leads to community structuring and patterns, has been widely adopted as a basic truth about nature. Thus, Ricklefs (1975) observed that "few ecologists doubt that competition is a potent ecological force or that it has guided the evolution of species relationships within communities," and Diamond (1979) concluded that "now, few ornithologists question the widespread role of competition." Other features of this paradigm seem to have become similarly rooted in our conceptions of and approaches to communities.

Another part of Kuhn's thesis, however, is that, inevitably, observations will begin to appear that are at odds with the views of the prevailing paradigm in a discipline. As such anomalies accumulate, the structure of the paradigm may be brought into question, and its dominance in the discipline weakened. Eventually, Kuhn suggested, disillusionment with the prevailing paradigm may reach a level at which it no longer provides a satisfying research tradition for the descipline, and it may be abandoned in favor of some new or developing paradigm that seems more capable of explaining the anomalous observations. A "scientific revolution" has then occurred.

It would be premature to suggest that such a series of events is now occurring in community ecology, but during the past few years the adquacy of the conceptual framework of the MacArthur paradigm has been increasingly questioned (e.g., Connell 1975, 1980; Wiens 1977b, in press; Simberloff 1980; Dayton and Oliver 1980), and observations that cannot be reconciled with prevailing theory have become more frequent (e.g., Rabenold 1978; Wiens and Rotenberry 1979, 1980a,b; Abbott 1980; Rotenberry and Wiens 1980a,b; Vuilleumier and Simberloff 1980). This may indicate that, despite the wide acceptance of the MacArthur approach to community studies, it is not necessarily correct. Some reassessment of this paradigm is thus in order.

In this chapter, I do not attempt a general review of the theories and facts of avian community ecology but instead consider some problems that seem to plague such studies and that may compromise our confidence in the utility of the theories and the veracity of the facts. I will do so from an explicity personal perspective, one gained from my own studies of bird communities in grassland and shrubsteppe environments over the past 20 years and from my mounting disillusionment with much of what I read in the scientific literature. I've come to the conclusion that, in spite of so much study of bird communities, we really know rather little about them, much less than we think we know. The remainder of this chapter will ask, in a variety of ways, why this is so.

Problems with logical procedures

Ecology and evolutionary biology, in general, seem to suffer from several basic problems inherent in the sort of logic that is frequently employed by workers in these fields (Peters 1976; Brady 1979; Clutton-Brock and Harvey 1979; Gould and Lewontin 1979; Simberloff 1980; Strong 1980). Although it would be inappropriate to delve deeply into the philosophy and logic of scientific explanation here, these problems are sufficiently pervasive in avian community ecology that they merit attention. The nature and effects of these abuses of logic will become apparent in some of the example that will be discussed in following sections of this chapter.

The problems

Perhaps the most prevalent problem with the logical approach of community ecology is that research is generally conducted within a circumscribed world view—a set of preconceptions about how nature *should* be—that leads us to expect certain patterns and processes to be prevalent and others inconsequential. By accepting the basic assumptions of a prevailing world view as true without question, we are compelled to interpret the observed patterns in a way that will support those basic truths. As Dayton and Oliver (1980) have observed, "preconceptions tend to flavor questions, determine research designs, and bias interpretation of the data. The preconceptions commonly result in an emphasis on verification rather than falsification of hypotheses, a process whereby counter examples are ignored, alternative hypotheses brushed aside, and existing paradigms manicured." Community ecology has been conducted largely with the preconceptions that the systems studied are likely to be highly ordered or deterministic and at (or reasonably close to) equilibrium with respect to their species composition, resource utilization, and allocation patterns. This is the foundation from which the search for patterns is launched and on which the explanations of the patterns rest. This perception of nature stems in part from a world view derived from Greek metaphysics, which proposes that nature must express an orderly reality (Simberloff 1980), and in part from our theory, which is largely founded upon equilibrium or near-equilibrium mathematics (Wiens, in press). This preconception of order and equilibrium of communities is often simply accepted as an article of faith (e.g., Cody 1974, 1981), with little concern for documenting whether or not this is actually so. The possibility that natural systems might frequently be nonequilibrial or only loosely ordered, with many of their features varying stochastically, is alien to the accepted world view and thus generally not considered.

Much of the recent literature of community ecology has made reference to "hypothesis testing" in one way or another. Unfortunately, most often only a single hypothesis is subjected to testing, by comparing observations of nature with the predictions of the hypothesis. At best, this leads to a situation in which nonagreement of observations with predictions can effectively falsify the hypothesis, but agreement does not necessarily indicate that the hypothesis is true. Other alternative hypotheses, which might account for the observations as well or even more parsimoniously, are generally not considered (Wiens and Rotenberry 1979). The prevalent attitude seems to be that if the observations agree with a hypothesis that matches the dictums of the accepted paradigm or our preconceptions of nature, there is really no need to consider alternatives. Furthermore, many of the hypotheses of community ecology are untestable, in that it is seemingly not possible to gather the observations that would lead to their falsification. For example, most hypotheses incorporating an historical element, such as the hypothesis that the structuring of a contemporary community is a consequence of competition operating in the past – what Connell (1980) has dubbed the "ghost of competition past" – cannot be falsified, yet they are nonetheless key features of the overall conceptual framework of the discipline. The most readily falsifiable hypotheses, of course, are appropriately structured null hypotheses (Strong 1980) or neutral models (Caswell 1976), yet such hypotheses are not a part of the research tradition fostered by the MacArthur paradigm, perhaps because they seem intentionally trivial.

Consideration of only a single hypothesis leads to a third logical problem, that of argument by consistency. This problem is especially severe if the hypothesis is untestable or can be falsified only with difficulty. Predications are generated from the hypothesis and then compared with the patterns of nature. If the observations are more or less as anticipated, they are taken to be clear confirmatory evidence for the legitimacy of the hypothesis and its associated causal explanation of the patterns, simply because they are consistent with the predictions. For example, we might document that two sympatric and ecologically similar bird species differ in the relative sizes of their bills. Such an observation is consistent with the hypothesis that competition in the past has selected for divergence in bill size between the species and that this difference now facilitates their coexistence. It is also consistent, however, with the alternative hypothesis that the two species are exposed to different selective regimes on bill size during their life histories quite apart from any interspecies interactions such as competition or with the alternative that the two species represent different branches of phylogeny and that the bill size differences are reflections of their phylogeny and not their immediate ecological relationships. Given such multiple consistency, this evidence contributes little to our ability to determine which hypothesis might be most likely. A rigorous form of scientific explanation would require that such evidence be not only consistent with the predictions of a given hypothesis but also inconsistent with the predictions of well-structured alternative hypotheses.

The force of our preconceptions about nature, combined with the testing of single hypotheses structured within the framework of the prevailing paradigm, produces an emphasis upon the verification rather than the falsification of hypotheses and theories. This disposition, in turn, leads to a tendency to consider only confirmatory evidence or observations in "testing" the hypotheses. If the treatment of observations is largely anecdotal, based upon carefully selected examples rather than statistical analyses, this problem is intensified. Negative observations or examples are generally not reported or even considered, and in some cases the evidence bearing upon the hypothesis is more circumstantial than direct (Strong 1980). It is little wonder that argument by consistency is so prevalent.

If one is not so selective in the treatment of evidence, however, it is likely that at least some of the observations will not match the predictions of the favored theory very well. In this case, various ad hoc explanations or adjustments of the hypothesis or theory are often undertaken to reconcile the observations with the predictions. If one is unable to find agreement between the ecological distribution of a species and the predictions of a model of direct interspecific competition, for example, one may postulate ad hoc that the competitive interactions are spread more diffusely over a large array of coexisting species and thereby reconcile the seemingly anomalous observations with the theory that competition does indeed determine the ecological distributions of species (e.g., MacArthur 1972; Diamond 1978). Resurection of the "ghost of competition past" (Connell 1980) is another frequently employed means of achieving an agreement between observations and theory. Making ad hoc adjustments of a hypothesis or theory when its predictions are not matched by reality is not necessarily fallacious – such revision of theory for subsequent retesting is a powerful tool of scientific explanation. However, when the adjustments are made to bring a set of anomalous observations into concordance with the theory and the observations are then used as confirmatory evidence without additional independent testing, the procedure becomes logically flawed. It becomes nothing more than an exercise based upon faith in the axioms of the theory and the truthfulness of the paradigm, and it does not contribute in any way to rigorous scientific explanation.

Scientific explanation, of course, involves linking an observed pattern of nature to a process that is likely to have produced the pattern –

coupling cause with effect. Unfortunately, the detection of pattern and the establishment of a process explanation for the pattern are tightly coupled in much of the recent work in avian community ecology. The predictions of some hypothesis of pattern are tested against observations from nature; if the observations are consistent with the predictions, not only is the hypothesis of pattern considered to be verified, but its underlying premises are also accepted as providing the process explanation for the observed pattern. It is commonplace, for example, to read of niche differences among coexisting species and to find these interpreted as consequences of competition; they are the means by which the species reduce competition sufficiently to permit their continuing coexistence. Usually, what has been established in such studies is that the species differ in some attributes that are potentially important ecologically. This represents a documentation of pattern. The conclusions regarding the role of competition in causing the differences, or the role of the divergences in permitting coexistence of the species, are only inferences. They have no direct empirical support but are simply assertions derived from an acceptance of the overall hypothesis that species coexist by virtue of niche differences, which are the product of selection mediated by competitive interactions. Finding agreement or consistency between the predictions of a hypothesis of pattern and observations may contribute to our confidence in the pattern hypothesis or its predictiveness, but it does not mean that we understand how the patterns are produced – we have not generated a process explanation. That requires formulation and testing of a separate hypothesis that explicitly deals with process. Discerning the patterns of natural systems and deriving process explanations for those patterns are separate, sequential phases of scientific activity (Osman and Whitlatch 1978; Eldridge and Cracraft 1980; Roth 1981; Wiens, in press).

An example: the assembly of communities

An example of the operation of such logical difficulties in analyses of ecological communities is provided by some recent attempts to conceptualize how communities of species are assembled (Figure 10.1). Ultimately, community attributes are delimited by the composition of the pool of species from which communities must be derived. The species pool of a region is enriched through speciation and diminished through extinction (see Rosenzweig 1975; Cracraft, unpub.), and although a large variety of factors can influence the rates of these pro-

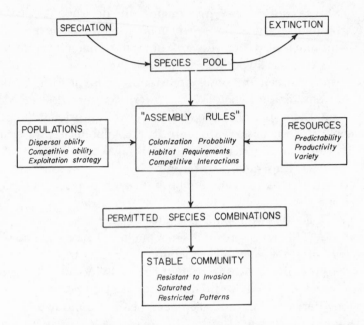

Figure 10.1. Flow diagram of the features that contribute the assembly of communities, according to current conceptualizations (e.g., Diamond 1975).

cesses, most are historical or biogeographic in nature. Intercontinental comparisons of tropical bird communities (Karr 1976) or communities in Mediterranean habitats (Cody 1975) reveal clear differences in patterns of species diversity, and these have been attributed, at least in part, to the influences of past history and of habitat and avifaunal biogeography on the species pools present in the different continents. The analyses by Vuilleumier and Simberloff (1980) and Haffer (1981) of South American avifaunas reinforce the importance of such influences. In the West Indies, the widespread occurrence of savannah and scrub woodland habitats during the Pleistocene glaciations has left an indelible mark on the contemporary avifauna, producing a species pool that consists mainly of shrub-adapted species despite the present scarcity of this habitat (Pregill and Olson 1981). The geological record provides ample evidence of environmental and biogeographic changes in the past and clearly indicates differences between different continental areas. Nonetheless, generating explicit hypotheses that relate historical events to the characteristics of the species pool of specific areas and that are testable in a rigorous fashion is extremely difficult.

It has recently been widely suggested that the generation of a community from a species pool proceeds according to some set of "assembly rules" (e.g., Diamond 1975; Cody 1978; Feinsinger and Colwell 1978; Herrera 1981; Brown 1981; Nudds et al. 1981). Given an environment that is characterized by a particular variety, abundance, and stability or predictability of resources and species in the pool that differ in features such as dispersal capability, competitive ability, or characteristics of their exploitation of resources, such "rules" attempt to define how particular species or species types will be assembled to form the community (Fig. 10.1). Thus, the biogeographic attributes of the species pool may interact with the dispersal characteristics of the species to determine the probability of colonization of a particular community by different species. These species, in turn, may or may not have the potential to become established, depending upon whether or not the environment satisfied their habitat or resource requirements. Finally, competitive interactions among the species will act to sort out those sets of species that can stably coexist from those that cannot. This results in sets of "permitted species combinations," distinctly nonrandom suites of species that will form stable communities. From the dictates of the assembly rules, these communities can be expected to be fully saturated with species and resistant to invasion by other species. Such communities will display a restricted set of patterns, and, to the degree that the environmental setting is similar in different areas, there will be an overall convergence in the patterns of the communities occupying them (Cody and Diamond 1975).

From our preconceptions of order and pattern in nature, we fully expect that actual communities should represent some nonrandom subset of the species available in a regional or continental species pool, and such a scenario of community assembly has an intuitive appeal. Diamond (1975) formalized such assembly rules in terms of a series of anticipated community patterns, using evidence from his studies of bird distributions on New Guinea and its satellite islands. Diamond's analysis, however, suffered from an ad hoc application of anecdotal examples as evidence to support the patterns. Moreover, Diamond used the seeming consistency of his observations with the predictions of the patterns to verify the process explanations (chiefly interspecific competition) embodied in the premises of his assembly rules. The existence of pattern was thus used to assert an underlying causal explanation. Connor and Simberloff (1979) analyzed Diamond's assembly rules and concluded that several rules were simple tautologies, in that the expected patterns followed inevitably from the definitions of terms such as "permissible combinations" and "forbidden combinations."

They suggested that Diamond assumed competition to be the primary determinant of the patterns and then sought to rationalize the observations in the light of this assumption, with no consideration of alternative null hypotheses. Connor and Simberloff structured a null hypothesis based upon a random distribution of species on the islands, subject to some constraints on dispersal and colonization, and found that several of Diamond's patterns were consistent with the predictions of their random model as well. This, of course, does not justify the conclusion that the species *are* randomly distributed (that would also represent fallacious verification of the hypothesis via consistency), but it should raise doubts about the logical rigor of Diamond's analysis and the necessity of postulating species interactions to account for the patterns.

Other features of this community assembly scenario also suffer logical deficiencies. Thus, although the community is held to develop within limits set by resource conditions, the resource levels and characteristics are rarely measured directly (Newton 1980). Rather, the consistency of observed patterns with those predicted by the theory is used as implicit confirmation that resources are indeed limiting. If resources instead actually vary between limiting and superabundant, the limits on species distributions and assembly patterns might be relaxed at times (e.g., Pulliam and Parker 1979; Pulliam, unpub), resulting in species combinations not permitted by theory. So long as "tests" of the theory rely on ex post facto generation of predictions to match observations gathered in the absence of any measures of resources, such anomalous observations are not likely to be detected.

Thus, despite our basic conviction that natural communities *must* be nonrandom assemblages of species and the intuitive appeal of such assembly rules, a logically rigorous approach to this problem has yet to be defined. Recently, Haefner (1981) proposed a model of community assembly based upon generative grammars, in which various ecological attributes of species are itemized and then related in a hierarchial fashion to specifications of the abiotic environment to determine which species can be inserted into a community. Some of these species are then removed from the community according to a set of "deletion rules," resulting in the final suite of "permitted" species that comprise the community. Unfortunately, the approach is constrained by the considerable amount of information required to describe each species ecologically, the reliance of most of the deletion rules upon some form of species interaction, and the deterministic nature of the model operations. Nonetheless, the procedures show promise, and they avoid some of the logical pitfalls that have plagued previous approaches.

I suggest that the logical problems discussed earlier are not of isolated occurrence but are prevalent modes of reasoning in studies of avian communities, perhaps largely because they are implicitly embodied in the procedures for doing science that have become established as part of the MacArthur paradigm of community ecology. Scientists are disposed to accept and follow the logic and procedures that seem to have been successful in providing answers (or at least insights) to the questions of current importance within a discipline. Certainly, an undiscriminating reading of the literature of avian community ecology over the past quarter century would lead one to conclude that this approach has been quite successful in defining and explaining the patterns of nature. Given this apparent success, it would perhaps seem rather easy to dismiss these logical problems as philosophical musings that are largely irrelevant to how science is actually done. However, there is a second class of problems that besets studies of avian communities that is more methodological in nature and that may impinge more directly upon the reality of the patterns that we study – the "facts" of community ecology. I now turn to a consideration of some of these patterns and how they may be influenced by methodology.

Problems in pattern detection

Much of the emphasis in avian community studies during the past two decades has been on patterns in species diversity, habitat relationships, niche differences, guild structuring of communities, and the like – the sorts of patterns that so interested MacArthur (e.g., 1971, 1972) and that became established as the focus of the community paradigm. To deal with these patterns adequately would require an entire volume; here, I will consider only a few selected features of community patterns that exemplify the possible influences of a variety of methodological or procedural problems.

Species diversity

Diversity is an attribute of any collection of species and individuals. Thus, comparisons of communities at any scale of spatial resolution may express some pattern in diversity variations. By convention (MacArthur 1965; Whittaker 1975, Cody 1975), different geographic scales of diversity are usually categorized as α-diversity of a community present within a single local habitat; β-diversity, the increased diversity that characterizes a large area containing a variety of habitats, each probably containing some species restricted to that habitat type; and γ-diversity, the variety of species found when one considers an area large enough for geographic species replacements to occur. Each of these scales of diversity implicitly carries with it an hypothesis of process. Thus, α-diversity is held to reflect the consequences of within-habitat interaction among competitors (as envisioned in assembly rules); β-diversity, the effects of adding different habitat patches (heterogeneity) to the area (thereby enhancing the coexistence of species that differ in habitat niche dimensions); and γ-diversity, the influences of range exclusions between competitors and biogeographic patterns in species distribution and speciation.

It seems evident that the processes that contribute to diversity should be different at different scales of spatial resolution. How species accumulate or populations are limited may be much more strongly influenced by stochastic environmental variations or details of population demography in local communities within a habitat, for example, than at a regional scale, where biogeographical processes may assume greater importance (Wiens 1981a). Thus, the relative contributions of richness and evenness to patterns of diversity will be likely to differ at different spatial scales, as found in the Pacific Northwest (Wiens 1981a). We cannot assume that biological processes operate without regard to spatial scale. One thus cannot extrapolate processes that operate at one level (e.g., competition among individuals in local populations) to explain patterns at another level (e.g., regional distributions) or employ evidence derived from patterns at one scale (e.g., variations in γ-diversity) to test process explanations at some other scale (e.g., local competitive interactions). Unfortunately, most studies of avian communities have been insensitive to such problems of spatial scale.

Investigations of diversity have also employed a variety of procedures to obtain values of species numbers and/or densities at the various scales of resolution. Patterns of γ-diversity, for example, have been defined using a broad array of data sources. MacArthur and Wilson (1967) and Bock and Lepthien (1975) derived counts of breeding land birds in large blocks of North America from the range maps given in field guides or from the results of North American Breeding Bird Surveys (see Robbins and Van Velzen 1969). The patterns derived in this manner indicate something of the nature of diversity variations on a continent-wide scale, but they are so general as to be almost useless. Short (1979) and Schall and Pianka (1978), on the other hand, plotted variations in avian diversity in North America with isopleths of species numbers. Short's analysis was based upon values recorded in Audubon

Breeding Bird Censuses conducted between 1936 and 1973, whereas Schall and Pianka considered only insectivorous birds ("possible competitors with lizards"), plotting the distributions gleaned from field guides in 1° latitude–longitude blocks. These isopleth mappings, and those of Rabinovich and Rapoport (1975) for Argentinian passerines, indicate at the very least how complex diversity patterns can be at this scale. Rabinovich and Rapoport suggested that the patterns they discerned depended strongly on local factors, implying that a clear demarcation between γ-diversity and α- or β-diversity may not be possible.

Patterns of α- and β-diversity (or, for that matter, of other community characteristics) are usually derived from comparisons among censuses or surveys taken, using a variety of techniques, in local areas of <5 to 50 ha. Despite the considerable attention that has been devoted to procedural (e.g., Ralph and Scott 1981) and statistical (e.g., Burnham et al. 1980) aspects of censusing birds, community ecologists have generally paid scant attention to the procedures used to determine the species present in an area or their actual or relative abundances. Diamond (e.g., 1975) used a variety of nonquantitative observational techniques (including interviews with local natives) to determine the distributions of species from which his patterns were derived, and Bédard (1976) characterized Cody's (1973) sampling of seabird distributions as "an unspecified number of transects over an unspecified number of days in unspecified sea conditions and in an unspecified number of localities." Karr (1976), on the other hand, explicitly recognized how methodological considerations can comprise the value of observations, warning his readers to ask continually, How will the sampling biases affect these results? A statement of Karr's warning should be required in the introduction to every study of avian communities.

The detection of community patterns rests upon comparisons, and serious errors can arise if the compared communities were censused with different methods. Franzreb (1981), for example, conducted both variable-strip and fixed-strip width transect surveys of the birds of a mixed coniferous forest; not only did the total density estimates differ by a factor of 4, but the species differed in their relative abundances or their presence or absence in the community, as recorded by the transect censuses. Hildén (1981) determined that single-line transect censuses detected only 46–49% of the breeding pairs determined by a mapping procedure to be present in an area, and DeSante (1981) suggested that variable circular plot censuses might underestimate the densities of different species by 2–70%! These problems of technique comparability are exacerbated if the censuses are conducted by different observers, as differences in their level of expertise with the census method, their ability to detect or identify birds, or their ability to estimate distances may introduce substantial bias into the census results (Scott et al. 1981; Ramsey and Scott 1981; Faanes and Bystrak 1981).

Most established avian census procedures were developed in north-temperate latitudes for application to breeding birds, and they therefore exhibit a bias toward the particular behavioral and populational attributes of temperate bird species. The application of such methods to bird communities in the tropics or south-temperate regions, such as Australia, produces further complications (Karr 1976, 1981; Recher 1981). Many birds in these regions are more secretive and sing less than their north-temperate counterparts; combined with denser vegetation, this makes detection of the species more difficult, and more are likely to be missed in a census. Many tropical species depart from the conventional temperate-zone territorial spacing system during breeding: some form display leks, others wander over large areas, and others occupy overlapping home ranges. Counts based on any aspect of territorial display (e.g., singing male counts, territory mapping) thus will differ in their applicability to different species and will generally be unsuitable for community investigations. Vocal mimicry is frequent among some tropical and Australian species, confounding species identifications and thus the determination of community composition. Lovejoy (1975), Karr (1976, 1981), and others have used mist nets to survey bird populations in tropical forests. This procedure circumvents some of the problems associated with using traditional temperate-derived procedures to census birds but carries with it other biases (see Karr 1976, 1981) and cannot be employed to survey entire avian communities. Karr (1981) expressed "grave doubts about the possibility of producing reliable census data across a wide spectrum of species" in the tropics. Collectively, these methodological difficulties would seem to decimate efforts to compare tropical and temperate bird communities (as, for example, in the determination of latitudinal diversity gradients) at even the most rudimentary level of quantitative rigor.

The temporal framework of sampling can also produce biases in the observations obtained. Best (1981), for example, documented substantial changes in the detectability of individuals (and thus the likelihood that they would be recorded in censuses) through the breeding season in temperate habitats. Because the seasonal changes in detectability were different for different species, surveys conducted over short periods at different times during the season would indicate different patterns of community composition. Here again, the problems are more severe in the tropics, where individual species have prolonged breeding periods but the tight seasonal synchrony of breeding among all members of the community found in temperate latitudes is lacking. Longer surveys, on the order of a year or more, are thus necessary (Karr 1976).

An example of the combination of several of these methodological problems is provided by the work of Tiainen (1980), who used two censuses from Poland, three from central Finland, and three from Finnish Lapland to examine regional patterns in breeding bird communities of mature pine forests. One of the censuses was his own, conducted during a 2-hour interval on one morning in Poland using a line transect count. The remaining seven data sets were obtained from the literature. They involved five (possibly nine) different observers, employed six different survey techniques, and were taken over periods ranging from 1 day to 3 years. Nest boxes were present in some areas but not others, and one data set included counts of flocks of two species. The censuses also differed in the area that was surveyed. Despite such heterogeneity in the data sets, Tiainen used them in comparisons to derive patterns in the distributions of species, community diversity and evenness, species dominance, biomass diversity, and year-round residency of species and individuals. In view of the methodological concerns just expressed, it is clear that none of these patterns can be considered valid.

It has long been known that the number of species that will be recorded in a census is a function of the area sampled (see Connor and McCoy 1979). Despite this, avian ecologists have generally sampled communities with little regard for such species–area effects. Cody (1968, 1975), for example, counted birds on plots of 1.66 to 4 ha, whereas we have used 10.6-ha plots in our studies in similar habitats (Wiens 1974; Rotenberry and Wiens 1980a); the Audubon Breeding Bird Censuses considered by James and Rathbun (1981) were conducted on plots of 6 to 60 ha. Even if one discounts the differences in species–area curves that occur between habitats or between regions, it should be apparent that a comparison of results from, say, a 6-ha census and a 60-ha census would be invalid. Short's (1979) analysis of diversity patterns across North America, for example, used censuses conducted on areas of 8 to 40 ha, although he standardized the results to areas of 40 ha. James and Rathbun (1981) advocated using the procedures of rarefaction to express data on a common spatial scale. In contrast to Short's standardization, this approach correctly adjusts estimates of densities or species numbers to the least censused unit. Rarefaction procedures, however, contain some inherent limitations (Tipper, 1979; James and Rathbun 1981). Such procedures assume that the spatial distribution of each species in the community is homogeneous. They can be applied properly only to censuses that have followed the same methods and that are conducted in similar habitats. Nonetheless, the approach does have potential, and it at least addresses the problem of area biases in censusing. Rarefaction, however, is not a technique that will remedy the ills of inadequate, biased, or incorrect sampling.

Censusing methodology is thus not a trivial matter that can be dismissed or taken lightly in the rush to get on with determining the patterns of nature and relating them to theory. To conduct science in a responsible, rigorous manner, we should be reasonably certain that the patterns of community attributes such as diversity are indeed real before we attempt to explain them. Improper censuses produce results that are faint, blurred, and incomplete images of the actual community. This is especially likely if ad hoc, nonquantitative census procedures are employed. When the results of such censuses are joined together in comparisons, they will produce "patterns" whose reality must be seriously questioned and vigorously challenged.

Habitat relations

The relations of birds to features of their habitats has been a major focus of study in avian ecology, perhaps because at least some aspects of habitat use are so conspicuous in birds. Some of MacArthur's earliest studies of bird communities concentrated on bird–habitat patterns, such as the differentiation of foliage use by foraging warblers (MacArthur 1958) or the relationship of bird community diversity (BSD) to the physical structuring of the vegetation, which MacArthur indexed as foliage height diversity (FHD) (MacArthur and MacArthur 1961; MacArthur 1965; MacArthur et al. 1966). Arguing from the premise that physical structuring of the habitat is important, in determining habitat selection in birds (Hildén 1965), MacArthur and his colleagues interpreted the close correlation between BSD and FHD that they demonstrated in several habitat types as a manifestation of an increased availability of niches or opportunities for resource subdivision in habitats that expressed greater vertical layering and patchiness. Several additional studies, in a variety of habitats, have confirmed the BSD–FHD correlation (e.g., Karr 1968; Recher 1969; Karr and Roth 1971; Haila et al. 1980; Rotenberry and Wiens 1980a). Other studies (e.g., Willson 1974; Tomoff 1974; Szaro and Balda 1979; Roth 1981;

Wiens and Rotenberry 1981), however, failed to demonstrate a relationship between diversity and FHD or vertical structuring, probably because some other variables were involved or because habitat structure was incorrectly measured or indexed (Karr 1980; Roth 1981).

Such departures from the BSD–FHD relationship demonstrated by MacArthur cast doubt upon the generality of this pattern in avian communities. Even where the relationship does obtain, its biological interpretation is not immediately apparent (Willson 1974; Haila et al. 1980; James and Rathbun 1981). An index like FHD includes a number of variables that may or may not be important to the birds, and different bird species may respond differently to various habitat features that are incorporated into a single index. The search for a single index to quantify avian habitat (Roth 1981) may not only be difficult (Rotenberry and Wiens 1980a) but misdirected.

An additional and more fundamental problem in interpreting the BSD–FHD pattern (or its absence) relates to the manner in which the relationship is established in the first place. In their initial work, Mac-Arthur and MacArthur (1961) calculated FHD by determining the proportion of vegetation present in three height strata. In fact, they initially compared diversity with various subdivisions of the vertical vegetation profile and selected the layers 0–0.6, 0.6–7.6, and >7.6m, because this subdivision "made the collection of points on the graph most orderly" (MacArthur and MacArthur 1961: 596). When the approach was extended (MacArthur et al. 1966), censuses from a tropical location (Panama) and from Puerto Rico did not fit the relationship; they then subdivided the foliage profile into four layers for the Panama data and two layers for Puerto Rico, preserving the BSD–FHD relationship for these habitats. Austin (1970) used still different height strata to achieve a good fit between BSD and FHD in Mohave Desert habitats, and more recently Moss (1978) obtained a strong correlation between BSD and FHD in British woodlands by adjusting the height strata that were used. In other words, it appears that considerations of the BSD–FHD relationship have been motivated from the outset by a belief that the pattern must be correct, and in many cases observations have been adjusted in an ex post facto, ad hoc fashion to preserve their consistency with the anticipated pattern. Moreover, MacArthur et al. (1966) used the Panama and Puerto Rico censuses to conclude that bird species in the tropics recognize finer vertical subdivisions of the habitat than birds in temperate habitats, whereas those in Puerto Rico subdivide habitats even less. Thus, their questionable generalization on pattern led to an inference on process.

MacArthur's (1958) work with coniferous-forest warblers pointed to the important role that habitat subdivision might have in facilitating the coexistence of potentially competing species, and numerous studies over the past quarter century have investigated habitat partitioning in this context (see Schoener 1974). The approach is seemingly straightforward. By measuring the behavioral aspects of habitat use and/or the features of habitat occurring in areas occupied by individuals of the coexisting species in a community, one may document the patterns of differences between the species. These differences are presumed to be the consequences of competition and are the mechanisms by which the species achieve stable coexistence. For example, "differences in habitat occupancy and utilization were seemingly adequate to circumvent direct competition and to explain the co-occupancy of the study area by the seven breeding species" (Wiens 1969:89). If the species have been found to be quite similar on the basis of such analyses, it has been a common practice to suggest, in an ad hoc fashion, that the species must differ on some other niche dimension that has not been measured (e.g., Cody 1974; Noon 1981) or perhaps that the habitat niche patterns are governed by diffuse competitive interactions that are not likely to be detected through pairwise calculations of similarity between species (Diamond 1975, 1978). Such contentions are generally untestable (Wiens 1977b, in press).

Documenting habitat partitioning by behavioral means has been especially popular in studies of birds, largely because at least some of their behavior is so easily observed. Stiles (1978), for example, used spot observations of the behavior and habitat use of individuals he encountered to determine patterns of niche overlap and niche breadth among breeding birds in alder forests. He concluded that in the more predictable environments birds feed in more specialized ways, allowing more species to coexist. A similar observational approach to documenting habitat use was followed by Ricklefs and Cox (1977) and Szaro and Balda (1979). Ricklefs and Cox obtained their observations of foraging movements of passerines on St. Kitts (samples of 8–37 foraging movements per species) during the mornings of 7 days in August. Coupling this data base with measures of morphology and an index of the breadth of habitat distribution for each species, they determined that habitat overlap among the species was unrelated to either behavioral or morphological similarity (which were strongly related to each other). Ricklefs and Cox found this pattern "not surprising considering the great breadth of habitat distributions of species on faunally depaupaerate islands . . . and the fact that small differences in morphology and feeding behavior may be sufficient for ecological segregation!"

Quite apart from the logical problems of such an argument, there are severe methodological problems that afflict such determinations of behavioral patterns. Use of spot observations of the behavior of individuals when they are first encountered, for example, is inevitably biased (Wiens 1969; but see Wagner 1981), as one is likely to record visible behaviors occurring in conspicuous or open habitat settings disproportionately often. This, in turn, can produce a systematic bias in the evaluation of habitat features associated with such behaviors. Cody (1978), for example, measured the habitat of species by first estimating the center of an individual's territory (determined by locating singing birds) and then taking vegetation profiles in each of four haphazardly selected directions from the center. It would seem unlikely that such sampling would produce an unbiased estimation of habitat features for territories as a whole.

Some of these biases can be circumvented by recording continuous or regularly spaced observations of individuals (Wiens 1969; Wiens et al. 1970; Altmann 1974). Such observations not only require some sort of a priori research plan, but they introduce other biases as well (Bradley and Bradley pers. comm.). If the strings of continuous observations are short, they will still contain a disproportionate representation of the conspicuous behaviors or habitat positions that led one to discover the individual and observe it. This problem can be circumvented by recording observations over long periods and then discarding the initial (biased) observations in the sequence. The successive observations of behaviors or habitat positions in such a sequence, however, are not independent of one another but are serially autocorrelated. Successive observations of an individual may thus not be used as data points in statistical analyses, as they violate the assumptions of independence of samples. Gathering reliable observations of behavioral patterns of habitat use is thus far more difficult than most ecologists have recognized, and the patterns that have been described must be regarded with caution.

Most studies of avian habitat relations have not included behavioral observations but have simply measured or recorded various features of habitats occupied by species in a community. Recently, multivariate statistical procedures have been widely used to analyze such data (see Capen 1981). All of these procedures carry statistical assumptions (which are not always considered), but some have critical biological implications as well. In particular, applications of discriminant function analysis (DFA) may pose important difficulties. Several key statistical assumptions (e.g., equal sample sizes and homogeneous variance–covariance matrixes for the groups being analyzed; see Green 1971) are rarely satisfied by ecological data. Furthermore, DFA may be mathematically predisposed toward support of hypotheses it purports to test. Thus, if one intends to test competition-based community theory by determining how coexisting species differ in niches (e.g., Cody 1978; Noon 1981), DFA is a biased procedure in that it is specifically designed to analyze a matrix of data so as to maximally differentiate between classes (species) (Rotenberry and Wiens 1980a). If there is any possibility of doing so, the procedure will discern but not create differences between the species, to which (following our expectations from theory) ecological meaning will then be assigned. Approaches using DFA to examine patterns of niche breadth and overlap of species such as that advocated by Dueser and Shugart (1979) may also be flawed. Dueser and Shugart (1978, 1979) conducted an a priori univariate screening of their data to produce a reduced data matrix that was then subjected to DFA. The screening procedure, however, specifically selected those variables that differed between the species, thus amplifying the predisposal of the DFA to discriminate between the species. Furthermore, their procedures for calculating niche breadths were biased by sample size differences in such a way that the common species inevitably emerge as being broad-niche generalists relative to rarer species (Van Horne and Ford 1982; Carnes and Slade 1982). Such procedures cannot be applied, for example, to tests of hypotheses relating relating abundance to niche breadth (e.g., McNaughton and Wolf 1970).

An analysis of habitat relationships of community members is obviously sensitive to what habitat features are recorded, when, and on what scale. Thus, although most avian studies seem to have assigned features of habitat structure overriding importance, there is mounting evidence that many bird species show close associations with specific plant species in their behavior or distribution (Tomoff 1974, Wiens and Rotenberry 1981, Holmes and Robinson 1981). The community patterns that are likely to be discerned will thus be quite different if one measures just a few preselected structural features of the habitat in contrast to a larger set of structural *and* floristic variables. Moreover, both habitats and bird populations change during the course of a breeding season, and it thus should be no surprise that characterizations of the habitat relations among species also change through the season (Wiens 1973; Whitmore 1979). Finally, the scale at which studies are conducted can have a major influence on the patterns that emerge. In our studies in grassland and shrubsteppe habitats (Rotenberry and Wiens 1980a; Wiens and Rotenberry 1981), for example, associations of species and of features of community structure with habitat variables were quite different when we conducted the analysis at a "continental" (Great Plains through Great Basin) versus a "regional" (Pacific Northwest) scale of resolution, It is quite apparent that by structuring a habitat measurement scheme that considers only a few variables that are expected to be important contributors to niche patterning (e.g., Cody 1968, 1974), without regard for the time at which the measurements are taken, or for the scale of the comparisons, one is likely to obtain misleading patterns.

No mention has been made of alternative hypotheses. One clear alternative to the view that species are arrayed in habitats following a pattern of niche differentiation driven by competition, for example, is that species occupy habitats each according to its own intrinsic ecological requirements, quite independently of one another and in the absence of species interactions. This, of course, is a resurrection of the neglected Gleasonian view of communities. Several recent studies (e.g., Bock et al. 1978; Haila et al. 1980; Lanyon 1981; Holmes and Robinson 1981) have concluded that the habitat distributions of species in the communities they investigated represented independent, species-specific responses to habitat characteristics, and our recent work with grassland and shrubsteppe bird communities (Rotenberry and Wiens 1980a,b; Wiens and Rotenberry 1981) has produced a considerable amount of evidence that is consistent with this hypothesis. This, of course, does not mean that such an interpretation is correct, but the inconsistency of our findings with the competition hypothesis strengthens the case.

Niche shifts

Niche shifts represent yet another community pattern, in which features of the niche breadth (e.g., range of habitat occupancy, variety of foraging behaviors) or the population density of a species change in apparent response to the presence or absence of some other species. According to the MacArthur paradigm, such niche shifts (or "ecological release") are among the most convincing evidence of the role of competition in structuring communities. Thus,

the accepted interpretation of this phenomenon is based on the fact that there are fewer competing species on the islands. On the mainland, one species may be excluded by competing species from habitats and vertical zones in which its competitors are superior. On islands where these competitors are absent, the species is able to occupy these habitats and zones. (Diamond and Jones 1980)

The existence of a large number of supposed niche shifts among sets of bird species on islands in the New Hebrides, for example, led Diamond and Marshall (1977) to conclude that competition was the proximate cause of the distributional limits for most of the species in this archipelago, and Diamond (1975, 1978) drew similar conclusions from his observations of niche shifts in the New Guinea region. In instances in which niche shifts cannot be attributed to the presence or absence of a single competitor, they are suggested to reflect diffuse competition with a set (usually unspecified) of competing species (e.g., Diamond 1978). Yeaton and Cody (1974), for example, used changes in territory sizes of Song Sparrows (*Melospiza melodia*) in various areas as a measure of niche shifts under conditions of varying competitive pressures (as gauged by the numbers of other species of small birds present in the areas). Territory size was found to increase with the number of competing species. They offered the following explanatory scenario:

In comparison to communities with few competitors, in bird communities with many species, the Song Sparrow becomes restricted, apparently to those parts of the potentially useable food resources on which it is most efficient. Thus, in order to have within the territory enough of these particular food resources for the maintenance of adults and the raising of young, the size of the territory needs to be increased. (Yeaton and Cody 1974).

After discounting a possible influence on territory size of variations in habitat or food supplies (which were measured only crudely), Yeaton and Cody (1974) concluded that "differences in interspecific competition remain the major cause of variation in Song Sparrow territory size." The possible impact of yearly or seasonal variations in territory size (e.g., Wiens 1969, 1981b) on the pattern they discussed was not considered.

Niche shifts demonstrate the importance of the ceteris paribus ("other things being equal") assumption to comparisons in community ecology perhaps more clearly than any other patterns. Thus, it is normally assumed that, were it not for competition, the niche characteristics of a species would be more or less the same throughout its range, even though there is evidence to the contrary (e.g., Nilsson and Allerstam 1976; Herrera and Hiraldo 1976; see also Grant 1972). Also, the comparisons that are used to document niche shift patterns involve different locations. Even using loose logic, one can infer that changes in the competitive mileu underlie the niche shifts only if the areas are assumed to be equivalent in all other respects (Abbott 1980; Vuilleumier and Simberloff 1980; Connell 1980). Resource levels or habitat characteristics are rarely measured in the compared areas, and the assumption is usually accepted as an article of faith. When features of the compared areas are measured (even crudely), the results may be ignored. Cox and Ricklefs (1977), for example, examined bird communities from several areas in the Caribbean for evidence of niche shifts in habitat occupancy. Their anlaysis of habitat structure showed that representative areas of the same habitat type sampled in different locations differed substantially in configuration. Because there was no clear trend to these habitat differences, however, they were ignored. Cox and Ricklefs (1977) then suggested that "the extent of ecological release in bird populations . . . is nearly complete in the habitats examined, *assuming that the carrying capacity of these habitats does not vary greatly from place to place*" (italics mine). Abbott (1976) did measure arthropod levels in an island–mainland comparison and found abundances to be greater on the island, which he attributed to a reduction in the densities of avian predators due to difficulties in colonizing the island. This could easily produce niche shifts paralleling those expected from competition theory in the absence of any species interactions if some types of sizes of prey increased more than others. Furthermore, because resources are more abundant on the island, population densities of the species that do occur there may be greater (e.g., Emlen 1979), producing an expansion in their niches as a consequence of the intensification of intraspecific competition (Svärdson 1949).

A particularly vivid example of the ways in which some of the logical problems noted earlier can lead to unjustified conclusions from the documentation of apparent niche shifts or ecological release is provided by Noon's (1981) recent investigation of the ecological relationships among thrush species distributed on elevational gradients. The study was conducted at one location in Vermont, where five species occurred, and at another location in the southern Appalachians, where only two species bred. In many respects, Noon's study exhibited an unusual attentiveness to the details of field and statistical methods and he avoided many of the methodological pitfalls discussed earlier. The patterns that he identified are probably correct. A discriminant function analysis of measures of habitat features in territories of the species at the northeastern site produced a clear and generally consistent pattern of separation among the species. Noon then used comparable measures of habitat occupancy for the two species occurring in the southeastern forests to determine the accuracy with which they were classified with the proper species group by the discriminate functions derived from the northeastern data sets. Finding a substantial difference from the classification functions for the same species at the northeastern site, he suggested that this was strong evidence of competitive release in the southeast – the two species occupied a broader and different range of habitats than they did in the five-species northeastern location. Other environmental features were thus assumed not to differ between the two locations, despite their considerable geographic separation. This documentation of ecological release, and the absence of evidence of direct (aggressive) competition between the thrushes in the northeast, was then used to infer that habitat selection (as revealed by DFA) is the mechanism that resolves interspecific competition in the northeastern forest. Habitat selection by the species was, in turn, suggested to be an evolved response to the pressures of past competition, largely on the basis of the apparent complementary in niche dimensions determined by DFA. This sequence of logic led to Noon's basic conclusions (1981): "the extensive ecological release of the thrush populations in allopatry . . . strongly implies that interspecific competition has (or had at some time in the past) a dominant role in forming the habitat affinities of this guild," and "although competition is apparently zero as a result of habitat selection, it remains very important. Presumably the observed patterns of specific habitat selection have evolved in response to pressures to avoid competition; if those pressures are removed, the habitat selection patterns would disappear, and interference competition would be obvious." Perhaps, as Williams (1981) suggested, discriminant function analysis may prove valuable for data exploration, but it is of limited value as an inferential procedure, especially when combined with faulty logic.

Guild structure

The species that occur in a community are generally not equally different from one another in their ecological attributes. Rather, species tend to be grouped by virtue of similarities in their manner of exploitation of the same type of resources. These "guilds" (Root 1967) group together species that overlap in niche requirements, regardless of their taxonomic affinities. The guild concept contains an explicit prediction of process: Interspecific competition is presumed to be intense within guilds but weak between members of different guilds (Root 1967; Cody 1974; Pianka 1980).

Because specific taxonomic affiliations of species are ignored, consideration of community structure in terms of guild patterns has been especially popular in broad-scale comparisons of communities. Schoener (1971), for example, sought to gain some understanding of the factors contributing to latitudinal diversity gradients by restricting attention to members of a single guild, and Karr and James (1975) used ecomorphological analyses to examine the patterns of guild structuring in several tropical and temperate habitats. There seems little doubt that at least some of the difference between species numbers in temperate and tropical areas is a consequence of the availability in the tropics of food resources, such as fruits, large insects, or nectar, that are not generally significant resources in most temperate habitats, which permits the addition of entire guilds of birds in the tropical habitats (Karr 1971, 1980). Such guilds differ in their patterns of response to seasonal variations (Karr 1976). These differences, considered with the demonstration by Holmes and Robinson (1981) that members of different guilds may differ in their responses to features of vegetation structure or floristics, argue that communities with different patterns of guild structuring will be quite likely to exhibit different dynamics.

It is critical to any community analysis based on guilds that the guild membership be correctly defined. Recently, multivariate statistical procedures have been applied to data matrixes that specify a variety of ecological, morphological, or behavioral attributes of the species in an attempt to identify suites of similar species in an objective fashion (e.g., Karr and James 1975; Holmes et al. 1979; Sabo 1980; Landres and MacMahon 1980). In general, these procedures have produced clusterings of species in multivariate space that agree with ecological intuitions, and, as long as proper attention is given to the statistical and methodological requirements of multivariate analysis (see the papers in Capen 1981), this approach holds considerable promise. Procedures that rely on one or another form of data clustering, however, suffer a potential bias in that, given any departure from uniformity or balance in the data matrix, the statistical procedures will generally produce clusters of the species. Some clusters may have marginal ecological relevance however (for example, the Chipping Sparrow (*Spizella passerina*), Red-shafted Flicker (*Colaptes auratus*), Mexican Jay (*Aphelocoma ultramarina*) grouping portrayed in a cluster diagram presented by Cody (1974). Such groupings may represent consequences of the clustering algorithm rather than bona fide ecological guilds. Strauss (1982) has developed procedures to determine which of the clusters in a dendrogram are statistically significant. As always, the interpretation of the results of multivariate procedures requires keen biological insights.

There is another problem that must be considered when one interprets the guild patterning of a community. Guilds are presumably founded upon resources. Some resources come in discrete categories (e.g., fruits, insects), whereas others may potentially exhibit a continuous distribution (e.g., sizes of insects or seeds). A recent simulation exercise (Bradley and Bradley, unpubl.), however, clearly indicates that species whose ecological attributes are assigned at random will nonetheless form distinct clusterings or "guilds" if there is any degree of discontinuity or unevenness in the availability of resources to them. In other words, environmental discontinuities may drive the apparent guild patterns. Similarly, unevenness in the patterns of branching in phylogeny will produce discontinuities between groups of species (which are recognized by higher taxonomic categories) – witness, for example, the strong phyletic component to the multivariate clusterings provided by Karr and James (1975). These points suggest that some sort of guild structuring of communities is really inevitable, given environmental discontinuities and the patterns of phyletic branching that occur in evolution. This does not mean that guild patterns are meaningless, but it does suggest that they may be a simple consequence of general features of environments and evolution and do not carry any a priori implications of species interactions.

Assumption of equilibrium and community saturation

It is a basic premise of the MacArthur community paradigm that natural communities are normally at or close to equilibrium in their structure and relations with the environmental resources. They are thus fully packed with species, and habitats likewise are fully saturated with individuals. Such a belief has been bolstered by various kinds of evidence, mostly inferential. The seeming consistency of observations of community patterns with the predictions of competition-based theory, for example, has been used to argue that the equilibrium/saturation premise must be correct. Cody (1966) and Recher (1971) interpreted the consistency of bird species diversity values among avifaunas occupying similar habitats in different continents as clear evidence of community saturation and equilibrium. The form of curves relating the species number within habitats to the size of the species pool available on several West Indies islands was used by Terborgh and Faaborg (1980) to substantiate their claim that the bird communities were indeed saturated. In view of such conclusions and the apparently widespread belief in the truth of the equilibrium/saturation assumption (Wiens, in press), it is appropriate to inquire whether or not there is evidence that communities may at times be nonequilibrial or environmental resources not saturated.

Certainly bird communities and populations in many habitats exhibit substantial fluctuations through time. Pulliam (1975), for example, documented a pattern of divergences among coexisting finch species that was consistent with the predictions of community theory during 1 year. He failed to confirm the theory in subsequent years (Pulliam, unpub.) and found instead that predictions based upon a model of random co-occurrence of species matched the observations about as well as those of community theory. Winternitz (1976) noted that only 78% of the 50 species recorded in her study area over a 5-year period were present in any 1 year, and Järvinen (1978) reported considerable annual change in the avifauna of an environmentally stable peatland habitat in Finland over a 20-year period. Some of these changes were associated with climatic variations: In the Finnish peatlands, rare species were much more frequent components of the breeding community in years with warm springs than in years with cool springs (Väisänen and Järvinen 1977). Such annual variations may involve localized extinction of species' populations in isolated habitat patches, especially on the margins of the species' range (e.g., Fritz 1980). Because recolonization of such habitat patches may be probabilistic, the effects of such variations on the composition of local communities are likely to persist for some time.

Järvinen (1979, 1980) and Myers et al. (in press) have attempted to consider such annual changes in a geographic context by searching for latitudinal gradients in the temporal stability of community diversity. Järvinen found a clear pattern of variation in stability with latitude in Europe, high-latitude areas being significantly more variable than those in central Europe and southern Scandinavia. Myers et al., on the other hand, found no relationship between latitude and variations in diversity in several habitat types in North America. They were unable to reconcile their findings with those of Järvinen, but it is possible that the temporal dynamics and structuring of bird communities in Europe simply differs from that in North America, as a comparison of the studies of Rabenold (1978) and Nilsson (1979a) would seem to suggest.

Avian communities are also characterized by change over longer time periods. Järvinen and his colleagues (Järvinen and Väisänen 1977, 1978, 1979; Järvinen 1978, 1980; Järvinen and Ulfstrand 1980) have examined changes in the avifaunas of various northern European habitats by comparing censuses taken decades apart. In all cases there were dramatic alterations, generally involving more additions than losses of species. These compositional changes in the communities led Järvinen (1980) to conclude that "it is thus impossible to conclude that a community is at equilibrium, unless it is studied for decades." Such investigations are not part of the research tradition of the MacArthur paradigm.

Järvinen attributed much of the avifaunal turnover that he documented to long-term changes in habitats, raising the possibility that the community changes represented a close tracking of changes in environmental resources to maintain a dynamic equilibrium. Cody (1980, 1981) has suggested that the apparent chaos in avian community fluctuations reflects failure to measure variations in resource levels that are closely tracked from year to year rather than an expression of nonequilibrium in the systems. His attempt to demonstrate such resource tracking (1981), however, is complicated by uneven and superficial censusing, coarse measures of resource levels, and the use of qualitative rather than quantitative data in the most critical year of the test. Other studies (e.g., Rotenberry and Wiens 1980b) have failed to document a relationship between substantial annual variations in resource conditions and the dynamics of bird popoulations or communities, despite intensive efforts. The expectation of close community tracking of varying resources requires that the consumers be able to match their resources on an annual basis with a precision such as suggested by Wynne-Edwards (1962). It seems likely, however, that such responses would be affected by time lags of various sorts, eroding a close tracking of environmental variations. Site fidelity, for example, might lead individuals to occupy previous breeding areas during successive years despite changes in habitats or resources (see Hilden 1965; Jedraszko-Dabrowska 1979), creating a sort of "community inertia." Additional complications to a tracking hypothesis arise from Järvinen's suggestion (Järvinen 1978; Järvinen and Ulfstrand 1980) that many of the long-term changes in breeding bird communities in northern Europe are consequences of habitat changes on the distant wintering grounds (see also Wolf and Gill 1980). Attempts to relate breeding-community fluctuations to proximate breeding habitat variations would therefore be misguided, largely through a neglect of the problems posed by geographic scale in such systems.

Determining the equilibrium or nonequilibrium status of a community requires that we be able to relate the patterns of community packing to the abundance of potentially limiting resources. The conditions of limited resources that imply equilibrium are most often inferred as consequences of the process explanations that are attached to a pattern, which is all that is actually observed. Thus, Dhondt (1977) demonstrated an inverse correlation between Blue Tit (*Parus caeruleus*) density and Great Tit (*Parus major*) reproduction. From this pattern, he inferred a process of interspecific competition for food, from which he in turn concluded that food must be limiting during the breeding season for these species. Ulfstrand (1977) followed similar logic to reach the same conclusion for communities of woodland birds. Unfortunately, assertions that food is superabundant (and the communities thus not fully saturated and not equilibrial) tend also to rest on inferences derived from patterns. In particular, the demonstration of unexpectedly great niche overlap among presumed competitors may be used to infer that food is superabundant (e.g., Herrera and Hiraldo 1976; Ford and Patton 1977). Baker (1977), for example, suggested that the considerable overlap in feeding habits among several Arctic shorebird species was perhaps associated with the abundance of food on the tundra during the breeding season, although he also raised the possibility that partitioning on other (unmeasured) niche dimensions might reduce the overall overlap. Rotenberry (1980), however, demonstrated not only that diet overlap was great during the breeding season among the members of the shrubsteppe bird community that he studied but that the species shifted diets through the season in a parallel fashion. Alatalo (1980) found similar parallel shifts in diet during summer among different species of a foliage-gleaning guild in Finland, although diet overlap was reduced in other seasons. Such patterns seem more consistent with an hypothesis of opportunistic utilization of superabundant (but changing) resources than with an hypothesis of competition over limited resources.

Environments vary, both seasonally and over longer time periods, and these variations may often create "feast or famine" conditions through pulses and bottlenecks in resource abundance (Wiens 1977b). If the increases in resource levels are rapid, bird populations may lag in their functional or numerical responses, creating situations of resource superabundance, removing competitive constraints on community organization, and leading to unsaturated, nonequilibrial communities (e.g., Abbott and Grant 1976; Rabenold 1979; Pulliam and Parker 1979; Wolf and Gill 1980; Alatalo 1980). The degree to which such seasonal or yearly fluctuations are "tracked" by immigration or emigration of migrants and/or transients in a manner that maintains community saturation and the patterns predicted by theory depends upon both the predictability and the magnitude of the resource changes. The flush of food abundance that occurs in temperate-zone woodlands during the breeding season, for example, may far surpass the.demands of consumers (Rabenold 1979; Alatalo 1980), altering both the patterns and the processes expressed in communities.

Unfortunately, at the present time arguments on both sides of this issue are founded largely upon the relative consistency of patterns with equilibrium or nonequilibrium expectations and upon assertions regarding resource limitation or superabundance. Diamond (1978) has observed that "doubters of the role of interspecific competition object that, until resource levels are actually measured, an essential link in the argument remains hypothetical and the 'overwhelming' evidence for competition remains circumstantial." He is right. Until avian ecologists begin the difficult task of measuring resource availability in relation to consumer demands, the arguments are likely to remain unresolved.

Process explanations of community patterns

Competition

Within the framework of the MacArthur paradigm, competition is viewed as the major, perhaps the only, process acting to determine community patterns. The logical and methodological limitations of this approach should be clear, but this does not mean that competition is unimportant or undeserving of study. By definition, interspecific competition occurs when individuals of different but co-occurring species overlap in their requirements for resources that are in limited supply in relation to their demands. One element of this definition, resource limitation, has been discussed earlier. A second element states that the species overlap in resource use. Unfortunately, we have no real means of specifying how much overlap or similarity between species is sufficient to promote competition between them, and theory provides only ambiguous guidance (Wiens 1977b). Thus, one may assert that resource scarcity (unmeasured) will increase overlap between species as they are restricted to common resources, which in turn produces competition (e.g., Ford and Patton 1977), or, alternatively, that resource scarcity will promote resource diversification between species, circumventing competition (e.g., Herrera and Hiraldo 1976). The propositions that competition might be expected to be intense among congeneric species (Lack 1971) or among members of the same guild (Root 1967) have intuitive appeal, but supporting evidence seems derived more from inferences based upon that appeal than from empirical studies. There is a widespread tendency to hypothesize that competition should be intense between such similar species. In some cases, such competitive effects may indeed be documented, but more often there is no clear evidence that it is occurring. Usually, attention is then drawn to the differences that exist between the species despite their overall similarity. The differences are interpreted as "coexistence mechanisms," and the ghost of competition past is called upon to explain them. Such arguments do not enhance the testability of the competition hypothesis.

The third key element of the definition of competition is that it occurs between individuals and that the effects of these interactions are most directly expressed in the structure and dynamics of local populations of species, which, in turn, comprise local communities. These local processes may contribute to the determination of patterns at a larger geographical scale (as, e.g., the determination of range boundaries), but at these large scales other factors, such as biogeography and past history, also influence patterns, perhaps overriding the evidence of competitive effects. Furthermore, because the nature, direction, and intensity of competition are likely to differ between different local populations of a species, it is unlikely that the amalgamation of many such local patterns to form a larger-scale pattern represents a simple additive process. The point is that "species" as units do *not* compete, and when one considers patterns at such a broad scale that species rather than populations are the basic units, it is unlikely that competitive processes will make major contributions to the patterns or that the patterns can be used to supply inferential "evidence" of the operation of competition.

Under what conditions should we expect competition to be in evidence? Diamond (1978) has suggested that evidence of competition should be clearest in the tropics, where species are more closely packed in ecological space and have tighter relationships with resources. Connell (1975, 1980), on the other hand, believes that competition is most likely to occur in moderately harsh physical environments. Under extremely harsh conditions, Connell suggests, populations are continually depressed below levels of resource limitation and competition; under benign conditions, "natural enemies" (predators, parasites, etc.) may be able to maintain sizable populations and thus keep populations in check and beneath competitive levels. Only under intermediate conditions will populations attain resource exploitation levels at which they compete. This may be so, although Connell provides little guidance to exactly what "harsh" or "benign" mean in the real world, and it is thus difficult to carry his prediction further.

An alternative approach is to look to the resources themselves as a focus of potential competitive interactions. Resources that are rather narrowly defined, that can be measured, and that support a suite of specialist species (e.g., floral nectar, nesting cavities) seem especially appropriate for analysis. Feinsinger and Colwell (1978) have observed that nectar-feeding hummingbirds of open tropical habitats may be the premier example of a food-limited guild, although Carpenter's (1978) studies suggest that the degree of such limitation may vary among communities. Nonetheless, nectar-based guilds seem nearly ideal for studies linking process to pattern: The birds obtain most of their energy from nectar; their behavioral interactions at the resources can easily be observed; use of the resource can be documented, not only behaviorally but in energetic terms; the availability, energy and nutrient content, and renewal rates of the resource can be measured; resource availablity can be related to the morphology of the species; and the dynamics of the community can be related to resource dynamics (Feinsinger and Colwell 1978; Gill and Wolf 1979; Wolf and Gill 1980).

Perhaps the surest way to demonstrate competition in natural systems is to conduct field experiments (Abbott 1980; Wiens, in press). Schoener (1974) and Cody (1974) have noted some of the problems that are inherent in such manipulations and have expressed their preference for deducing competition from correlations in "natural experiments." Connell (1975, 1980), however, has argued vehemently in support of experiments and has provided some general guidelines for their design. Such experiments generally manipulate either local populations of one of a pair of potentially competing species (e.g., Davis 1973; Williams and Batzli 1979; Högstedt 1980; Minot 1981) or the abundance or availability of resources (e.g., Hogstad 1975; Slagsvold 1978; Dhondt and Eyckerman 1980). Although some of the results of these experiments are open to alternative interpretations, several (e.g., Högstedt 1980; Dhondt and Eyckerman 1980; Minot 1981) clearly indicate the operation of resource-based competition between species and its effects upon breeding densities or reproductive success. Such studies strengthen our confidence that competition over limited resources does in fact occur and that this may in turn influence community structure. They do not indicate how frequent or widespread such competitive interactions might be, however, and thus do not substantiate the ubiquity of competition as a process determining community patterns.

Other processes

Diamond (1978) noted that some authors suggest that processes other than competition may contribute to community patterns such as niche shifts and object to interpreting them as consequences of competition. This objection, Diamond (1978) claimed, "strains one's credulity." Nonetheless, processes such as predation, disturbance, and chance do occur in nature, and they are thus likely to have some effects upon species' populations and, thence, on the patterns of communities (Connell 1975; Menge and Sutherland 1976; Glasser 1979; see also Paine 1981; Newman and Stanley 1981). Predation, for example, may depress population levels below competitive levels, permitting greater overlap among the coexisting species and enhancing diversity, and recurrent disturbance of environments may prevent communities from attaining equilibrium and thus from expressing the patterns expected from theory. Such processes have received little consideration in studies of bird communities, perhaps because they are difficult to document or because of the dominance of the competition explanation. Fretwell (1972) and Tomialojc and Profus (1977) suggested that the density of bird communities may depend upon the density of species that are nest predators. Nilsson (1979b) found that individuals of six seed-eating species were concentrated in habitat patches where seed density was greatest during one winter, but in the following winter only two of the species showed such a distribution. Nilsson attributed this change in resource use patterns to a greater abundance of predators during the second winter. A more complex situation was reported by Slagsvold (1980): The presence of a nest predator species (*Corvus corone*) apparently contributed to an increase in community diversity. Diversity was reduced, however, when another species (*Turdus pilaris*), whose colonies seemed to offer some protection against predation for other species nesting in their midst, was present.

Disturbance creates obvious changes in community structuring through alterations in habitat features. The members of a three-species breeding bird community that we studied in a shrubsteppe habitat, for example, underwent dramatic changes in their densities relative to one another on a plot that suffered a range fire, which removed most of the shrub cover (Rotenberry and Wiens 1978). The response, however, was not immediate but occurred gradually, lagging over several years. More subtle (and potentially more important) disturbances are those that create quite localized patches within a habitat. Such disturbances produce a mosaic of patches of differing sizes and ages since disturbance in an area; if the disturbance rate is high, most patches will be young, and the system may be chronically nonequilibrial and unsaturated (Osman and Whitlatch 1978). In forests in Panama, Schemske and Brokaw (1981) found that diversity was greater in the gaps created

by treefalls than in adjacent undisturbed areas, a pattern they attributed to the greater structural heterogeneity generated in the forest by
the patches. There are thus likely to be variations within and between
forests in community features according to the frequency of treefall,
the size- and age-class distribution of treefall gaps, and the regeneration time of the gaps.

The specific timing and location of treefalls within a forest, and the
specific pattern of regeneration that occurs within the patches so
created, are to a large degree matters of chance (Hubbell 1979). Because the traditional approach to understanding community patterns
has relied almsot exclusively on deterministic explanations, however,
such stochastic effects have rarely been considered (but see Wolf and
Gill 1980). As part of a recent move advocating more explicit consideration of alternative hypotheses, however, null hypotheses that incorporate strong elements of randomness and chance have been advanced
to test various aspects of avian community patterns (e.g., Simberloff
1978; Connor and Simberloff 1978, 1979; Strong et al. 1979; Strong
1980; Ricklefs and Travis 1980; Grant and Abbott 1980; Coleman et al.
1982). Usually, such null models generate patterns that can be compared with observations using some algorithm to draw species or
species attributes from a species pool, subject to various constraints
(refer again to Figure 10.1). The results are thus quite sensitive to the
way in which the species pool is specified and the particular attributes
of the randomization procedures used. Depending upon how these a
priori operations are performed, it is possible to generate "null" models
that will inevitably show the observed pattern to be highly deterministic
or to be indistinguishable from random. The methodology of structuring and testing such random null hypotheses still must be refined, but
the approach offers a clear alternative to the conventional mode of
testing process hypotheses in community ecology.

Conclusions

It should be apparent from the foregoing that the study of
community patterns and processes is not the simple, straightforward
matter that many avian ecologists seem to have thought. It is beset with
a host of logical and methodological problems, which collectively generate considerable doubt about what we really know of the patterns of
avian communities, much less the processes that have produced them.
Some of the community patterns that we have taken to be true may

thus be epiphenomena resulting from attributes of individuals or populations (Simberloff 1980) or artifacts of inappropriate or incomplete sampling, and many of our process explanations are likely to be more myth than reality (Wiens, in press). This state of affairs leads to a consideration of several critical questions.

1. Can short-term studies tell us much about communities? Probably not, unless the communities are clearly equilibrial. Otherwise, short-term studies will give us only glimpses of communities as they exist at a particular time (Roth 1976), and glimpses at different times may well reveal quite different patterns (Wiens 1981c). Roth's (1976) hope that "such glimpses can tell us something of the way species fit together as communities – if we look at enough of them to permit generalizations" is quixotic.

2. Can comparisons of observations obtained by different observers using different methods be used to determine community patterns? Probably not, except at the most general scale of comparison. Robert MacArthur, in fact, once told me that one should trust only one's own observations in drawing comparisons!

3. At what scale should we pursue the study of community patterns and processes? Ricklefs and Travis (1980) and Brown (1981) have suggested that the complexity of interactions in local communities and the difficulties involved in assembling such components to produce an overview of the general features of community organization argue for studying communities on a broad scale. Holmes et al. (1979), however, have proposed that an understanding of avian community structuring and resource partitioning is likely to emerge only from intensive studies of a restricted set of species or guilds in a local setting. I believe that the problems inherent in determining pattern and inferring causation at broad geographic scales clearly indicate the appropriateness of intensive, local studies.

4. Should we expect to be able to detect general patterns and ubiquitous processes in communities? Certainly not at our current level of understanding, and certainly not if the systems are nonequilibrial. Despite the lust for general explanations of general patterns that characterizes community ecology (and any science, really), it seems apparent that, when we really know so little, an emphasis on generality is premature and can only delude us. Local communities should be investigated on their own terms, with the objective of attempting to determine exactly how a complex interplay of processes may produce the patterns that are expressed.

5. Should we follow a formal hypotheticodeductive approach to in-

vestigating communities, adhering closely to the tenets espoused by Popper (1968a) in his early (and most frequently cited) works? Probably not, at least in any rigid, formalistic fashion. Certainly we should be concerned with testing hypotheses or posing questions about the communities we study rather than retreating into detailed but sterile descriptions. To insist on pure testability of our hypotheses, however, is probably premature – too many important aspects of community structure cannot be framed in explicitly testable terms. Should they thus be ignored and perhaps forgotten? Perhaps the most appropriate philosophical pathway to follow lies somewhere between the formalism of early Popper and the more relaxed "anything goes" views of Feyerabend (1978).

6. Are the bird communities that we have been studying real biological entities or artifacts that we have created for ease of study? MacArthur (1971) defined a community as "any set of organisms currently living near each other and about which it is interesting to talk." To MacArthur, the question of the actual existence of such communities was of little consequence. Such a laissez-faire approach to community definition, however, fosters the practice of using the apparent existence of some pattern in the "community" to justify the conclusion that the "community" is in fact real in a biological sense and the patterns (and inferred processes) thus of biological relevance.

A more meaningful approach would be to define a community on the basis of common use of some set of resources by its members, somewhat akin to a "superguild." When this is done, it immediately becomes apparent that "bird communities" are arbitrarily defined taxonomic subsets of such a community. Nectar, for example, is the primary energy source for a variety of insects and bats, and to restrict attention to its use by birds and study avian "community" patterns without consideration of interrelationships with such other consumers is simply incomplete. Recently, the artificiality of taxonomic boundaries on community membership and routes of possible interactions has received increasing attention (e.g., the papers in the symposium introduced by Reichman, 1979), although (predictably) almost entirely in the context of intertaxa competition. Nonetheless, such studies clearly cast doubt upon the reality, as biological entities, of bird communities (Wiens 1980).

The tone of my comments in this chapter has been largely negative. I believe that, although the studies conducted within the research tradition established by MacArthur and his colleagues have posed many interesting questions and presented an array of plausible arguments and neat studies about community structuring, the general lack of scientific rigor in the approach has left us unable to determine which patterns and which processes have real merit and which are mythological. Perhaps the use of informal methods and relaxed logical procedures was justified two decades ago when this research tradition became established. Perhaps it still is an acceptable way to do science, and my detailing of problems in this chapter can be dismissed as symptomatic of an excessive idealism that is inappropriate to the study of complex ecological systems. However, I maintain that if avian community ecology is to merit respect and gain credibility, it *must* adopt rigorous methodology and follow sound logical procedures.

Despite my obvious skepticism about the recent past of avian community ecology, I have guarded optimism for its future. It *is* possible to gain reliable knowledge about assemblages of organisms. We must do this, I feel, by turning our attention to the diversity of factors that may influence individuals and populations that are related through their common modes of resource use, attempting to measure the resources, including manipulative experiments in a carefully designed research protocol that addresses alternative hypotheses, concentrating our studies on local assemblages, and maintaining a clear separation between our attempts to define patterns and the generation of process explanations of those patterns. This may be a more tortuous route to gaining knowledge about communities, but I believe it is a more certain one.

Literature cited

Abbott, I. 1976. Comparisons of habitat structure and plant, arthropod and bird diversity between mainland and island sites near Perth, Western Australia. *Aust. J. Ecol. 1:*275–280.

– 1980. Theories dealing with the ecology of landbirds on islands. In A. Macfadyen (ed.). *Advances in ecological research*, Vol. 11, pp. 329–371. New York: Academic Press.

Abbott, I., and P. R. Grant. 1976. Nonequilibrial bird faunas on islands. *Am. Nat. 110:*507–528.

Alatalo, R. V. Seasonal dynamics of resource partitioning among foliage-gleaning passerines in northern Finland. *Oecologia 45:*190–196.

Altmann, J. 1974. Observational study of behavior; sampling methods. *Behaviour 49:*227–267.

Austin, G. T. 1970. Breeding birds of desert riparian habitat in southern Nevada. *Condor 72:*431–436.

Baker, M. C. 1977. Shorebird food habits in the eastern Canadian Arctic. *Condor 79:*56–62.

Beals, E. W. 1960. Forest bird communities in the Apostle Islands of Wisconsin. *Wilson Bull. 72:*156–181.

Bédard, J. 1976. Coexistence, coevolution and convergent evolution in seabird communities: a comment. *Ecology 57:*177–184.

Best, L. B. 1981. Seasonal changes in detection of individual bird species. *Stud. Avian Bio. 6:*252–261.

Bock, C. E., and L. W. Lepthien. 1975. Patterns of bird species diversity revealed by Christmas counts versus breeding bird surveys. *West. Birds 6:*95–100.

Bock, C. E., M. Raphael, and J. H. Bock. 1978. Changing avian community structure during early post-fire succession in the Sierra Nevada. *Wilson Bull. 90:*119–123.

Bond, R. R. 1957. Ecological distribution of breeding birds in upland forests of southern Wisconsin. *Ecol. Monogr. 27:*351–384.

Brady, R. H. 1979. Natural selection and the criteria by which a theory is judged. *Syst. Zool. 28:*600–621.

Brown, J. H. 1981. Two decades of homage to Santa Rosalia: toward a general theory of diversity. *Am. Zool. 21:*877–888.

Burnham, K. P., D. R. Anderson, and J. L. Laake. 1980. Estimation of density from line transect sampling of biological populations. *Wildl. Monogr. 72:*1–202.

Capen, D. E. 1981. *The use of multivariate statistics in studies of wildlife habitat,* USDA Forest Service General Technical Report RM-87. Fort Collins, Colo.: Rocky Mountain Forest and Range Experiment Station.

Carnes, B. A., and N. A. Slade. 1982. Some comments on niche analysis in canonical space. *Ecology 63:*888–893.

Carpenter, F. L. 1978. A spectrum of nectar-eater communities. *Am. Zool. 18:*809–819.

Caswell, H. 1976. Community structure: a neutral model analysis. *Ecol. Monogr. 46:*327–354.

Clutton-Brock, T. H., and P. H. Harvey. 1979. Comparison and adaptation. *Proc. R. Soc. London Ser. B 205:*547–565.

Cody, M. L. 1966. The consistency of intra- and inter-continental grassland bird species counts. *Am. Nat. 100:*371–376.

– 1968. On the methods of resource division in grassland bird communities. *Am. Nat. 102:*107–147.

– 1973. Coexistence, coevolution and convergent evolution in seabird communities. *Ecology 54:*31–44.

– 1974. *Competition and the structure of bird communities.* Princeton, N.J.: Princeton University Press.

– 1975. Towards a theory of continental species diversities: bird distributions over Mediterranean habitat gradients. In M. L. Cody and J. M. Diamond (eds.). *Ecology and evolution of communities,* pp. 214–257. Cambridge, Mass.: Belknap Press.

– 1978. Habitat selection and interspecific territoriality among the sylviid warblers of England and Sweden. *Ecol. Mongr. 48:*351–396.

– 1980. Species packing in insectivorous bird communities: density, diversity, and productivity. In R. Nöhring (ed.). *Acta XVII Congressus Internationalis Ornithologici*, pp. 1071–1077. Berlin: Verlag der Deutschen Ornithologen-Gesellschaft.

– 1981. Habitat selection in birds: the roles of vegetation structure, competitors, and productivity. *BioScience 31:*107–113.

Cody, M. L., and J. M. Diamond (eds.). 1975. *Ecology and evolution of communities.* Cambridge, Mass.: Belknap Press.

Coleman, B. D., M. A. Mares, M. R. Willig, and Y.-H. Hsieh. 1982. Randomness, area, and species richness. *Ecology 63:*1121–1133.

Connell, J. H. 1975. Some mechanisms producing structure in natural communities: a model and evidence from field experiments. In M. L. Cody and J. M. Diamond (eds.). *Ecology and evolution of communities, pp. 460–490. Cambridge, Mass.: Belknap Press.*

– 1980. Diversity and the coevolution of competitors, or the ghost of competition past. *Oikos 35:*131–138.

Connor, E. F. and E. D. McCoy. 1979. The statistics and biology of the species-area relationship. *Am. Nat. 113:*791–833.

Connor, E. F., and D. Simberloff. 1978. Species number and compositional similarity of the Galápagos flora and avifauna. *Ecol. Mongr. 48:*219–248.

– 1979. The assembly of species communities: chance or competition? *Ecology 60:*1132–1140.

Cox, G. W., and R. E. Ricklefs. 1977. Species diversity and ecological release in Caribbean land bird faunas. *Oikos 28:*113–122.

Curtis, J. T. 1959. *The vegetation of Wisconsin.* Madison: University of Wisconsin Press.

Davis, J. 1973. Habitat preferences and competition of wintering juncos and Golden-crowned Sparrows. *Ecology 54:*174–180.

Dayton, P. K., and J. S. Oliver. 1980. An evaluation of experimental analyses of population and community patterns in benthic marine environments. In K. R. Tenore and B. C. Coull (eds.). *Marine benthic dynamics*, pp. 93–120. Columbia: University of South Carolina Press.

DeSante, D. F. 1981. A field test of the variable circular-plot censusing technique in a California coastal scrub breeding bird community. *Stud. Avian Biol. 6:*177–185.

Dhondt, A. A. 1977. Interspecific competition between Great and Blue Tit. *Nature (London) 268:*521–523.

Dhondt, A. A., and R. Eyckerman. 1980. Competition between the Great Tit and the Blue Tit outside the breeding season in field experiments. *Ecology 61:*1291–1296.

Diamond, J. M. 1975. Assembly of species communities. In M. L. Cody and J. M. Diamond (eds.). *Ecology and evolution of communities*, pp. 342–444. Cambridge, Mass.: Belknap Press.

– 1978. Niche shifts and the rediscovery of interspecific competition. *Am. Sci. 66:*322–331.

– 1979. Population dynamics and interspecific competition in bird communities. *Fortschr. Zool. 25* (2/3):389–402.

Diamond, J. M., and H. L. Jones. 1980. Breeding land birds of the Channel Islands. In D. M. Power (ed.). *The California Islands: proceedings of a multidis-*

ciplinary symposium, pp. 597–612. Santa Barbara, Calif.: Santa Barbara Museum of Natural History.

Diamond, J. M., and A. G. Marshall. 1977. Niche shifts in New Hebridean birds. *Emu 77:*61–72.

Dueser, R. D., and H. H. Shugart, Jr. 1978. Microhabitats in a forest-floor small mammal fauna. *Ecology 59:*89–98.

– 1979. Niche pattern in a forest-floor small-mammal fauna. *Ecology 60:*108–118.

Eldridge, N., and J. Cracraft. 1980. *Phylogenetic patterns and the evolutionary process.* New York: Columbia University Press.

Emlen, J. T. 1979. Land bird densities on Baja California islands. *Auk 96:*152–167.

Faanes, C. A., and D. Bystrak. 1981. The role of observer bias in the North American Breeding Bird Survey. *Stud. Avian Biol. 6:*353–359.

Feinsinger, P. A., and R. K. Colwell. 1978. Community organization among Neotropical nectar-feeding birds. *Am. Zool. 18:*779–795.

Feyerabend, P. 1978. *Against method.* London: Verso.

Ford, H. A., and D. C. Patton. 1977. The comparative ecology of ten species of honeyeaters in South Australia. *Aust. J. Ecol. 2:*399–407.

Franzreb, K. E. 1981. The determination of avian densities using the variable-strip and fixed-width transect surveying methods. *Stud. Avian Biol. 6:*139–145.

Fretwell, S. D. 1972. *Populations in a seasonal environment.* Princeton, N.J.: Princeton University Press.

– 1975. The impact of Robert MacArthur on ecology. *Annu. Rev. Ecol. Syst. 6:*1–13.

Fritz, R. S. 1980. Consequences of insular population structure: distribution and extinction of Spruce Grouse populations in the Adirondack Mountains. In R. Nöhring (ed.). *Acta XVII Congressus Internationalis Ornithologici,* pp. 757–763. Berlin: Verlag der Deutschen Ornithologen-Gesellschaft.

Gill, F. B., and L. L. Wolf. 1979. Nectar loss by Golden-winged Sunbirds to competitors. *Auk 96:*448–461.

Glasser, J. W. 1979. The role of predation in shaping and maintaining the structure of communities. *Am. Nat. 113:*631–641.

Gleason, H. A. 1917. The structure and development of the plant association. *Bull. Torrey Bot. Club 44:*463–481.

Gould, S. J., and R. C. Lewontin. 1979. The spandrels of San Marco and the Panglossian paradigm: a critique of the adaptationist programme. *Proc. R. Soc. London Ser. B 205:*581–598.

Grant, P. 1972. Convergent and divergent character displacement. *Biol. J. Linn. Soc. 4:*39–68.

Grant, P., and I. Abbott. 1980. Interspecific competition, island biogeography and null hypotheses. *Evolution 34:*332–341.

Green, R. H. 1971. A multivariate statistical approach to the Hutchinsonian niche: bivalve molluscs of central Canada. *Ecology 52:*543–556.

Gutting, G. (ed.). 1980. *Paradigms and revolutions. Applications and appraisals of Thomas Kuhn's philosophy of science.* Notre Dame, Ind.: University of Notre Dame Press.

Haefner, J. W. 1981. Avian community assembly rules: the foliage-gleaning guild. *Oecologia 50:*131–142.

Haffer, J. 1981. Aspects of Neotropical bird speciation during the Cenozoic. In G. Nelson and D. E. Rosen (eds.). *Vicariance biogeography. A critique,* pp. 371–394. New York: Columbia University Press.

Haila, Y., O. Järvinen, and R. A. Väisänen. 1980. Habitat distribution and species associations of land bird populations on the Åland Islands, SW Finland. *Ann. Zool. Fenn. 17:*87–106.

Herrera, C. M. 1981. Combination rules among western European *Parus* species. *Ornis Scand. 12:*140–147.

Herrera, C. M., and F. Hiraldo. 1976. Food-niche and trophic relationships among European owls. *Ornis Scand. 7:*29–41.

Hildén, O. 1965. Habitat selection in birds: a review. *Ann. Zool. Fenn. 2:*53–75.

– 1981. Sources of error involved in the Finnish line-transect method. *Stud. Avian Biol. 6:*152–159.

Hogstad, O. 1975. Quantitative relations between hole-nesting and open-nesting species within a passerine breeding community. *Norw. J. Zool. 23:*261–267.

Högstedt, G. 1980. Prediction and test of the effects of interspecific competition. *Nature* (London) *283:*64–66.

Holmes, R. T., and S. K. Robinson. 1981. Tree species preferences of foraging insectivorous birds in a northern hardwoods forest. *Oecologia 48:*31–35.

Holmes, R. T., R. E. Bonney, Jr., and S. W. Pacala. 1979. Guild structure of the Hubbard Brook bird community: a multivariate approach. *Ecology 60:*512–520.

Hubbell, S. P. 1979. Tree dispersion, abundance, and diversity in a tropical dry forest. *Science 203:*1299–1309.

James, F. C., and S. Rathbun. 1981. Rarefaction, relative abundance, and diversity of avian communities. *Auk 98:*785–800.

Järvinen O. 1978. Är nordliga fågelsamhällen mättade? [Are northern bird communities saturated?] *Anser Suppl. 3:*112–116.

– 1979. Geographical gradients of stability in European land bird communities. *Oecologia 38:*51–69.

– 1980. Dynamics of north European bird communities. In R. Nöhring (ed.). *Acta XVII Congressus Internationalis Ornithologici,* pp. 770–776. Berlin: Verlag der Deutschen Ornithologen-Gesellschaft.

Järvinen, O. and R. A. Väisänen. 1977. Long-term changes of the north European land bird fauna. *Oikos 29:*225–228.

– 1978. Recent changes in forest bird populations in northern Finland. *Ann. Zool Fenn. 15:*279–289.

– 1979. Climatic changes, habitat changes, and competition: dynamics of geographical overlap in two pairs of congeneric bird species in Finland. *Oikos 33:*261–271.

Järvinen, O., and S. Ulfstrand. 1980. Species turnover of a continental bird fauna: northern Europe, 1850–1970. *Oecologia 46:*186–195.

Jedraszko-Dabrowska, D. 1979. Rotation of individuals in breeding populations of dominant species of birds in a pine forest. *Ekol. Pol. 27:*545–569.

Karr, J. R. 1968. Habitat and avian diversity on strip-mined land in east–central Illinois. *Condor 70:*348–357.

– 1971. Structure of avian communities in selected Panama and Illinois habitats. *Ecol. Monogr. 41:*207–233.

– 1976. Within- and between-habitat avian diversity in African and Neotropical lowland habitats. *Ecol. Monogr. 46:*457–481.

– 1980. History of the habitat concept in birds and the measurement of avian habitats. In R. Nöhring (ed.). *Acta XVII Internationalis Congressus Ornithologici*, pp. 991–997. Berlin: Verlag der Deutschen Ornithologen-Gesellshaft.

– 1981. Surveying birds in the tropics. *Stud. Avian Biol. 6:*548–553.

Karr, J. R., and F. C. James. 1975. Eco-morphological configurations and convergent evolution in species and communities. In M. L. Cody and J. M. Diamond (eds.). *Ecology and evolution of communities*, pp. 258–291. Cambridge, Mass.: Belknap Press.

Karr, J. R., and R. R. Roth. 1971. Vegetation structure and avian diversity in several New World areas. *Am. Nat. 105:*423–435.

Kuhn, T. S. 1970. *The structure of scientific revolutions*, 2nd ed. Chicago: University of Chicago Press.

Lack, D. 1971. *Ecological isolation in birds.* Cambridge, Mass.: Harvard University Press.

Lakatos, I., and A. Musgrave (eds.). 1970. *Criticism and the growth of knowledge.* Cambridge: Cambridge University Press.

Landres, P. B., and J. A. MacMahon. 1980. Guilds and community organization: analysis of an oak woodland avifauna in Sonora, Mexico. *Auk 97:*351–365.

Lanyon, W. E. 1981. Breeding birds and old field succession on fallow Long Island farmland. *Bull. Am. Mus. Nat. Hist. 168:*1–60.

Lovejoy, T. E. 1975. Bird diversity and abundance in Amazon forest communities. *Living Bird 13:*127–191.

MacArthur, R. H. 1958. Population ecology of some warblers of northeastern coniferous forests. *Ecology 39:*599–619.

– 1965. Patterns of species diversity. *Biol. Rev. 40:*510–533.

– 1971. Patterns of terrestrial bird communities. In D. S. Farner and J. R. King (eds.). *Avian biology*, Vol. 1, pp. 189–221. New York: Academic Press.

– 1972. *Geographical ecology.* New York: Harper & Row.

MacArthur, R. H., and J. H. Connell. 1966. *The biology of populations.* New York: Wiley.

MacArthur, R. H., and J.W. MacArthur. 1961. On bird species diversity. *Ecology 42:*594–598.

MacArthur, R. H., H. Recher, and M. Cody. 1966. On the relation between habitat selection and species diversity. *Am. Nat. 100:*319–332.

MacArthur, R. H., and E. O. Wilson. 1967. *The theory of island biogeography.* Princeton, N.J.: Princeton University Press.

McIntosh, R. P. 1975. H. A. Gleason–"Individualistic ecologist" 1882–1875: his contributions to ecological theory. *Bull. Torrey Bot. Club 102:*253–273.

McNaughton, S. J., and L. L. Wolf. 1970. Dominance and the niche in ecological systems. *Science 167:*131–139.

Menge, B. A., and J. P. Sutherland. 1976. Species diversity gradients: synthesis of the roles of predation, competition, and temporal heterogeneity. *Am. Nat. 110:*351–369.

Mertz, D. B., and D. E. McCauley. 1980. The domain of laboratory ecology. *Synthese 43:*95–110.

Minot, E. O. 1981. Effects of interspecific competition for food in breeding Blue and Great Tits. *J. Anim. Ecol. 50:*375–385.

Moss, D. 1978. Diversity of woodland song-bird populations. *J. Anim. Ecol. 47:*521–527.

Myers, J. P., R. L. Mumme, and F. A. Pitelka. in press. Components of breeding bird community stability in relation to diversity, habitat, and latitude. *Oecologia.*

Newman, W. A., and S. M. Stanley. 1981. Competition wins out overall: reply to Paine. *Paleobiology 7:*561–569.

Newton, I. 1980. The role of food in limiting bird numbers. *Ardea 68:*11–30.

Nilsson, S. G. 1979a. Density and species richness of some forest bird communities in south Sweden. *Oikos 33:*392–401.

– 1979b. Seed density, cover, predation and the distribution of birds in a beech wood in southern Sweden. *Ibis 121:*177–185.

Nilsson, S. G., and T. Allerstam. 1976. Resource division among birds in north Finnish coniferous forest in autumn. *Ornis Fenn. 53:*15–27.

Noon, B. R. 1981. The distribution of an avian guild along a temperate elevational gradient: the importance and expression of competition. *Ecol. Monogr. 51:*105–124.

Nudds, T. D., K. F. Abraham, C. D. Ankney, and P. D. Tebbel. 1981. Are size gaps in dabbling- and wading-bird arrays real? *Am. Nat. 118:*549–553.

Osman, R. W., and R. B. Whitlatch. 1978. Patterns of species diversity: fact or artifact? *Paleobiology 4:*41–54.

Paine, R. T. 1981. Barnacle ecology: is competition important? The forgotten roles of disturbance and predation. *Paleobiology 7:*553–560.

Peters, R. H. 1976. Tautology in evolution and ecology. *Am. Nat. 110:*1–12.

Pianka, E. R. 1980. Guild structure in desert lizards. *Oikos 35:*194–201.

– 1981. Competition and niche theory. In R. M. May (ed.). *Theoretical ecology. Principles and applications,* 2nd ed., pp. 167–196. Sunderland, Mass.: Sinauer Associates.

Popper, K. R. 1968a. *The logic of scientific discovery,* 2nd ed. New York: Harper & Row.

– 1968b. *Conjectures and refutations: the growth of scientific knowledge,* 2nd ed. New York: Harper & Row.

Pregill, G. K., and S. L. Olson. 1981. Zoogeography of West Indian vertebrates in relation to Pleistocene climatic cycles. *Annu. Rev. Ecol. Syst. 12:*75–98.

Pulliam, H. R. 1975. Coexistence of sparrows: a test of community theory. *Science 184:*474–476.

Pulliam, H. R., and T. A. Parker III. 1979. Population regulation of sparrows. *Fortschr. Zool. 25:*(2/3):137–147.

Rabenold, K. N. 1978. Foraging strategies, diversity, and seasonality in bird communities of Appalachian spruce-fir forests. *Ecol. Monogr. 48:*397–424.

– 1979. A reversed latitudinal diversity gradient in avian communities of eastern deciduous forests. *Am. Nat. 114:*275–286.

400   J. A. Wiens

Rabinovich, J. E., and E. H. Rapoport. 1975. Geographical variation of diversity in Argentine passerine birds. *J. Biogeogr. 2:*141–157.

Ralph, C. J., and J. M. Scott (eds.). 1981. Estimating numbers of terrestrial birds. *Stud. Avian Biol. 6:*x, 1–630.

Ramsey, F. L., and J. M. Scott. 1981. Tests of hearing ability. *Stud. Avian Biol. 6:*341–345.

Recher, H. F. 1969. Bird species diversity and habitat diversity in Australia and North America. *Am. Nat. 103:*75–80.

– 1971. Bird species diversity: a review of the relation between species number and environment. *Proc. Ecol. Soc. Aust. 6:*135–152.

– 1981. Introductory remarks: environmental influences. *Stud. Avian Biol. 6:*251.

Reichman, O. J. 1979. Introduction to the symposium: competition between distantly related taxa. *Am. Zool. 19:*1027.

Ricklefs, R. E. 1975. Review of M. L. Cody, "Competition and the structure of bird communities." *Evolution 29:*581–585.

Ricklefs, R. E., and G. W. Cox. 1977. Morphological similarity and ecological overlap among passerine birds on St. Kitts, British West Indies. *Oikos 29:*60–66.

Ricklefs, R. E., and J. Travis. 1980. A morphological approach to the study of avian community organization. *Auk 97:*321–338.

Robbins, C. S., and W. T. Van Velzen. 1969. *The breeding bird survey, 1967 and 1968,* U.S. Fish Wildlife Service, Special Scientific Report on Wildlife 124. Washington, D.C.: Government Printing Office.

Root, R. B. 1967. The niche exploitation pattern of the Blue-gray Gnatcatcher. *Ecol. Monogr.* 317–350.

Rosenzweig, M. L. 1975. On continental steady states of species diversity. In M. L. Cody and J. M. Diamond (eds.). *Ecology and evolution of communities,* pp. 121–140. Cambridge, Mass.: Belknap Press.

Rotenberry, J. T. 1980. Dietary relationships among shrub-steppe passerine birds: competition or opportunism in a variable environment? *Ecol. Monogr. 50:*93–110.

Rotenberry, J. T., and J. A. Wiens. 1978. Nongame bird communities in northwestern rangelands. In R. M. DeGraaf (ed.). *Proceedings of the workshop on nongame bird habitat management in the coniferous forests of the western United States,* pp. 32–46. USDA Forest Service General Technical Report PNW-64, pp. 32–46. Portland, Oreg.

– 1980a. Habitat structure, patchiness, and avian communities in North American steppe vegetation: a multivariate analysis. *Ecology 61:*1228–1250.

– 1980b. Temporal variation in habitat structure and shrubsteppe bird dynamics. *Oecologia 47:*1–9.

Roth, R. R. 1976. Spatial heterogeneity and bird species diversity. *Ecology 57:*773–782.

– 1981. Vegetation as a determinant in avian ecology. In *Proceedings of the 1st Welder Wildlife Foundation Symposium,* pp. 162–174.

Roth, V. L. 1981. Constancy in the size ratios of sympatric species. *Am. Nat. 118:*394–404.

Sabo, S. R. 1980. Niche and habitat relations in subalpine bird communities of the White Mountains of New Hampshire. *Ecol. Monogr. 50:*241–259.

Schall, J. J., and E. R. Pianka. 1978. Geographical trends in numbers of species. *Science 201:*679–686.

Schemske, D. W., and N. Brokaw. 1981. Treefalls and the distribution of understory birds in a tropical forest. *Ecology 62:*938–945.

Schoener, T. W. 1971. Large-billed insectivorous birds: a precipitous diversity gradient. Condor 73:154–161.

– 1972. Mathematical ecology and its place among the sciences (Review of R. H. MacArthur, "Geographical ecology"). *Science 178:*389–391.

– 1974. Resource partitioning in ecological communities. *Science 185:*27–39.

Scott, J. M., F. L. Ramsey, and C. B. Kepler. 1981. Distance estimation as a variable in estimating bird numbers from vocalizations. *Stud. Avian Biol. 6:*334–340.

Short, J. J. 1979. Patterns of alpha-diversity and abundance in breeding bird communities across North America. *Condor 81:*21–27.

Simberloff, D. 1978. Using island biogeographic distributions to determine if colonization is stochastic. *Am. Nat. 112:*713–726.

– 1980. A succession of paradigms in ecology: essentialism to materialism and probabilism *and* Reply. *Synthese 43:*3–39, 79–93.

Slagsvold, T. 1978. Competition between the Great Tit *Parus major* and the Pied Flycatcher *Ficedula hypoleuca:* an experiment. *Ornis Scand. 9:*46–50.

– 1980. Habitat selection in birds: on the presence of other bird species with special regard to *Turdus pilaris. J. Anim. Ecol. 49:*523–536.

Stiles, E. W. 1978. Avian communities in temperate and tropical alder forests. *Condor 80:*276–284.

Strauss, R. E. 1982. Statistical significance of species clusters in association analysis. *Ecology 63:*634–639.

Strong. D. R., Jr. 1980. Null hypotheses in ecology. *Synthese 43:*271–285.

Strong, D. R., Jr., L. A. Szyska, and D. S. Simberloff. 1979. Tests of community-wide character displacement against null hypotheses. *Evolution 33:*897–913.

Svärdson, G. 1949. Competition and habitat selection in birds. *Oikos 1:*157–174.

Szaro, R. C., and R. P. Balda. 1979. Bird community dynamics in a ponderosa pine forest. *Stud. Avian Biol. 3:*1–66.

Terborgh, J. W., and J. Faaborg. 1980. Saturation of bird communities in the West Indies. *Am. Nat. 116:*178–195.

Tiainen, J. 1980. Regional trends in bird communities of mature pine forests between Finland and Poland. *Ornis Scand. 11:*85–91.

Tipper, J. C. 1979. Rarefaction and rarefiction – the use and abuse of a method in paleoecology. *Paleobiology 5:*423–434.

Tomialojc, L., and P. Profus. 1977. Comparative analysis of breeding bird communities in two parks of Wroclaw and in an adjacent Querco-Carpinetum forest. *Acta Ornithol. 16:*117–177.

Tomoff, C. S. 1974. Avian species diversity in desert scrub. *Ecology 55:*396–403.

Ulfstrand, S. 1977. Foraging niche dynamics and overlap in a guild of passerine birds in a south Swedish coniferous woodland. *Oecologia 27:*23–45.

Väisänen, R. A., and O. Järvinen. 1977. Structure and fluctuation of the breeding bird fauna of a north Finnish peatland area. *Ornis Fenn. 54:*143–153.

Van Horne, B., and R. G. Ford. 1982. Niche breadth calculation based on discriminant analysis. *Ecology 63:*1172–1174.

Vuilleumier, F., and D. Simberloff. 1980. Ecology versus history as determinants of patchy and insular distributions in high Andean birds. In M. K. Hecht, W. C. Steere, and B. Wallace (eds.). *Evolutionary biology*, Vol. 12, pp. 235–379. New York: Plenum.

Wagner, J. L. 1981. Visibility and bias in avian foraging data. *Condor 83:*263–264.

Whitmore, R. C. 1979. Temporal variation in the selected habitats of a guild of grassland sparrows. *Wilson Bull. 91:*592–598.

Whittaker, R. H. 1975. *Communities and ecosystems*, 2nd ed. New York: Macmillan.

Wiens, J. A. 1969. An approach to the study of ecological relationships among grassland birds. *Ornithol. Monogr. 8:*1–93.

– 1973. Interterritorial habitat variation in Grasshopper and Savannah Sparrows. *Ecology 54:*877–884.

– 1974. Habitat heterogeneity and avian community structure in North American grasslands. *Am. Midl. Nat. 91:*195–213.

– 1976. Review of M. L. Cody, "Competition and the structure of bird communities." *Auk 93:*396–400.

– 1977a. Review of M. L. Cody and J. M. Diamond, "Ecology and evolution of communities." *Auk 94:*792–794.

– 1977b. On competition and variable environments. *Am. Sci. 65:*590–597.

– 1980. Concluding comments: Are bird communities real? In R. Nöhring (ed.). *Acta XVII Congressus Internationalis Ornithologici*, pp. 1088–1089. Berlin: Verlag der Deutschen Ornithologen-Gesellschaft.

– 1981a. Scale problems in avian censusing. *Stud. Avian Biol. 6:*513–521.

– 1981b. Avian consumers. In P. G. Risser (ed.). *The true prairie ecosystem*, pp. 214–264. Stroudsburg, Pa.: Hutchinson Ross.

– 1981c. Single-sample surveys of communities: are the revealed patterns real? *Am. Nat. 117:*90–98.

– in press. On understanding a nonequilibrium world: myth and reality in community patterns and processes. In D. R. Strong and D. Simberloff (eds.). *Ecological communities: conceptual issues and the evidence*. Princeton, N.J.: Princeton University Press.

Wiens, J. A., and J. T. Rotenberry. 1979. Diet niche relationships among North American grassland and shrubsteppe birds. *Oecologia 42:*253–292.

– 1980a. Patterns of morphology and ecology in grassland and shrubsteppe bird populations. *Ecol. Monogr. 50:*287–308.

– 1980b. Bird community structure in cold shrub deserts: competition or chaos? In R. Nöhring (ed.), *Acta XVII Congressus Internationalis Ornithologici*, pp. 1063–1070. Berlin: Verlag der Deutschen Ornithologen-Gesellschaft.

– 1981. Habitat associations and community structure of birds in shrubsteppe environments. *Ecol. Monogr. 51:*21–41.

Wiens, J. A., S. G. Martin, W. R. Holthaus, and F. A. Iwen. 1970. Metronome timing in behavioral ecology studies. *Ecology 51:*350–352.

Williams, B. K. 1981. Discriminant analysis in wildlife research: theory and applications. In D. E. Capen (ed.). *The use of multivariate statistics in studies of wildlife habitat,* pp. 50–71. USDA Forest Service General Technical Report RM-87. Fort Collins, Colo.: Rocky Mountain Forest and Range Experiment Station.

Williams, J. B., and G. O. Batzli. 1979. Competition among bark-foraging birds in central Illinois: experimental evidence. *Condor 81:*122–132.

Willson, M. F. 1974. Avian community organization and habitat structure. *Ecology 55:*1017–1029.

Winternitz, B. L. 1976. Temporal change and habitat preference of some montane breeding birds. *Condor 78:*383–393.

Wolf, L. L., and F. B. Gill. 1980. Resource gradients and community organization of nectarivorous birds. In R. Nöhring (ed.). *Acta XVII Congressus Internationalis Ornithologici,* pp. 1105–1113. Berlin: Verlag der Deutschen Ornithologen-Gesellschaft.

Wynne-Edwards, V. C. 1962. *Animal dispersion in relation to social behavior.* New York: Hafner.

Yeaton, R. I., and M. L. Cody. 1974. Competitive release in island Song Sparrow populations. *Theor. Popul. Biol. 5:*42–58.

Commentary

JAMES R. KARR

Avian community ecology has been one of the most active areas of ornithological research for over two decades. Indeed, its impact, for better or worse, has gone well beyond the field of ornithology. Although I do not disagree significantly with any of the points made by Wiens in Chapter 10, I address several points in this comment from a slightly different perspective. In addition, I comment on several points treated only briefly or not at all by Wiens.

On MacArthurian ecology

Wiens clearly and forcefully demonstrates several weaknesses of the "MacArthur approach to community studies." These weaknesses developed due to the desire to discover broad generalizations that account for ecological patterns in all habitats and for all organisms. The search for generalizations had led many disciples of the MacArthur approach to seek repeated patterns while divergences from such patterns among habitats were ignored. Real understanding of pattern and process in ecological communities depends on analysis and comprehension of both similarities and differences. MacArthur (1972:1) recognized this when he wrote "a bird pattern would only be expected to look like that of a *Paramecium* if birds and *Paramecium* had the same morphology, economics, and dynamics, and found themselves in environments of the same structure." Similarly, factors regulating bird communities differ in space and time, with consequences for realized community patterns. The search for *the* cause of latitudinal gradients in species diversity, an early effort in community ecology, was doomed to failure because this fact was ignored or overlooked. Most biologists now recognize the diversity of physical and biotic processes that regulate biotic communities.

Recently, the search for a few generalizations has been replaced by more careful examination of the complexity of natural systems. Although the search for general principles, the primary goal of science, persists, the simplicity of those principles is more reasonably examined. This shift coincides with a more cautious treatment of the central dogma of modern theoretical ecology. That dogma, the assumption of equilibrium in biological communities, has fallen slowly. An early recognition came with Hutchinson's paradox of the plankton and later in studies of marine intertidal, tropical forest tree, and coral reef communities. Indeed, the significance of spatial and temporal variability in bird communities is clear from several recent studies. Without equilibrium, the "competition is king" approach to community ecology is undermined. I emphasize, as does Wiens, that the existence of competition is not in question. Its pervasiveness in space and time must be more carefully documented than it has been in the past.

I end this section with recognition of a positive legacy of the MacArthur era in community ecology. MacArthur and his colleagues stimulated us to think more analytically, to explore the why of a phenomenon rather than simply to describe it. That was a positive accomplishment and one for which we owe a great debt. The strength of the MacArthur philosophy is clear in the first sentence of *Geographical Ecology* (1972): "To do science is to search for repeated pattern, not simply to accumulate facts. . . ."

On birds as study organisms

The appeal of birds as study organisms is widely known. Their diurnal habits, a tendency to advertise their presence through song and plumage, the extensive documentation of basic natural histories, and their being well known taxonomically have led many to use birds as tools for exploration of ecological principles.

Optimism about these attributes, however, may have led us to ignore certain problems. First and foremost are the difficulties associated with conducting accurate censuses or of accumulating unbiased foraging data. At the very least, many classical procedures do not document differences in populations and communities with sufficient accuracy and precision to distinguish the ecological differences that must be detected to understand ecological patterns. Some innovative approaches are being developed that sample subsets of avian communities, but these are often discounted, incorrectly in my view, because they fail to deal with the entire community. We should, on the whole, be more concerned with what we can do with precision and accuracy and less with estimating densities of all avian species.

A second problem is the difficulty involved in sampling relevant factors such as the food base on predation rates. Despite repeated efforts, few have effectively treated food resources of insectivorous birds. Problems may arise due to our inability to sample those resources of real significance as food for birds, because other complicating and controlling variables are not adequately treated or perhaps because theoretical foundations, such as optimal foraging theory, are inadequate. Predation presents another array of intractable problems. Birds are clearly finely tuned to avoid predation, but most ornithologists ignore predation, not because of their intuition but because it is too difficult to gather sufficient relevant data. Resolution of these and other similar problems requires careful research design.

A third critical weakness is that studies of avian communities continue to be largely correlational. Few have attempted to utilize carefully controlled experimental procedures in the complex world of natural avian communities. A few studies have attempted to adjust resource densities (especially seeds and fruits), avian population densities, or vegetation structure. These commendable efforts should be continued, but we should also recognize that experimental manipulations are not always possible or practical with avian studies. (I have initiated several research efforts on fish communities in the last decade for precisely this reason.)

However, I would not argue for abandonment of studies of avian communities as some have done. First, birds must be studied if we are to understand the details of pattern and process in bird assemblages. Considering MacArthur's comments quoted earlier, attempts to generalize results among taxa should be made with caution. However, we must also admit problems to prevent future ornithologists from pursuing exceedingly complex problems with a simplistic approach and perspective. Sophisticated studies of birds *can* yield insight into community pattern and process as is demonstrated by progress made in the past two decades.

One of the strengths of modern ecology is the continuing feedback that results from the taxonomic breadth of contributions to ecological research. The last thing we need is to develop an equivalent of the fruit fly (genetics) or white rat (physiology and behavior) in ecology with a focus on some single unit as a tractable research "organism," either species, higher taxon, or community type. Some might argue that many problems of community ecology may have been precipitated by a strict adherence to results of studies on birds, but overreaction to avoid studies of birds would be equally counterproductive.

On community metrics

Resource managers often seek simplistic principles and convenient metrics to guide their decisions. Theoreticians and basic biologists do the same. However, simple metrics (e.g., species diversity) are inadequate to account for the complexity of both pattern and process in natural systems, especially the dynamics of communities. The search for a single metric to measure all pattern is like the search for the single factor responsible for diversity gradients noted earlier.

Historically, the primary habitat variable of interest in studies of habitat selection in birds has been vegetation structure (vertical and horizontal pattern) but the importance of other variables (tree species, moisture conditions, etc.) is now clear. If we are to understand the factors that influence communities and their dynamics, we must comprehend how each species responds to the many physical and biological aspects of its environment and how, in the aggregate, they produce communities of particular attributes. Average vegetation structure is a meaningless concept in organisms that exhibit sex, age, seasonal, and year-to-year variation in attributes of habitat selected for short- or long-term occupancy. Indeed, bottlenecks may exist in species life histories as a consequence of short-term habitat needs.

Simple metrics cannot describe this complexity. Thus, we should strive to understand that complexity before developing more tractable metrics to depict it and thereby guide the decisions of resource managers.

On space, time, and scale

Environmental heterogeneity is a fundamental characteristic of nature with significance in both spatial and temporal dimensions. This important point has been clearly demonstrated by research on birds in shrubsteppe, eastern deciduous forest, and tropical wet forest environments. Incorporation of that knowledge into ornithological research, however, is hampered by the difficulty of reconciling human and avian perception of environmental scale and by a reluctance to abandon old dogmas.

With respect to scale, Wiens advocates the use of intensive, local-scale study of community pattern. I dissent from this opinion. Scale phenomena, as they affect avian communities, clearly form a continuum of levels, a hierarchy. To suggest that one level is paramount is to reduce understanding of community ecology to a simplistic one-dimensional perspective. For example, avian species richness in a forest island can be predicted with some accuracy if island size is known. Pattern at three other scales, however, may also be important in regulating species richness: habitat heterogeneity within the island, regional mosaic of forest islands (and other habitats), and continental setting (e.g., North America, Africa, Australia). Historical time scales were brought into focus by the refuge theory of Haffer, a contribution that has provided insights for a number of other major taxa. Understanding of community pattern depends upon consideration of the full hierarchy of scales. Indeed, the use of α-, β-, and γ-diversity was an early recognition of the importance of heterogeneity and scale, although I urge caution in the uncritical use of such discrete classifications.

On the reluctance to abandon old dogma, the equilibrium concept is an excellent example treated by Wiens. He outlines assumptions of the equilibrium concept noting that systems are presumed to be "highly ordered or deterministic and at (or reasonably close to) equilibrium." In addition, he notes that the equilibrium assumption results in communities that are very stable or, in cases where environmental conditions vary through time, species populations "track" resource variation to maintain optimal community structure and composition. In this scenario, suitable habitat is saturated.

Two alternatives presented by Wiens, at least implicitly, are (a) perfect tracking with population saturation and (b) random variation in community structure and composition independent of the specific context of current environmental conditions. Presentation as "If not a, then b" derives from too simplistic a perspective. Clearly, communities do not vary at random; patterns in richness, composition, population dynamics, trophic structure, and other attributes are too regular. Conversely, communities are too dynamic and influenced by too many variables to be solely deterministic. We must seek solutions in the middle ground between these extremes.

I prefer to begin with the assumption that some variation in avian communities in space and time is traceable to specific factors, whereas, other variation is due to random or stochastic processes. Furthermore, it is important to distinguish between variation that is truly stochastic and that which is due to a specific cause-and-effect relationship that we have not been able to identify. For example, variability in population and, thus, community attributes at a site over time may be due to changes in weather or climate, food availability, or the success of populations on geographically distant wintering grounds. It would be easy to throw up our hands in despair and attribute variation due to these factors (or their interactions) to stochastic processes. In the long term, however, that would be counterproductive.

For the same reasons that birds and *Paramecium* do not exhibit the same attributes, avian communities of different habitats (grassland vs. forest) are probably not regulated by the same balance of variables. Even at a single site the role of local (wet vs. dry period at time of territory establishment) and distant (survivorship on geographically remote wintering areas) factors varies with time.

Finally, the choice of metrics used to assess community attributes is also important. One assemblage attribute may be quite plastic over time, whereas others are very stable. Rather than seek to determine whether a community is stable or not, we should seek to understand which attributes are variable over time and space and which are not. Our goal should be to comprehend the pattern and process of these dynamics.

On the guild concept

The concept of guilds of species with similar foraging ecology developed in the heydey of competition and the equilibrium concept in community ecology. As a result, the concept includes the explicit assumption that competition produces ecological pattern within guilds. The time has come for a more open consideration of what shapes the pattern of species within guilds. Perhaps, for example, similarity among species is simply the result of species being subjected to the same selective pressures involved in exploiting a specific nonlimiting resource. Thus, if food is not limiting, similarities among species within a guild might be interpreted as resulting from numerous selective pressures, perhaps not even including competition. Some might argue that the guild concept should not be so modified. If not, we should develop a new concept that more accurately reflects current understanding of determinants of assemblage attributes.

Another difficulty involves the tendency to force very complex systems to comply with simplistic models. In the study of nectarivorous birds, bill lengths and flower morphology are commonly interpreted solely on the basis of interspecific competition or on coevolutionary grounds with the plant. However, hermits (*Phaethornis*) and their Old World counterparts, the spiderhunters (*Arachnothera*), commonly feed on arthropods such as those in spiderwebs. Longer bills give access to arthropods while avoiding the problem of entanglement. Selective pressures other than competition may be instrumental in molding bill and flower morphology, a fact that should be represented in interpretation of these adaptive complexes.

Summary

The central problem in community ecology in the past two decades has been a dependence on a simplistic body of theory that is incapable of accounting for pattern in an exceedingly complex world. The failure of that theory/dogma to account for pattern has precipitated a revolution in the approach to study of avian and other communities. Recognition of the weaknesses of the old approach is easy with the benefit of 20–20 hindsight. Indeed, without that approach and its weaknesses, we would not be where we are now.

An important lesson from the study of avian communities is the extent to which a philosophical approach to science can dominate a science. Another danger is replacement of that philosophical approach with another equally narrow one. Our science will be more stimulating and less likely to stagnate if we tolerate a diversity of approaches. Diverse approaches act as checks on one another and foster progress toward understanding as they limit the dominance of the field by an approach favored over the short term. Avian community ecology has stimulated ornithologists as well as biologists interested in other taxa. I fully expect that stimulation to continue as we move into the second century of the American Ornithologists' Union. If the past two decades can be taken as an indication, we have learned a great deal but there is much to be learned.

Literature cited

MacArthur, R. H. 1972. *Geographical ecology*, New York: Harper & Row.

11 Biogeography: the unification and maturation of a science

DANIEL SIMBERLOFF

One may define *biogeography* as the geographic arrangement of living organisms, the history of how a particular arrangement is achieved, and an analysis of possible reasons for the arrangement. Biogeography thus overlaps broadly with ecology (as defined by Krebs 1978) and with large sections of systematics, paleontology, and geology. These overlaps are mirrored in a schism between historical biogeography and ecological biogeography, the former grounded in evolutionary systematics and the earth sciences, the latter in ecology. Literatures in the two traditions are largely separate, and Rosen (1978) feels that this schism is inevitable and likely permanent. I predict that the differences between the two approaches will lessen. They are more a function of different foci: long time frames and large geographic scales for historical biogeography and short periods and local geography for ecological biogeography (Simberloff et al. 1981). Furthermore, a complete understanding of current biogeographical patterns requires both approaches, and I feel that the two traditions are beset by the same main problem: frequent absence of falsifiable hypotheses and of statistical tests that can help to choose between competing theories. Ball (1975) suggests that all sciences pass through a descriptive stage (when patterns are noted), a narrative stage (when explanations are proposed for the patterns but are simply coherent narratives), and an analytical stage (in which competing hypotheses are rigorously tested) and that biogeography is on the threshold of the analytical stage. I hope to show that Ball is correct and that the anticipated progress stems partly from more rigorous hypothesis tests in ecological biogeography and from the introduction of these methods into historical biogeography.

There are several important and active areas of biogeography that I will not treat here or will treat only tangentially as they bear on my main subjects of vicariance and dispersal, equilibrium island biogeography, and community composition. Perhaps foremost among the areas given very short shrift is intraspecific geographic variation. Such re-

411

search is in turn but a subset of the larger question of how geographic variation in habitats and resources is reflected in geographic distributions of species' abundances and phenotypes. Ultimately the explanations for the patterns that vicariance and equilibrium biogeographers feel they perceive will have to come from reductionist research by physiologists on what is adaptive or maladaptive about a given phenotype in a particular environment and by geneticists on what actually determines a phenotype, given an environment and a genotype. So my omission of this area does not imply denigration, only limited space and expertise. Simlarly, avian paleontology will have to underlie or at least be consistent with any valid precept of historical biogeography, yet it has been downplayed here. This again reflects the constraints of this chapter, not the importance of the field.

Historical biogeography

The dispersal paradigm

Although various papers that might now be assigned to the descriptive stage of biogeography appeared in the eighteenth and early nineteenth centuries (Nelson 1978), the first and (until recently) dominant coherent narrative biogeographic explanation arose with Darwin's and Wallace's elucidation of evolution (Ghiselin 1969). Their general view was that large-scale patterns, such as the division of the earth into clearly defined provinces (*e.g.*, Sclater's, 1858, six ornithological regions), the disharmony of island biotas, or the endemicity on islands of an archipelago, could be traced to differences in dispersal ability among taxa plus occasional changes in sea level. The sea level changes would erect and sunder land bridges in the regions of shallow seas, and the dispersal differences would allow different taxa to use with differing frequencies land bridges (when they existed) or long-distance dispersal over barriers like oceans or mountains in order to colonize new sites. Having reached a new site, populations would undergo genetic differentiation, leading to speciation. The distinct regions, then, are generated by barriers to dispersal; island biotas are disharmonious because some organisms (e.g., birds) disperse better than do others (e.g., amphibians); endemicity on archipelagoes reflects the isolation of populations on different islands. This dispersal paradigm was passed more or less intact to Matthew (1915) and then to Darlington (1957, 1965) and Simpson (1953, 1965), although with some variation of detail, such as location of major areas of radiation. Perhaps the most notable present adherent of these views is Carlquist (1974).

This approach produces narrative explanations: No matter how plausible one may be, it is difficult to conceive of direct evidence that could falsify a claim that, for example, a particular species colonized a particular site at some time in the past by long-distance dispersal. The dispersalist approach is probablistic, resting on the notion that events that appear extraordinarily unlikely for one individual at one time – for example, a seed's being carried for thousands of miles on a bird's foot (Camp 1948) – are plausible given large populations and geological time spans. Simpson (1952) was explicit about this idea, but the only generalization that can be drawn is that from fixed probabilities of dispersal each year, one can easily calculate probabilities for very long periods of time, and they are much larger. Since one can rarely if ever calculate even the probability for 1 year, and it is unlikely that this probability is constant, it is difficult to see how a falsifiable hypothesis can be constructed.

The vicariance paradigm

It is perhaps this inability to pose statistically falsifiable dispersalist hypotheses that most incurs the wrath of advocates of vicariance biogeography (Croizat 1964, 1981; Croizat et al. 1974; Rosen 1974a,b; Cracraft 1980). The competing vicariance paradigm states that "on a global basis the general features of modern biotic distribution have been determined by sudivision of ancestral biotas in response to changing geography" (Croizat et al. 1974). That is, long-distance dispersal from centers of origin is insignificant, and instead one envisions an ancestral form gradually and continuously diffusing on a generation-by-generation basis until it occupies all available area that is not precluded by either barriers or inimical physical environment. Subsequent splitting of this widespread biota is achieved by tectonic plate movement, mountain building, or other epeirogenic events that create barriers to gene flow, and the very same event that generates the new geographic distribution also allows for allopatric speciation. Dispersal of one of the new sister species into the range of the other may then produce sympatry. Mayr's propagule isolated by long-distance dispersal (Bond 1948) has no role in such a scheme.

Its adherents contend that vicariance, unlike dispersal, allows scientific statements that can be falsified statistically (Croizat et al. 1974; Rosen 1974a,b; Nelson 1975; Cracraft 1980), but their earliest efforts, delineating "generalized tracks," fall short of this goal (McDowall 1978; Simberloff et al. 1981). A track is a line connecting members of a monophyletic group (like species in a genus), and a generalized track is a group of tracks that coincide approximately. For example, Croizat et al. (1974) depict the tract of the *Columba fasciata/albilinea* complex, a line running primarily along the Pacific coast from British Columbia to Chile, with an extension to Trinidad. How closely must another track approximate this one in order to comprise a generalized track? What of tracks that do not coincide? Appealing to probabilistic intuition, Croizat et al. (1974) say that a generalized track "becomes statistically . . . significant," presumably meaning that it would be unlikely that a group of taxa dispersing individually would by chance achieve identical tracks, so one must seek a common force like a tectonic event acting simultaneously on all of them. However, when one considers that if there are, say, 50 genera to examine, there are 1,225 pairs of genera, 19,600 trios, and so forth, it no longer seems surprising that by chance alone two or three tracks should approximately coincide, even if all were established independently. The contention that a generalized track was, in fact, produced by a common geological force would be buttressed if the track was rationally related to an independently elucidated geological history, but it would still be difficult to attach probabilities to the biogeography.

McDowall (1978) raises a number of cogent objections to the vicariance paradigm even if it could be improved to allow statistical tests. In essence, he objects to the implicit invocation of parsimony in seeking common geological causes for coincident tracks and also to the notion that a falsifiable vicariance would be preferable to a dispersal theory simply by virtue of its falsifiability. He notes that the White-faced Heron (*Ardea novaehollandiae*) has an Australia–New Zealand track coincident with that of numerous invertebrate taxa, yet the heron is known to have dispersed to New Zealand in this century, although there is strong independent evidence that at least some of the invertebrate tracks were established in the Mesozoic by continental drift. Furthermore, the happenstance that nine birds found on New Zealand's subantarctic islands (e.g., the Eurasian Blackbird, *Turdus merula*, and the Common Redpoll, *Carduelis flammea*) are European and have been in New Zealand itself for at most a century allows us to be certain that colonization occurred by over-water dispersal. Borrowing a leaf from Croizat's common-patterns-demand-common-causes credo, McDowall asks if this pattern does not imply that Australian and New Zealand birds found on these islands also dispersed there over water.

Also, the number of nonresident vagrant bird species recorded in New Zealand each year similarly proves that long-distance dispersal is frequent, even if there is usually no clear way to show that a given colonization did or did not occur this way. If there is no way to falsify a general dispersal hypothesis, should one ignore evidence of dispersal on formal philosophical grounds alone?

Because most birds fly, and many regularly migrate long distances, birds have traditionally been viewed (e.g., Darlington 1957) as perhaps the least likely taxon to leave traces of the pathways by which current biogeographic patterns were achieved and the most likely taxon to establish new populations in the manner envisioned by dispersalists. Cracraft (1980), however, views the generalized tracks of early vicariance models as the appropriate null model for birds as for other taxa, and only "strong evidence" (of unspecified nature) as capable of falsifying it in favor of dispersal. Although this choice seems grounded in the unfalsifiability of a dispersal hypothesis, Cracraft presents no clearer criterion of falsifiability for generalized tracks than did Croizat et al. (1974). Working primarily at the level of orders divided into families, Cracraft generates plausible narrative explanations for the tracks of Galliformes, Anseriformes, Gruiformes, Caprimulgiformes, and Piciformes, most of which conform reasonably well to the hypothesized order of the breakup of Pangaea into present landmasses (e.g., Hallam 1981). The explanations rest on the correctness of phylogenetic cladograms for each of these orders, as does a similar scheme for the ratites (Cracraft 1973). That is, although the similarity of the individual ordinal tracks may argue for a parsimonious vicariance explanation, it is not the generalized tracks per se that most persuasively support Cracraft's contention. Rather, it is the similarity of the systematic cladograms to a "geographic cladogram" of the hypothesized historical relationships of the landmasses.

The original advocates of vicariance have similarly abandoned generalized tracks as providing falsifiable hypotheses and sought better evidence in the relation between cladistic and biogeographic analyses. Hennig (1966) originated this approach, although this entomological pedigree has not been acknowledged by subsequent enthusiasts. Hennig's *multiple sister-group rule* (Ashlock 1974) states that if several sister groups within a monophyletic taxon all span the same two regions (Figure 11.1), a previous direct connection is implied. Again, no statistics attach to this observation, but one's probabilistic intuition suggests the basis for this rule: It would be unlikely for many individually dispersed taxa all to achieve identical tracks. As noted earlier, one still

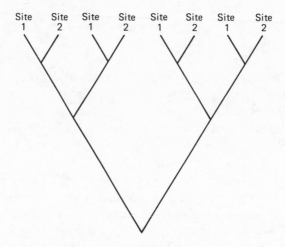

Figure 11.1 Cladogram of several sister groups within one monophyletic taxon all spanning the same two regions.

wishes to know how to assess the statistical significance of the sister groups *not* spanning the two regions. Brundin (1966) and Edmunds (1972), also working with insects, elaborated the multiple sister group rule to a *drift sequence rule* by examining multiple sister groups at a series of taxonomic levels, thereby producing a full cladogram. They do not attempt to attach a probability to their cladograms, but the number of sister groups plus the accordance with plate tectonic evidence is impressive. Edmunds's suggested cladogram for the breakup of Gondwanaland (Figure 11.2), for example, is based wholly on mayfly systematics except for the position of India and the order in which Australia and South America separated from Antarctica. It is identical to that suggested by geological evidence (e.g., Hallam 1981).

Rosen (1978) for poeciliid fishes was the first to attempt to attach probabilities to instances of multiple sister groups, and Platnick and Nelson (1978) to the matching of cladistic and geographic cladograms. Observing that two poeciliid genera share five sites and that their geographic cladograms (that is, the systematic cladograms with sites substituted for species) are identical, Rosen (1978) reasoned that since there are 105 topologically distinct five-member cladograms, the null probability of his observation is 1/105 = 0.0095 and concluded that the genera were unlikely to have achieved their present distributions by individual dispersal. Since the turtle genus *Terrapene* shares three sites with the two poeciliid genera and the three three-member geographic

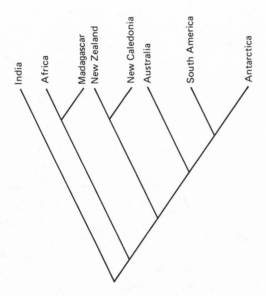

Figure 11.2 Cladogram suggested by Edmunds (1972) for the breakup of Gondwanaland.

cladograms of these sites are identical, Rosen says that this event has null probability $(1/3)^2 = 0.111$, since there are three topologically distinct three-member cladograms. He suggests scanning other North and Middle American taxa to find other cladograms topologically identical to these. He claims that such cladograms can only lower the null probability and thereby strengthen the inference that individual dispersal cannot account for the similarity of the biogeographic ranges. Platnick and Nelson (1978) describe an approach very similar to Rosen's. The key difference is that they first erect a geographic cladogram based on geological evidence, then transform taxonomic cladograms into geographic ones for comparison to the geologically derived cladogram. They contend that any taxonomic cladograms that coincide with the geological ones can only support the geological hypothesis, as well as the hypothesis that the speciation and biogeography of the taxa whose cladograms match were caused by the geological events.

Simberloff et al. (1981), who level a number of criticisms at these most recent cladistic variants of vicariance biogeography, feel that the method is capable of yielding falsifiable hypotheses but that the statistics used to date are invalid and even with valid statistics the approach would not apply to many taxa. First, Rosen (1978) and Platnick and Nelson (1978) restrict their effort to taxa endemic to their present sites.

These are exactly those whose geographic distributions would be most likely to be geologically determined and for whom dispersal would be least likely to occur. It is instructive that Rosen's application is to fresh-water fishes, the vertebrate class traditionally viewed (e.g., Darlington 1957) as having the most difficulty dispersing long distances. Few bird taxa would have many members so narrowly limited in range.

Second, Rosen (1978) and Platnick and Nelson (1978) do not consider the size of the universe of cladograms in which they seek matches; this problem is exactly analogous to the one already discussed for tracks. The null probability that *some* pair or trio of cladograms coincides topologically may be quite large if there are 50 cladograms to draw from, since there would then be 1,225 pairs and 19,600 trios possible. So the match of three three-member cladograms (the two poeciliids and the turtle) need not be striking. Had these three genera been specified in advance, before their topologies were examined, the null probability of identity would indeed be $(1/3)^2 = 0.111$. However, there are many vertebrate genera in North and Middle America, and one could not simply scan them to find matches to one another (Rosen 1978) or to an area cladogram (Platnick and Nelson 1978); one would have to know the sampling universe. Even then, a statistical test would likely have to be by simulation (Simberloff et al. 1981), since matches of pairs are not independent. If we know genera A and B have identical cladograms and that A and C do also, we know that B and C must match.

Third, the null probability to assign to any particular cladogram, and, therefore, to a match, is no trivial matter. For four or more members, there are not only a limited number of cladograms but a distinct number of "types" (Simberloff et al. 1981); Figure 11.3 depicts the three possible types of a five-member cladogram. The number of distinguishable cladograms is not evenly apportioned among types (Figure 11.3), so if one views types and not distinguished cladograms as equiprobable, each distinguishable cladogram does not have identical probability. Furthermore, the Markovian model of cladogenesis (Raup et al. 1973; Gould et al. 1977), which states that every extant lineage is as likely to split in the next instant as every other extant lineage, predicts yet a third distribution of distinguishable cladograms among types, and Savage (in press) has recently shown that published cladograms from many taxa are consistent with the Markovian model and not with the hypotheses of equiprobable types or equiprobable distinguishable cladograms. As an example of the differences in null probability of identity that the models generate, for two five-member cladograms the

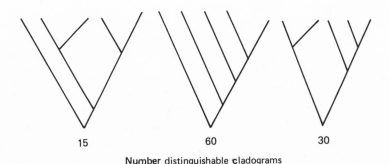

Figure 11.3 Three possible topological types of five-member cladograms, with number of distinguishable cladograms of each type.

probability is 0.0095 for equiprobable cladograms, 0.0130 for equiprobable types, and 0.0120 for the Markovian model.

Null probabilities of an observed number of matches among a set of cladograms, or beween a geologically derived geographic cladogram and a set of taxonomically derived geographic cladograms, were calculated with an algorithm for randomly generating sets of cladograms (once one chooses a Markovian or other model) to see if the observed set produced an extraordinary number of matches relative to a random arrangement that would correspond to an individual dispersal hypothesis (Simberloff et al. 1981). It is unlikely that sufficient systematic and geographic information exists for any taxon to apply this rigorous test to the vicariance hypotheses. Pielou (1981a) suggests the further restriction that systematists and geologists must construct their cladograms independently of one another, analogously to a double-blind procedure. It is perhaps too much to ask a systematist to desist from reading the geological literature in the course of what may be the major part of a lifetime's work. Endler (1982) also notes that falsification of vicariance models has not been attempted and offers a speciation simulation that suggests that concordant cladograms should not generally reflect common sequences of geographic vicariant events.

Haffer's (1974) refuge model for the biogeography of Amazon bird species and related models for African birds by Moreau (1963, 1966) and for Andean birds by Vuilleumier (1969), Haffer (1970), and Vuilleumier and Simberloff (1980) may be viewed as vicariance models in which splitting of widespread ancestral taxa is achieved not by epeirogenic events but by changing climates (Haffer 1981). Glaciations may reduce widespread lowland forest habitat to isolated refugia in which allopatric speciation may occur, whereas interglacials allow expansion and secondary contact as forest again becomes continuous (Simpson and Haffer 1978; Haffer 1981). Similarly, paramos during interglacials are isolated on mountaintops and allopatric speciation may occur, whereas in glacial periods this habitat will descend further and further down the mountainsides, ultimately coalescing and establishing a secondary contact among new species (Vuilleumier and Simberloff 1980). As stated, these models are at best plausible narratives, reasonable by virtue of accordance with palynological and geological evidence on Quaternary habitats but not framed so as to be falsifiable.

The lowland scheme, at least, might be tested against a null hypothesis that species' ranges are established independently, not simultaneously. Haffer's probabilistic evidence for the model consists of (1) the claimed congruence of refugia that he determines for birds with those of other workers on other taxa (e.g., Lynch, 1979, for amphibians) and with rainfall centers and (2) the clustering of the boundaries of numerous pairs of closely related species into narrow secondary contact zones. To me, the congruence depicted for different taxa by, for example, Simpson and Haffer (1978) does not look very close, but a simulation of random placing of refugia of fixed areas subject to relevant constraints, followed by a comparison of observed overlap to simulated overlap, might convincingly demonstrate congruence. Similarly, it is disputable that the boundaries of species pairs are remarkably clustered relative to the pattern one would see if each species pair's boundary were located independently with respect to other pairs' boundaries. Underwood (1978) has published a statistical test of exactly this problem in his attempt to see if intertidal zones are real constructs or whether the clustering of species' boundaries are no more pronounced than if they were randomly and independently located on the shore. This test could be applied, mutatis mutandis, to the Amazon bird boundaries. Endler (1981) has similarly questioned whether species' boundaries are in fact clustered between postulated African bird refugia.

The dynamic equilibrium model

The basic model
Nineteenth-century plant biogeography, beginning with de Candolle (1820) and continued by such writers as Grisebach (1884) and Schimper (1898), was primarily ecological, seeking explanations for biogeographic patterns in the physical climate of today and not the

evolutionary past of taxa. Although most zoologists quickly followed Darwin's and Wallace's lead in seeking evolutionary interpretations of biogeography, animals were also occasionally classified into ecological rather than evolutionary groups (e.g., Allen's, 1871, division of birds into inhabitants of circumpolar zones). Such an approach tends toward description, lists of which species are in which climates, and such narrative explanations as are produced usually revolve around which physical factors are of prime importance for which groups of species. It is difficult to see what further developments this thrust could produce, and it generates neither the evolutionary questions that have rent historical biogeography nor the search for adaptive significance of various traits that has motivated modern population ecology.

Perhaps the resulting stagnation of ecological biogeography explains the instant enthusiasm that greeted the dynamic equilibrium model of island biogeography (Preston 1962; MacArthur and Wilson 1963, 1967). This model was a logical intersection of two ecological foci: the dynamics of single-species population growth and pairwise interactions (especially of insects) and the distributions of population sizes in nature (primarily of birds) (Simberloff 1978a). It proposes that the number of species on an island (and, by implication, any insular habitat) is a dynamic equilibrium between immigration of species not on the island and local extinction of species already there. This idea was adumbrated by Darwin, Wallace, and other early biogeographers, if not to the extent of positing an equilibrium, then at least in a concern with the numbers of species at different sites and recognition that different immigration rates must play a role. In the *Origin,* for example, Darwin (1896) ponders why large and habitat-diverse New Zealand has scarcely more angiosperm species that the little island of Anglesea or the "uniform" county of Cambridge. For this and similar examples of the depauperation of distant islands, he concludes that the rarity of immigration is key. Even as a young naturalist on the *Beagle,* Darwin (1845) concerned himself with the differences in numbers of bird species between the island of Georgia (now South Georgia) and Iceland. Wallace (1869) recognized, as do current island biogeographers, that island isolation and size determine the number of bird species on islands of the Malay Archipelago.

By its simple mathematics, use of a familiar statistic (species number) and conception of nature as a dynamic entity that can be grasped by virtue of its division into subunits, the equilibrium model quickly captured the attention of ecologists and systematists and spawned a plethora of papers not only on oceanic islands but on a host of insular microcosms (references in Simberloff 1974) as well as entire continents (e.g., Rosenzweig 1975; Webb 1976). Lack (1969, 1976) objected that bird community richness is not a dynamic equilibrium but a rather static entity almost wholly determined by which habitats an island contains. However, his views in this area were quickly rejected (e.g., Grant 1977; Ricklefs 1977; Williamson 1981).

The early euphoria among equilibrium adherents (Simberloff 1974) has recently dissolved upon careful consideration of the demands and predictions of the model and the data available to test it (Lynch and Johnson 1974; Abbott and Grant 1976; Simberloff 1976a, in press-a, -b; Slud 1976; Abbott 1980; Gilbert 1980). At a minimum, the model requires that (1) number of species at a site remain constant (equilibrium), (2) the species comprising the community continually change (immigration and extinction), (3) the immigration and extinction constitute population processes, not just transient intrapopulation movement, and (4) the dynamic species richnesses thus produced be rational with respect to variation in island size and isolation (the area and distance effects).

As for the existence of an equilibrium, reports conflict. Abbott and Grant (1976), for avifaunas of 14 southwest Pacific islands, find number of nonpasserine species approximately constant over the last century but the number of passerines increasing. Haila et al. (1979) report land bird species numbers have increased in the Åland Archipelago over the last 50 years. Reed (1980) says there has been a slow increase in the number of bird species on Britain and its satellite islands. Using four censuses, Terborgh and Faaborg (1973) felt that Mona Island birds were equilibrial, whereas Diamond drew the same conclusion from two censuses of Karkar (1971) and the California Channel Islands (1969). For the latter archipelago, Jones and Diamond (1976) provide additional census data. It is striking that equilibrium has traditionally been a subjective judgment. For example, Figure 11.4 depicts the numbers of species versus time for Farne Islands birds and Skokholm Island passerines. The former were felt to be equilibrial by Diamond and May (1977), the latter readily judged nonequilibrial by Abbott and Grant (1976). Ad hoc views on how much variation in species numbers can be tolerated by an equilibrium are perplexing: Diamond (1969) views variation of 16% or less equilibrial, and other authors state other critical percentages. As Abbott and Grant (1976) observe, these criteria are arbitrary. Diamond and May (1977) demand a coefficient of variation less than 0.20 and Abbott and Black (1980) a coefficient of variation less than 0.05. Again these values seem arbitrary. One needs a

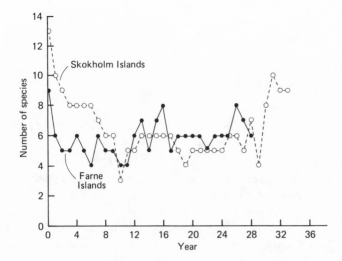

Figure 11.4 Changes through time in numbers of breeding landbird species on Skokholm (dashed line) and Farne Islands (solid line).

falsifiable hypothesis if "equilibrium" is not to embrace every conceivable data set. For one island, there must be some sequence of censuses that one would *not* call equilibrial if "equilibrium" is to have scientific meaning.

It is clear that for just two data points there is no statistical hypothesis possible, and whether one calls Karkar's 43 lowland birds in 1917 and 49 species in 1969 equilibrial (Diamond 1971) is a matter of taste. With longer sequences, however, statistics may be brought to bear. One might suggest that since species number ($\hat{S}$) is bounded by 0 and P (the number of species in the pool), it must be equilibrial. This suggestion reduces the equilibrium to an unfalsifiable truism and surely does not accord with MacArthur and Wilson's (1967) expectation of more limited temporal variation. On the other hand, since the equilibrium is a probabilistic one, compounded of many species-specific rates plus possible interaction terms (Simberloff 1969, in press-b), some variation must be entertained. A simple model is

$$\hat{S} = \sum_{k=1}^{P} \frac{i_k}{i_k + e_k}$$

where the i_k and e_k are species-specific immigration and extinction rates, respectively (Simberloff 1969; Gilpin and Diamond 1981). The rates are averages, and there will be variation over time. Simberloff (in press-b) suggests that the equilibrium model connotes regulation of species number and that even if we cannot say what the equilibrium is, we ought to be able to detect regulation (vs. a null hypothesis of no regulation) by a runs test, since there should be a tendency for an increase in species number to be redressed by a decrease, and vice versa. Tail probabilities for observed numbers of runs or fewer (Swed and Eisenhart 1943) show that neither the Farne Islands ($Pr = 0.227$) nor Skokholm ($Pr = 0.704$) has unusually many runs, so that neither avifauna depicts a clear regulatory equilibrium (Simberloff, in press-b). The major complication would be that if species number ever drew very near to 0 or P, there would be a high probability of an increase or decrease, respectively, in the next time interval by virtue of the definitions of immigration and extinction, and direction of change would no longer be independent of the present value of S, as demanded by the null hypothesis of a runs test. A second deficiency of this method is that it cannot assign probabilities to observations of "no change" between two successive censuses, and the existence of a large number of these would have to be viewed as supporting a hypothesis of regulatory equilibrium.

If the evidence for an equilibrium is equivocal, the evidence that it is dynamic and that substantial turnover occurs is even less supportive. Avifaunas of the Tres Marias Islands (Grant and Cowan 1964), Cocos Island (Slud 1976), the Faroe Islands (Salomonsen 1976), and Christmas Island and Raine Island (Abbott 1980) have approximately constant species number and virtually no turnover, so they do not confirm the dynamic equilibrium model. Even when turnover appears substantial, its interpretation is not always straightforward. First, estimated turnover rates will be lower the longer the intervals between censuses, and statistical attempts to correct for this tendency (Simberloff 1969; Diamond and May 1977) do not have confidence limits. Second, the nature of even a valid estimate cannot be determined without further information (Simberloff 1976a). Smith (1975) pointed out that if one monitored robins coming to and leaving a tree, a turnover rate could be calculated: Every time the tree emptied, an extinction would be recorded. This transient, intrapopulation movement, however, does not constitute the population extinction and immigration envisioned by the dynamic equilibrium model. What one must know is what fraction of an island population's recruitment derives from breeding as opposed to invasion, and this datum can rarely be collected (Simberloff 1976a).

Thus, Lynch and Johnson (1974) dispute Diamond's (1969) claim of substantial turnover of Channel Island birds on grounds of frequent interisland movement. Since the Farne Islands comprise but 80 acres total and are less than 2 miles from Britain, the documented turnover (Diamond and May 1977) is beclouded by the likelihood that many of the island birds are parts of mainland populations. A similar reservation must be expressed for the Skokholm Island bird turnover (Abbott and Grant 1976). Haila et al. (1979) believe that most bird "populations" on single islands in the Åland Archipelago are parts of larger populations that encompass other islands and the mainland as well. By contrast, T. Reed (pers. comm.) has recently amassed data for some of the Hebrides showing that individuals of at least certain bird species breed on the same island year after year. The avifaunal turnover reported by Abbott and Grant (1976) for isolated Pacific islands cannot result from transient movement.

Equilibrium turnover also connotes no secular environmental change, yet reported turnover may be caused by anthropogenous and other modifications. Lynch and Johnson (1974) contend that much of the avian turnover noted on Mona Island by Terborgh and Faaborg (1973) and the Channel Islands by Diamond (1969) is caused by human activities. Abbott and Grant (1976) believe that at least half of the avian turnover on their Pacific islands (admittedly not in equilibrium) is anthropogenous, whereas Reed (1978) explains much of the nonequilibrial Hebridean bird turnover as bound up with vegetational change. Simberloff and Abele (1976b) point out that the oft-cited nonequilibrial turnover of Barro Colorado Island land birds has occurred against a background of drastic successional vegetational change.

The *area effect* (that larger islands have more species than smaller ones) and *distance effect* (that isolated islands are depauperate) are well known in the ornithological literature (Hamilton et al. 1964), and the evidence does not require summarization here. Among recent examples, Higuchi's (1980) study of woodpeckers on 98 Japanese and neighboring islands shows clear distance and area effects, whereas Slud (1976) depicts several striking species–area relationships. Both the area and distance effects have ready interpretations in equilibrium theory terms. Larger islands should have lower extinction rates, therefore, larger equilibrium numbers of species (Figure 11.5a), whereas distant islands should have lower immigration rates, therefore, lower numbers of species (Figure 11.5b). Many papers have cited an observed area effect as evidence for the equilibrium model (references in Gilbert 1980).

However, neither the distance effect nor the area effect is uniquely predicted by the equilibrium model (Simberloff 1974, 1978a), so the observation of either effect cannot be construed as strong evidence in

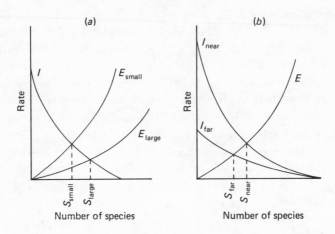

Figure 11.5 Area effect (a) and distance effect (b) predicted by equilibrium theory of island biogeography.

its support, though lack of either effect should be taken as falsification. In particular, Connor and McCoy (1979) suggest that at least three forces that are not mutually exclusive may lead to a classical species–area relationship. First, larger islands tend to have more habitats, and each habitat has at least partially distinct species. Second, a set of islands need not be insular with respect to some taxon, and each island community may just be a sample (proportional to the island's size) of a larger community. Third, extinction–immigration dynamics may contribute, as the equilibrium model envisions. Statistical techniques cannot apportion observed species–area effects among these three potential causes. Increasing area has often been interpreted as simply an index of increasing habitat diversity, as in Power's (1972) study of California Channel Island birds, Watson's (1964) work on passerines of Aegean islands, and the study of Hamilton et al. (1964) on birds of three island groups. Because of the high correlation between habitat diversity (however indexed) and area, experimental rather than statistical means will be required to determine whether turnover rates contribute to the species–area effect: Area must be modified while habitat diversity is maintained. Only for arboreal arthropods of Florida Keys mangrove islands has such manipulation been performed, and there it demonstrated a small but clear effect of area independently of habitat diversity (Simberloff 1976b).

The basic equilibrium model is therefore not confirmed. Observed species–area relationships are at best consistent with the model, and direct observation of species richness changes is obscured by habitat change and the possibility of transient movement. Some islands appear to have no turnover whatever, and perhaps the model is best viewed as a testable hypothesis for each archipelago and taxon. Gilbert's (1980) overall review and Abbott's (1980) review of the avian literature agree that, as a universally applicable model of nature, the model has been falsified.

It is also apparent that the equilibrium model, although spawned (see earlier) from two approaches to the study of population sizes, takes no direct account of population sizes and focuses instead on presence or absence of species. It is therefore not an appropriate indicator of some kinds of community change. For example, a species whose population sizes at a site in a sequence of annual censuses was 0–1–0–1–0–1 would contribute substantially to immigration and extinction rates, whereas a species whose population sizes over the same period were 5–100–5–100–5–100 would not contribute to changes in species number. Yet most ecologists would surely view the changes in the second species' populations as more interesting and important than those in the first species' populations. Haila and Järvinen (1981) have stressed the importance of population sizes, and not just presence/absence, in island bird studies.

Refinements of equilibrium theory

Qualitative refinements of the basic eqilibrium model are numerous (Pielou 1979). Perhaps most importantly, Brown and Kodric-Brown (1977) proposed that extinction is not independent of isolation, since sufficiently high invasion rates can repeatedly rescue a species from extinction on an island. Although the operation of this "rescue effect" has been demonstrated only for Brown and Kodric-Brown's insects on flowers, it is likely that local extinction has been forestalled for many taxa by intermittent immigration.

An attempt to quantify the equilibrium theory – that is, to predict not just the existence of an equilibrium and the shape of a colonization curve, but the actual equilibrium number and trajectory of species number through time – is Diamond's (1972) *relaxation model*. This model postulates that each island and taxon has a dynamic equilibrium number of species and that whenever the number present is not this equilibrium, it will "relax" toward the equilibrium by the equation

$$S(t) = \hat{S} + [S(0) - \hat{S}]e^{-t(K_i + K_e)}$$

where $S(t)$ = the number of species at time t, $\hat{S}$ is the equilibrium number, and K_i and K_e are constant coefficients of immigration and extinction, respectively. Diamond's test of this model consisted of the comparison of species–area relationships for birds of New Guinea satellite islands that were recently defaunated by volcanoes or modified by rising sea levels with islands not so modified. Abele and Connor (1979) show that (1) the species–area relationships of the modified islands do not differ from those of the other islands, except perhaps as a function of distance (which is not part of Diamond's model), and (2) the estimates of how many species were originally present are inflated (a criticism also raised by Abbott, (1980).

Terborgh (1974) proposed a similar model for bird species loss in the West Indies as sea levels rose over the past 10,000 years and created islands out of former mainland sections. His estimates of number of species originally present are based on interpolation in recent mainland species–area curves and so are not as likely to be inflated as Diamond's were, but his model allows for no immigration, and there is no way of falsifying it. A key difference from Diamond's model, in addition to no immigration coefficient, is that Terborgh's extinction coefficient is multiplied by the square of species number and so extinction rate is not a constant fraction of number of species present.

Gilpin and Diamond (1976) similarly propose relaxing the constraints that extinction rate be linearly related to number of species present on an island and that immigration rate be linearly related to number of species not present. MacArthur and Wilson (1967) had also suggested nonlinearity; this suggestion is reflected in their concave immigration and extinction curves. The rationale is that gradients of invasion and persistence abilities among the available species pool would dictate that the good dispersers would likely be the first species on an island, and the good persisters the last species to be extinguished (cf. MacArthur 1972; Pielou 1979). Gilpin and Diamond (1976) tried 13 different equilibrium models (i.e., 13 pairs of expressions for immigration and extinction as functions of island species number, isolation, and area), with up to seven fitted parameters per model. They attempted then to fit the models to data on lowland birds of the Solomon Islands and were able to explain well over 90% of the variation in species number with several models. This curve-fitting tour de force serves as a textbook example of how not to test a hypothesis; it is an elaborate narrative. First, one cannot use the same set of data to produce a model and then to test it statistically (Selvin and Stuart 1966; Pielou and Pielou 1967; Pielou 1979). At the very least, the models would have to be tested against data from another archipelago. Second, with up to seven fitted constants there must surely be a high a priori likelihood of constructing a curve that fits 52 data points very well. In fact, Gilpin and Diamond produce a five-parameter model, where the parameters have no ready biological interpretation, that explains as much of the variation in species number as a function of area and distance as does their best-fit deductive model with seven parameters. Third, the submodels that generate the particular seven parameters underpinning their deductive equations are untested, and some are not even falsifiable with realizable data. Finally, Selvin and Stuart (1966) observe that it is difficult to attach probabilities to the fit of data to a model when a series of related models have been tried in order to find the best one. All that one can say with assurance is that the null probability that *some* model will not be rejected is far greater than the null probability that any particular model will not be rejected.

The *molecular theory of island biogeography* (Gilpin and Diamond 1981) builds on the Solomon Islands analysis and attempts to determine for any island an equilibrium number of species plus the shape of the immigration and extinction curves from the properties of individual species' immigration and extinction probabilities (the i_k and e_k of Simberloff, 1969). That an equilibrium is obtained given that these probabilities are constant was demonstrated by Simberloff (1969), who pointed out that their values and constancy would be very difficult to assess with real data. Crowell (1973) and Simberloff (1981) have attempted to gather such data for rodents and insects, respectively, but both efforts are insufficient to estimate the equilibrium. Gilpin and Diamond (1981) assume the i_k and e_k are both constant and log normally distributed among species within a pool. This assumption is based partly on the estimation that Solomon Islands extinction rates vary with area more than immigration rates do, but, as noted earlier, the Solomon models are unsupported by data and are deduced. No extinctions or immigrations were documented. Further support for log-normal distributions of the e_k comes from the observation that extinction rates are inversely related to population size, and population sizes are usually log normally distributed. At best, this correlation suggests approximate log normality for the e_k, and the log-normal parameters cannot be estimated without knowledge of the extinction rates. Since extinctions on the islands in question (the Bismarcks and Solomons) have not been well documented, their rates are unknown. Again a theory reduces to an untestable narrative. Given that the i_k and e_k are log normal (or of any known distribution), it is a straightforward matter to combine them in various combinations (e.g., as suggested by Simberloff, 1969, and MacArthur, 1972) to estimate island-wide immigration and extinction curves. For log-normal distributions, these curves are concave (Gilpin and Diamond 1981), as MacArthur and Wilson (1967) and MacArthur (1972) predicted they would be for any realistic distributions of i_k and e_k. Simberloff (1969) noted that since the i_k and e_k are not known for any pool, there is no advantage to be gained from pursuing this exercise further. This state of affairs persists.

The equilibrium model and conservation

It is surprising that a model whose applicability is as limited as that of the dynamic equilibrium model should spawn an applied sub-discipline, biogeographic refuge design, and more surprising still that its even less successful relaxation offshoot should contribute to it. Terborgh (1974, 1975), Moore and Hooper (1975), Diamond (1975a), Diamond and May (1976), Galli et al. (1976), Forman et al. (1976), Nilsson (1978), Faaborg (1979), Sampson (1980), and Gilpin and Diamond (1980) have all invoked the equilibrium model and the notion of relaxation in bird conservation schemes. Two particular recommendations for maximizing species richness are frequently repeated:

1.   One large refuge is better than two or more smaller ones of equal total area (Wilson and Willis 1975; Diamond and May 1976)
2.   A round shape is best (Wilson and Willis 1975; Diamond 1975a; Diamond and May 1976; Butcher et al. 1981)

In fact, recommendation 1 does not derive from the equilibrium theory (Simberloff and Abele 1976a,b; Simberloff in press-b), and what data have been examined in this context indicate exactly the opposite trend: A series of small sites contain at least as many species as one large one does (Simberloff and Abele 1982; Simberloff in press-b). Equilibrium theory is neutral with respect to whether one large refuge or several small ones should contain more species, and the key datum for any particular system is empirical: the gradient of colonizing among species in the pool. Simberloff and Abele (1976a) and Higgs (1981) have pointed out that the stronger this gradient, the more likely it is that one large refuge will be better than several small ones. Simberloff and Abele (1976a) also noted that minimum population sizes for different species would limit how small refuges could be, but these are also an empirical matter. Several criticisms of these conclusions (Diamond 1976; Terborgh 1976; Whitcomb et al. 1976) did not address the arguments (Simberloff and Abele 1976b; Kitchener et al. 1980; Cole 1981). The only direct experiment on the question of one large versus several small refuges is on insects of Florida mangrove islands, and results to date are equivocal (Simberloff and Abele 1982). Lovejoy (1980) has set up such an experiment on birds of the Brazilian rainforest. All other reports proceed by compiling lists for small and large sites, then merging the lists for a group of small sites for comparison to that of one large site of equal total area. This procedure has been carried out for numerous taxa in many regions and invariably has shown that several small sites have as many species as one large site or even more. Among these attempts are three on birds: The data of Forman et al. (1976) and Forman (pers. comm.) on New Jersey forest birds and of Nilsson (1978) on birds of Swedish islands show no statistical difference between one large and several small sites (Simberloff, in press-b). Gilpin and Diamond (1980) find for New Hebrides lowland and forest birds that two small islands generally contain 5–10% more species than a single large one of equal area. Butcher et al. (1981) do not actually compare one large to two small sites but, based on their long-term study of one Connecticut study site, agree with Simberloff and Abele (1976a) that the equilibrium theory does not dictate single large refuges and find that species number on their site supports neither an equilibrium nor a relaxation interpretation.

A number of nonequilibrium theory considerations must, of course, come into play with respect to whether one large or several small refuges would be better, the most important of which would likely be the problems of idiosyncratic catastrophes (like fires, epizootics, or introduced predators) and inbreeding depression (Simberloff and Abele 1976a, 1982). In general, the former would be vitiated by many small reserves, the latter by fewer and larger ones, although there have been recent arguments (e.g., Drury 1974; Christiansen 1974, 1975; Chesser et al. 1980) that polymorphisms would be protected by subdivision into small populations so long as the small populations are not so small that inbreeding depression within them causes extinction. Several avian examples come to mind that exemplify the interplay of catastrophic and genetic forces, and all of them seem to argue for several small refuges and not one large one. Simberloff (1982) describes the extinction of the Heath Hen (*Tympanuchus cupido cupido*) in New England, in which restriction to a single Martha's Vineyard refuge, subsequently devastated by a series of catastrophes, appears to have played a key role. Drury (1974), comparing the fates of single and subdivided populations of the Laughing Gull (*Larus atricilla*), concludes that subdivision is the better conservation strategy. Simberloff (1978b) notes the series of anthropogenous environmental insults that have devastated the Seychelles Islands and observes that the fact that there are about 30 small islands instead of one large one was probably instrumental in restricting extinction to just 2 of 17 species, the green parakeet (*Psittacula* [*eupatria*] *wardi*) and the Chestnut-flanked White-eye (*Zosterops mayottensis semiflava*).

The second contention, that equilibrium theory dictates a round refuge, is equally specious. This view seems to stem from the "peninsula effect," first observed by Simpson (1964) for North American mammals and subsequently by Cook (1969) for North American birds. The observation is that species richness decreases from the base of a peninsula to its tip. Both Simpson and Cook suggested that decreased immigration of mainland individuals to a peninsular tip might be the reason and called for further study. Three studies that have heeded that call – Wamer (1978) on Florida birds, Taylor and Regal (1978) on Baja California rodents, and Seib (1980) on Baja California reptiles – all assign a primary role to habitat variation along the length of the respective peninsulas, and none envisions an immigration–extinction equilibrium. Furthermore, Game (1980) observes that a long, thin shape maximizes perimeter, which might increase immigration rate into a refuge, and that, ceteris paribus, long, thin refuges might well encompass greater habitat diversity than round ones would. Finally, the greater perimeter might enhance the "edge effect" (Williamson 1975), which would likely lead to increased species richness but at the same time act against species restricted to central habitats (cf. Butcher et al. 1981). In sum, equilibrium theory has no lessons for conservation. Such characteristics as refuge shape and contiguity may affect individual species, but species number, the statistic the equilibrium theory purports to explain, does not vary consistently with these characteristics in such a way as to support the theoretical recommendations.

Biogeographic patterns and community composition

Species/genus ratios and related statistics

The major thrust of ecological biogeography in the wake of equilibrium theory is a search for rules that constrain community composition within the equilibrium species richness (Simberloff 1980). Attempts to show statistically that morphology and biogeography of island birds are determined by interspecific competition have taken three directions: taxonomic relations among coexisting species, morphological relationships, and numbers of combinations of coexisting species. Two of these approaches preceded equilibrium theory.

Elton (1946) interpreted the fact that species/genus (S/G) ratios for many local animal and plant communities in Britain were lower than those for all of Britain as caused by "existing or historical effects of competition between species of the same genus, resulting in a strong tendency for species of any genus to be distributed as ecotypes in different habitats, or if not, to be unable to co-exist permanently on the same area of the same habitat." Williams (1951) used Moreau's data (1948) on birds of 32 East African habitats to suggest that the lower local S/G ratios were likely artifactual, since even random subsets drawn from the entire avifauna had lower S/G ratios than did the entire avifauna. In fact, there tended to be more congeneric pairs and trios of African birds coexisting within habitats than one would have expected for random subsets. Williams (1951) felt that similar habitat requirements among congeners would override intense congeneric competition, even if the latter occurred.

Grant (1966, 1968) performed an analysis similar to Elton's on the land birds of the Tres Marias Islands, using percentage congeneric species (which is highly correlated with S/G ratio), and concluded as Elton had that the lower values on the islands than on a nearby patch of mainland indicate more intense competition. Simberloff (1970, in press-c) performed a random draw procedure similar to Williams' to show that the islands tend actually to have more congeners than would random subsets drawn from the entire avifaunas of either a restricted mainland and patch near the islands or all of Mexico within 300 km of the islands. Grant's (1966) observation of a monotonic increase in percentage congeners with area for 18 islands is also likely an artifact of the fact that species richness tends to increase with area and S/G ratios (or percentage cogeners) are increasing functions of species richness, even for random draws (Simberloff 1970). Juvik and Austring (1979) noted this correlation for Hawaiian birds. Isopleths of species/family (S/F) ratios for land birds of North America, which Cook (1969) interpreted as a reflection of structural habitat complexity and, therefore, likelihood that competing confamilials could coexist, are probably also an artifact, for the S/F isopleths appear highly correlated with species richness isopleths (Simberloff 1980). That the patterns may be statistical artifacts is not to say that competition is not more intense between closely related species. It only says that the ratio statistics do not manifest such heightened competition if it is occurring.

It appears that not only the Tres Marias Islands and the African habitats but also islands generally usually have slightly higher S/G ratios (more congeneric species) for land birds than random subsets of the same size would have (Simberloff 1970). Abbott (1975, 1980) finds the same trend for avifaunas of islands around Australia. Simberloff (1970, in press-b) and Grant and Abbott (1980) agree that interpretation of this trend, and of exceptions to it, are probably impossible because of the multitude of forces that act on S/G ratios. Even if intensified competition were to affect certain genera, its effect on S/G ratios in the biota as a whole would be diluted by unaffected genera and beclouded by other factors affecting the ratios.

Morphological relationships

The hypothesis of competitive character displacement (Brown and Wilson 1956) – competition in sympatry selects for greater morphological differences than obtain between allopatric populations of two species – has similarly motivated a biogeographic search for rules constraining which species can comprise an equilibrium number. Hutchinson (1959) contends that in order for species in the same trophic level to coexist, they must differ in size of trophic appendages by more than the critical ratio of 1.3, and a host of other workers, mostly on birds, have repeated this contention, albeit with various critical ratios (references in Simberloff and Boecklen 1981). Furthermore, the same competitive force is often believed to have led to extraordinarily constant morphological ratios when three or more competitors coexist (references in Simberloff and Boecklen 1981). Finally, morphological ratios on islands are said to exceed those on mainland because a poorer insular resource base intensifies competition (Schoener 1965; Grant 1966, 1968).

Grant (1972) feels that the evidence on most published examples of character displacement admits of other interpretations, and the three coexistence patterns based on character displacement have similarly been questioned. Roth (1981) suggests that the literature generally does not confirm a critical coexistence ratio of 1.3 and that a particular claimed example of ratio constancy for fruit pigeons (*Ptilinopus* and *Ducula*) (Diamond 1978), is at best marginally significant when tested against a null hypothesis of independently distributed sizes. Simberloff and Boecklen (1981), using different statistical tests, conclude that many data sets cited as evidence for either minimum or constant ratios cannot be statistically tested because the data have been selected (Selvin and Stuart 1966) and that those that are sufficiently complete to allow tests generally do not support the contention that ratios are extraordinarily large or constant relative to hypothesis of independently distributed sizes. Certain individual data sets do support one or the other contention, and neither Roth (1981) nor Simberloff and Boecklen

(1981) can be construed as denying that interspecific competition exists or even that it affects sizes. However, both papers show that the size distributions themselves do not require such an explanation.

As for the size ratios of birds coexisting on islands as opposed to mainland, Strong et al. (1979) found for several archipelagoes that island ratios *were* larger, but no more so than one would have expected given the fact that islands have fewer species. Even random proper subsets of a mainland species pool are expected to have larger ratios than the entire pool. Grant and Abbott (1980) and Hendrickson (1981) criticize this paper on a number of statistical grounds, but Strong and Simberloff (1981) and Simberloff (in press-b) show that even incorporating their criticisms one finds virtually the same results; few if any observations differ significantly from those expected for random subsets of birds.

When there is no relevant mainland pool from which to draw random sets of birds for comparison to the observed set, Strong et al. (1979), Ricklefs and Travis (1980), and Hendrickson (1981) produced an artificial species pool by adding together all the species in the archipelago. They next randomly drew species sets from this artificial pool to compare to observed sets, asking generally whether observed sets show greater morphological differences than would have been expected were species (or races) assigned to islands independently of morphology. Grant and Abbott (1980) and Grant and Schluter (unpub.) object that the species on the islands could already have been morphologically selected, so that the pooling procedure is circular. I disagree; it exemplifies the statistical technique of randomization (e.g., Bradley 1968), and we are not claiming that competition had no morphological effects, only asking if the morphologies of coexisting species (or races) provide additional evidence of such competition. For Galapagos tree finches (*Camarhynchus*), no such evidence is forthcoming, but for ground finches (*Geospiza*), at least some bill features seem to be displaced among coexisting species or races, relative to random species sets (Simberloff, in press-b), exactly as a competition hypothesis would have predicted. This is also the key result of Grant and Schluter's (unpub.) study of the same birds.

Ricklefs and Travis (1980) similarly compared Cody's (1974) 11 temperate-zone scrub bird communities to randomly drawn communities and concluded that distance and regularity of spacing between morphological nearest neighbors in the natural communities did not differ significantly from random expectations. Their random communities were constituted in two ways: (1) Species were drawn randomly from a total pool of all species in all of Cody's lists (the randomization procedure described earlier) and (2) random species were synthesized by randomly drawing morphological characters from the morphological hyperspace generated by the ranges observed, in the real communities, of eight traits: wing length, beak depth, and so forth. This second approach ignores the extensive literature (e.g., Seilacher 1970; Raup 1972; Oxnard 1978; Mosimann and James 1979) demonstrating that sizes and shapes of different morphological traits do not vary independently and are selected jointly. Consequently, the vast majority of "species" randomly constructed by drawing within a morphological hyperspace do not correspond to anything that nature could conceivably produce.

As with the equilibrium model, morphological ratios between coexisting species have traditionally been considered independently of the species' population sizes. So long as species are present, their size ratios are computed and rationalized. However, if the sizes are actually selected for by competition between the species, it is clear that one need not expect the same ratios to obtain between rare species as between common ones. If two species are so rare that they have no effect on their common resource, exploitation competition between them would not cause their sizes to diverge. James and Boecklen (in press) have sought correlations between degree or morphological displacement between pairs of species on the one hand and compensatory population size fluctuations on the other. That is, when one species' population increases, do the populations of morphologically similar species decrease? For one well-studied bird community, they found little evidence of such a pattern.

Of course, sizes of sympatric species have long been studied from a viewpoint very different from that of character displacement. Grant (1972) feels that most published claims of character displacement, including the classic example of the rock nuthatches *Sitta neumayer* and *Sitta tephronota* (Vaurie 1951, Grant 1975), can more parsimoniously be explained as simple clinal variation (Endler 1977) of one or both of the putatively displaced species, and the clines need not be attributed to competitive interactions between the two species. In this spirit, James (1970) has shown that complex patterns of clinal variation in sizes of 12 bird species in the eastern and central United States are concordant. All of these species' sizes increase and decrease together geographically, and these changes are highly correlated with various statistics that reflect the combined effects of temperature and humidity. One can infer, then, that the overriding influence on size (and shape) of birds is weather, perhaps through its effect on heat loss or energy budgets, and that any character displacement generated by exploitation competition will constitute but a minor adjustment to the climatically determined main effect. James (1970) observes that her conclusions are a variant of Bergmann's Rule (1847) and that similar patterns have been observed by, among others, Rensch (1939) for a variety of birds.

Species combinations

The numbers of combinations of species found on a set of islands or mainland sites have also been used to support the contention that biogeographic patterns are heavily influenced by competition (references in Simberloff and Connor 1981). The basic data consist of a site × species binary matrix, with "1" representing a species' presence at a particular site, and "0" its absence. Pielou and Pielou (1968) first treated this problem for arthropods and bracket fungi, by proposing as a null hypothesis that each species is arranged among available sites independently of where other species are. That is, they asked whether, if each species were present in a simulated matrix exactly as many times as it is observed to be present in nature and were situated independently of all other species, would there be anything about the frequencies of the various combinations of species that would differ from what is actually observed. Since the most obvious way to approach this problem – a χ^2 test comparing observed and expected frequencies of the combinations – was impossible because the expected frequencies would be too low, Pielou and Pielou (1968) instead first simulated filling the matrix, then scanned the matrix to see how many species combinations were present. They performed the operation repeatedly to generate an expectation and variance, then asked how the observed number of combinations compared to the expected. For their arthropods, the observations were generally close to the expectation, but Pielou and Pielou (1968) caution that even had there been fewer combinations then expected, one could not from the pattern data alone conclude that species interactions were the cause, since different preferences among the species for different sites (e.g., if the sites were physically different in any way) would have produced the same observation. That is, if interspecific competition among the species of some combinations would prevent those combinations from occurring, one would see the same thing as if different species ordinated site suitability differently – too few combinations relative to a random, independent arrangement.

It is worth emphasizing that neither Pielou and Pielou (1968) nor Connor and Simberloff (1979) and Simberloff and Connor (1981) in similar simulation approaches are addressing the question of whether the differing frequencies among the species—that some species are found at few sites and others at many—might have been generated by interspecific competition. All they are asking is whether the *joint* arrangements—which species are found together and how frequently—are compatible with a hypothesis of interspecific competition.

Siegfried (1976) attempted an *adjusted χ^2 analysis* of the frequencies of combinations of four diving duck species in 395 ponds in southern Manitoba. Since he did not record how many ponds had no ducks, he normalized the expected combination frequencies that his individual species frequencies predicted so that they summed to 395. When he compared his observations to the adjusted expectations he found that the four classes of pond with only one species all are more frequent than expected whereas 9 of the 11 multispecies combinations were less frequent than expected. He consequently concluded that the ducks exclude one another competitively. However, Siegfried's method of adjusting the expected frequencies is invalid and is highly biased toward producing expecteds that are less than observed for single species and greater than observed for multispecies combinations (Simberloff and Connor 1981). It is likely that nothing short of a complicated simulation can yield a valid interpretation of data from which empty sites are omitted. Abbott (1981), using Siegfried's method, concludes that three passerine birds found in various combinations on southwestern Australian islands tend to exclude one another competitively. Since the method is incorrect, this part of his analysis cannot support his hypothesis.

Abbott et al. (1977), following Abbott's (1977) analysis for plants on Australian islands, examine combinations of Galapagos *Geospiza* finches by Maxwell–Boltzmann statistics (Feller 1950), which give the probability of exactly m missing species combinations as

$$\frac{1}{n^r} \begin{bmatrix} n \\ m \end{bmatrix} \sum_{i=0}^{n-m} (-1)^i \begin{bmatrix} n-m \\ i \end{bmatrix} (n - m - i)^r$$

where there are n possible combinations and r islands. Corrected tail probabilities are given by Simberloff and Connor (1981) and Grant and Schluter (unpub.). Some uncertainty exists about exactly which species breed on which islands, but the general result is that there appear to be fewer combinations of one to five species present than one would have expected if the species colonized independently of one another.

As Simberloff and Connor (1981) note, the Maxwell–Boltzmann model is not appropriate for testing whether *joint* frequencies, over and above frequencies of individual species occurrences, are improbable for an independent colonization hypothesis, since the Maxwell–Boltzmann model assumes that all combinations (even those composed of rare species) are equiprobable. Simberloff and Connor (1981) thus suggested a *weighted model* for independent colonization, analogous to the Maxwell–Boltzmann model except that the null probability that any island is colonized by a given combination is proportional to the product of the frequencies of the component species. By this test, the tail probabilities for one- and two-species combinations in the Galapagos do not cause rejection of the independent colonization hypothesis, but there are still too few three-, four-, and five-species combinations (Simberloff and Connor 1981; Simberloff, in press-c).

Abbott et al. (1977) and Grant (in press) observe that two *Geospiza* pairs (*conirostris–fortis* and *conirostris–scandens*) are found together on no islands. Not only do they never coexist together on two-species islands, but they are never embedded in larger combinations. These pairs are suggested to be prime candidates for competitive exclusion, but one would likely expect at least two missing combinations even if all species were independently arranged on the islands (Connor and Simberloff 1979). *Conirostris* occurs on the fewest islands, so missing combinations that include *conirostris* are to be expected. This is not to say that *conirostris*'s rarity is not caused by competition, only that the absence of certain combinations that include it does not constitute additional support for a competition hypothesis. Grant (in press) feels that these two missing pairs resemble one another very closely in bill size and that this resemblance strengthens the implication of competition. Simberloff (in press-c), however, shows that the missing pairs are not particularly different in bill morphologies relative to all possible species pairs.

In addition to the fact that there are surprisingly few three-, four-, and five-species combinations, aspects of *Geospiza* morphology combined with combination data are consistent with a competition hypothesis and inconsistent with an independence hypothesis (Simberloff, in press-c; Grant and Schluter, unpub.). For example, 11 of 13 sympatric *Geospiza* species pairs show smaller morphological differences between randomly drawn allopatric races than between races that are actually sympatric (Simberloff, in press-c). This pattern is consistent with the hypothesis of competitive character displacement. One's confidence in this interpretation should be mitigated, though, by the realization that a battery of correlated statistical tests have been run on a host of correlated characters, and only certain ones showed significant departures from expectations (Simberloff, in press-c). This is exactly the circumstance in which it is especially difficult to assess the significance of the result in toto (Selvin and Stuart 1966).

Higuchi (1980) has similarly combined observations of how many combinations are found and not found, respectively, with a consideration of the morphology of the constituent species of the two kinds of combination to infer that interspecific competition structures Japanese island woodpecker communities. Eleven species arranged on 98 islands produce only five two-species combinations (on 13 islands). Using only the pool of six species found on two-species islands (cf. Abbott 1977), one finds that 15 combinations are possible, and the weighted model of Simberloff and Connor (1981) shows that a result as extreme as that observed (five combinations or fewer) would have obtained 0.200 of the time even had the species been independently arranged on the islands, but with the observed individual species' frequencies. Higuchi, however, notes that the missing combinations tend to comprise pairs that are very similar morphologically, whereas the pairs that one finds consist of species that are morphologically very different. Thus, I find for culmen length ratio that the 5 observed pairs exceed the 10 missing ones, with probability less than 0.01 that the observed ranks would be as different by chance alone (one-tailed Wilcoxon rank sum test). However, this is probably a statistical artifact. The five observed pairs all consist of one other species plus *Picoides* (=*Dendrocopus*) *kizuki,* which is not only the most frequent species (half of all occurrences by all species on two-species islands) but also by far the smallest bird. Its mean culmen length, for example, is only 54% of that of the next smallest of the six species, *Picoides* (=*Dendrocopus*) *major.* Consequently, morphological ratios of every species pair that includes *P. kizuki* will perforce be very large, and since it is also by far the most frequent species one would have expected most of the observed combinations to include it. None of this is to say that its small size does not alleviate competition with other species and that this circumstances is not what allows it to be most frequent, only that the size ratios of the observed combinations do not constitute an additional piece of evidence that competition is the key force here.

Diamond (1975b) asserted that the composition of bird communities on islands is governed by seven "assembly rules," which state that some pairs and larger groups of species, both related and unrelated to one another, will not be found coexisting an any island within an archipelago and that smaller islands will have particularly few species combinations. These rules, which he felt satisfactorily explain patterns in the Bismarck Archipelago, he attributed primarily to the workings of interspecific competition. Connor and Simberloff (1979) showed that three of the rules are tautologies, and a fourth is generally untestable. The remaining three simply state, respectively, that some combinations of related species are nowhere found, that some pairs of species (related or not) are nowhere found, and that some large combinations are nowhere found even though subcombinations are found. Connor and Simberloff (1979) note that all three of these patterns would obtain even if species were independently distributed subject only to the constraints of species frequencies and island richnesses, so that observing such patterns does not provide additional evidence of competition beyond whatever insight one can bring to bear on the causes of species' differing frequencies. To find whether observed data on any of these patterns would be inconsistent with an independence hypothesis, one would have to generate the relevant expected values for random, independent arrangements. When Connor and Simberloff (1979) generated these expectations by computer simulation for the birds of the New Hebrides and West Indies, they found that a large fraction of the missing species combinations would have obtained even had species individually and randomly colonized the islands, although for the West Indies the differences between observed and expected are statistically significant. The assignment of competition as the cause of even this statistically significant fraction is questionable, however, in light of the fact that exactly the same pattern would be produced by allopatric speciation (Connor and Simberloff 1979) and differences among islands in suitability for different birds (Pielou and Pielou 1968; Simberloff and Connor 1981).

Diamond and Gilpin (1982) and Gilipin and Diamond (1982) argue against the Connor–Simberloff analysis of assembly rules on several grounds. First, they suggest that the rules are meant only for application within guilds. Connor and Simberloff (unpub.) reply that the data used to assign the Bismarck birds to guilds have not been gathered but that the guilds instead appear to be delineated by species' joint distributions. Second, Diamond and Gilpin (1982) object that Connor–Simberloff analysis of the New Hebrides data was destined to confirm an independence hypothesis, because only a very few matrixes could have resulted given the observed species frequencies and island richnesses. Connor and Simberloff (unpub.) show that this objection is ill-founded, since there are many possible matrixes. Third, Diamond and Gilpin (1982) feel that the Connor–Simberloff matrix simulation method is unable to detect arrangements known to have been produced by competition; in short, the power of the method may be low. This last criticism is quite possibly valid, but its validity has not been demonstrated nor has the power of an alternative test by Diamond and Gilpin (1982) been shown to be greater (Connor and Simberloff unpub.). Finally, Diamond and Gilpin (1982) propose an analytic alternative test that they feel identifies which species combinations are nonrandomly distributed. When they compute the relevant test statistics for the observed Bismarck bird patterns, they find a great excess of positively associated species pairs (the opposite to the result that interspecific competition should have produced) and are unable to say which pairs are in the tails of the test statistic because of chance–after all, 5% would be in the 5% tail for this reason alone–and which are there for deterministic reasons (Connor and Simberloff, unpub.), so their claims of the overarching importance of competition and of the insights provided by their analysis are not sustained.

Haefner (1981) has suggested constructing algorithms by which island communities may be assembled, then testing observed combinations to see if they match those predicted by the algorithms. He conducts this exercise for foliage-gleaning birds of New England islands (Morse 1977) and avoids the problem of testing the model against the very data from which it was constructed by using 4 years' data to construct the model and the next 4 years' data to test it. However, Haefner's algorithms are phenomenological rather than deductive. For example, several algorithms delete species from habitats if they are similar in size to species already there, and these algorithms are simply restatements of observed patterns in the first run of data. Consequently, as Haefner concedes, the entire exercise is correlative and cannot attempt to determine causal relationships; it is an elaborate narrative description. The test is a test only of whether the second run of data on the same community shows the same species combinations present and absent as the first run shows. There is no test of a hypothesis.

A new method of analyzing biogeographic species × site matrixes has been proposed by Whittan and Siegel-Causey (1981), who apply multiway contingency tables and log–linear models (Fienberg 1970) to distributional patterns of Alaskan seabirds in a search for species interactions. The simple log–linear model is $\ln e_{ijklm} = u + u_A + u_B + u_C + u_D + u_E$, where e_{ijklm} is the expected number of colonies with a particular combination of species and i, j, k, l, and m are 1 or 0 depending on whether species A, B, C, D, or E, respectively, are present or absent. The overall mean effect (average of logarithms of the expectations of all combinations) is u, whereas the main effect of each species t (u_t, where $t = A$ through E) is the mean over all combinations including species t minus u.

The method is of course generalizable to any number of species, not just five. If the species are mutually independent in their biogeographic patterns, the e_{ijklm} can be calculated from the totals of each species, just as one calculates expectations in a two-way contingency table. If these expectations do not match observations well, additional interaction terms may be added. There are 10 two-way interaction terms possible, 10 three-way terms, and so forth. One adds as many of these as necessary to get a good fit, but what constitutes "good" and the order in which interaction terms are added are arbitrary.

Whittam and Siegal-Causey (1981) constructed four groups of five species such that the species in each group had similar nest site preferences. For these four groups of 40 two-way interactions, 24 were used to achieve a good fit: 22 positive interactions and 2 negative ones. The latter are the only ones consistent with a competition hypothesis. Of 40 three-way interactions, 2 were added. In both, a negative interaction between two species is exacerbated in the presence of a third species, and Whittam and Siegel-Causey view these as consistent with a hypothesis of diffuse competition (MacArthur 1972). Finally, in order to estimate what fraction of the added interactions might be statistical artifacts, a "dummy" group was constructed of five birds, no two of which have similar nesting requirements. In the log–linear model of this group are seven two-way interactions of which six are positive. So it is not yet clear that there is more evidence of competitive structure in the groups chosen to maximize competition than in a group chosen to minimize it.

Prospect

The various null models I have discussed for vicariance, equilibrium theory, and the biogeography of community composition were all inspired by Popper's (1959, 1963) conclusion that scientific advancement is best achieved by posing falsifiable hypotheses and then attempting to falsify them, rather than by seeking confirmatory evidence for a theory. I do not intend that the only valid biogeographic statement is a falsifiable hypothesis, even if I would agree with Ball (1975) that this approach should dominate a mature biogeography. For one thing, many interesting biogeographic patterns may be sufficiently localized that they are singular phenomena, and a statement about their causes cannot be tested by data from another site nor, of course, are many of the patterns of historical biogeography and some of those ecological biogeography replicable in time or space. It may be that a narrative explanation is the best that we can aspire to in such a situation, though Popper (1959, 1980) is explicit that singular and historical events are not precluded from explanation by falsifiable hypotheses simply by virtue of their singularity and past occurrence. I am impressed by McDowall's (1978) argument that much evidence exists implicating dispersal as an important force, even though we may never be able to frame a falsifiable dispersal hypothesis. If this argument is true, we should be appropriately modest about the generality of biogeographic statements and the rigor with which they are derived relative to those of the experimental physical sciences. We must admit we are propounding local narratives and not general theories.

Dispersal hypotheses are not unfalsifiable by virtue of the impossibility of conceiving of an observation that they cannot explain. For example, the data that Patterson (1981) marshals in his discussion of marsupial biogeography could be explained by dispersal only at the cost of so many ad hoc subsidiary hypotheses that the overall explanation would be extremely suspect. Rather, the difficulty in generating a falsifiable causal hypothesis with dispersal having a central role is that one cannot yet attach any statistical significance level to an observation evaluated in the light of such a hypothesis. On the other hand, it is not clear (see earlier) that any observation could ever be taken by certain observers to falsify the equilibrium theory, statistics or no. In my opinion, this is a far more serious defect. It turns a theory into a construct that, like a religion (Lewontin 1972), is not susceptible to any sort of test whatever and, therefore, renders a search for evidence irrelevant since it must always succeed. Diffuse competition (MacArthur 1972) has almost achieved this status. Abbott (1980) has noted that it is widely advocated in the avian ecology literature in spite of very weak evidence, and Abbott and Grant (1976) point out that the hypothesis of diffuse competition is usually invoked as an ad hoc response to failure to demonstrate competitive exclusion and is virtually unfalsifiable with available data.

It is interesting that in the wake of criticism based on null hypotheses, at least some biogeographers have suggested that falsifying null hypotheses is perhaps not an appropriate modus operandi for biogeography. Farris (1981), for example, suggests that perhaps estimation is a more appropriate statistical approach to the vicariance-versus-dispersal debate, whereas Gilpin and Diamond (1982) view Popper as jejune and null models as inappropriate for island biogeographic data. To me, this view seems too pessimistic, and the value of null models in ecology seems so well established (MacFadyen 1975; Strong 1980) that much progress will likely derive from a search for data to test further null models and more powerful and/or elegant methods, especially statistical ones, for assessing them. Whittam and Siegel-Causey's (1981) effort seems a good example of such statistical progress, and the agreement of all concerned parties that species/genus and related ratios can tell us nothing about interspecific interactions is an example of how tests of null hypotheses and their refinement generated a broad consensus where there had been decades of strife. Endler's (1982) test of the vicariance paradigm appears also to be a promising step.

Many null models of the sort I have discussed have been assailed (e.g., Grant and Abbott 1980; Farris 1981) as lacking realism and likely having high type II error rates. The latter charge may be a fair one, and I can only justify the approach by saying that a null hypothesis with low power tests is better than none and perhaps a springboard for better null hypotheses. One must constantly emphasize that failure to reject a null hypothesis does not mean its acceptance. Rather, it means that the available data do not allow us to reject it in favor of the alternative. In general, in ecological biogeography, the null hypothesis approach has aimed to show that pattern cannot tell us much about causal mecahnisms (Pielou and Pielou 1968; Simberloff, in press-a) and that if possible an experimental approach is required. Abbott (1980) and Grant and Abbott (1980) have similarly pleaded that avian ecological biogeography requires a far greater experimental effort, and Abbott (1980; cf. Connell 1975) rejects the "natural experiment" (e.g., Diamond 1978) as far too uncontrolled and to be replaced by manipulation if at all possible. In historical biogeography, Eldredge (1981) has similarly distinguished between pattern and process and emphasized the need, if pattern is to be used to infer causal mechanisms, to frame the proposed mechanism as a falsifiable hypothesis. However, there seems little likelihood of making historical biogeography an experimental discipline.

Increasing realism of hypotheses, however, is a mixed blessing (Simberloff, in press-a). Unrealistic hypotheses have been as useful as realistic ones in ecology, simply by virtue of generating falsifiable hypotheses and aiding in the perception of discrepancies (Pielou 1981b). Adding parameters or constraints to a model to make it more realistic eventually becomes counterproductive for two reasons. First, any model is an abstraction and is not faithful to nature in many particulars. Operationally, this abstractness need not be debilitating, so long as those aspects of the model that are unrealistic do not affect the model's performance with respect to the issues of interest (Simberloff, in press-c).

Second, the more realistically a model matches a certain system, the less likely it is to match any other system, and eventually it becomes untestable, except perhaps by more reductionist methods. It is no longer a generalized abstraction of elements common to many systems but just an elaborate narrative. If such a narrative spawns research at a more fundamental level it can contribute substantially to scientific progress. If instead it is enshrined as canon, it usually ceases to be useful. James (1970) makes this point particularly well by observing that the frequent adherence of geographic size variation in birds to Bergmann's Rule is well demonstrated as a generality, but the biological bases for this pattern are unknown. What is required now, I feel, are falsifiable hypotheses of exactly why and to what degree wing length differences of 0.5 to 4.0 mm, out of 100 m, are associated with fitness differences. The sorts of research that such hypotheses will require will, I predict, be quite different from those practised by most ecologists, evolutionists, and biogeographers today. Similarly, the species–area relationship is a well-established generality and several possible causes, foremost among them the effects of habitat diversity, have been proposed (Connor and McCoy 1979). Consequently, it is no longer a major contribution to show that a particular avifauna adheres more or less closely to a species–area curve with specified parameters. What is needed is evidence on the mechanism by which the relationship is achieved. If it is habitat diversity that is responsible, what exactly is it about a particular habitat or suite of them that causes a given bird species to be present? Even establishing a correlation between a species and a habitat is not a sufficient reduction. We must try to learn why the correlation obtains.

Literature cited

Abbott, I. 1975. Coexistence of congeneric species in the avifauna of Australian islands. *Aust. J. Zool.* 23:487–494.
– 1977. Species richness, turnover, and equilibrium in insular floras near Perth, Western Australia. *Aust. J. Bot.* 25:193–208.
– 1980. Theories dealing with the ecology of landbirds on islands. *Adv. Ecol. Res.* 11:329–371.
– 1981. The composition of landbird faunas of islands round south-western Australia: is there evidence for competitive exclusion? *J. Biogeogr.* 8:135–144.
Abbott, I., L. K. Abbott, and P. R. Grant. 1977. Comparative ecology of

Galápagos ground finches (*Geospiza* Gould): evaluation of the importance of floristic diversity and interspecific competition. *Ecol. Monogr. 47:*151–184.

Abbott, I., and R. Black. 1980. Changes in species composition of floras on islets near Perth, Western Australia. *J. Biogeogr. 7:*399–410.

Abbott, I., and P. R. Grant. 1976. Non-equilibrium bird faunas on islands. *Am. Nat. 110:*507–528.

Abele, L. G., and E. F. Connor. 1979. Application of island biogeography theory to refuge design: making the right decision for the wrong reasons. In R. M. Linn (ed.). *Proceedings of the First Conference on Scientific Research in the National Parks,* Vol. 1, pp. 89–94. Washington, D.C.: Department of the Interior.

Allen, J. A. 1871. Mammals and winter birds of east Florida, and a sketch of the bird faunae of eastern North America. *Bull. Mus. Comp. Zool. 2:*161–450.

Ashlock, P. D. 1974. The uses of cladistics. *Annu. Rev. Ecol. Syst. 5:*81–99.

Ball, I. R. 1975. Nature and formulation of biogeographic hypotheses. *Syst. Zool. 24:*407–430.

Bergmann, C. 1847. Ueber die Verhältnisse der Wärmeökonomie zu ihrer Grösse. *Gottinger Studien,* Vol. 3, Pt. 1, pp. 595–708.

Bond, J. 1948. Origin of the bird fauna of the West Indies. *Wilson Bull. 60:*207–229.

Bradley, J. V. 1968. *Distribution-free statistical tests.* Englewood Cliffs, N.J.: Prentice-Hall.

Brown, J. H., and A. Kodric-Brown. 1977. Turnover rates in insular biogeography: effect of immigration on extinction. *Ecology 58:*445–449.

Brown, W. L., and E. O. Wilson. 1956. Character displacement. *Syst. Zool. 5:*49–64.

Brundin, L. 1966. Transantarctic relationships and their significance. *K. Sven. Vetenskapsakad. Hand.* Ser. 4 *11:*1–472.

Butcher, G. S., W. A. Niering, W. J. Barry, and R. H. Goodwin. 1981. Equilibrium biogeography and the size of nature preserves: an avian case study. *Oecologia 49:*29–37.

Camp, W. H. 1948. *Rhipsalis* – and plant distributions in the Southern Hemisphere. *J. N.Y. Bot. Gar. 49:*88–91.

Carlquist, S. 1974. *Island biology.* New York: Columbia University Press.

Chesser, R. K., M. H. Smith, and I. L. Brisbin, Jr. 1980. Management and maintenance of genetic variability in endangered species. *Int. Zoo Yearb. 20:*146–154.

Christiansen, F. B. 1974. Sufficient conditions for protected polymorphism in a subdivided population. *Am. Nat. 108:*157–166.

– 1975. Hard and soft selection in a subdivided population. *Am. Nat. 109:*11–16.

Cody, M. L. 1974. *Competition and the structure of bird communities.* Princeton, N.J.: Princeton University Press.

Cole, B. J. 1981. Colonizing abilities, island size, and the number of species on archipelagoes. *Am. Nat. 117:*629–638.

Connell, J. H. 1975. Some mechanisms producing structure in natural communities: a model and evidence from field experiments. In M. L. Cody and J. M. Diamond (eds.). *Ecology and evolution of communities,* pp. 460–490. Cambridge, Mass.: Harvard University Press.

Connor, E. F., and E. D. McCoy. 1979. The statistics and biology of the species-area relationship. *Am. Nat. 113*:791–833.

Connor, E. F., and D. Simberloff. 1979. The assembly of species communities: chance or competition? *Ecology 60*:1132–1140.

Cook, R. E. 1969. Variation in species density of North American birds. *Syst. Zool. 18*:63–84.

Crowell, K. L. 1973. Experimental zoogeography: introductions of mice to small islands. *Am. Nat. 107*:535–558.

Cracraft, J. 1973. Continental drift, paleoclimatology, and the evolution and biogeography of birds. *J. Zool. 169*:455–545.

– 1980. Avian phylogeny and intercontinental biogeographic patterns. In R. Nöhring (ed.). *Acta XVII Congressus Internationalis Ornithologici*, pp. 1302–1308. Berlin: Verlag der Deutschen Ornithologen-Gesellschaft.

Croizat, L. 1964. *Space, time, form: the biological synthesis.* Caracas: Author.

– 1981. Biogeography: past, present, future. In G. Nelson and D. E. Rosen (eds.). *Vicariance biogeography: a critique*, pp. 510–523. New York: Columbia University Press.

Croizat, L., G. Nelson, and D. E. Rosen. 1974. Centers of origin and related concepts. *Syst. Zool. 23*:265–287.

Darlington, P. J., Jr. 1957. *Zoogeography: the geographical distribution of animals.* New York: Wiley.

– 1965. *Biogeography of the southern end of the world: distribution and history of far-southern life and land, with an assessment of continental drift.* Cambridge, Mass.: Harvard University Press.

Darwin, C. 1845. *Journal of researches into the natural history and geology of the various countries visited during the voyage of H. M. S. Beagle bound round the world.* London: Murray.

– 1896. *The origin of species by means of natural selection or the preservation of favoured races in the struggle for life*, 6th ed. New York: Appleton.

de Candolle, A. P. 1820. *Essai Élémentaire de Géographie Botanique.* Strasbourg.

Diamond, J. M. 1969. Avifaunal equilibrium and species turnover rates on the Channel Islands of California. *Proc. Natl. Acad. Sci. USA 64*:57–63.

– 1971. Comparison of faunal equilibrium turnover rates on a tropical and a temperate island. *Proc. Natl. Acad. Sci USA 68*:2742–2745.

– 1972. Biogeographic kinetics: estimation of relaxation times for avifaunas of southwest Pacific islands. *Proc. Natl. Acad. Sci. USA 69*:3199–3203.

– 1975a. The island dilemma: lessons of modern biogeographic studies for the design of natural reserves. *Biol. Conserv. 7*:129–146.

– 1975b. Assembly of species communities. In M. L. Cody and J. M. Diamond (eds.). *Ecology and evolution of communities*, pp. 342–444. Cambridge, Mass.: Harvard University Press.

– 1976. Island biogeography and conservation: strategy and limitations. *Science 193*:1027–1029.

– 1978. Niche shifts and the rediscovery of interspecific competition. *Am. Sci. 66*:322–331.

Diamond, J. M., and M. E. Gilpin. 1982. Examination of the "null" model of Connor and Simberloff for species co-occurrences on islands. *Oecologia 52*:64–74.

Diamond, J. M., and R. M. May. 1976. Island biogeography and the design of natural reserves. In R. M. May (ed.). *Theoretical ecology*, pp. 163–186. Philadelphia: Saunders.

– 1977. Species turnover rates on islands: dependence on census interval. *Science 197:*266–270.

Drury, W. H. 1974. Rare species. *Biol. Conserv. 6:*162–169.

Edmunds, G. F., Jr. 1972. Biogeography and evolution of Ephemeroptera. *Annu. Rev. Entomol. 17:*21–42.

Eldredge, N. 1981. Discussion (of paper by Udvardy). In G. Nelson and D. E. Rosen (eds.). *Vicariance biogeography: a critique*, pp. 34–38. New York: Columbia University Press.

Elton, C. 1946. Competition and the structure of ecological communities. *J. Anim. Ecol. 15:*54–68.

Endler, J. A. 1977. *Geographic variation, speciation, and clines.* Princeton, N.J.: Princeton University Press.

– 1981. Pleistocene forest refuges: fact or fancy? In G. Prance (ed.). *Biological diversification in the tropics*, pp. 641–657. New York: Columbia University Press.

– 1982. Problems in distinguishing historical from ecological factors in biogeography. *Am. Zool. 22:*441–452.

Faaborg, J. 1979. Qualitative patterns of avian extinction on Neotropical landbridge islands: lessons for conservation. *J. Appl. Ecol. 16:*99–107.

Farris, J. S. 1981. Discussion (of paper by Simberloff et al.). In G. Nelson and D. E. Rosen (eds.). *Vicariance biogeography: a critique*, pp. 73–84. New York: Columbia University Press.

Feller, W. 1950. *An introduction to probability theory and its applications*, Vol. 1. New York: Wiley.

Fienberg, S. E. 1970. The analysis of multiway frequency tables. *Ecology 51:*419–433.

Forman, R. T. T., A. E. Galli, and C. F. Leck. 1976. Forest size and avian diversity in New Jersey woodlots with some land use implications. *Oecologia 26:*1–8.

Galli, A. E., C. F. Leck, and R. T. T. Forman. 1976. Avian distribution patterns in forest islands of different sizes in central New Jersey. *Auk 93:*356–364.

Game, M. Best shape for nature reserves. *Nature (London) 287:*630–632.

Ghiselin M. T. 1969. *The triumph of the Darwinian method.* Berkeley: University of California Press.

Gilbert, F. S. 1980. The equilibrium theory of island biogeography: fact or fiction? *J. Biogeo. 7:*209–235.

Gilpin, M. E., and J. M. Diamond. 1976. Calculation of immigration and extinction curves from the species-area-distance relation. *Proc. Natl. Acad. Sci. USA 73:*4130–4134.

– 1980. Subdivision of nature reserves and the maintenance of species diversity. *Nature (London) 285:*567–568.

– 1981. Immigration and extinction probabilities for individual species: relation to incidence functions and species colonization curves. *Proc. Natl. Acad. Sci. USA 78:*393–396.

– 1982. Factor contributing to nonrandomness in species co-occurrences on islands. *Oecologia 52:*75–84.

Gould, S. J., D. M. Raup, J. J. Sepkoski, Jr., T. J. M. Schopf, and D. Simberloff. 1977. The shape of evolution: a comparison of real and random clades. *Paleobiology 3:*23–40.

Grant, P. R. 1966. Ecological compatibility of bird species on islands. *Am. Nat. 100:*451–462.

– 1968. Bill size, body size and the ecological adaptations of bird species to competitive situations on islands. *Syst. Zool. 17:*319–333.

– 1972. Convergent and divergent character displacement. *Biol. J. Linn. Soc. 4:*39–68.

– 1975. The classical case of character displacement. *Evol. Biol. 8:*237–337.

– 1977. Review of Lack's *Island Biology. Bird-Banding 48:*296–300.

– in press. The role of interspecific competition in the adaptive radiation of Darwin's Finches. In R. I. Bowman and A. E. Leviton (eds.). *Patterns of evolution in Galapagos organisms.* San Francisco: American Association for the Advancement of Science.

Grant, P. R., and I. Abbott. 1980. Interspecific competition, island biogeography and null hypotheses. *Evolution 34:*332–341.

Grant, P. R., and I. McT. Cowan. 1964. A review of the avifauna of the Tres Marias Islands, Nayarit, Mexico. *Condor 66:*221–228.

Grisebach, A. H. R. 1884. *Die Vegetation der Erde nach ihrer klimatischen Anordnung,* 2nd ed. Leipzig: Engelmann.

Haefner, J. W. 1981. Avian community assembly rules: the foliage-gleaning guild. *Oecologia 50:*131–142.

Haffer, J. 1970. Geological-climatic history and zoogeographic significance of the Urabá region in northwestern Colombia. *Caldasia 10:*603–636.

– 1974. *Avian speciation in tropical South America, with a systematic survey of the toucans (Ramphastidae) and jacamars (Galbulidae).* Cambridge, Mass.: Nuttall Ornithological Club.

– 1981. Aspects of Neotropical bird speciation during the Cenozoic. In G. Nelson and D. E. Rosen (eds.). *Vicariance biogeography: a critique,* pp. 370–394. New York: Columbia University Press.

Haila, Y., and O. Järvinen. 1981. The underexploited potential of quantitative bird censuses in insular ecology. *Stud. Avian Biol.,* No. 6:559–565.

Haila, Y., O. Järvinen, and R. A. Väisänen. 1979. Effect of mainland population changes on the terrestrial bird fauna of a northern island. *Ornis Scand. 10:*48–55.

Hallam, A. 1981. Relative importance of plate movements, eustasy, and climate in controlling major biogeographical changes since the early Mesozoic. In G. Nelson and D. E. Rosen (eds.). *Vicariance biogeography: a critique,* pp. 303–330. New York: Columbia University Press.

Hamilton, T. H., R. H. Barth, Jr., and I. Rubinoff. 1964. The environmental control of insular variation in bird species abundance. *Proc. Natl. Acad. Sci. USA 52:*132–140.

Hendrickson, J. A., Jr. 1981. Community-wide character displacement reexamined. *Evolution 35:*794–809.

Hennig, W. 1966. *Phylogenetic systematics*. Urbana: University of Illinois Press.

Higgs, A. J. 1981. Island biogeography theory and nature reserve design. *J. Biogeogr. 8:*117–124.

Higuchi, H. 1980. Colonization and coexistence of woodpeckers in the Japanese Islands. *Misc. Rep. Yamashina Inst. Ornithol. 12:*139–156.

Hutchinson, G. E. 1959. Homage to Santa Rosalia, or why are there so many kinds of animals? *Am. Nat. 93:*145–159.

James, F. C. 1970. Geographic size variation in birds and its relationship to climate. *Ecology 51:*365–390.

James, F. C., and W. J. Boecklen. in press. Interspecific morphological relationships and the densities of birds.

Jones, H. L., and J. M. Diamond. 1976. Short-time-base studies of turnover in breeding bird populations on the California Channel Islands. *Condor 78:*526–549.

Juvik, J. O., and A. P. Austring. 1979. The Hawaiian avifauna: biogeographic theory in evolutionary time. *J. Biogeogr. 6:*205–224.

Kitchener, D. J., A. Chapman, J. Dell, B. G. Muir, and M. Palmer. 1980. Lizard assemblage and reserve size and structure in the Western Australian wheatbelt – some implications for conservation. *Biol. Conserv. 17:*25–62.

Krebs, C. J. 1978. *The experimental analysis of distribution and abundance*, 2nd ed. New York: Harper & Row.

Lack, D. 1969. The numbers of bird species on islands. *Bird Study 16:*193–209.

– 1976. *Island biology illustrated by the land birds of Jamaica*. Oxford: Blackwell Scientific Publications.

Lewontin, R. C. 1972. Testing the theory of natural selection. *Nature (London) 236:*181–182.

Lovejoy, T. E. 1980. Discontinuous wilderness: minimum areas for conservation. *Parks 5:*13–15.

Lynch, J. F. 1979. The amphibians of the lowland tropical forests. In W. E. Duellman (ed.). *The South American Herpetofauna: its origin, evolution and dispersal*, pp. 189–215. Lawrence: Museum of Natural History, University of Kansas.

Lynch, J. F., and N. K. Johnson. 1974. Turnover and equilibria in insular avifaunas, with special reference to the California Channel Islands. *Condor 76:*370–384.

MacArthur, R. H. 1972. *Geographical ecology*. New York: Harper & Row.

MacArthur, R. H., and E. O. Wilson. 1963. An equilibrium theory of insular zoogeography. *Evolution 17:*373–387.

– 1967. *The theory of island biogeography*. Princeton, N.J.: Princeton University Press.

McDowall, R. M. 1978. Generalized tracks and dispersal in biogeography. *Syst. Zool. 27:*88–105.

MacFadyen, A. 1975. Some thoughts on the behaviour of ecologists. *J. Anim. Ecol. 44:*351–363.

Matthew, W. P. 1915. Climate and evolution. *Ann. N.Y. Acad. Sci. 24:*171–318.

Moore, N.W., and M. D. Hooper. 1975. On the number of bird species in British woods. *Biol Conserv. 8:*239–250.

Moreau, R. E. 1948. Ecological isolation in a rich tropical avi-fauna. *J. Anim. Ecol. 17*:113–126.

– 1963. Vicissitudes of the African biomes in the Late Pleistocene. *Proc. Zool Soc. London 141*:395–421.

– 1966. *The bird faunas of Africa and its islands.* New York: Academic Press.

Morse, D. H. 1977. The occupation of small islands by passerine birds. *Condor 79*:399–412.

Mosimann, J. E., and F. C. James. 1979. New statistical methods for allometry with application to Florida Red-winged Blackbirds. *Evolution 33*:444–459.

Nelson, G. 1975. Historical biogeography: an alternative formalization. *Syst. Zool. 23*:555–558.

– 1978. From Candolle to Croizat: comments on the history of biogeography. *J. Hist. Biol. 11*:269–305.

Nilsson, S. G. 1978. Fragmented habitats, species richness and conservation practice. *Ambio 7*:26–27.

Oxnard, C. E. 1978. One biologist's view of morphometrics. *Annu. Rev. Ecol. Syst. 9*:219–242.

Patterson, C. 1981. Methods of paleobiogeography. In G. Nelson and D. E. Rosen (eds.). *Vicariance biogeography: a critique,* pp. 446–489. New York: Columbia University Press.

Pielou, D. P., and E. C. Pielou. 1968. Association among species of infrequent occurrence: the insect and spider fauna of *Polyporus betulinus* (Bulliard) Fries. *J. Theor. Biol. 21*:202–216.

Pielou, E. C. 1979. *Biogeography.* New York: Wiley.

– 1981a. Crosscurrents in biogeography. *Science 213*:324–325.

– 1981b. The usefulness of ecological models: a stocktaking. *Q. Rev. Biol. 56*:17–31.

Platnick, N. I., and G. Nelson. 1978. A method of analysis for historical biogeography. *Syst. Zool. 27*:1–16.

Popper, K. 1980. Letter to editor. *New Sci. 87*:611.

Popper, K. R. 1959. *The logic of scientific discovery.* London: Hutchinson.

– 1963. *Conjectures and refutations: the growth of scientific knowledge.* New York: Harper & Row.

Power, D. M. 1972. Numbers of bird species on the California Islands. *Evolution 26*:451–463.

Preston, F. W. 1962. The canonical distribution of commonness and rarity. *Ecology 43*:185–215, 410–432.

Raup, D. M. 1972. Approaches to morphologic analysis. In T. J. M. Schopf (ed.). *Models in paleobiology,* pp. 28–45. San Francisco: Freeman.

Raup, D. M., S. J. Gould, T. J. M. Schopf, and D. S. Simberloff. 1973. Stochastic models of phylogeny and the evolution of diversity. *J. Geol. 81*:525–542.

Reed, T. 1978. Immigrations and extinctions: a Hebridean context. *Hebr. Nat. 1*:11–21.

Reed, T. M. 1980. Turnover frequency in island birds. *J. Biogeogr. 7*:329–336.

Rensch, B. 1939. Klimatische Auslese von Grössenvarianten. *Arch. Naturgesch. 8*:89–129.

Ricklefs, R. E. 1977. Review of Lack's *Island Biology. Auk 94*:794–797.

Ricklefs, R. E., and J. Travis. 1980. A morphological approach to the study of avian community organization. *Auk 97:*321–338.

Rosen, D. E. 1974a. The phylogeny and zoogeography of salmoniform fishes, and the relationships of *Lepidogalaxias salamandroides. Bull. Amer. Mus. Nat. Hist. 153:*265–326.

– 1974b. Space, time, form: the biological synthesis (review of Croizat 1964). *Syst. Zool. 23:*288–290.

– 1978. Vicariant patterns and historical explanation in biogeography. *Syst. Zool. 27:*159–188.

Rosenzweig, M. L. 1975. On continental steady states of species diversity. In M. L. Cody and J. M. Diamond (eds.). *Ecology and evolution of communities,* pp. 121–140. Cambridge, Mass.: Harvard University Press.

Roth, V. L. 1981. Constancy in the size ratios of sympatric species. *Am. Nat. 118:*394–404.

Salomonsen, F. 1976. The main problems concerning avian evolution on islands. In *Proceedings of the 16th International Ornithological Congress,* pp. 585–602.

Sampson, F. B. 1980. Island biogeography and the conservation of nongame birds. In *Transactions of the 45th North American Wildlife and National Resources Conference,* pp. 245–251.

Savage, H. M. in press. The shape of evolution: systematic tree topology. *Biol. J. Linn. Soc.*

Schimper, A. F. W. 1898. *Pflanzen-Geographie auf physiologische Grundlage.* Jena: Fischer.

Schoener, T. W. 1965. The evolution of bill size differences among sympatric species of birds. *Evolution 19:*189–213.

Sclater, P. L. 1858. On the general geographical distribution of the members of the Class Aves. *J. Proc. Linn. Soc. London (Zool.) 2:*130–145.

Seib, R. L. 1980. Baja California: a peninsula for rodents but not for reptiles. *Am. Nat. 115:*613–620.

Seilacher, A. 1970. Arbeitskonzept zur Konstruktions-Morphologie. *Lethaia 3:*393–396.

Selvin, H. C., and A. Stuart. 1966. Data-dredging procedures in survey analysis. *Am. Stat. 20:*20–23.

Siegfried, W. R. 1976. Segregation in feeding behavior of four diving ducks in southern Manitoba. *Can. J. Zool. 54:*730–736.

Simberloff, D. 1969. Experimental zoogeography of islands: a model for insular colonization. *Ecology 50:*296–314.

– 1970. Taxonomic diversity of island biotas. *Evolution 24:*22–47.

– 1974. Equilibrium theory of island biogeography and ecology. *Annu. Rev. Ecol. Syst. 5:*161–182.

– 1976a. Species turnover and equilibrium island biogeography. *Science 194:*572–578.

– 1976b. Experimental zoogeography of islands: effects of island size. *Ecology 57:*629–648.

– 1978a. Colonisation of islands by insects: immigration, extinction, and diversity. In L. A. Mound and N. Waloff (eds.). *Diversity of insect fauna,* pp. 139–153. Oxford: Blackwell Scientific Publications.

– 1978b. Islands and their species. *Nat. Conserv. News 28:*4–10.

– 1980. Dynamic equilibrium island biogeography: the second stage. In R. Nöhring (ed.). *Acta XVII Congressus Internationalis Ornithologici*, pp. 1289–1295. Berlin: Verlag der Deutschen Ornithologen-Gesellschaft.

– 1981. What makes a good island colonist? In R. F. Denno and H. Dingle (eds.). *Insect life history patterns: habitat and geographic variation*, pp. 195–206. New York: Springer.

– 1982. Ecological generalizations and wildlife refuge design. *Nat. Hist.* 91(4):6–14.

– in press-a. Island biogeographic theory and the design of wildlife refuges. *Ékologiya*.

– in press-b. Biogeographic models, species' distributions, and community organization.

– in press-c. Properties of coexisting bird species in two archipelagoes.

Simberloff, D. S., and L. G. Abele. 1976a. Island biogeography theory and conservation practice. *Science 191:*285–286.

– 1976b. Island biogeography and conservation: strategy and limitations. *Science 193:*1032.

– 1982. Refuge design and island biogeographic theory: effects of fragmentation. *Am. Nat. 120:*41–50.

Simberloff, D., and W. Boecklen. 1981. Santa Rosalia reconsidered: size ratios and competition. *Evolution 35:*1206–1228.

Simberloff, D., and E. F. Connor. 1981. Missing species combinations. *Am. Nat. 118:*215–239.

Simberloff, D., K. L. Heck, E. D. McCoy, and E. F. Connor. 1981. There have been no statistical tests of cladistic biogeographical hypotheses. In G. Nelson and D. E. Rosen (eds.). *Vicariance biogeography: a critique*, pp. 40–63. New York: Columbia University Press.

Simpson, B. B., and J. Haffer. 1978. Speciation patterns in Amazonian forest biota. *Annu. Rev. Ecol. Syst. 9:*497–518.

Simpson, G. G. 1952. Probabilities of dispersal in geologic time. *Bull. Am. Mus. Nat. Hist. 99:*163–176.

– 1953. *Evolution and geography*. Eugene: Oregon State System of Higher Education.

– 1964. Species diversity of North American Recent mammals. *Syst. Zool. 13:*57–73.

– 1965. *The geography of evolution*. Philadelphia: Chilton Books.

Slud, P. 1976. Geographic and climatic relationships of avifaunas with special reference to the comparative distribution in the Neotropics. *Smithson. Contrib. Zool. 212:*1–149.

Smith, F. E. 1975. Ecosystems and evolution. *Bull. Ecol. Soc. Am. 56:*2–6.

Strong, D. R. 1980. Null hypotheses in ecology. *Synthèse 43:*271–286.

Strong, D. R., and D. Simberloff. 1981. Straining at gnats and swallowing ratios: character displacement. *Evolution 35:*810–812.

Strong, D. R., L. A. Szyska, and D. Simberloff. 1979. Tests of community-wide character displacement against null hypotheses. *Evolution 33:*897–913.

Swed, F. S., and C. Eisenhart. 1943. Tables for testing randomness of grouping in a sequence of alternatives. *Ann. Math. Stat. 14:*66–87.

Taylor, R. J., and P. J. Regal. 1978. The peninsular effect on species diversity and the biogeography of Baja California. *Am. Nat. 112:*583–593.

Terborgh, J. 1974. Preservation of natural diversity: the problem of extinction-prone species. *BioScience 24:*715–722.

– 1975. Faunal equilibria and the design of wildlife preserves. In F. Golley and E. Medina (eds.). *Tropical ecological systems: trends in terrestrial and aquatic research*, pp. 369–380. New York: Springer-Verlag.

– 1976: Island biogeography and conservation: strategy and limitations. *Science 193:*1029–1030.

Terborgh, J., and F. Faaborg. 1973. Turnover and ecological release in the avifauna of Mona Island, Puerto Rico. *Auk 90:*759–779.

Underwood, A. J. 1978. The detection of non-random patterns of distribution of species along a gradient. *Oecologi 36:*317–326.

Vaurie, C. 1951. Adaptive differences between two sympatric species of nuthatches (*Sitta*). *Proceedings of the Tenth International Ornithological Congress*, pp. 163–166.

Vuilleumier, F. 1969. Pleistocene speciation in birds living in the high tropical Andes. *Nature (London) 223:*1179–1180.

Vuilleumier, F., and D. Simberloff. 1980. Ecology versus history as determinants of patchy and insular distributions in high Andean birds. *Evol. Biol. 12:*235–379.

Wallace, A. R., 1869. *The Malay archipelago.* London: Macmillan.

Wamer, N. O. 1978. Avian diversity and habitat in Florida: an analysis of peninsular diversity gradient. M.S. thesis, Florida State University, Tallahassee.

Watson, G. 1964. Ecology and evolution of passerine birds on the islands of the Aegean Sea. Ph.D. thesis, Yale University.

Webb, S. D. 1976. Mammalian faunal dynamics of the great American interchange. *Paleobiology 2:*220–234.

Whitcomb, R. F., J. F. Lynch, P. A. Opler, and C. S. Robbins. 1976. Island biogeography and conservation: strategy and limitations. *Science 193:*1030–1032.

Whittam, T. S., and D. Siegel-Causey. 1981. Species interactions and community structure in Alaskan seabird colonies. *Ecology 62:*1515–1524.

Williams, C. B. 1951. Intra-generic competition as illustrated by Moreau's records of East African birds. *J. Anim. Ecol. 20:*246–253.

Williamson, M. 1975. The design of wildlife preserves. *Nature (London) 256:*519.

– 1981. *Island populations.* Oxford: Oxford University Press.

Wilson, E. O., and E. O. Willis. 1975. Applied biogeography. In M. L. Cody and J. M. Diamond (eds.). *Ecology and evolution of communities*, pp. 522–534. Cambridge, Mass.: Harvard University Press.

Commentary

JOEL CRACRAFT

Biogeography is, next to systematics, the most fundamental discipline underlying evolutionary biology. Systematics is primary because biogeography cannot be undertaken without it. Given the present emphasis on the multifarious field of population biology, the assertion that systematics and biogeography are the prime components of evolutionary biology may sound anachronistic. The statement is not meant to polarize or challenge population biology but merely to reaffirm the indispensability of systematics and biogeography and to suggest that their current role in evolutionary biology must be enlarged. The reason for this is simple. The goals of systematic and biogeographic analysis are, first, to define the units of evolution – any monophyletic group from the subspecific level on up – and, second, to describe patterns of form (including phenotype and genotype) through space and time. In other words, without systematic and biogeographic analysis, evolutionary biology would have little, if anything, to explain. The focus of evolutionary biology is ultimately the description and explanation of diversity, which can take two forms: patterns of taxonomic diversity and patterns of diversity in form. Systematic and biogeographic analysis reveals those patterns and thereby lays the groundwork for their explanation.

In its broadest sense, biogeography is the study of the distributions of organisms, but such a simple definition is not really informative. Biogeography today is usually seen from two perspectives. The first, *historical biogeography*, encompasses much of the subject matter classically included in the discipline for the last 150 years. The second perspective, usually called *island biogeography* (or *equilibrium biogeography*) has a much more recent inception (MacArthur and Wilson 1963, 1967).

It is instructive to consider the relationship between these two perspectives of biogeography. Island biogeographic theory is often viewed – almost in a paradigmatic sense – as having transformed classical biogeography into a more modern, scientific discipline:

[Island biogeography] represents an effort to take biogeography out of its present natural-history setting and place it in a new one, more conducive to experimentation and more amenable to theoretical analysis. (Hamilton 1968:71)

[Island biogeographic theory has] revolutionized biogeography . . . spawned a mass of research which has given biogeography general laws of both didactic and predictive power. (Simberloff 1974:163)

Is this an accurate evalution? Has island biogeographic theory, in fact, transformed classical biogeography? The answer probably depends upon one's point of view. Nevertheless, although granting the significance and widespread influence of island biogeographic theory within ecology, it is also possible to argue that the approach has had little, if any, influence on the traditional domains of biogeographic thinking. Fundamental conceptual advances within a discipline usually have two major effects. They provide a theoretical basis for solving diverse problems already indigenous to the field, but for which there is as yet no unifying solution. Also, they permit the older data to be seen in a different light by using them to answer entirely new kinds of questions. Although island biogeographic theory may satisfy these criteria within ecology, it has not done so within biogeography.

Some workers have suggested that biogeography is becoming unified as a discipline, with historical and island biogeographic analyses occupying the ends of a continuous methodological spectrum, distinguished only by differences in the time and geographic scales under investigation (Simberloff 1974; Simberloff et al. 1981; Chapter 11). Thus, historical biogeography is said to concern itself with *evolutionary time* and broad geographic scales, whereas island biogeography applies to *ecological time* and local geographic situations. For those accepting this view, perhaps it is reasonable to see a synthesis just over the hill and to perceive that island biogeographic theory represents a conceptual advance within the discipline of biogeography. I believe this view is false, however, not only from a historical standpoint but also when examined in terms of contemporary research in biogeography. The primary reason why the equilibrium theory does not constitute an advance within biogeography is simple: When formulated, it was directed toward a question that never was a primary concern to classical biogeography (see also Rosen 1978).

Three observations persist in the writings of biogeographers of the last 200 years (Nelson 1978; Nelson and Platnick 1981), and each leads to specific questions.

1. *Taxa are not distributed at random nor are they found everywhere: What are the areas of endemism or provinciality?* Early in the history of biogeography, it was discovered that most taxa are very localized in their distributions and that distributions are clustered. Much of nineteenth-century biogeographic analysis was directed toward defining these areas of endemism. The concept of zoogeographic realms (Sclater 1858), for example, was one of many attempts to do this. Despite a deemphasis by contemporary ecologically oriented biogeographers on delimiting areas

of endemism and understanding their significance, in recent years the study of endemism has taken on increasing importance.

2. *An area of endemism often exhibits more similarity to a second area than either does to a third: What are the interrelationships of areas of endemism?* It was an expectation of many early biogeographers that as long as the climate and topography of areas were more or less similar, so too would be their biotas. This expectation was not confirmed upon careful investigation, and it was soon realized that historical continuity, not climate and topography, was the most important determinant of biotic similarity. A measure of similarity among biotas has two components: the numbers of shared widespread species and the numbers of species whose close relatives are distributed in the areas under consideration. Historical relationships among areas cannot be established by comparison of the shared widespread species but only by the degrees of relationship among closely related species endemic in those different areas (Platnick and Nelson 1978; Nelson and Platnick 1978, 1981).

3. *Areas of endemism often exhibit more similarity to areas that are far away than they do to those relatively closer: How could such similarities have developed?* This third observation was derived not only from analyses of the biotas of numerous areas of endemism but also from systematic study of individual groups (the ratite birds, for instance).

These three observations and questions about areas of endemism have by no means formed the content of all biogeographic studies, but they represent classes of observations calling for general explanations. Without a doubt, most work in historical biogeography has been dominated by a concern for "tracing the distributional history" of individual groups of organisms. The highly influential works of Darwin, Matthew, Darlington, Mayr, and Simpson characterize this viewpoint. Nevertheless, even if biogeographic analysis is directed toward understanding the geographic history of individual groups rather than the interrelationships of entire biotas, both approaches focus on history, and the core concept of historical biogeography is the problem of endemism: Where is each taxon distributed? Do these distributions cluster to form shared areas of endemism? What is the history of these taxa and their areas? How is that history of endemism related to earth history? How has that history of endemism influenced the taxonomic composition of biotas through time? Whereas it is entirely possible to interpret distributional history from either a dispersalist or vicariance point of view – although such interpretations will be significantly different – the theme of historical analysis still remains the problem of endemic taxa and areas of endemism.

Conversely, the literature pertaining to equilibrium biogeography seldomly addresses the question of endemism. MacArthur and Wilson (1967), for example, mention endemism only once and then with regard to an illustration of the percentage of endemics on islands. MacArthur's (1972) highly influential book, *Geographical Ecology*, does not even raise the issue; neither does Simberloff (1974) in his extensive review. This may or may not be significant, depending upon whether one thinks the subject of endemism plays a relevant role in the phenomena that island biogeography is supposed to explain.

One of the most important questions addressed by island biogeography has been, How can the number of species on islands (or "patches") be explained? Expressed somewhat differently, the question becomes, How is the taxonomic and ecological structure of communities assembled? As originally formulated, the question was directed toward understanding the species richness of biotas, independent of their specific taxonomic composition. In other words, endemism and patterns of relationship among the taxonomic components can be ignored. Species number, then, is said to be a function of immigration and extinction rates, which are, in turn, a function of island size and distance from the "source" area (MacArthur and Wilson 1967). At some point in time, each island reaches an equilibrium number of species, that number being characterized by a dynamic balance between immigration and extinction to produce a given turnover rate.

It is not the purpose of this commentary to present an extensive criticism of island biogeographic theory. That there are exceedingly few empirical data, after nearly 20 years of research, to support its predictions has been forcefully argued by others (Simberloff 1976; Connor and McCoy 1979; Gilbert 1980). It is sufficient to note that one of the central questions posed by island biogeographic theory – How can the numbers of species on islands be explained? – has not been a pressing problem within biogeographic research for the last 200 years. Thus, in a sense, the question of whether there are two approaches to biogeography is a non sequitur: Island biogeography can be considered a major subdiscipline of biogeography only if one accepts its central question as being substantially important within the field as a whole, which it has not been.

What is it about the biotas of "islands" – whether oceanic or continental – that attracts our attention? From an evolutionary perspective, it is that these biotas designate centers of differentiation or areas of endemism. Upon reflection, that must be the crucial observation, *if* our perspective is evolutionary. Of course, it is possible to look upon these biotas in terms of the widespread taxa they may or may not share with other areas, but by definition these taxa have undergone no evolutionary change and, from that standpoint at least, are not terribly noteworthy. In terms of its research strategy, the field of geographical ecology typically emphasizes the distributional patterns of widespread species, and the taxonomic composition of biotas is interpreted in terms of the presumed ecological characteristics of the component species (Diamond 1975; Gilpin and Diamond 1982). In the words of Diamond and Gilpin (1982:65), "The reasons why particular combinations were observed to exist in nature, while others did not, were interpretable in terms of ecological attributes of the species, such as their resource overlap, dispersal ability, proneness to local extinctions, and distributional strategy."

Connor and Simberloff (1979) have criticized this approach, claiming that the particular species combinations found by Diamond (1975) also could have been assembled even if the species were distributed at random. It would appear there is indeed some cause for worry when virtually any distributional pattern can be "explained" by invoking *dispersal ability, proneness to local extinction,* or *distributional strategy,* and it is unclear how such explanatory tools could be applied in a manner that is not inherently ad hoc. Furthermore, it does not help to respond to this apparent problem by advocating the replacement of a *falsificationist* philosophy of science with a *pluralist* philosophy (Diamond and Gilpin 1982:73), a move seemingly designed to justify an approach to distributional data that is less then rigorous. The "null" hypothesis formulations of Simberloff and his colleagues cannot be equated with a falsificationist approach to scientific explanation, and one can reject the former while accepting the latter. Instead of doing this, Diamond and Gilpin give the appearance of rejecting both, seemingly because they do not want to be forced to relinquish hypotheses in the face of conflicting data. Surely the point is not that hypotheses must always be abandoned when faced with conflicting data. Rigid falsificationism need not be followed if we admit that patterns in nature, and their explanations, will always be less than perfectly rendered; some conflicting data are always to be expected. On the other hand, an admission of this sort does not justify a pluralist position when formulating and testing hypotheses; if hypotheses are not susceptible to critical evaluation and rejection, then any progress these hypotheses seem to confer may be only illusionary.

Although proponents of rigorous testing have raised valid criticisms about much of the literature of geographic ecology, it is questionable whether a null hypothesis evaluation of biotic distribution (Simberloff

1978; Connor and Simberloff 1978, 1979, Chapter 11), in which a "deterministic" explanation is judged relative to a "random" alternative, will necessarily prove enlightening in deciphering distribution patterns (see Grant and Abbott 1980; Farris 1981; Diamond and Gilpin 1982; Gilpin and Diamond 1982). Questions of statistical treatment aside, two problems arise with the use of random null hypotheses. The first is the dubious nature of the philosophical argument that claims these hypotheses have "logical primacy" in attempts to evaluate a particular deterministic hypothesis (Strong et al. 1979:910). Others have noted that (1) the identification of a hypothesis as being null is not altogether a simple and easily justifiable procedure, (2) the designation of a hypothesis as being truly random depends upon a host of assumptions, and (3) preference for a random hypothesis over alternative deterministic hypotheses as the null hypothesis is not clear (Grant and Abbott 1980; Farris 1981; Diamond and Gilpin 1982). The second problem relates to the latter point just raised. Because many of the assumptions often underlying discussions of island biogeographic theory (geographic ecology) lack biological realism (e.g., random dispersal, community structure rigidly controlled by competition), null hypotheses based on those assumptions scarcely can possess greater validity than do deterministic hypotheses.

In their attempts to explain the distributional patterns of species, geographic ecologists often allude to a contrast between the ecological and historical determinants of those patterns (e.g., MacArthur 1972; Vuilleumier and Simberloff 1980). Not surprisingly, it is usually concluded that nature is complex, with distributions being an intricate combination of short-term ecological factors and long-term historical events. The problem, it is claimed, is with nature, not with the nature of our science: "The reason why we have some trouble disentangling ecology from history (or from chance) is not so much necessarily due to a methodological hurdle as perhaps to the very nature of the 'problem'" (Vuilleumier and Simberloff 1980:344). If we accept this assessment, then surely there is little we can do except catalog nature's complexity and construct scenarios to explain it. If, as it is claimed, our methods of analysis are essentially correct, then there is little or no hope our approach to the problem will be changed: It will always remain at the level of narrative scenario, which in fact is exactly what the methods of geographic ecology often entail. In testing hypotheses within a given theoretical–methodological framework, we are restricted to the same theory and methodology used to construct the hypotheses in the first place; new data are interpreted the same way (although they do not necessarily have to give the same results, in which case the hypothesis is cast into doubt).

If we accept, however, that the theory and methodology is wanting for one or more reasons, it is then possible to approach the problem in an altogether different fashion. This is what I would like to do in the remainder of this commentary.

The problem we wish to investigate is how the taxonomic and ecological structure of an avian community might have been assembled. The analytical procedures could apply either to the entire avifauna or to selected taxonomic or ecological components of that community. To begin, it is assumed the following basic data have been collected for the sample of species: (1) A species list has been compiled, and the taxonomic status of each differentiated form (at the species or subspecies level, whichever is appropriate) has been established, (2) the distribution of each form is mapped, as are those of all taxa potentially closely related to the taxa of the sample, and (3) the relevant ecological parameters of each species have been noted.

Using these basic data, it is now possible to outline a research strategy capable of generating testable hypotheses about the history of community assembly:

1. From the taxonomic and distributional data, determine the centers of endemism not only for the taxa in the community but also for their close relatives in other areas.

2. Perform cladistic analyses on individual clades within the community sample (Eldredge and Cracraft 1980; Nelson and Platnick 1981; Wiley 1981), then reconstruct the general area cladogram for the centers of endemism (Rosen 1978; Platnick and Nelson 1978; Nelson and Platnick 1981; Platnick 1981; Wiley 1981). This area cladogram constitutes a general historical hypothesis for all the taxa endemic in the areas of endemism under consideration.

3. Compare the distributions of the species found in the community being investigated with the locations of the areas of endemism. Then, taking into account the general area cladogram and the phylogenetic relationships of the species being studied, establish which of those species had an autochthonous origin (by vicariance) within the area of endemism containing the community and which species had an allochthonous origin and subsequently dispersed into the area.

4. After characterizing the ecological structure of each area of endemism, use the general area cladogram (and the paleogeographic and paleoclimatic data used to explain it) to construct a hypothesis for the historical changes in ecology for these areas. This procedure is a logical

extension of using cladistic analysis to infer historical changes in ecology within a clade (Cracraft 1974; Morse and White 1979; Andersen 1979; Andrews 1982). Combined, these two procedures make it possible to erect a hypothesis about the history of species interactions by comparing the ecologies of the autochthonous species with those of the allochthonous species. If the comparison is extended to communities in other areas, testable hypotheses can be constructed regarding the historical pattern of species interactions when different combinations of species are sympatric in different areas. To the extent that we understand the phylogenetic relationship and vicariance biogeography of the species of any community, we can piece together the ecological history of that community.

This historical component has been missing from the discipline of geographic ecology. Needless to say, the difficulty of such an analysis increases as more species are included, therefore, in practice we may not be able to apply it to the entire avifauna of a community. On the other hand, the procedure is particularly feasible when analyzing a specific ecological substructure of a community such as different guilds (e.g., all foliage gleaners or frugivores) or of a taxonomic subset such as all antbirds, warblers, or tanagers.

As I have argued, the key to understanding both historical biogeography and geographic ecology is the history of areas of endemism. These areas can arise in two primary ways. First, areas can be subdivided by a geological or climatic event, with subsequent taxonomic differentiation producing two or more descendant areas of endemism. Second, an area may already by separated from a "source" area and receive colonists by long-distance dispersal *across a preexisting barrier*, with colonizers then differentiating. Thus, there are two modes of dispersal, each having significantly different biological implications, that can produce autochthonous species within a community. The first is the gradual enlargement of species ranges, a process regulated by various ecological rules acting at the *populational* level. The second mode is long-distance dispersal across barriers, which, at least as presently understood, is not governed by any identifiable, predictable ecological regularities but instead seems to reflect chance-based characteristics of individual organisms. It is this second kind of dispersal that is invoked by geographic ecologists and dispersalist biogeographers to explain the "ecological" component of areas of endemism.

Critics of vicariance biogeography have failed universally to understand the biological differences and systematic implications of these two types of dispersal (e.g., McDowall 1978; Briggs 1981; Mayr 1982;

Chapter 11). The first type of dispersal produces widespread species until they become geographically subdivided; the second may also, but more often it produces isolated, random occurrences of species in places of compatible habitat. Leaving this distinction aside, there is another aspect of this controversy in need of emphasis. Science attempts to explain the nonrandomness of nature. Given a single observation, one cannot assert whether it is part of a more general pattern or simply "noise," an observation that has been shaped by its own unique contingent circumstances. In biogeography, as in all science, it is first necessary to identify the general patterns, for only then can we specify the unique component within the system. Geographic ecologists and dispersalist biogeographers advocate a methodology that assumes that patterns of nature are statistical summations of many unique species histories (i.e., they assume biotas are assembled primarily as a result of long-distance dispersal). Epistemologically, this assumption will not permit biologists to reconstruct the history of biotas or community assemblages.

Considerable theoretical work has been undertaken on the analysis of area cladograms (see references cited earlier), and much more certainly will follow in the future. One potentially influential critique of these studies is the work of Simberloff and his colleagues (Simberloff et al. 1981; Chapter 11). They attempt to establish a null hypothesis for evaluating the postulated congruence of area cladograms by treating areas of endemism as if they were random draws from a metaphorical biological urn of all possible cladograms. For example, given four areas of endemism, what is the probability that the cladistic patterns we observe could not have been randomly drawn if we assume there are 15 different cladistic patterns for four areas and X number of clades to be examined? The problem with this approach is that the null hypothesis is weakly formulated, because it omits the primary systematic data supporting the original cladistic hypotheses, that is, the character distributions. The null hypothesis constructed by Simberloff et al. apparently assumes either that each possible cladistic hypothesis is devoid of supporting data or that each possible cladistic hypothesis has equal empirical support. I am unaware of a single study in which either assumption holds. Furthermore, I cannot imagine any serious vicariance biogeographer postulating general patterns on this basis as it would be biologically unrealistic.

To illustrate this point more emphatically, consider the following. Assume there are three areas of endemism, A–C, and assume we have three clades with species endemic in each area. Suppose the first clade exhibits a pattern of relationships (A + B) + C, the second (A + B) + C, and the third (A + C) + B. From these results, we might postulate that (A + B) + C is a general pattern. Simberloff in Chapter 11 notes that if there were 50 clades, there would be 19,600 trios to choose from, consequently the congruence of our first two clades may not be "statistically" significant. If one accepts this method of establishing a null hypothesis, then Simberloff may have a point. However, assume there are seven well-defined derived characters linking A and B in the first clade, five linking A and B in the second clade, and only one linking A and C in the third. Given these additional data, how would we construct a null hypothesis?

Lest I be misunderstood, let me state that it is not the use of null hypothesis that is at issue. Philosophically speaking, a null hypothesis does not have to be random. If, however, such hypotheses are so simplified as to make comparisons with empirically supported hypotheses unrealistic, then the efficacy of the "null" hypothesis is impaired. At the present, null hypotheses in biogeography appear incapable of providing a realistic standard against which to compare competing hypotheses.

To summarize, we now have the theoretical and analytical tools to reconstruct the history of island and continental biotas (for a didactic example, see Cracraft 1982). In the past, biologists have viewed island biotas as calling for different methods of analysis, presumably based on the assumption that the only way endemics could have arisen is through long-distance dispersal. Given what we now know about earth history, this is clearly an unwarranted assumption. Many, if not most, of the islands of the world probably have shared a contiguous geographic history with continental mainlands or with nearby islands. It is a mistake to assume that the positions, size, or the numbers of present-day islands have remained unchanged and, consequently, that their biotas may not be older than previously assumed. If we are to decipher these complexities, we must adopt a research program that directs attention first to the general patterns of endemism and then to second-order hypotheses about the unique components of these biotas. Coupled with evidence from paleogeography, paleoclimatology, and ecology, these methods can provide a basis for hypotheses about community evolution.

Acknowledgments

I am most grateful to Alan H. Brush, Jared M. Diamond, Norman I. Platnick, and John A. Wiens for their helpful comments on an earlier draft. This work was supported by NSF Grant DEB 79-21492.

Literature cited

Andersen, N. M. 1979. Phylogenetic inference as applied to the study of evolutionary diversification of semiaquatic bugs (Hemiptera: Gerromorpha). *Syst. Zool. 28:*554–578.

Andrews, P. 1982. Ecological polarity in primate evolution. *Zool. J. Linn. Soc. 74:*233–244.

Briggs, J. C. 1981. Do centers of origin have a center? *Paleobiology 7:*305–307.

Connor, E.F., and E.D. McCoy. 1979. The statistics and biology of the species-area relationship. *Am. Nat. 113:*791–833.

Conner, E. F., and D. Simberloff. 1978. Species number and compositional similarity of the Galapagos flora and avifauna. *Ecol. Monogr. 48:*219–248.

– 1979. The assembly of species communities: chance or competition? *Ecology 60:*1132–1140.

Cracraft, J. 1974. Phylogeny and evolution of the ratite birds. *Ibis 116:*494–521.

– 1982. Geographic differentiation, cladistics, and vicariance biogeography: reconstructing the tempo and mode of evolution. *Am. Zool. 22:*411–424.

Diamond, J. M. 1975. Assembly of species communities. In *Ecology and evolution of communities*, pp. 342–444. M. L. Cody and J. M. Diamond (eds.). Cambridge, Mass.: Harvard University Press.

Diamond, J. M. and M. E. Gilpin. 1982. Examination of the "null" model of Connor and Simberloff for species co-occurrences on islands. *Oecologia 52:*64–74.

Eldredge, N., and J. Cracraft. 1980. *Phylogenetic patterns and the evolutionary process.* New York: Columbia University Press.

Farris, J. S. 1981. Discussion (of paper by Simberloff, Heck, McCoy, and Connor). In *Vicariance biogeography: a critique*, pp. 73–84. G. Nelson and D. E. Rosen (eds.). New York: Columbia University Press.

Gilbert, F. S. 1980. The equilibrium theory of island biogeography: fact or fiction? *J. Biogeogr. 7:*209–235.

Gilpin, M. E., and J. M. Diamond., 1982. Factors contributing to non-randomness in species co-occurrences on islands. *Oecologia 52:*75–84.

Grant, P. R., and I. Abbott. 1980. Interspecific competition, island biogeography and null hypotheses. *Evolution 34:*332–341.

Hamilton, T. H. 1968. [Review of] *The theory of island biogeography* by R. H. MacArthur and E. O. Wilson. *Science 159:*71–72.

MacArthur, R. H. 1972. *Geographical ecology.* Cambridge, Mass.: Harvard University Press.

MacArthur, R. H. and E. O. Wilson. 1963. An equilibrium theory of island biogeography. *Evolution 17:*373–387.

– 1967. *The theory of island biogeography.* Princeton, N.J.: Princeton University Press.

Mayr, E. 1982. [Review of] *Vicariance biogeography. Auk 99:*618–620.

McDowall, R. M. 1978. Generalized tracks and dispersal in biogeography. *Syst. Zool. 27:*88–104.

Morse, J. C., and D. F. White, Jr. 1979. A technique for analysis of historical biogeography and other characters in comparative biology. *Syst. Zool. 28:*356–365.

Nelson, G. J. 1978. From Candolle to Croizat: comments on the history of biogeography. *J. Hist. Biol. 11*:269–305.

Nelson, G. J. and N. I. Platnick. 1978. The perils of plesomorphy: widespread taxa, dispersal, and phenetic biogeography. *Syst. Zool. 27*:474–477.

– 1981. Systematics and biogeography. New York: Columbia University Press.

Platnick, N. I. 1981. Widespread taxa and biogeographic congruence. In V. A. Funk and D. R. Brooks, (eds.). *Advances in cladistics*, pp. 223–227. Bronx, N.Y.: New York Botanical Garden.

Platnick, N. I., and G. J. Nelson. 1978. A method of analysis for historical biogeography. *Syst. Zool. 27*:1–16.

Rosen, D. E. 1978. Vicariant patterns and historical explanation in biogeography. *Syst. Zool. 27*:159–188.

Sclater, P. L. 1858. On the general geographical distribution of the members of the Class Aves. *J. Linn. Soc. Zool. 23*:130–145.

Simberloff, D. 1974. Equilibrium theory of island biogeography and ecology. *Annu. Rev. Ecol. Syst. 5*:161–182.

– 1976. Species turnover and equilibrium island biogeography. *Science 194*:572–578.

– 1978. Using island biogeographic distributions to determine if colonization is stochastic. *Am. Nat. 112*:713–726.

Simberloff, D., K. L. Heck, E. D. McCoy, and E. F. Connor. 1981. There have been no statistical tests of cladistic biogeographical hypotheses. In G. Nelson and D. E. Rosen (eds.). *Vicariance biogeography: a critique*, pp. 41–63. New York: Columbia University Press.

Strong, D. R., Jr., L. A. Szyska, and D. Simberloff. 1979. Tests of community-wide character displacement against null hypotheses. *Evolution 33*:897–913.

Vuilleumier, F., and D. Simberloff. 1980. Ecology versus history as determinants of patchy and insular distributions in high Andean birds. *Evol. Biol. 12*:235–379.

Wiley, E. O. 1981. *Phylogenetics: the theory and practice of phylogenetic systematics.* New York: Wiley.

Commentary

DENNIS M. POWER

Philosophically, the message Simberloff offers is sound. Biogeography can be strengthened by examining falsifiable hypotheses and employing appropriate statistical tests – a Popperian scientific approach. Where much of the published criticism has been directed is in the way Simberloff and his colleagues have constructed specific random models against which hypotheses are tested. In most cases, the criticism centers on the ways specific random models increase the risk of making a type

II error—accepting a null hypothesis when it is, in fact, false. When a null hypothesis is accepted, it may mean one of three things: (1) the null hypothesis is true, (2) the data are inadequate as supporting evidence for the alternative hypothesis, or (3) there may be something faulty about the procedure for generating the expected distribution with which the observed distribution is compared. It is item 3 that worries many critics.

In some island biogeographic studies, there is a test of whether or not certain pairs of species occur together more or less often than would be expected by chance. If they occur together less often than expected, then competitive exclusion may have taken place. In certain simulations used by Simberloff et al. (1981), the assumption is made that any species may occur on any island with a probability that depends only on the number of islands occupied by the species and the number of species found on each island. Terborgh (1981) has pointed out that the notion any species may occur on any island in an archipelago will not be true unless the species pool has come to equilibrium with respect to dispersal, a situation that may not obtain where colonization proceeds down an island chain or where a group of islands is exposed to invasion from more than one source. To provide evidence that dispersal has not come to equilibrium in some archipelagos, Terborgh graphs the percentage of breeding land birds in the Virgin Islands and Lesser Antilles and shows, among other things, the South American and Greater Antillian faunal components decrease rapidly with distance from the source. We cannot factor out the effects of interspecific competition or the effects of limited powers of dispersal of species emanating from both sources, and we should not test against a model of randomness that makes the assumption that species are equally likely to occur on any island. Nonetheless, Simberloff et al. (1981) do not like visual inspection of graphs and require statistical tests against some null hypothesis. Simberloff et al. (1981:85–86), state, "Our point is that there must be *some* null hypotheses with stated criteria for rejection, or biogeography will be hopelessly restricted to perpetual advocacy of subjective . . . schemes." In such cases testing against an unreasonable random model may be more "scientific," but it may not add to our knowledge.

In looking at the potential effects of competition, Strong et al. (1979) generated expected ratios of beak sizes among birds on certain archipelagos. They calculated ratios from observed species combinations and compared them with ratios from randomly paired species. Grant and Abbott (1980) discussed several ways that the procedures of Strong et al. favor acceptance of the null hypothesis when it is false. The first of these is pooling species regardless of the degree of relationship. Species not in the same genera, guild, or feeding group are not expected to compete; including them may obscure real competitive effects. Second, when comparing island with mainland ratios, the use of large mainland species pools (e.g., states, countries, or areas that are much larger than the island assemblage) will reduce the chances of detecting nonrandom colonization when there is geographic variation in community membership and/or beak size within the mainland area. Third, in order to obtain expected species/genus ratios on islands, equiprobability of dispersal among all species in the mainland pool has been assumed. As I have already mentioned, Terborgh (1981) indicates this is a problem, and it is also admitted by those constructing stochastic models (e.g., Connor and Simberloff 1978; Simberloff 1978). Stochastic models that allow for different dispersal abilities are admittedly difficult to construct. Fourth, when there is no identifiable source area and species pool, the solution has been to treat the archipelago as a universe from which random combinations are drawn. As a result, observed and expected samples are not independent and are artificially more similar by the inclusion of the same data in each. Again, Simberloff (1978) and Connor and Simberloff (1978) acknowledge the difficulties of determining the source pool.

Wright and Biehl (1982) have commented on the inappropriateness of the analyses proposed by Simberloff (1978) and Connor and Simberloff (1978, 1979) to determine if island distributions have resulted from random colonizations. Wright and Biehl propose a set of null hypotheses that they claim are better able to discriminate between random and nonrandom distributions of species among islands. If this claim is true, then many of the objections previously raised about construction of the null model may be moot. Wright and Biehl also emphasize that in studies of competitive effects on distributions, rather than simultaneously analyzing the distributions of all birds, the analysis should be restricted to subsets of potential competitors.

Simberloff also says that explanations involving dispersal or generalized tracks in vicariance models have not provided falsifiable hypotheses and are therefore not part of the analytical science stage of biogeography. The first attempts by vicariance biogeographers to attach probabilities to instances of multiple sister groups, and to match organismic and geographic cladograms, are criticized in Chapter 11 and in Simberloff et al. (1981). Rebuttals and comments regarding the 1981 paper have been written by Terborgh (1981), Smith (1981), and Farris (1981). The greatest problem seems to be in agreeing on what constitute appropriate criteria for the null hypotheses. As Simberloff shows, null probability differs with equiprobable cladograms, equiprobable types of cladograms, and Markov models.

Simberloff also criticizes many of the flaws that have been uncovered regarding the dynamic equilibrium model. I concur with all of the criticisms and especially support the view that because of the almost universally high correlation between habitat diversity and island area, experimental rather than statistical means will be required to determine which of these features contribute to extinction–immigration dynamics. In addition, it is likely that area *and* distance affect both extinction and immigration rates (Brown and Kodric-Brown 1977; Simberloff 1974, 1976). Similarly, predictions for the design of wildlife refuges are best made by empirical studies and ecological knowledge of the specific communities to be conserved rather than the now conflicting arguments about the dynamic equilibrium model.

Species/genus (S/G) ratios have been used to show that geographic variation and distribution of some island birds are determined by interspecific competition. Simberloff summarizes evidence that interpretations are difficult because of the multitude of forces acting on the S/G ratios. Even if interspecific competition were to affect certain genera, its effect in the biota as a whole is diluted by unaffected genera. In addition, I have always been bothered by the fact that the genus is an artificial taxonomic construct and is not a suitable constant. It is possible that noncongeneric birds might be in greater competition in an environment with limiting resources than would congeners in a rich environment.

Aside from the difficulties already discussed, there are other complications. For example, competition hypotheses may not be adequately tested by comparing S/G ratios on islands with those on the adjacent mainland (Grant and Abbott 1980:339): "Species within a genus may have similar propensities to disperse and colonize islands. If these are greater within some genera than other . . . the species/genus ratio on islands would be in equilibrium between two opposing processes – differential dispersal among genera, tending to elevate the ratio, and intrageneric competitive exclusion tending to lower it."

In addition to species-specific dispersal characteristics, there are ecological matters to contend with – a species has to be tolerant of ecological conditions in order to have the opportunity to disperse. For example, Marshall et al. (1982) applied island biogeography to the great interchange of mammals between North and South America during the late Cenozoic. After the formation of the Panamanian land bridge, tropical areas may have acted as barriers to dispersal so that only families with some constituent species distributed in tropical or subtropical areas took part in the interchange, whereas families with only temperate species did not.

There is also a need for a better historical perspective in many ecologic models of biogeography. For example, data on late Pleistocene climates and sea levels support a hypothesis that extinction of a considerable number of vertebrate species in the West Indies was a result of climatic changes since the end of the Pleistocene (Pregill and Olson 1981). Without a historical perspective, one may propose biogeographic models that, in fact, explain very little.

One rather important ecologic/biogeographic problem that is not touched on by Simberloff is the taxon cycle. Originally developed by Wilson (1961) and applied in many cases since (e.g., Ricklefs and Cox 1972, 1978), the taxon cycle considers species to follow a typical pattern of range expansion, differentiation, contraction, and extinction. Although the progression is supposed to be continuous rather than disjunct, it has been useful to consider discrete stages:

1. Species with widespread or continuous distributions through an archipelago with little or no geographic variation
2. Species widespread with considerable differentiation into subspecies
3. Species with fragmented and reduced distributions and well-differentiated populations
4. Species endemic to single islands

Pregill and Olson (1981) point out problems with the taxon cycle concept and relate these to the distribution of West Indian birds. The stages are simply criteria that describe distributional data. Any species would provide a close fit to one or the other categories, and with a collection of species from an archipelago it would be tempting mistakenly to visualize species evolving from stage 1 to stage 4. There is often no evidence that each species progressed through the stages during its history.

Another difficulty in ecological biogeography is rarely considered; it goes back to the basic data that are used to record the presence or absence of a species on an island or in an insular habitat. Species richness, the species/genus ratio, and morphological differences all assume that the presence of a species is important regardless of population size. Population size, however, is especially important when dealing with potential effects of competition. It is usually assumed that if a species is present, it occurs in large enough numbers to have an evolutionary or ecological impact on other species in the community. This assumption is rarely tested.

I must admit, I have a bias in favor of the use of descriptive statistical procedures (e.g., regression analysis), all of which have built-in tests against null models. Simberloff relegates hypothesis-generating statistical analyses to the marginally desirable status of "elaborate narratives." Very often, biogeographic patterns of interest are compared with other patterns to gain insight into causal relations. For example, I compared similarity among avifaunas of the Galapagos Islands with similarity among the islands based on vegetation, area, distance, and so forth (Power 1975). Simberloff is primarily concerned with determining whether the primary pattern that emerges is due to chance (e.g., similarity among avifaunas), not whether secondary relationships are (e.g., the relation of similarity among avifaunas to similarity among floras). However, there is heuristic value in looking at the relations among patterns in nature. Consider, for example, Haffer's (1981) problem of relating centers of evolution to Pleistocene refugia and present-day centers of rainfall in South America. Presumably, a large geographic region could be subdivided into numerous quadrats and the presence or absence of species recorded. A similarity matrix among quadrats could then be generated. A similar environmental matrix could be generated for the same quadrats based on annual rainfall. A statistically significant correlation between avifaunal similarity and rainfall similarity between quadrats would indicate a relationship between the two, suggesting one may be on the right track and prompting further investigation for a causal relationship.

The title of Chapter 11 indicates that biogeography is becoming unified and mature as a science. Personally, I find biogeography more diverse than unified these days. It also seems more rejuvenated than on the maturational continuum to menopause and senescence. Nonetheless, Simberloff and his colleagues are doing a tremendous service to biogeography by having created a faction that is jousting with all other factions (island biogeographers, historical biogeographers, dynamic equilibrium modelers, vicariance advocates, dispersalists, etc.). Wherever there is controversy there is excitement. I am not sure I would go so far as to say that historical biogeography and ecological biogeography are on the threshold of unification due to entering the analytical stage of science, but I have no doubt that the next few years will see stimulating advances in the field.

Literature cited

Brown, J. H., and A. Kodric-Brown. 1977. Turnover rate in insular biogeography: effect of immigration on extinction. *Ecology 58:*445–449.

Connor, E. F., and D. Simberloff. 1978. Species number and compositional similarity of the Galapagos flora and avifauna. *Ecol. Mongr. 48:*219–248.

– 1979. The assembly of species communities: chance or competition? *Ecology 60:*1132–1140.

Farris, J. S. 1981. Discussion. In G. Nelson and D. E. Rosen (eds.). *Vicariance biogeography: a critique,* pp. 73–84. New York: Columbia University Press.

Grant, P. R., and I. Abbott. 1980. Interspecific competition, biogeography and null hypotheses. *Evolution 34:*332–341.

Haffer, J. 1981. Aspects of Neotropical bird speciation during the Cenozoic. In G. Nelson and D. E. Rosen (eds.). *Vicariance biogeography: a critique,* pp. 370–394. New York: Columbia University Press.

Marshall, L. G., S. D. Webb, J. J. Sepkoski, Jr., and D. M. Raup. 1982. Mammalian evolution and the great American interchange. *Science 215:*1351–1357.

Power, D. M. 1975. Similarity among avifaunas of the Galapagos Islands. *Ecology 56:*616–626.

Pregill, G. K., and S. L. Olson. 1981. Zoogeography of West Indian vertebrates in relation to Pleistocene climatic cycles. *Annu. Rev. Ecol. Syst. 12:*75–98.

Ricklefs, R., and G. W. Cox. 1972. Taxon cycles in the West Indian avifauna. *Am. Nat. 106:*195–219.

– 1978. Stage of taxon cycle, habitat distribution, and population density in the avifauna of the West Indies. *Am. Nat. 112:*875–895.

Simberloff, D. 1974. Model in biogeography. In T. J. M. Schopf (ed.). *Models in paleobiology,* pp. 160–191. San Francisco: Freeman.

– 1976. Experimental zoogeography of islands: effects of island size. *Ecology 57:*629–648.

– 1978. Using island biogeographic distributions to determine if colonization is stochastic. *Am. Nat. 112:*713–726.

Simberloff, D., K. L. Heck, E. D. McCoy, and E. F. Connor. 1981. There have been no statistical tests of cladistic biogeographical hypotheses. In G. Nelson and D. E. Rosen (eds.). *Vicariance biogeography: a critique,* pp. 40–63, 85–98. New York: Columbia University Press.

Smith, C. L. 1981. Discussion. In G. Nelson and D. E. Rosen (eds.). *Vicariance biogeography: a critique,* pp. 69–72. New York: Columbia University Press.

Strong, D. R., Jr., L. A. Szyska, and D. S. Simberloff. 1979. Tests of community-wide character displacement against null hypotheses. *Evolution 33:*897–913.

Terborgh, J. 1981. Discussion. In G. Nelson and D. E. Rosen (eds.). *Vicariance biogeography: a critique,* pp. 66–68. New York: Columbia University Press.

Wilson, E. O. 1961. The nature of the taxon cycle in the Melanesian ant fauna. *Am. Nat. 95:*169–193.

Wright, S. J., and C. C. Biehl. 1982. Island biogeographic distributions: testing for random, regular, and aggregated patterns of species occurrence. *Am. Nat. 119:*345–357.

12 Bird song learning: theme and variations

P. J. B. SLATER

Perhaps the greatest attraction of bird song to scientist and layman alike is its immense variety. In some birds, songs are brief and discrete, whereas in others they can continue for long periods without a break. In some, each male has but a single phrase that is repeated in identical form; in others, the repertoire of phrases is so large that it is difficult to measure exactly. In some species, neighbors tend to have similar songs; in others, they do not. There are species in which the female sings as well as the male, there are those that sing at times of year far removed from breeding, there are those that sing loudly from conspicuous perches and those that sing quietly from the deepest undergrowth. With all this variety, generalizations about bird song are not easy to come by, even if one considers only the songbirds (Oscines), and the knowledgeable will think of exceptions to every rule.

One likely reason for this variation is that the function of song differs among species. Three main functions for song have been suggested: that it repels rivals, that it attracts mates, and that it stimulates the female to lay eggs. There is some experimental evidence for each of these (e.g., Kroodsma 1976; Wasserman 1977; Falls 1978). However, there is no reason why, in a particular species, song should not serve more than one of these functions at the same time, and there are cases in which there is experimental evidence that both territorial defense and mate attraction occur in the same species (Krebs 1977; Krebs et al. 1981). Song may vary from species to species due to differences in the relative importance of each function, and each function demands that song should be rather different, tending to pull it in opposite directions. It might also differ among species because their social organization is rather different. In some species, the territorial function of song might be primarily through its utility in repelling incursions by neighbors; in others, its role may be more to advertize occupancy to wandering intruders looking for vacant territories to occupy. Song after pairing might cease to be advantageous for mate attraction in species where polygyny and stolen matings are rare but continue to serve this function where this is not the case. In fact, differences among species in many aspects of their way of life may have implications for features of their singing behavior, and the variety of song is not so surprising when these possible ramifications are considered.

It is not simple to devise experiments to examine the functions of song. Those that have been carried out have been concerned with simple qualitative questions, determining whether or not there is evidence in favor of a particular function, rather than quantitative assessment of the importance of different functions. Yet, if song can function in more than one way, some idea of the balance among these uses is essential to understanding its variations. An alternative approach to the experimental approach is that involving correlations: Can one relate variations in different aspects of song to each other and to variations in the way of life of different species? Although correlations can arise for many reasons without the variables being directly related to one another, this approach can provide useful pointers on functional questions. For instance, it brought to light the relationship between continuity and versatility in bird song (Hartshorne 1956, 1973), which appears to be a genuine phenomenon (Kroodsma 1978a) even if it is open to a variety of interpretations (Slater 1981). A similar correlational approach is one that will be used, in a rather informal way, in this chapter.

The development of bird song is one aspect that has received a great deal of attention in the past decades. It has become apparent that learning has a part to play in the ontogeny of song in all or nearly all songbirds. Yet its exact role and the detailed schedule of development varies considerably from one species to another. Enough laboratory studies have now been completed for this variation to be examined and related to differences among species in singing behavior in the wild. In this chapter, I shall briefly review the literature on song learning, consider local and geographic variations in song in those species whose song development has received most attention, and end by discussing the link between song learning and the variation of song in the wild to explore the functional significance of these phenomena.

Song learning

One of the simplest ways to determine whether learning has a role in the development of bird vocalizations is to exclude the possibil-

ity of copying by raising young birds in isolation from other individuals. In most such "Kaspar Hauser" experiments, the effects have been
profound. Isolate songs are often described as slower (e.g., Immelmann 1969; Kroodsma 1977; Ewert 1979), including fewer elements
(e.g., Poulsen 1954; Lemon and Scott 1966; Rice and Thompson
1968), and being more variable (e.g., Marler et al. 1962; Lanyon
1979). In some cases, the basic structure of the song approximates
that typical of the species despite lack of opportunity for copying this
(Guttinger 1979; Lanyon 1979), but in others the arrangement of
song is abnormal (Marler et al. 1972; Thielcke-Poltz and Thielcke
1960; Ewert 1979). Occasionally isolated birds develop songs that appear to lie within the limits of variability of individuals in the wild, as
suggested by Thorpe (1964) for single Corn and Reed Buntings (*Emberiza calandra* and *Emberiza schoeniclus*). However, subtle differences
may still exist. Mulligan (1966) reported that the songs of Song Sparrows (*Melospiza melodia*) reared by Canaries (*Serinus canarius*) were approximately normal; however, Kroodsma (1977) repeated the experiment and found the birds developed songs that were longer, of lower
frequency range, with longer but simpler syllables and a slower rate of
syllable repetition, and abnormal in organization. The discrepancies
therefore had many of the same characteristics as those found in
other species. In another experiment, Lanyon (1979) found the young
Wood Thrushes (*Hylocichla mustelina*) that he raised sang near-normal
songs but that wild birds did not respond to their playback, suggesting
that the slight differences were important. Few species have been
reared in isolation from the egg, and unless this is done, learning
before isolation could contribute to similarities between normally
reared and Kaspar Hauser birds. It is probably true to say that no
bird reared in isolation has developed fully normal song (Marler
1981). Experience has also been found to be necessary for the normal
development of some passerine call notes, particular among cardueline finches (Poulsen 1954; Mundinger 1970; Marler and Mundinger
1975), although not of others (Lanyon 1960) nor of many nonpasserine sounds that have been studied (Konishi 1978). In the Ring Dove
(*Streptopelia risoria*) and the domestic fowl (*Gallus gallus*), calls develop
normally even in deafened birds (Konishi 1963; Nottebohm and Nottebohm 1971). Thus, although learning is important in song development, it does not influence all bird vocalizations, and its involvement
does not correlate simply with complexity, for some quite simple
sounds are learned (Mundinger 1979).

The opportunity to hear adults singing is only one aspect of experience that may contribute to the development of song. Often young birds reared in groups develop songs that are more normal than those of totally isolated individuals (Thorpe 1958; Marler 1967). In some such cases, grouped birds clearly copy each other and develop songs that are closely similar (Thorpe 1958; Waser and Marler 1977). However, this is not always so (Marler 1967), and another reason why song in grouped birds may be more normal is that they stimulate each other to sing and so gain more practice through which their song improves even without copying. Practice is an important aspect of song learning. Indeed, song development can often be split into two phases between which some species show no temporal overlap: a *memorization phase* in which the young bird memorizes sounds that it hears and a *motor phase* in which it matches its output to those sounds. This conceptualization of bird song development was stated most clearly by Marler (1970, 1976). He referred to young birds as being born with a "crude template" of what their species song should sound like. On its own, this is insufficient to enable them to produce normal song, as the songs of isolated birds show, but it acts to exclude the copying of totally inappropriate sounds. In other words, the bird has a rough idea of what it should learn and memorizes (at least in a form accessible for later production) only those sounds that fit in with this. As a result of learning, the crude template becomes an exact template. This is often well before the bird begins to sing itself, so that the memorization appears to involve the storing of an auditory image rather than a motor program. When the bird starts to sing, its output is matched to this image, in some cases with startling precision. Young birds that are deafened before they start to sing produce highly abnormal songs compared to those which are simply isolated at the same stage (Konishi 1964a, 1965a,b; Nottebohm 1968; Marler and Waser 1977). This suggests that it is necessary for them to hear the sounds of their own voices to match their output to the structure of songs they have learned.

The general picture put forward by Marler is a theme on which there are a great many variations. The age at which the memorization phase of learning occurs is varied, being throughout life in some species but limited to a brief sensitive period in others. The extent to which the crude template limits what can be learned also varies from species to species: Some are constrained to a very small range of sounds, whereas others have remarkable mimetic ability. The accuracy of learning is a third feature that varies, only the general features of song being picked up in some species, whereas even the finest details are learned and accurately reproduced in others. These variations in song learning deserve some attention here, for they may give us some insight into its functional significance and into the reasons for variations among birds in wild populations.

Timing

The age at which song is learned can be studied in the laboratory by training otherwise isolated individuals with tape-recorded songs at specific times and seeing whether any features of these songs are later reproduced. Field data, particularly of marked individuals, can also help by indicating whether or not the song of known birds changes over time. In most species, song is a seasonal phenomenon, and young birds, although they may be capable of learning at other times, as Thorpe (1958) showed in Chaffinches (*Fringilla coelebs*), will normally only have the opportunity to learn when adults are singing. Song is usually fully developed by the time a young bird is 1 year old. In most species, this means that learning can only occur in the first few weeks of life, unless adult males have stopped singing at that stage of the season, or in the following spring when song commences again, and the young birds are starting to sing themselves. It is also possible that new songs can be learned later in life after song is fully developed.

The extent to which learning occurs at these different stages varies from species to species. Attempts to train birds as nestlings have so far been unsuccessful (Thielcke-Poltz and Thielcke 1960; Marler 1970; Kroodsma 1978b). This is perhaps not surprising given the brevity of the nestling period and the early stage of development of hatchlings in altricial species. Nevertheless, it is possible that songs heard at this time may have a generalized effect without being copied exactly. By contrast, the evidence that some species can learn as fledglings is good; phrases heard at that stage are memorized and sung later. In the Zebra Finch (*Taeniopygia guttata*), learning is gradual and is completed by around day 66 of life, shortly before the young birds of this rapidly maturing species are fully adult (Immelmann 1969). In the White-crowned Sparrow (*Zonotrichia leucophrys*), memorization is almost entirely in the first 2 months of life, long before the birds sing themselves (Marler 1970), and the sensitive period for the Swamp Sparrow (*Melospiza georgiana*) is similar (Marler and Peters 1977). Marler and Peters (1981b) have shown clearly that this species can produce syllables memorized 8 months earlier without the need for rehearsal.

In other species, learning can take place both in juveniles and in young adults the following spring (e.g., Poulsen 1954; Lanyon 1960; Lemon and Scott 1966). In the Chaffinch, Poulsen (1951) found that two birds caught in the autumn and isolated thereafter sang songs like those of Kaspar Hauser individuals. However, those caught by Thorpe (1958) at an equivalent time sang songs that were less abnormal although still different from those of wild birds. He suggested that the basic features of song were learned in the summer and fine details added the next spring. More recently, Slater (Slater and Ince 1982; Slater and Clements, unpub.) has found some young birds to produce near-perfect copies of songs that they only heard as fledglings as well as others they heard the following spring. Adult Chaffinches cease to sing in July, at a stage when some young birds are still in the nest. Thus, young birds caught in the autumn and subsequently isolated may have had very different amounts of experience from each other, and this will certainly influence whether or not they sing normally in the spring. The influence of hatching date is well illustrated by the results of Kroodsma and Pickert (1980). They raised Marsh Wrens (*Cistothorus palustris*) on short and long days, simulating conditions that would have been experienced by birds hatched in August and those hatched in June. Those kept on long days molted later, showed more autumn subsong, and learned more songs in their first autumn than those on short days. On the other hand, those on shorter daylengths after hatching were more prepared to learn the following spring. Thus, it is likely that the sensitive period and preparedness to learn of birds in the wild are influenced by daylength, which in turn depends on their hatching date.

A proximate factor influencing the timing of sensitive periods in young birds may be their hormone levels. Testosterone stimulates song in most bird species (e.g., Poulsen 1951; Thielcke-Poltz and Thielcke 1960), and Nottebohm (1969a) found that a male Chaffinch castrated in its first winter would learn songs in its second spring, when it began to sing after injection with this hormone. Normally, Chaffinches will not learn new songs after 13 months of age (Thorpe 1958). It is thus possible that testosterone influences not only the amount of song produced but the timing of the sensitive period for learning as well. Higher testosterone levels in birds on longer days might be a reason for Kroodsma and Pickert's results. As testosterone levels can be affected by many environmental factors, abnormal titers stemming from laboratory housing conditions may be a reason why individuals vary in their preparedness to learn songs in the laboratory.

The species considered earlier can learn songs either as juveniles or in the following spring, and which of these stages they learn at probably depends on the opportunities they have for learning. In other species, however, learning seems not to occur during the fledgling period and to be primarily in the spring when the young birds are setting up territories and can learn from neighbors. In a long-term study of wild Saddlebacks (*Philesturnus carunculatus*), Jenkins (1978) found young birds to sing songs identical with those of their neighbors but unlike those of their father and birds neighboring on his territory. Rice and Thompson (1968) showed that Indigo Buntings (*Passerina cyanea*) were still prepared to learn at 18 months of age, and Payne (1981b) found that their song was abnormal if they were isolated from 60 days onward. Thus, fledglings do not appear to learn in this case either, and the sensitive period is mainly when the birds are beginning to sing themselves.

Once a young bird is in full song, it is common for no further learning to take place. In such species, laboratory attempts to modify the songs of adults with tutors or with tape recordings are ineffective (e.g., Barrington 1773; Poulsen 1954; Thorpe 1958; Lemon and Scott 1966), and the songs of wild birds do not change from year to year (e.g., Blase 1960; Dowsett-Lemaire 1979). However, in other species, the capacity to learn new songs is not lost in adulthood, and birds may change their repertoires during the season or from 1 year to the next (e.g., Laskey 1944; Guttinger 1979). This capacity to change songs in adulthood has been described for several thrush species (Hall-Craggs 1962; Farkas 1969; Marler et al. 1972) and is also true of Canaries (Nottebohm and Nottebohm 1978). In the Village Indigobird (*Vidua chalybeata*), all the birds in a display area have similar song, and if a bird moves to a new area, it changes its song appropriately (Payne 1981a).

From this brief review, it will be apparent that strong species differences exist in the timing of the memorization phase of song learning. Most species that have been studied can be placed in one or other of four categories: (1) those in which learning is restricted to a brief sensitive period at the fledgling stage, before the bird begins to sing itself; (2) those in which learning is primarily in young adults at the time that they are beginning to sing themselves; (3) those in which learning can occur at both these stages but is restricted to the first $1-1\frac{1}{2}$ years of life; (4) those in which learning can occur throughout life.

Constraints on what is learned

Within the sensitive period for song learning, many factors may influence whether or not a bird learns a song that it hears. How often a particular song is heard may be one of these, although recognizable copying has been reported with remarkably few repetitions. Thielcke-Poltz and Thielcke (1960) found one male European Blackbird (*Turdus* merula) to produce an element that it had heard only 12 times, whereas some other Blackbird sounds heard much more frequently were not imitated. Other results suggest that the number of repetitions, beyond some minimum, may be less important than other factors (e.g., Kroodsma 1978b). Even with extensive training at a time when others would learn, some birds persist in singing songs more typical of isolates (Thielcke-Poltz and Thielcke 1960; Marler 1970; personal observation). The importance of daylength and of hormone levels have already been suggested as possible factors influencing the timing of the sensitive period. Social factors have also been found to influence preparedness to learn in some species. Zebra Finches learn the songs of their Bengalese Finch (*Lonchura striata*) foster father even if members of their own species are singing in the same room (Immelmann 1969). Indigo Buntings learn from males with which they can interact rather than those from which they are visually isolated (Payne 1981b). In Nightingales (*Luscinia megarhynchos*), visual contact with a tutor is required for complete songs to be learned (Todt et al. 1979). Kroodsma (1978b) found that Sedge Wrens (*Cistothorus platensis*) would not learn from tape recordings and that, although Marsh Wrens would do so, they sang only the songs of live males if they heard these later but still within the sensitive period. Thus, social factors may lead young birds to select particular models in preference to others in the wild.

Experiments to determine whether or not a bird has learned a song that it was exposed to depend on whether or not it subsequently produces that song or one based upon it. For this to be possible, the bird must be able both to memorize the sound and to produce it, and failure to reproduce a particular song may depend on an incapacity in either or both of these respects. As far as memorization is concerned, there is some evidence that birds may learn more sounds than they later produce. Both wild and hand-reared Chaffinches sometimes sing perfect copies of song types during quiet subsong that they do not produce in full song (Slater and Ince, unpub.). In the Swamp Sparrow, males may perfect many more syllables in plastic song than are retained in full song (Marler and Peters 1981b). It is not known why certain sounds are rejected and others accepted from among those that have been learned.

Considering the motor function of song learning, the syrinxes of passerine birds are all remarkably similar, and it is unlikely that morphological differences can account for differences among species in what songs birds can be taught to produce (Marler 1970). The idea of a rough template acting as a filter to constrain what is learned is much more plausible as the major factor limiting what is reproduced. The specificity of this filter varies enormously. At one extreme are the vocal mimics, such as Grey Parrots (*Psittacus erithacus*), Indian Hill Mynahs (*Gracula religiosa*), and Mockingbirds (*Mimus polyglottos*) in which constraints appear to be slight (Laskey 1944; Bertram 1970; Todt 1975), so that in some cases even such unbirdlike sounds as human speech can be reproduced. A remarkable example of interspecific mimicry in the wild is provided by the Marsh Warbler (*Acrocephalus palustris*), in which Dowsett-Lemaire (1979; Lemaire 1974) has shown each wild individual to mimic an average of 76 other species. Altogether 212 different species were detected as mimicked by Marsh Warblers. The learning appears to occur entirely in the first autumn and winter when adult Marsh Warblers are not singing, and it involves copying of birds from both Europe and Africa learned before, during, and after migration. In this case, the main constraint may well be a motor one: Some of the sounds that they hear may not be copied simply because they are outside the frequency range of the bird's syrinx.

In the great majority of bird species, constraints on song learning ensure that they only produce the songs of their own species in the wild, although there are occasional instances of mimicry in species that do not normally mimic (e.g., Baptista 1972; Conrads 1977). Laboratory training can lead a wider variety of species to mimic. Barrington (1773) showed this in several European species; Thorpe (1958) succeeded in training a Chaffinch to sing Tree Pipit (*Anthus trivialis*) song; Marler et al. (1972) found that Red-winged Blackbirds (*Agelaius phoeniceus*) would learn Northern Oriole (*Icterus galbula*) song, but only if they were isolated without females of their own species. Training with songs that have been altered experimentally can indicate some of the rules that constrain learning. Swamp Sparrows will only learn syllables of their own species and not those from Song Sparrows and will learn them even if they are edited to give Song Sparrow temporal organization (Marler and Peters 1977). The equivalent is not true of Song Sparrows: They will not learn ordinary Swamp Sparrow song but will learn syllables from it organized like those of a Song Sparrow or Song Sparrow syllables edited into the Swamp Sparrow pattern (Marler and Peters 1981a). The cues that birds use to exclude particular patterns from learning presumably depend on how likely they would be to be misled in their normal auditory environment. In some cases these rules are very strict, and, despite extensive training at appropriate times, young birds will not learn songs of alien species even if these are similar to their own (Marler 1970).

The song of isolated birds gives some indication of the form of the crude template that constrains their learning. In Chaffinches, for example, isolate song is approximately the normal length and in the correct frequency range; it also tends to descend in frequency like normal song, although it is grossly aberrant in many other ways (Thorpe 1958; Nottebohm 1968). However, neither isolate song nor the crude template is simply the lowest common denominator of song in a particular species. Isolate songs vary considerably from bird to bird (Nottebohm 1968; Konishi 1978), and the birds can be trained with edited tapes to produce songs quite different from those of their species in the wild. Thus, the basic whistle followed by trill form of White-crowned Sparrow song does not always appear in isolate song nor in tutored birds (Konishi 1978). Chaffinch song in the wild always consists of a trill followed by a complex end phrase, yet Thorpe (1958) succeeded in training a bird to learn a song in which the end phrase had been transposed to the middle.

In summary, the main constraint on song learning seems not to be motor but to be provided by a filter the specificity of which varies considerably from one species to another. Some birds will mimic a wide variety of bird and other sounds, whereas in others the restrictions are so tight that they are limited to learning a small variety of sounds similar to those normally produced by their own species. In some species, there appear to be further limitations within these general constraints, dictating that some individuals are more likely to be copied than others and that some sounds that have been memorized are more likely to be included in full song than are others.

Accuracy

Whether a bird learns only the song of its own species or will mimic a wide range of other sounds, it may copy with a greater or lesser degree of accuracy. In some species, learning is normally extremely accurate, as shown by the sharing of songs in more or less identical form between wild individuals, as well as the results of training experiments. Laboratory experiments on White-crowned Sparrows (Marler 1970) and on Chaffinches (Thorpe 1958; Slater and Ince 1982) suggest that in these species the exact details of complete songs can be learned. That this normally occurs in the field is indicated by the close similarities found between the songs of different individuals in an area (Marler and Tamura 1964; Slater and Ince 1979). Slater et al. (1980) estimate that song learning in wild Chaffinches is precise on at least 85% of occasions, the same song type being passed on without alteration from one individual to another. Many of the song types within an area are similar to each other, suggesting that they have arisen from one another by inaccurate copying (Slater and Ince 1979), and, consistent with this idea, the song types present in an area change with time (Ince et al. 1980). The songs of laboratory-reared birds can also differ from those on which they were tutored (Slater and Ince 1982).

The very precise copying of complete songs may be rather unusual. In the Indigo Bunting, syllable sequences are passed on from one individual to another both in the wild (Payne et al. 1981) and in the laboratory (Payne 1981b), and this may involve complete songs. Again, the sequences present in an area change over time (Payne et al. 1981). Neighboring Winter Wrens (*Troglodytes troglodytes*) tend to have many syllables in common, and some of these are in the same sequences when they are shared (Kroodsma 1980). Song Sparrows trained on different song types often construct novel types using syllables from various different training models (Marler 1981). In these cases, syllables, or sequences of a few syllables, are copied accurately but not complete songs.

Some laboratory studies point to improvisation as being an important source of variety in bird songs. It has been considered a major influence on song development in Dark-eyed Juncos (*Junco hyemalis*) by Marler et al. (1962) and might also be the case in Song and Swamp Sparrows as these species invent syllables even when trained on a variety of models (Marler 1981). This may explain the great individual variation found in the songs of wild Song Sparrows (Harris and Lemon 1972). However, it is not easy to be certain of the role of improvisation in the wild as songs sung by one individual might have been copied from others unknown to the observer. Although laboratory experiments suggest that species differ in the accuracy with which they copy from others and the extent to which they can improvise, here too one must be cautious. Laboratory conditions and training procedures are usually quite different from those the birds would experience in the wild, and this may lead to a greater or lesser degree of accuracy than is found in nature. Because of this, it is essential to study the distribution and use of different songs in the wild if the functional significance of song learning is to be understood.

Natural variations

Variation of song in the wild can be examined on several different levels. Do neighbors share syllables, phrases, or songs? Can similarities be found between more distant individuals within a population? Are there differences between populations? Are these greater than those between individuals within a population? Do the differences between populations appear random or can they be related to environmental variations?

There are substantial differences between species in the variations that are found in the wild. As studies of geographic variation have recently been reviewed by Krebs and Kroodsma (1980), they will only be dealt with briefly here. A number of factors contribute to the exact pattern shown by a particular species, including the variations in song learning reviewed earlier. By improvisation, reordering, and recombining of subunits, birds can develop songs that differ from those to which they have been exposed, and this will lead the songs of individuals within a population to differ. If young birds learn songs as fledglings and then disperse some distance to their breeding sites, there may also be differences among neighbors even with accurate learning. On the other hand, similarities within a population are likely to be found if learning occurred in fledglings, was very accurate, and was followed by little movement before breeding. This could also arise if birds learned accurately from their territorial neighbors after dispersing, in which case the songs of neighbors would be expected to have more in common than those of more distant individuals within the population. The major determinants of song sharing within a population are therefore the accuracy of learning, the distance of dispersal, and the timing of dispersal in relation to that of learning.

Generally speaking, male birds do not disperse as far as females, and males may often nest on or close to the territory of their birth (Greenwood 1980). It is also common for adults to nest on the same territory in successive years. Given this pattern of dispersal, it is not easy to determine whether similarities between the songs of males nesting near each other arise from learning as fledglings or as young adults. Less close similarity within a population may also arise for two reasons: greater dispersal or lesser accuracy in learning. These considerations indicate that there are several ways in which the individuals in an area might come to sing the way they do and that the exact schedule of song development and dispersal may be an important factor in the pattern of geographic variation in song found in a particular species (Kroodsma 1974; Krebs and Kroodsma 1980).

Some species exhibit a very high degree of song sharing among individuals on neighboring territories. In Marsh Wrens, for example, Verner (1975) found five males that averaged 112 song types each to have only 127 types among them. Such a high degree of sharing appears, however, to be more common in birds with small repertoires, perhaps because larger repertoires are more often built up by improvisation. In many species, the songs of neighbors are more similar to each other than are those of more distant birds within a population (e.g., Kroodsma 1974; Avery and Oring 1977; Payne 1978; Bitterbaum and Baptista 1979). However, there is sometimes only chance association between the songs of neighbors compared with nonneighbors (Payne and Budde 1979; Slater and Ince 1982). Neighboring White-eyed Vireos (*Vireo griseus*) actually share less than expected by chance (Bradley 1981), and in both Nightingales (Hultsch and Todt 1981) and Cirl Buntings (*Emberiza cirlus*) (Kreutzer 1979), sharing is also greater between more distant individuals than it is between neighbors. All these findings involve comparison with chance expectation. In other studies, where this has not been done, strong variation among individuals in a population has been reported, suggesting either that birds copy inaccurately or that individuals with similar repertoires show no tendency to nest on adjacent territories (e.g., Konishi 1964b, 1965a).

The ontogeny of song has been examined in some detail in two of the species that share more than expected with neighbors. Interestingly, the reason for the similarity with neighbors appears to differ between them. In the Indigo Bunting, young males learn primarily from territorial neighbors when they are starting to sing themselves, and Payne (1981b) has found a tendency for them to learn the songs of individuals with which they are in visual contact and can interact. The similarity of songs is thus because birds learn from their neighbors. In the White-crowned Sparrow, by contrast, the sensitive period for song learning is much earlier, and the fact that neighbors usually have identical songs seems to stem from the very strong tendency of young males to breed in the same area as that in which they hatched. In the White-crowned Sparrow, as in some other species (Grimes 1974; McGregor 1980), similar songs are found in areas with more or less sharp boundaries between them (Baptista 1975; Orejuela and Morton 1975), a mosaic distribution of song types for which the word "dialect" is sometimes reserved. As Payne (1978) suggests, this is most likely to arise where songs are learned by birds when they settle on their territories, although one would only expect this to lead to sharp boundaries between dialects if there were geographic barriers to copying. These are present at some borders between dialect areas (Baptista 1975) but certainly not in all cases (McGregor 1980). Another possibility is that secondary contact between previously isolated groups has led to sharp divisions

(Baker 1975; Jenkins 1978). In species such as the White-crowned
Sparrow, where there are sharp boundaries despite learning early in
life, the dialect boundary probably acts as a barrier to dispersal (Baker
and Mewaldt 1978), although there is conflicting evidence on this (Petrinovich et al. 1981; Baker and Mewaldt 1981).

The mosaic pattern of song distribution seems to correlate with small
repertoire size. The Splendid Sunbird and European Redwing have
but one song type (Grimes 1974; Bjerke and Bjerke 1981), and the
White-crowned Sparrow usually has only one (Marler and Tamura
1964); in the Corn Bunting, each male has two, both of which show the
same dialect boundaries (McGregor 1980). The Chaffinch has a repertoire that is often as low as this, averaging 2.9 song types (Slater et al.
1980), but the distribution of song in this species is quite different.
Neighbors in a group of 42 territories shared no more than would be
expected at random (Slater and Ince 1982), and each population contains many song types. Up to half the individuals in a small area may
sing the most common of these; in other woods some distance away,
this song may be found rarely but another will be very common (Slater
and Ince 1979). In other words, there is a gradual shift in the frequencies of different song types with distance rather than sharp boundaries.
A song type that is rare in a particular area, perhaps being sung by
only one bird, may be so for two reasons. It may be a song that is
common elsewhere but that is, in this area, on the edge of its distribution, or it may be one that has recently been formed by inaccurate
copying. Slater et al. (1980) argue that 15% of the songs sung in the
wood they studied have either been inaccurately copied or introduced
from elsewhere. The song distribution of Chaffinches seems therefore
to arise because birds often learn their songs before they set up territories but, unlike White-crowned Sparrows, quite frequently learn them
inaccurately. They also seem not to choose to settle next to individuals
with similar songs. Thus, although the songs of Chaffinches do fall into
discrete types, there are no discontinuities that could be used to define
dialect areas. The lack of sharp boundaries is also found in some other
species despite the sharing of phrases of songs between individuals
(e.g., Nottebohm 1969b; Bitterbaum and Baptista 1979). The pattern
of song distribution and development in Cardinals (*Cardinalis cardinalis*)
appears similar to that in Chaffinches (Lemon 1975), although neighbors probably share song types to a greater extent (Lemon 1968).

Perhaps the most important question about variation in song from
one area to another is whether it is of any functional significance. It has
been argued that natural selection might favor song learning partly

because it leads to such variations and that these in turn are advantageous because they enable individuals to mate assortatively (Nottebohm 1972). Nottebohm (1969b) found song to change most with distance in the Rufous–collared Sparrow (*Zonotrichia capensis*), where there were also sharp habitat changes, and suggested that birds mating within a dialect group might be at an advantage as this would preserve adaptations to that particular habitat. Consistent with this, there is some evidence that birds within a dialect area are more similar genetically than individuals from different areas, both for this species (Nottebohm and Selander 1972) and for the White-crowned Sparrow (Baker 1974, 1975). One interpretation of these results is that the dialect of a male provides a means whereby a female can choose a mate well matched to her. However, it is equally possible that gene flow between populations is restricted for other reasons and that dialects have formed because birds learn song within the population from which they come. An alternative form of assortative mating would be that which ensured outbreeding. Jenkins (1978) suggested this as a reason for dialects in the Saddleback, a species in which the young males moved to a different area from that in which they hatched and learned the songs of their neighbors on arrival.

Another possibility is that geographic variation in song is a by-product of vocal learning and is of no functional significance in its own right (Andrew 1962). This is unlikely for species where there are sharp dialect boundaries unless these correspond to geographic barriers. However, such sharp boundaries have been described in only a few species: They should not be presumed to be widespread simply because they occur in the White-crowned Sparrow, a species Baptista (1975) refers to as "the white rat of the ornithological world." Even in this species, sharp boundaries do not occur in all populations (DeWolfe et al. 1974). Where song variation from one area to another is gradual and not obviously correlated with ecological variables, it seems plausible to suggest that it has no functional significance. If features of song are learned from other individuals, if birds do not move long distances between learning and singing, and if, as seems inevitable with any learning process, the copying is not always exact, some geographic variation would appear to be inevitable. Slater et al. (1980) found that random inaccuracies in copying could have been responsible for the distribution of song types found in a population of Chaffinches and that a similar rate of inaccurate copying would account for the changes over time described by Ince et al. (1980) in the same species. Changes over time have also been described by Payne et al. (1981), on the basis of a long-term study of Indigo Buntings. Like changes with distance, they would be predicted if song learning is not always totally accurate regardless of whether or not they have any functional significance.

In summary, song learning in some species may have been selected for because it leads to geographic variation, this in turn being advantageous because song is a cue in assortative mating. However, in many other species, geographic variation, although a consequence of song learning, is not likely to be a reason for it. As we shall see in the next section, the question of why these latter birds learn their songs is very much an open one.

Why learn?

The significance of song learning is a fundamental issue that has attracted a good deal of interest, because it may be relevant to discussions about why vocal learning arose in our own species (Marler 1970, 1976). Nottebohm (1972) suggested that the development of vocalizations without learning was probably the primitive situation in birds and that learning must have arisen subsequently through natural selection. He suggested several reasons why this change might have been favored in addition to the idea that it would promote mating within populations through the formation of dialects that act as barriers to outbreeding. These and other possibilities will be discussed briefly here. It is important to remember that several selective forces may have been acting simulateously on the same species and that different species may have been subject to different forces. Especially given the variety of song, no single answer to the question of why it is usually learned seems likely.

The different explanations for song learning that have been proposed fall into two general classes. First is the suggestion that learning may be required if something as complicated as song is to be passed from one generation to another with any precision. Second is the idea that the young bird may be better able to fit its song to some aspect of its animate or inanimate environment if it is learned. The first of these is rather more speculative and not easy to put to the scientific test. The most likely reason why song is sometimes very complex is through sexual selection, males with more elaborate songs being more successful in obtaining mates or territories, as shown by Howard (1974) and Yasukawa et al. (1980). Nottebohm (1972) suggests that complexity may be difficult to code beyond a certain point and that selection may therefore have favored learning, and this seems plausible given that selection seems to favor economy of genetic material (Williams 1966). As Marler (1981) puts it, "The interspecies complexities in bird song exceed what one usually thinks of as limits to the competence of genetic mechanisms."

If all that selection required was complexity, this could probably be achieved most simply and economically by improvisation, but a striking feature of song learning in many species is its precision, fine details being passed very exactly from one individual to another. It is commonly argued that behavior that is learned is likely to be more variable than that which is passed from one generation to the next without learning, but Slater and Ince (1982) have found that variation between individual Chaffinches is most marked in features of song that appear not to be easily modified by learning. They suggest that this may have arisen through high genetic variance being tolerated where precision of song is not required and that learning, far from leading to differences among individuals, may have been selected for because of the precision of copying that it allows. Thus, as well as being necessary for the generation of large and complex repertoires, learning might also enable individuals to achieve songs that match each other more precisely than would otherwise be possible.

The general idea that learning is necessary if song is to be precisely matched to that of other individuals or to the ideal for the environment in which the bird finds itself has been proposed a number of times. Many simple vocalizations are learned and it is unlikely that this is because they are too complex to be developed without learning. Instead, theories of this sort propose that the exact form of a particular vocalization that is best for an individual is unpredictable and therefore cannot be programmed without learning. For example, different environments have different sound transmission characteristics (Morton 1975; Wiley and Richards 1978), and there is evidence in the Rufous-collared Sparrow (Nottebohm 1975), the White-throated Sparrow (*Zonotrichia albicollis*) (Wassermann 1979), and the Great Tit (Hunter and Krebs 1979) that song features correlate with these environmental differences. Hansen (1979) has suggested that song learning may be the means whereby this is brought about, young birds preferentially learning those sounds that reach them through the environment in which they learn. Consistent with this, Gish and Morton (1981) have shown that songs of Carolina Wrens (*Thyrothorus ludovicianus*) degrade less with distance at their site of recording than at other sites. However, in many species, the differences in song from one area or habitat to another do not seem in any way systematic, and the variation in an area approaches that found between distant sites (e.g., Lemon and Harris 1974; Martin 1979). Thus, although the hypothesis that learning matches song to the environment deserves to be examined in more species, the extent of its applicability is doubtful.

An alternative type of matching is not to the physical environment but to other individuals of the same species, and this again may necessitate learning if the songs to be matched cannot be predicted. In some species where learning occurs very early in life, young birds are most likely to learn from their fathers (Nicolai 1959; Immelmann 1969). In such cases, song may be used as a cue to mate choice to achieve an optimal degree of outbreeding (see Bateson 1978), and learning may have become involved because it leads to more rapid divergence than would be possible with genetic control. Learning also makes it possible for the song of a son to match that of his father despite the fact that only half his genes are from that source. This would ease the discrimination between relatives and nonrelatives. These points are, however, purely theoretical: There is at present very little evidence on how song influences mate choice.

Vocalizations learned at a later stage, at around the time of territory establishment, may be important in relationships between neighboring males (Payne 1978, 1981a). Matched countersinging is common in species where neighbors share songs (e.g., Lemon 1968; Krebs et al. 1981) and is likely to be important in their relations with each other (Kroodsma 1979; Smith and Norman 1979). If young birds cannot predict where they will settle, learning from neighbors is the best way of ensuring that their songs match. The learning of particular songs in this way may enable birds to address their songs to specific neighbors (Armstrong 1963; Lemon 1968), to provide a graded signal, the gradation being based on the amount of matching they show (Krebs et al. 1981), or to take over territories more easily by matching songs that had been sung by the previous occupant (Payne 1978; Slater 1981). Learned songs may therefore facilitate communication between neighbors and thus help in territory maintenance. Learned signals may also assist in communication by identifying an individual or the group to which it belongs. Several species of cardueline finches have call notes that are pair specific and are learned from each other within the pair (Mundinger 1979): The role of learning here is most likely to be that it facilitates partner recognition. Sharing of songs or calls more widely within a group of individuals (e.g., Bertram 1970) may enable members of the group to identify each other and so exclude intruders more easily, as suggested by Feekes (1977).

There are thus many reasons why natural selection might have favored a role for learning in the development of bird vocalizations. Two of these perhaps deserve some stress because they may go some way toward explaining the variety both of song itself and of its patterns of development and, within that variety, to account for the correlation between continuity and versatility first described by Hartshorne (1956). Where song is primarily a mate attractant signal, sexual selection may have led to learning as a means of achieving an elaborate repertoire. Continuous singing is possible here because the individual is not listening for a reply. On the other hand, where song is mainly used as a signal between territorial males, less variety is required, but learning may have arisen to allow individuals to match their songs accurately to those of their neighbors. The gaps between songs that accompany low variety and give rise to Hartshorne's correlation may arise simply because countersinging males, having sung, must pause and listen for a reply (Slater 1981). Similar ideas have been independently elaborated by Catchpole (in press) with a view to accounting for differences in song complexity among *Acrocephalus* warblers.

These arguments suggest that there may be a continuum of species between those with relatively simple and stereotyped songs that are copied precisely and sung discontinuously and those with elaborate and varied songs produced more continuously and copied less precisely and that the position of a species on this continuum depends on the relative importance of mate attraction and territory maintenance in shaping its song. This may account for some of the variety of song, but there are undoubtedly species in which it also serves other functions (see, e.g., West et al. 1981; Brenowitz 1981) and in which the form and ontogeny of song are best viewed as compromises shaped by several different selective forces pulling in different directions. With this in mind, it is perhaps less surprising that song varies so much between species and that efforts at making simple generalizations tend to be accompanied by long lists of exceptions.

Acknowledgments

I am grateful to F. A. Clements, J. R. Krebs, and R. H. Wiley for comments. The author's research is financed by a grant from the Science and Engineering Research Council, U.K.

Literature cited

Andrew, R. J. 1962. Evolution of intelligence and vocal mimicking. *Science* *137*:585–589.

Armstrong, E. A. 1963. *A study of bird song*. London: Oxford University Press.

Avery, M., and L. W. Oring. 1977. Song dialects in the Bobolink (*Dolichonyx oryzivorus*). *Condor 79:*113–118.

Baker, M. C. 1974. Genetic structure of two populations of White-crowned Sparrows with different song dialects. *Condor 76:*351–356.

– 1975. Song dialects and genetic differences in White-crowned Sparrows (*Zonotrichia leucophrys*). *Evolution 29:*226–241.

Baker, M. C., and L. R. Mewaldt. 1978. Song dialects as barriers to dispersal in White-crowned Sparrows (*Zonotrichia leucophrys nuttalli*). *Evolution 32:*712–722.

– 1981. Response to "Song dialects as barriers to dispersal: a re-evaluation." *Evolution 35:*189–190.

Baptista, L. F. 1972. Wild House Finch sings White-crowned Sparrow song. *Z. Tierpsychol. 30:*266–270.

– 1975. Song dialects and demes in sedentary populations of the White-crowned Sparrow (*Zonotrichia leucophrys nuttalli*). *Univ. Calif. Berkeley Publ. Zool. 105:*1–52.

Barrington, D. 1773. Experiments and observations on the singing of birds. *Phil. Trans. R. Soc. 63:*249–291.

Bateson, P. P. G. 1978. Early experience and sexual preferences. In J. B. Hutchison (ed.). *Biological determinants of sexual behaviour*, pp. 29–53. New York: Wiley.

Bertram, B. 1970. The vocal behaviour of the Indian Hill Mynah *Gracula religiosa*. *Anim. Behav. Monogr. 3:*81–192.

Bitterbaum, E., and L. F. Baptista. 1979. Geographical variation in songs of California House Finches (*Carpodacus mexicanus*). *Auk 96:*462–474.

Bjerke, T. K., and T. H. Bjerke. 1981. Song dialects in the Redwing, *Turdus iliacus*. *Ornis Scand. 12:*40–50.

Blase, B. 1960. Die Lautäusserungen des Neuntöters (*Lanius c. collurio* L.), Freilandbeobachtungen und Kaspar-Hauser-Versuche. *Z. Tierpsychol. 17:*293–344.

Bradley, R. A. 1981. Song variation in a population of White-eyed Vireos (*Vireo griseus*). *Auk 98:*80–87.

Brenowitz, E. A. 1981. 'Territorial song' as a flocking signal in Redwinged Blackbirds. *Anim. Behav. 29:*641–642.

Catchpole, C. K. in press. The evolution of bird songs in relation to mating and spacing behavior. In D. Kroodsma and E. H. Miller (eds.). *Acoustic communication in birds*. New York: Academic Press.

Conrads, K. 1977. Entwicklung einer Kombinationstrophe des Buchfinken (*Fringilla c. coelebs* L.) aus einer Grunlings-Imitation und arteigenen Elementen im Frieland. *Ber. Nat. Ver. Bielefeld. 5:*91–101.

DeWolfe, B. B., D. D. Kaska, and L. J. Peyton. 1974. Prominent variations in the songs of Gambel's White-crowned Sparrows. *Bird-Banding 45:*224–252.

Dowsett-Lemaire, F. 1979. The imitative range of the song of the Marsh Warbler, *Acrocephalus palustris*, with special reference to imitations of African birds. *Ibis 121:*453–468.

Ewert, D. N. 1979. Development of song of a Rufous-sided Towhee raised in acoustic isolation. *Condor 81:*313–316.

Falls, J. B. 1978. Aggression, dominance and individual spacing. In L. Krames, P. Pliner, and T. Alloway (eds.). *Advances in the study of communication and affect*, Vol. 4, pp. 61–89. New York: Plenum.

Farkas, T. 1969. Notes of the biology and ethology of the Natal Robin *Cossypha natalensis*. *Ibis 111:*281–292.

Feekes, F. 1977. Colony specific song in *Cacicus cela* (Icteridae, Aves): the pass-word hypothesis. *Ardea 65:*197–202.

Gish, S. L., and E. S. Morton. 1981. Structural adaptations to local habitat acoustics in Carolina Wren songs. *Z. Tierpsychol. 56:*74–84.

Greenwood, P. J. 1980. Mating systems, philopatry and dispersal in birds and mammals. *Anim. Behav. 28:*1140–1162.

Grimes, L. G. 1974. Dialects and geographical variation in the song of the Splendid Sunbird *Nectarinia coccinigaster*. *Ibis. 116:*314–329.

Guttinger, H. R. 1979. The integration of learned and genetically programmed behaviour; hierarchical organisation in songs of Canaries, Greenfinches, and their hybrids. *Z. Tierpsychol. 49:*285–303.

Hall-Craggs, J. 1962. The development of song in the Blackbird, *Turdus merula*. *Ibis. 104:*277–300.

Hansen, P. 1979. Vocal learning: its role in adapting sound structures to long-distance propagation and a hypothesis on its evolution. *Anim. Behav. 27:*1270–1271.

Harris, M. A., and R. E. Lemon. 1972. Songs of Song Sparrows: individual variation and dialects. *Can. J. Zool. 50:*301–309.

Hartshorne, C. 1956. The monotony threshold in singing birds. *Auk 83:*176–192.

– 1973. *Born to sing*. Bloomington: Indiana University Press.

Howard, R. D. 1974. The influence of sexual selection and interspecific competition on Mockingbird song (*Mimus polyglottos*). *Evolution 28:*428–438.

Hultsch, H., and D. Todt. 1981. Repertoire sharing and song-post distance in Nightingales (*Luscinia megarhynchos* B.). *Behav. Ecol. Sociobiol. 8:*183–188.

Hunter, M. L., and J. R. Krebs. 1979. Geographical variation in the song of the Great Tit (*Parus major*) in relation to ecological factors. *J. Anim. Ecol. 48:*759–785.

Immelmann, K. 1969. Song development in the Zebra Finch and other estrildid finches. In R. A. Hinde (ed.). *Bird vocalisations* pp. 61–74. Cambridge: Cambridge University Press.

Ince, S. A., P. J. B. Slater, and C. Weismann. 1980. Changes with time in the songs of a population of Chaffinches. *Condor 82:*285–290.

Jenkins, P. F. 1978. Cultural transmission of song patterns and dialect development in a free-living bird population. *Anim. Behav. 28:*50–78.

Konishi, M. 1963. The role of auditory feedback in the vocal behaviour of the domestic Fowl. *Z. Tierpsychol. 20:*349–367.

– 1964a. Effects of deafening on song development in two species of juncos. *Condor 66:*85–102.

– 1964b. Song variation in a population of Oregon Juncos. *Condor 66:*423–436.
– 1965a. Effects of deafening on song development in American Robins and Black-headed Grosbeaks. *Z. Tierpsychol. 22:*584–599.
– 1965b. The role of auditory feedback in the control of vocalization in the White-crowned Sparrow. *Z. Tierpsychol. 22:*770–778.
– 1978. Auditory environment and vocal development in birds. In R. D. Walk and H. L. Pick (eds.). *Perception and experience,* pp. 105–118. New York: Plenum.
Krebs, J. R. 1977. Song and territory in the Great Tit *Parus major.* In B. Stonehouse and C. Perrins (eds.). *Evolutionary ecology,* pp. 47–62. London: Macmillan Press.
Krebs, J. R., R. Ashcroft, and K. van Orsdol. 1981. Song matching in the Great Tit *Parus major* L. *Anim. Behav. 29:*918–923.
Krebs, J. R., M. Avery, and R. J. Cowie. 1981. Effect of removal of mate on the singing behaviour of Great Tits. *Anim. Behav. 29:*635–637.
Krebs, J. R., and D. E. Kroodsma. 1980. Repertoires and geographical variation in bird song. *Adv. Study Behav. 11:*143–177.
Kreutzer, M. 1979. Etude du chant chez le Bruant zizi (*Emberiza cirlus*) le répertoire, caractéristiques et distribution. *Behaviour 71:*291–321.
Kroodsma, D. E. 1974. Song learning, dialects and dispersal in the Bewick's Wren. *Z. Tierpsychol. 35:*352–380.
– 1976. Reproductive development in a female songbird: differential stimulation by quality of male song. *Science 192:*574–575.
– 1977. Re-evaluation of song development in the Song Sparrow. *Anim. Behav. 25:*390–399.
– 1978a. Continuity and versatility in bird songs: support for the monotony-threshold hypothesis. *Nature (London) 274:*681–683.
– 1978b. Aspects of learning in the ontogeny of bird song: where, from whom, when, how many, which and how accurately. In G. M. Burghardt and M. Bekoff (eds.). *The development of behavior: comparative and evolutionary aspects,* pp. 215–230. New York: Garland.
– 1979. Vocal dueling among male Marsh Wrens: evidence for ritualized expressions of dominance/subordinance. *Auk 96:*506–515.
– 1980. Winter Wren singing behavior: a pinnacle of song complexity. *Condor 82:*357–365.
Kroodsma, D. E., and R. Pickert. 1980. Environmentally dependent sensitive periods for avian vocal learning. *Nature (London) 288:*477–479.
Lanyon, W. E. 1960. The ontogeny of vocalizations in birds. In W. E. Lanyon and W. N. Tavolga (eds.). *Animal sounds and communication,* pp. 321–347. Washington D.C.: American Institute of Biological Science.
– 1979. Development of song in the Wood Thrush (*Hylocichla mustelina*), with notes on a technique for hand rearing passerines from the egg. *Am. Mus. Novit.,* No. 2666, 1–27.
Laskey, A. R. 1944. A Mockingbird acquires his song repertoire *Auk 61:*211–219.
Lemaire, F. 1974. Le chant de la Rousserolle verderolle (*Acrocephalus palustris*): étendue du répertoire imitatif, construction rhythmique et musicalité. *Gerfaut 64:*3–28.

Lemon, R. E. 1968. The relation between organization and function of song in Cardinals. *Behaviour 32*:158–178.

– 1975. How birds develop song dialects. *Condor 77*:385–406.

Lemon, R. E., and M. Harris. 1974. The question of dialects in the songs of White-throated Sparrows. *Can. J. Zool. 52*:83–98.

Lemon, R. E., and D. M. Scott. 1966. On the development of song in young Cardinals. *Can. J. Zool. 44*:191–197.

Marler, P. 1967. Comparative study of song development in sparrows. In *Proceedings of the XIV International Ornithological Congress*, pp. 231–244.

– 1970. A comparative approach to vocal learning: song development in White-crowned Sparrows. *J. Comp. Physio. Psychol. 71* (Suppl.): 1–25.

– 1976. Sensory templates in species-specific behavior. In J. C. Fentress (ed.). *Simpler networks and behavior*, pp. 314–329. Sunderland, Mass.: Sinauer Associates.

– 1981. Birdsong: the acquisition of a learned motor skill. *Trends Neurol. Sci. 3*:88–94.

Marler, P., M. Kreith, and M. Tamura. 1962. Song development in hand raised Oregon Juncos. *Auk 79*:12–30.

Marler, P., and P. C. Mundinger. 1975. Vocalizations, social organisation and breeding biology of the Twite *Acanthus flavirostris*. *Ibis 117*:1–17.

Marler, P., P. Mundinger, M. S. Waser, and A. Lutjen. 1972. Effects of acoustical stimulation and deprivation on song development in Redwinged Blackbirds (*Agelaius phoeniceus*). *Anim. Behav. 20*:586–606.

Marler, P., and S. Peters. 1977. Selective vocal learning in a sparrow. *Science 198*:519–521.

– 1981a. Birdsong and speech: evidence for special processing. In P. Eimas and J. Miller (eds.). *Perspectives in the study of speech*, pp. 75–112. Hillsdale, N.J.: Erlbaum.

– 1981b. Sparrows can learn adult song and more from memory. *Science 213*:780–782.

Marler, P., and M. Tamura. 1964. Song "dialects" in three populations of White-crowned Sparrows. *Science 146*:1483–1486.

Marler, P., and M. S. Waser. 1977. Role of auditory feedback in Canary song development. *J. Comp. Physiol. Psychol. 91*:8–16.

Martin, D. J. 1979. Songs of the Fox Sparrow. II. Intra- and interpopulation variation. *Condor 81*:173–184.

McGregor, P. K. 1980. Song dialects in the Corn Bunting (*Emberiza calandra*). *Z. Tierpsychol. 54*:285–297.

Morton, E. S. 1975. Ecological sources of selection on avian sounds. *Am. Nat. 109*:17–34.

Mulligan, J. A. 1966. Singing Behavior and its development in the Song Sparrow *Melospiza melodia*. *Univ. Calif. Berkeley Publ. Zool. 81*:1–73.

Mundinger, P. C. 1970. Vocal imitation and individual recognition of finch calls. *Science 168*:480–482.

– 1979. Call learning in the Carduelinae: ethological and systematic considerations. *Syst. Zool. 28*:270–283.

Nicolai, J. 1959. Familientradition in der Gesangsentwicklung des Gimpels (*Pyrrhula pyrrhula* L.). *J. Ornithol. 100*:39–46.

Nottebohm, F. 1968. Auditory experience and song development in the Chaffinch *Fringilla coelebs*. *Ibis 110*:549–568.

– 1969a. The "critical period" for song learning in birds. *Ibis 111*:385–387.

– 1969b. The song of the Chingolo (*Z. capensis*) in Argentina: description and evaluation of a system of dialects. *Condor 71*:299–315.

– 1972. The origins of vocal learning. *Am. Nat. 106*:116–140.

– 1975. Continental patterns of song variability in *Zonotrichia capensis:* some possible ecological correlates. *Am. Nat. 109*:605–624.

Nottebohm, F., and M. Nottebohm. 1971. Vocalizations and breeding behaviour of surgically deafened Ring Doves. *Anim. Behav. 19*:313–327.

– 1978. Relationship between song repertoire and age in the Canary *Serinas canarius*. *Z. Tierpsychol. 46*:298–305.

Nottebohm, F., and R. K. Selander. 1972. Vocal dialects and gene frequencies in the Chingolo Sparrow (*Zonotrichia capensis*). *Condor 74*:137–143.

Orejuela, J. E., and M. L. Morton. 1975. Song dialects in several populations of mountain White-crowned Sparrows (*Zonotrichia leucophrys*) in the Sierra Nevada. *Condor 77*:145–153.

Payne, R. B. 1978. Microgeographic variation in songs of Splendid Sunbirds *Nectarinia coccinigaster:* population phenetics, habitats and song dialects. *Behaviour 65*:282–308.

– 1981a. Population structure and social behavior: models for testing the ecological significance of song dialects in birds. In R. D. Alexander and D. W. Tinkle (eds.). *Natural selection and social behavior,* pp. 108–120. New York: Chiron Press.

– 1981b. Song learning and social interaction in Indigo Buntings. *Anim. Behav. 29*:688–697.

Payne, R. B., and P. Budde. 1979. Song differences and map distances in a population of Acadian Flycatchers. *Wilson Bull. 91*:29–41.

Payne, R. B., W. L. Thompson, K. L. Fiala, and L. L. Sweany. 1981. Local song traditions in Indigo Buntings: cultural transmission of behavior patterns across generations. *Behaviour 77*:199–221.

Petrinovich, L., T. Patterson, and L. Baptista. 1981. Song dialects as barriers to dispersal: a re-evaluation. *Evolution 35*:180–188.

Poulsen, H. 1951. Inheritance and learning in the song of the Chaffinch *Fringilla coelebs* L. *Behaviour 3*:216–228.

– 1954. On the song of the Linnet (*Carduelis cannabina* (L.)). *Dansk. Ornithol. Foren. Tidsskr. 48*:32–37.

Rice, J. O., and W. L. Thompson. 1968. Song development in the Indigo Bunting. *Anim. Behav. 16*:462–469.

Slater, P. J. B. 1981. Chaffinch song repertoires: observations, experiments and a discussion of their significance. *Z. Tierpsychol. 56*:1–24.

Slater, P. J. B., and S. A. Ince. 1979. Cultural evolution in Chaffinch song. *Behaviour 71*:146–166.

– 1982. Song development in Chaffinches: what is learnt and when? *Ibis 124*:21–26.

Slater, P. J. B., S. A. Ince, and P. W. Colgan. 1980. Chaffinch song types: their frequencies in the population and distribution between the repertoires of different individuals. *Behaviour 75*:207–218.

Smith, D. G., and D. O. Norman. 1979. "Leader-follower" singing in Red-winged Blackbirds (*Agelaius phoeniceus*). *Condor 81*:83–84.

Thielcke-Poltz, H., and G. Thielcke. 1960. Akustisches Lernen verschieden alter schallisolierter Amseln (*Turdus merula* L.) und die Entwicklung erlernter Motive ohne und mit kunstlichem Einfluss von Testosteron. *Z. Tierpsychol. 17*:211–244.

Thorpe, W. H. 1958. The learning of song patterns by birds, with especial reference to the song of the Chaffinch *Fringilla coelebs*. *Ibis 100*:535–570.

– 1964. The isolate song of two species of *Emberiza*. *Ibis 106*:115–118.

Todt, D. 1975. Social learning of vocal patterns and modes of their application in Grey Parrots (*Psittacus erithacus*). *Z. Tierpsychol. 39*:178–188.

Todt. D., H. Hultsch, and D. Heike. 1979. Conditions affecting song acquisition in Nightingales (*Luscinia megarhynchos* L.). *Z. Tierpsychol. 51*:23–35.

Verner, J. 1975. Complex song repertoire of male Long-billed Marsh Wrens in eastern Washington. *Living Bird 14*:263–300.

Waser, M. S., and P. Marler. 1977. Song learning in Canaries. *J. Comp. Physiol. Psychol. 91*:1–7.

Wassermann, F. E. 1977. Mate attraction function of song in the White-throated Sparrow. *Condor 79*:125–127.

– 1979. The relationship between habitat and song in the White-throated Sparrow. *Condor 81*:424–426.

West, M. J., A. P. King, and D. H. Eastzer. 1981. The cowbird: reflections on development from an unlikely source. *Am. Sci. 69*:56–66.

Wiley, R. H., and D. G. Richards. 1978. Physical constraints on acoustic communication in the atmosphere: implications for the evolution of animal vocalizations. *Behav. Ecol. Sociobiol. 3*:69–94.

Williams, G. C. 1966. *Adaptation and natural selection.* Princeton, N.J.: Princeton University Press.

Yasukawa, K., J. L. Blank, and C. B. Patterson. 1980. Song repertoires and sexual selection in the Red-winged Blackbird. *Behav. Ecol. Sociobiol. 7*:233–238.

Commentary

LUIS F. BAPTISTA

All disciplines in biology usually begin with an observation/descriptive stage that then proceeds to an experimental/hypothesis-testing stage. The descriptive literature on bird song is vast, and we are now in an exciting era when evolutionary principles are being tested using bird song. Although a few excellent studies have appeared utilizing nonpasserines as subjects (e.g., reviewed in Abs 1980), most developmental or experimental work still tends to be with oscines.

Song learning

Slater's excellent review of the literature on song ontogeny in passerines reveals that in many species song is learned at an early "sensitive phase." Tape-recorded songs are played to isolates at specific ages, and their vocal output at maturity is compared with the taped songs. Using this technique, Marler (1970) has estimated the sensitive phase for the White-crowned Sparrow (*Zonotrichia leucophrys*) to be approximately between 10 and 50 days.

Sedge wrens (*Cistothorus platensis*), creepers (*Certhia* spp.), and Zebra Finches (*Poephila guttata*) do not learn from tutor tapes but require social interaction with adults (Thielcke 1970; Kroodsma 1978; Price 1979). Tape-recorded songs may be copied by some species; however, individuals exposed to live tutors learn more syllables (Waser and Marler 1977; Kroodsma 1978), demonstrating that social interaction is important in normal song development. Fifty- to fifty-nine-day-old (hand-raised) White-crowned Sparrows exposed to live adult conspecifics successfully learned their songs. Controls not so exposed sang typical isolate songs (Baptista, Petrinovich, and Morton, unpub.). These data suggest that (1) the sensitive phase is more variable than previously thought or (2) a live male in a natural setting is a more effective stimulus than a taped song, enabling birds to learn later in life. White-crowns disperse from their natal area at between 35 and 48 days of age (Blanchard 1941) and must be able to learn songs after that time.

Slater discussed Marler's (1970, 1976) idea that birds possess an innate auditory specification for conspecific song, the "template," which guides the fledgling to select conspecific song and reject alien sounds during the learning process. This theory can be interpreted in two ways: (1) As in Zebra Finches, birds *can* produce allospecific songs but given a choice will select conspecific songs (Immelmann 1969). (2) Birds *cannot* produce allospecific sounds and thus only learn conspecific sounds, as reported for Swamp Sparrows (*Melospiza georgiana*), Song Sparrows (*Melospiza melodia*), and White-crowned Sparrows (Mulligan 1966; Marler 1970; Marler and Peters 1977). The data on Swamp Sparrows are convincing. However, both Song Sparrows and White-crowns may learn alien songs from live tutors (Kroodsma 1977; Eberhardt and Baptista 1977; Baptista and Morton 1981). White-crowns also learn tape-recorded Mexican Junco (*Junco hyemalis*) songs (Petrinovich and Baptista, unpub.). White-crowned Sparrows placed in cages where they could see and hear Strawberry Finches (*Amandava aman-*

dava) learned their song and ignored White-crown songs in the same room (Baptista, Petrinovich, and Morton, unpub.).

Slater noted that in a number of species, including White-crowns, song learning is completed long before birds sing themselves. These data are usually based on birds raised in isolation boxes. We found that White-crowns raised in a room where they could hear ambient sounds began subsong at 16 to 30 days of age (Baptista, DeWolfe, and Petrinovich, unpub.). Subsong in the wild may be heard in June or July (Blanchard 1941; Baptista 1975; Baptista, unpub.). Wild juveniles may sing fully developed songs by the end of September or early October at as early as 120 days of age, long before sexual maturity (Baptista, DeWolfe, and Petrinovich, unpub.). These data suggest that: (1) The onset of singing (the *motor* phase of Slater) overlaps the learning period (the *memorization* phase of Slater). (2) In nature, where juveniles continually interact with adults, song development proceeds at a faster rate than in the laboratory. We noted many incidences of juveniles countersinging with adults. Both age classes responded to playback with song and displays in the fall and winter, suggesting year-round defense of territory.

Learning to respond

In addition to (1) perceptual learning, when birds receive and store songs, and (2) motor learning, when birds perfect their vocal output, playback studies indicate that dispositions to respond are also learned and may be coupled with song learning. Treecreepers (*Certhia familiaris*) occasionally learn songs of their sympatric sibling species (*Certhia brachydactyla*) and will respond to playback of those songs (Thielcke 1972; Becker 1977). However, song and response to playback may be learned separately. *Acrocephalus* warblers will respond to playback of sympatric congenerics without learning to produce their songs (Catchpole 1978). Rufous-sided Towhees (*Pipilo erythrophthalmus*) occasionally learn songs of Carolina Wrens (*Thryothorus carolinensis*). Neighboring towhees, but not nonneighbors, will respond to playback of wren songs, apparently due to associative learning (Richards 1979).

Isolating mechanisms

Are birds learning wrong (allospecific) songs prevented from pairing with conspecific females? Indigo (*Passerina cyanea*) and Lazuli

Buntings (*Passerina amoena*), sympatric in Nebraska, compete interspe-cifically, often learn each others' songs, and respond to them in play-back. Nonetheless, few hybrid pairings occur probably because females use other clues in mate selection (Emlen et al. 1975). In this as in other biological phenomena, for example, orientation behavior (Emlen 1969), several "fail-safe" mechanisms (redundancies) may occur in the system. Avian responses to playback tend to be aggressive rather than sexual (but see Payne 1973). Data from playback studies to test for isolating mechanisms, either between species or dialectal populations, must be treated with caution.

Natural variations

Earlier, I called attention to the importance of social (male/male) interaction in song learning. Amount of social interaction in na-ture is probably directly related to population density and must affect song variation, notably of obligate social learners, (e.g., *Certhia* spp.; Thielcke 1970). Accuracy of learning is proportional to amount of exposure to the model (Petrinovich and Baptista, unpub.). In areas of high creeper density with frequent male/male interaction, songs vary little. In regions with dispersed habitat and little male/male interaction, song variation is high, and song may be simpler in structure (Thielcke 1965; Baptista and Johnson 1982).

Slater contrasts the two best-studied species, Indigo Buntings and White-crowned Sparrows. Our data suggest that they are quite similar. Nearest neighbors sing the most similar songs in both species (Figure 5 in Baptista 1975, unpub.; Payne 1981a). In both species juveniles pre-fer to learn the songs of their social tutor (Payne 1981a; Baptista, Petrinovich, and Morton, unpub.). Few White-crowns remain to breed in their natal area (Petrinovich et al. 1980; Baptista and Morton 1982), and males may learn after 50 days of age. Thus, White-crowned Spar-rows may learn songs from males at settling sites, as Payne (1981a) has shown for Indigo Buntings. Some birds may learn songs at their natal area and then disperse and learn new songs at the settling site. Al-though the most intensely studied species indicate that learning songs at dispersal/settling time may be a widespread phenomenom, it is not necessarily the general rule. Slater and Ince (1982) noted that sharing by neighbors in Chaffinches (*Fringilla coelebs*) is only a chance phe-nomenon. Moreover, theme sharing may vary between populations within the same species (Wiens 1982).

Why learn?

Slater discusses the hypothesis that learning matches song to the environment. He is not convinced of its general applicability, because in many species differences in song from one area or habitat to another do not appear to be consistent. Most studies on geographic variation in song treat variation in syllable morphology, rhythm, and/or pitch. However, it is the pitch–amplitude envelope that appears to be correlated with habitat characteristics. Bowman (1979) found island differences (dialects) within species and convergences between species in these characteristics. Pitch-amplitude was such as to minimize transmission loss in each habitat. We clearly need more such studies before the generality of the phenomenon may be properly evaluated.

Slater discusses Nottebohm's (1969) hypothesis that birds select mates on the basis of home dialect and thus perpetuate genes adapting them to that habitat. My colleagues and I tested this with two races of White-crowned Sparrows (Z. l. nuttalli and Z. l. oriantha) by inducing song in females with testosterone and then comparing their songs with those of their mates. Songs of few females match those of their consorts (Petrinovich et al. 1980, unpub.; Baptista and Morton 1982). Moreover, birds learning alien dialects were often paired and bred successfully (Baptista 1974; Baptista and Morton 1982).

I agree with Slater that gene flow between dialectal populations may be restricted for reasons other than song. Baker's (1975) biochemical data on White-crowned Sparrows from Colorado lend support to Slater's argument, as changes in three loci frequencies occurred within a dialect area covering about 7.5 km. Payne's (1981b) examination of the evidence indicates that the amount of cross-dialect dispersal appears great enough to prevent significant genetic differentiation.

If dialects are not involved in assortative mating, why are they learned? Brown (1975:674) believes that the most important role of song is territoriality. Slater develops Payne's (1978) idea that learning songs of neighbors at sites settled may be important in territory establishment and maintenance. The energetic costs of territorial defense may be lessened if territory holders respond less to songs of known neighbors (often encountered) than to strangers (rarely encountered). This has been demonstrated recently in several species (Falls 1969; Baker et al. 1981). Females learn songs because they may aid males in defense of territory with song early in the breeding season (Baptista and Morton, unpub.).

Concluding remarks

Results from field and laboratory may differ because the laboratory milieu often does not provide the adequate quantity or quality of social stimulation necessary for proper song development. This is especially true if taped songs are used as learning stimuli. All recent reviews indicate that few studies are conducted with marked birds in natural populations. Both field and concurrent laboratory studies, with good sample sizes, are necessary if we are to better understand (1) the length of sensitive phases, (2) time of onset of song, (3) time of crystallization of song, (4) whether songs are learned from the natal area or from sites settled, and (5) amount of dispersal across dialect boundaries.

Literature cited

Abs, M. 1980. Zur Bioakustik des Stimmbruchs bei Vögeln. *Zool. Jahrb. Abt. Allg. Zool. Physiol. 84:*289–382.

Baker, M. C. 1975. Song dialects and genetic differences in White-crowned Sparrows (*Zonotrichia leucophrys*). *Evolution 29:*226–241.

Baker, M. C., D. B. Thompson, and G. L. Sherman. 1981. Neighbor-stranger song discrimination in White-crowned Sparrows. *Condor 83:*265–267.

Baptista, L. F. 1974. The effects of songs of wintering White-crowned Sparrows on song development in sedentary populations of the species. *Z. Tierpsychol. 34:*147–171.

– 1975. Song dialects and demes in sedentary populations of the White-crowned Sparrow (*Zonotrichia leucophrys nuttalli*) *Univ. Calif. Berkeley Publ. Zool. 105:*1–52.

Baptista, L. F., and R. B. Johnson. 1982. Song variation in insular and mainland California Brown Creepers. *J. Ornithol. 123:*131–144.

Baptista, L. F., and M. L. Morton. 1981. Interspecific song acquisition by a White-crowned Sparrow. *Auk 98:*383–385.

– 1982. Song dialects and mate selection in montane White-crowned Sparrows. *Auk 99:*537–547.

Becker, P. H. 1977. Verhalten auf Lautäusserungen der Zwillingsart, interspecifische Territorialität and Habitatansprüche von Winter und Sommergoldhähnchen (*Regulus regulus, R. ingnicapillus*). *J. Ornithol. 118:*233–260.

Blanchard, B. D. 1941. The White-crowned Sparrows (*Zonotrichia leucophrys*) of the Pacific Seaboard: environment and annual cycle. *Univ. Calif. Berkeley Publ. Zool. 46:*1–178.

Bowman, R. I. 1979. Adaptive morphology of song dialects in Darwin's Finches. *J. Ornithol. 120:*353–389.

Brown, J. L. 1975. *The evolution of behavior.* New York: Norton.

Catchpole, C. K. 1978. Interspecific territorialism and competition in *Acrocephalus* warblers as revealed by playback experiments in areas of sympatry and allopatry. *Anim. Behav. 26:*1072–1080.

Eberhardt, C., and L. F. Baptista. 1977. Intraspecific and interspecific song mimesis in California Song Sparrows. *Bird-Banding 48:*193–205.

Emlen, S. T. 1969. Migration: orientation and navigation. In D. S. Farner and J. R. King (eds.). *Avian biology*, Vol. 5, pp. 129–219. New York: Academic Press.

Emlen, S. T., J. D. Rising, and W. L. Thompson. 1975. A behavioral and morphological study of sympatry in the Indigo and Lazuli Buntings of the Great Plains. *Wilson Bull. 87:*145–179.

Falls, J. B. 1969. Functions of territorial song in the White-crowned Sparrow. In R. A. Hinde (ed.). *Bird vocalizations*, pp. 207–232. Oxford University Press.

Immelmann, K. 1969. Song development in the Zebra Finch and other estrildid finches. In R. A. Hinde (ed.). *Bird vocalizations*, pp. 61–74. Oxford University Press.

Kroodsma, D. E. 1977. A re-evaluation of song development in the Song Sparrow. *Anim. Behav. 25:*390–399.

– 1978. Aspects of learning in the ontogeny of bird song: where, from whom, when, how many, which, and how accurately. In G. Burghart and M. Bekoff (eds.). *Ontogeny of behavior*, pp. 215–230. New York: Garland.

Marler, P. 1970. A comparative approach to vocal learning: song development in White-crowned Sparrows. *J. Comp. Physiol. Psychol.* 71 (2):1–25.

– 1976. Sensory template in species-specific behavior. In J. C. Fentress (ed.). *Simpler networks and behavior*, pp. 314–329. Sunderland, Mass.: Sinauer Associates.

Marler, P., and S. Peters. 1977. Selective vocal learning in a sparrow. *Science 198:*519–521.

Mulligan, J. A. 1966. Singing behavior and its development in the Song Sparrow, *Melospiza melodia. Univ. Calif. Berkeley Publ. Zool. 81:*1–76.

Nottebohm, F. 1969. The song of the Chingolo, *Zonotrichia capensis*, in Argentina: description and evaluation of a system of dialects. *Condor 71:*299–315.

Payne, R. B. 1973. Vocal mimicry of the Paradise Whydah (*Vidua*) and response of female Whydahs to the songs of their hosts (*Pytilia*) and their mimics. *Anim. Behav. 21:*762–771.

– 1978. Microgeographic variation in songs of Splendid Sunbirds *Nectarinia coccinigaster:* population phenetics, habitats, and song dialects. *Behavior 65:*282–308.

– 1981a. Song learning and social interaction in Indigo Buntings. *Anim. Behav. 29:*688–697.

– 1981b. Population structure and social behavior: models for testing the ecological significance of song dialects in birds. In R. D. Alexander and D. W. Tinkle (eds.). *Natural selection and social behavior*, pp. 108–120. New York: Chiron.

Petrinovich, L. F., T. L. Patterson, and L. F. Baptista. 1980. Song dialects as barriers to dispersal; a re-appraisal. *Evolution 35:*180–188.

Price, P. H. 1979. Developmental determinants of structure in Zebra Finch song. *J. Comp. Physiol. Psychol. 93:*260–277.

Richards, D. G. 1979. Recognition of neighbors by associative learning in Rufous-sided Towhees. *Auk 96:*688–693.

Slater, P., and S. A. Ince. 1982. Song development in the Chaffinch: what is learned and when? *Ibis 124*:21–26.

Thielcke, G. 1965. Gesangsgeographische Variation des Gartenbaumläufers (*Certhia brachydactyla*) im Hinblick auf das Artbildungsproblem. *Z. Tierpsychol. 22*:542–566.

— 1970. Lernen vor Gesang als möglicher Schrittmacher der Evolution. *Z. Zool. Syst. Evolutionsforsch. 8*:309–320.

— 1972. Waldbaumläufer (*Certhia familiaris*) ahmen artfremdes Signal nach und reagieren darauf. *J. Ornithol. 113*:287–295.

Waser, M. S., and P. Marler. 1977. Song learning in Canaries. *J. Comp. Physiol. Psychol. 91*:1–7.

Wiens, J. A. 1982. Song pattern variation in the Sage Sparrow (*Amphispiza belli*): dialects or ephiphenomena? *Auk 99*:208–229.

Commentary

DONALD E. KROODSMA

The study of bird vocalizations has progressed far beyond the relevant thoughts offered by Nice (1933) 50 years ago at the semicentennial anniversary of the American Ornithologists' Union. Nice could at that time dispel the notion that bird song was a "little prayer of thankfulness . . . [sent] . . . straight up to heaven, . . . [a] sweet and sincere little petition . . . [expressing] . . . simple faith and trust" (Pearson 1917:52), and through her own and others' careful field work paved the way for a scientific interpretation of the significance of bird vocalizations. Since then, with the advent of modern technical equipment, exciting progress has been made in a number of fields, not only in that of ontogeny and microgeographic variation so thoroughly reviewed by Slater but also in the following areas: neural control of vocalization; both the production and perception of sounds; the influence of both the physical and biological environment on the structure of vocalizations; duetting and interspecific mimicry; the evolution of complex vocal repertoires; individuals, population, and species recognition; and last, but certainly not least, the conceptual framework within which we view the communicative process (see relevant chapters in Smith 1977; Kroodsma and Miller, in press). Progress in all of these areas deserves recognition, but I must be selective; hence, I will first address the topics introduced by Slater and then comment briefly on several other areas previously listed.

Ontogeny

A by-product of Slater's very thorough review of vocal development is the revelation that no two investigators design experiments in quite the same way. Controls and stimuli are never the same, and interpretations certainly vary. As with the study of imprinting, results may even be irreproducible, for they depend "on all sorts of conditions which the experimenter may vary sometimes knowingly and sometimes unwittingly" (Bateson, 1979:473).

It is clear that learning plays an important role in the vocal development of songbirds. Sensitive periods in experimental birds are undoubtedly more flexible than once imagined, with the timing and duration of the learning phase depending not only on the quantity and the quality of the song models but also on factors such as the photoperiod and housing conditions of the subjects (e.g., individual or group isolation; see references in Chapter 12). Unfortunately, data for a single species (or population) are difficult to interpret meaningfully. It certainly is dangerous to infer from laboratory data the timing and location of events occurring in nature, and methodological differences among investigators preclude many interspecies comparisons. However, comparable information from other selected species, preferably collected by the same investigator, are needed to place developmental data in perspective.

The comparative approach that I advocate here is certainly not new. It has been exploited by Marler (e.g., 1967; Marler and Peters 1977) for years, but it needs reemphasis. Features of vocal development must coevolve with other life history strategies, and our best window on this integrative process is not to fixate on absolute measures for aspects of vocal development in a single population or species but rather to compare different treatment groups from the same population, different populations of the same species, or different yet closely related species. Furthermore, in order to understand the evolution of a behavior, experiments must be designed so that *ecologically relevant* questions are addressed. Thus, (1) does the sensitive period differ between the sexes, which tend to disperse different distances (Greenwood 1980), (2) does it vary among birds of the same population that hatch at different times of the season and hence experience different environments, or (3) does it vary between resident and migratory populations of the same species? Are differences in the learned song repertoire size among populations of the same species determined by the immediate environment or by genetic differences that have accrued in those environments? Does the model of vocal development vary with the degree of philopatry expressed by populations or species? The significance of vocal learning and the evolution of dialects (see also the next section) in the White-crowned Sparrow may be understood less by further study of that species than by a close examination of the biology of a congener, such as the White-throated Sparrow (*Zonotrichia albicollis*), where dialects, if they exist (Lemon and Harris 1974), are certainly quite different.

Perhaps I belabor the point. Briefly, we need not despair at the "immense variety" of experimental results catalogued by Slater; rather, we must exploit that diversity and design thoughtful experiments and comparisons that reveal how vocal development has coevolved with other life history parameters.

Natural variations

Slater is wisely careful in his use of the term "dialect," for few investigators would agree to its definition and applicability. Such disagreement, however, need not deter debate on the significance of the microgeographic variation in bird song, which is so easily and frequently documented among the songbirds.

The Marler–Nottebohm–Baker hypothesis states, in a nutshell, that song dialects promote assortative mating. It is at once both highly attractive and ultra controversial. In my opinion, one reason this hypothesis is often rejected is because questions being asked are too global and not sufficiently focused with an evolutionary perspective. For example, progress will be slow if we continue to ask, What is (are) the function(s) of dialects? Instead, we must ask how survival and fitness are influenced when particular individuals select various available options. For example, we must ask not whether females that pair with males of an alien dialect *can* breed, but rather we must *compare the relative fitnesses* of two females, one of which pairs with a male that sings her own song and one of which pairs with a male of an alien dialect. We must ask not whether a male can disperse from one dialect to another and still breed successfully; rather, we must compare the relative fitnesses of two males, one of which remains in the natal dialect and the other of which disperses to another.

In short, neither dispersal across dialect boundaries nor successful breeding by mates with nonmatching songs nor genetic change over distance within a dialect is sufficient to dismiss this Marler–

Nottebohm–Baker hypothesis if dialect boundaries do in fact inhibit dispersal from what would be expected by chance alone, if relative fitnesses of individuals remaining in the natal dialect are greater than for those crossing the boundary, or if genetic changes between dialects are maintained and are greater than those within dialects. Evolution works with probabilities, not with absolute values, and selection coefficients may be rather small and perhaps even unmeasurable by our gross field techniques; very important, though, is that data to test any hypothesis be collected in habitats where the system evolved – to pay tribute to one of Lack's (1965) frequent admonitions. Sufficient data on these points are not available and will be extremely difficult to obtain. Meanwhile, the best research strategy is to *focus* on the important questions, do *field work,* and maintain an *open* but *critical mind.*

Neural control of song

During the last decade, some of the most exciting discoveries involving bird vocalizations have been made in the neural control of oscine song. These discoveries include (1) neural lateralization for song control, (2) discrete song control nuclei in the brain, (3) large sexual differences in brain anatomy correlated with sexual differences in vocal behavior, and (4) correlations between the volume of song control nuclei in the brain and the total song repertoire size of an individual (reviewed in Nottebohm 1980; Arnold in press). These spectacular findings have made the songbird an important model system for assessing the relationship between brain structure and behavior, and we can expect many important studies from this field for years to come.

Other areas

Many vocalizations provide results of broad interest. For example, Gaunt and Gaunt (1980) are challenging traditional views of how the avian syrinx produces sounds, and further work there will undoubtedly clarify how birds are able to produce and accurately reproduce complex sounds. Vocalizations and other sounds must be perceived and interpreted, and the behavioral and neurophysiological data now available for owls are superb (Knudson 1980). Song synthesis and ingenious experimentation have led to a better understanding of the design features of population- and species-specific song characteristics (e.g., Becker, in press). A reemphasis of the delicate balance between informing and deceiving in animal communication, which is a reemphasis of selection at the level of the individual, has helped to place the functions of vocalizing in perspective (Dawkins and Krebs 1978).

Concluding remarks

Progress in these and other areas, such as the influence of sound propagation and the avian community on the amplitude–frequency–time envelope of vocalizations, will be rapid during the coming years. Technological advances will allow what may now be only a dream. Yet there will be no substitute for careful design of research programs at both the organismal and suborganismal level, using both experimental and observational approaches (the latter of which is a necessary complement and not a distant second to experimentation). Thus, in present-day terms, a continuous spectrum analyzer does not by itself a good research program make. The greatest threat, no doubt, is to the evolutionary biologist; habitat destruction on breeding but especially nonbreeding grounds (Keast and Morton 1980) will have ramifications that make hypothesis testing in the field increasingly difficult.

Literature cited

Arnold, A. P. in press. Neural control of passerine bird song. In D. E. Kroodsma and E. H. Miller (eds.). *Acoustic communication in birds*, Vol. 1. New York: Academic Press.

Bateson, P. 1979. How do sensitive periods arise and what are they for? *Anim. Behav. 27:*470–486.

Becker, P. H. in press. The coding of species-specific characteristics in bird sounds. In D. E. Kroodsma and E. H. Miller (eds.). *Acoustic communication in birds*, vol. 1. New York: Academic Press.

Dawkins, R., and J. R. Krebs. 1978. Animal signals: information or manipulation? In J. R. Krebs and N. B. Davies (eds.). *Behavioural ecology: an evolutionary approach*, pp. 282–309. Sunderland, Mass.: Sinauer Associates.

Gaunt, S. L. L., and A. S. Gaunt. 1980. Phonation of the Ring Dove: the basic mechanism. *Am. Zool. 20:*757.

Greenwood, P. J. 1980. Mating systems, philopatry and dispersal in birds and mammals. *Anim. Behav. 28:*1140–1162.

Keast, A. and E. S. Morton (eds.). 1980. *Migrant birds in the Neotropics: ecology, behavior, distribution, and conservation.* Washington, D.C.: Smithsonian Institution.

Knudson, E. I. 1980. Sound localization in birds. In A. N. Popper and R. R.

Ray (eds.). *Comparative studies of hearing in vertebrates,* pp. 289–322. New York: Springer-Verlag.

Kroodsma, D. E., and E. H. Miller (eds.). in press. *Acoustic communication in birds,* Vols. 1, 2. New York: Academic Press.

Lack, D. 1965. Evolutionary ecology. *Anim. Ecol. 34:*223–231.

Lemon, R. E., and M. Harris. 1974. The question of dialects in the songs of White-throated Sparrows. *Can. J. Zool. 52:*83–98.

Marler, P. 1967. Comparative study of song development in sparrows. *Proceedings of the XIV International Ornithological Congress,* pp. 231–244.

Marler, P., and S. Peters. 1977. Selective vocal learning in a sparrow. *Science 198:*519–521.

Nice, M. M. 1933. The theory of territorialism and its development. In F. M. Chapman and T. S. Palmer (eds.). *Fifty years' progress of American ornithology 1883–1933,* pp. 89–100. Lancaster, Pa.: American Ornithologists' Union.

Nottebohm, F. 1980. Brain pathways for vocal learning in birds: a review of the first ten years. *Prog. Psychobio. Physiol. Psychol. 9:*85–124.

Pearson, T. G. 1917. *Birds of America,* Vol. III. New York: New York University Society.

Smith, W. J. 1977. *The behavior of communicating: an ethological approach.* Cambridge, Mass.: Harvard University Press.

13 Bird navigation

CHARLES WALCOTT AND ANTHONY J. LEDNOR

The question of how birds find their way during migration and homing has interested the lay public and ornithologists alike. Yet despite all the work that has been done, there are still no definite answers to the question. We know many pieces of the puzzle in the form of cues that birds seem to use, but so far these pieces fail to form a coherent picture. Indeed, there are such large gaps in the puzzle that it seems more than likely that we may not even have all the pieces yet!

During the past 25 years, there has been a great increase in the experimental work on bird orientation and consequently a great proliferation of research papers and reviews. The most useful of these reviews are the ones by Emlen (1975) on migratory birds, Keeton (1974a) on homing in pigeons, Schmidt-Koenig (1979) for a general review on bird orientation and navigation, and Able (1980a) for the general mechanisms of orientation in animals. Given this proliferation of excellent reviews, the only excuse we can offer for yet another is that the field is moving so rapidly that a brief overview of new developments would complement the existing literature and might be helpful to the nonspecialist.

In thinking about bird navigation, there are two major kinds of phenomena we seek to explain. The first is the twice yearly migration of birds, often over distances of thousands of miles. The second is the homing behavior of birds – the Manx Shearwaters (*Puffinus puffinus*) released in Boston that were back in their burrows off the coast of England $12\frac{1}{2}$ days later (Mazzeo 1953; Matthews 1953; see Matthews 1968 for other examples) or the homing pigeons that return to their lofts from distances of hundreds of miles. How different these phenomena may be is not clear.

Homing, at least in pigeons, seems to involve two processes: a "map" and a compass. The displaced bird must first determine the direction to home. This action corresponds to Kramer's (1953b) "map step." Once it has determined a direction to fly, a pigeon can use a compass to find and continue in that direction.

These two processes, map and compass, correspond to Griffin's (1955) type III or "true navigation" and type II or "compass orientation," respectively. Whether migratory birds use this two-step process is unknown. There is a growing body of data that suggests that for young birds, at least, fall migration is largely a matter of flying in a certain direction for a certain length of time. Both direction and distance seem to be under endogenous control (Gwinner and Wiltschko 1978; for a review, see Able 1980a: 346). There seems to be no convincing evidence of true navigation during normal migration, although the consistent return of birds to specific nests or wintering areas certainly suggests that some form of navigation exists.

Whether migratory birds navigate or not, a major issue is the sensory basis of the map or the compass. Recent years have brought the realization that a search for a single sensory mechanism to explain the map or compass is foredoomed to failure. In fact, birds seem remarkably like people – they use a variety of cues in their travels and different species seem to weigh them differently. This makes the investigation of what birds are using especially difficult, because as the investigator interferes with one cue, the bird may promptly switch to another. In addition to this problem, birds may even use several cues simultaneously, checking one against the other. Although this process offers the (human male) equivalent of using both belt and suspenders, it makes the investigator's job even harder! The recognition of the complexity of the phenomena we wish to investigate is, in itself, a big advance. It means that there may be several answers to the question of how birds navigate. Yet, as we hope to show in what follows, we still do not have even *one* wholly adequate explanation for the navigational feats performed by migrating or homing birds (Griffin 1973).

Weather

Weather might be said to be the cause of migration: Birds move from one habitat to another to find suitable climatic conditions. Whether this is the reason for the origin of migration is a matter that has been debated almost endlessly and will not be discussed further here (see Dingle 1980 for a summary).

In the general sense, the proximate cause of migration seems to be related to two factors: First, the changing daylength seems to entrain a circannual rhythm that is responsible for fat deposition, periods of migratory restlessness, and the like. Gwinner (1973) has shown recently that this circannual rhythm behaves in many ways like a circadian rhythm in that it is apparently endogenous and free-running, but, in normal circumstances, it is synchronized to the changing length of day, thus maintaining the birds' physiology and synchrony with the natural conditions (Gwinner 1973).

Given the general physiological readiness to migrate, the immediate cue for passerine migration seems to be the arrival of the appropriate pressure pattern producing favorable winds, that is, tailwinds for the birds' intended migratory direction. Along the east coast of North America, spring migration patterns seem to be associated with low-pressure areas resulting in southwesterly winds (for a comprehensive review, see Richardson 1978). Exactly how birds detect these pressure patterns, whether the wind direction acts as a cue or whether the birds are able to sense the changes in barometric pressure, is still very much an open question. It's worth pointing out, however, that Kreithen and Keeton (1974b) have shown that homing pigeons are extraordinarily sensitive to very small changes of pressure, responding in some cases to changes of less than 4 m in altitude. This sensitivity would suffice to act as a barometer. Given favorable winds, the heaviest volume of migration usually occurs under clear skies. However, there are also nights as shown by radar and radiosonde data in which birds migrate under heavy cloud cover (cited in Emlen 1975), and, indeed, Griffin (1973) has provided evidence that birds can actually orient within cloud layers themselves. Only heavy rain seems to be a significant deterrent to migration (Eastwood 1967).

Clearly then, birds are highly dependent upon the wind in their migration. Presumably, this is an energy-saving device. The birds are effectively getting a free ride by using the wind blowing in favorable directions. Indeed, Timothy and Janet Williams (1978) have shown that many migratory birds leaving the east coast of the United States head much further east than one would expect. They pass southeast over Bermuda and continue in that direction until they meet the trade winds blowing from the east, which take them to South America. They speculate that the reason for this rather indirect route is to save energy. If we remember that most small birds have an airspeed of 30–40 km hour^{-1} and that winds at the altitude where they are flying are commonly this speed or greater, the advantage in speed and savings in energy is obvious (see also Richardson 1980).

Finally, it should be pointed out that there is some evidence that birds use the wind direction itself as a directional cue. Gauthreaux and Able (1970) and Able (1974) have demonstrated that under overcast and clear skies, particularly in the southeastern United States, passerine migrants often actively fly in the downwind direction during their periods of migration (Able 1978). Cochran et al. (1967) tracking radio-tagged thrushes reported birds often fly downwind under overcast, and even homing pigeons with frosted contact lenses behave the same way in strong winds (Schmidt-Koenig, pers. comm.). Able (1980a:323) points out that in the southeastern United States winds during the migration season tend to blow in roughly the appropriate migratory direction. Thus, a bird that simply suspended itself in the air mass long enough would end up being displaced in the appropriate direction. In the Northeast, such a strategy might result in the bird being blown out to sea. Unfortunately, we cannot find any quantitative analysis of wind data to support this argument, although it is intuitively obvious that birds migrating inland would experience less risk from crosswinds than birds along the coast.

How and why birds actively fly downwind seems mysterious. Among many other difficulties, how could the birds know which direction the wind was blowing when they themselves are flying in the moving air mass in a cloud without visual reference cues? Also, if the birds had that information, one would certainly think they would tend to fly in the opposite direction to prevent their being blown off course (for a useful discussion, see Able 1980a).

We know that under some wind conditions birds *are* blown off course, or at least groups of birds tend to congregate and land on areas that are downwind (Able 1977, 1980a, among others). There are many accounts of birds under unfavorable conditions landing on ships, islands, and other places to one side of their normal migratory route, suggesting they have been deflected by strong crosswinds. It's not clear whether birds displaced in this way can compensate and return to their normal breeding grounds, although the work of Gauthreaux (1978) in watching the flight directions of birds the morning after a major migration certainly suggests that they can. More convincing evidence comes from experiments in which *Zonotrichia* Sparrows returned from a displacement from California to Maryland, far outside of their normal winter range (Mewaldt 1964). The data from a variety of displacement experiments show that some species of migratory birds can, in fact, home (for a summary, see Emlen 1975; Schmidt-Koenig 1979). This is a particularly interesting finding since, as we will see, we have no idea what migratory birds use as a navigational mechanism to find their way back to either their breeding area or wintering area, nor, indeed, whether they normally use navigation at all in their migrations!

The available cues

Topographic features

Bird migration does not occur as a uniform, solid movement of birds southward across the United States in the fall. Rather, migration tends to be concentrated along certain flyways that are generally associated with various topographic features. Birds seem to fly along the coast in the eastern United States or often follow major rivers and valleys all in a generally north–south direction. The obvious suggestion is that birds are paying attention to topographic features and simply following these valleys or other landmarks to the south (for a summary, see Emlen 1975:148). Yet the literature based on radar data gives little support to this obvious idea. We've mentioned already the birds heading east from New England following favorable winds, and Lack (1962), Eastwood (1967), Bruderer (1980), and others have shown that in Europe, although birds may respond to major topographic features, there is no simpleminded following of a series of topographic cues.

In studying the homing of pigeons, it has become quite clear that topography plays only a minor role (Michener and Walcott 1967). Wagner (1972), in Switzerland, followed pigeons in helicopters and showed that they tended to keep in the valleys unless so doing put them far off course in which case they did not hesitate to fly up and over ridges and low mountains. Schlichte (1973), Schmidt-Koenig and Schlichte (1972), and later Schmidt-Koenig and Walcott (1978) followed homing pigeons equipped with frosted contact lenses, which should have prevented the pigeons from using any topographic information. These pigeons oriented just as well as controls and often managed to arrive within 1 km or so of the home loft. These experiments certainly suggest that detailed form vision is not essential for pigeon homing and that whatever navigational system the pigeon is using is accurate to within 1–2 km of the loft. Able (pers. comm.) has used frosted contact lenses on White-throated Sparrows (*Zonotrichia albicollis*). When these birds, equipped with lenses, were released from balloon-borne boxes and tracked with radar, they actively flew downwind.

All of this is not to say that landmarks cannot be used and are not important to birds. For example, homing pigeons flying over the land seem to be able to compensate for crosswinds and are not blown off course (Michener and Walcott 1967). Presumably, the use of landmarks and topographic features enables pigeons to change their heading so as to minimize wind drift. For migratory birds, Bingman et al. (1982) have shown that the Hudson River can serve as a geographic cue and that birds flying incorrect courses tend to correct them as they approach this prominent landmark.

All in all, however, it appears that migratory birds place relatively small reliance upon topographic features and that they must be using some other source of information on their migratory journeys.

Stars

That the stars were used as a major source of directional information is an old idea, but it was only relatively recently that the pioneering work of Emlen (1967a,b) put it on a solid basis. Emlen showed that Indigo Buntings (*Passerina cyanea*) were able to choose the appropriate migratory direction using star patterns. By putting adult buntings in a planetarium, he demonstrated that their orientation depended on stars located within about 35° of the pole star. He showed also that juvenile buntings learned the star patterns by observing the rotation of the sky; if the rotation of the planetarium sky was altered so that the axis of rotation was centered around Betelgeuse in the constellation Orion, the buntings' orientations would be correspondingly shifted under the normal sky (Emlen 1970; reviewed in Emlen 1975).

Finally, by altering the physiological conditions of birds and bringing them into the fall migratory condition in the spring, Emlen (1969) and Miller and Weise (1978) were able to demonstrate that Indigo Buntings oriented in a direction corresponding to their internal condition, not that of the external cues. Emlen concluded that the Indigo Buntings learned the direction of north by watching the rotation of the sky pattern and that their choice of whether to fly north or south was determined by their internal physiological condition, not by any set of exterior cues. Thus, although there is good evidence that the stars serve as a compass reference for buntings, Emlen was not able to find any evidence that buntings shown a star pattern characteristic of an area far to the west of Ithaca, New York, were able to compensate for this apparent westward displacement. A few other species (see Emlen 1975 for a summary) have been shown to use star patterns in their orientation, but surprisingly few species have been studied carefully.

Star patterns cannot be the only cue used by migrating birds for at least two reasons. First, a transequatorial migrant starting off in South America or Antarctica and heading to Canada, for example, would not be able to see all the stars within 30° of the pole star until it reached a latitude of 30° N. If such birds used the stars, to what patterns are they paying attention in the initial part of their journey? As far as is known, there is no answer to this question. A second reason is, of course, the observation mentioned earlier that birds were able to migrate under conditions of total overcast, and, presumably, they cannot see the stars through the clouds. Yet the radar data (Emlen 1975) indicate that such birds are often well oriented, implying at the very least that they are making use of some other cue system to keep them on course. Thus, star patterns, important as they may be for some species, can only be *one* of the cues the birds are using on their migrations.

The sun

It was the work of Kramer (1951, 1953a) and Hoffman (1954) who showed that European Starlings (*Sturnus vulgaris*) made use of the sun as a compass. At the time, this seemed like an outrageous suggestion. The sun, after all, appears to move through the sky, and any animal using it as a compass reference must compensate for this apparent movement. This implies that the animal would have to possess an internal clock as well as having to deal with daily and seasonal variations in the rate of change of the sun's azimuth. By training experiments, Kramer and Von St. Paul (1950) were able to show that starlings did indeed use the sun as a compass reference, and Hoffmann (1954, 1965) was able to demonstrate the relationship of the sun compass to the birds' internal clock.

In the case of homing pigeons, the clock-shifting experiments have proved to be an enormously fruitful experimental technique, and Schmidt-Koenig (1979), Keeton (1974a), and others have shown that pigeons, whose internal clocks are 6 hours out of phase with the real day, vanish approximately 90° away from the direction chosen by control birds released at the same site. If homing pigeons are clock-shifted but released on an overcast day, Keeton (1969) demonstrated that both shifted and control birds were well oriented toward home. These are important experiments, because they demonstrate that pigeons appear to use the sun as a compass when it's available and that the clock-shifting procedure has had no effect upon the pigeons' ability to determine in which direction home lies. This is shown because, as pointed out earlier, the 90° shift is always with respect to the roughly homeward orientation of the control pigeons. Furthermore, under overcast skies when the pigeons are not using the sun, both the shifted and unshifted birds are clearly homeward oriented. Clock shifting has become one of the clearest and most repeatable experiments. Clearly, pigeons use the sun as a compass and are well able to compensate for its apparent movement through the sky.

In the 1950s, G. V. T. Matthews (see Matthews 1968 for a summary) proposed that pigeons also use the sun as a navigational reference. He proposed that pigeons used the sun to determine where they had been released relative to the home loft in a way rather analogous to that of a human navigator. Unfortunately, there is little direct evidence in support of this ingenious hypothesis. It turns out that birds given short clock shifts corresponding to modest displacements either east or west of the loft both orient and home normally, and birds released under total overcast skies also show good homeward orientation. Thus, at present, there is no evidence that supports the idea that birds are able to use the sun for positional information (see Keeton 1974a for a review).

Vleugel (1954) proposed, and Bingman and Able (1979), F. Moore (1978, 1980), and Emlen (1980) have shown, that some migrating species pay attention to sunset. They apparently use either the light glow, polarized light patterns (Able, in press), or some other feature of sunset to choose the direction of their nightly journey. Again, this appears to be an alternative compass mechanism. Once birds have chosen the appropriate direction, they may use the star patterns to maintain it, although, as pointed out before, they can even maintain orientation under overcast skies. To summarize then, the sun appears to act as a compass reference but not as a source of positional information, and homing pigeons and some migratory birds appear to use this source of information in preference to the stars or other directional cues.

Geomagnetism

Since migratory birds are able to maintain a constant course in the absence of stellar information and at night when the sun is not available, one must ask if there is an additional compass that they may be using. The pioneering work of Merkel et al. (1964) and Wiltschko and Wiltschko (1972) has shown that migratory birds, particularly European Robins (*Erithacus rubecula*) and a number of European warblers, are able to use the earth's magnetic field as a compass. This conclusion was drawn from a long series of experiments in which birds in migratory condition were placed in a cage with eight radial perches and their activity was monitored over the course of a night. Generally, it was found that each night's perch-hopping activity did not lead to significant orientation in any direction, but by pooling the nightly means over a number of nights, significant results were obtained. Furthermore, the direction chosen by the robins was in the appropriate migratory direction, namely, north in the spring and roughly south in the fall. By changing the direction of the magnetic field around the orientation cage, the mean direction was altered in accordance with the magnetic field. Thus, although one may object to the statistical techniques employed in these experiments, the results seem to be biologically meaningful, and the fact that changes in the magnetic field lead to predictable changes in the orientation lends further support to the idea that birds are using magnetic information. One may argue that the large scatter of bearings in the orientation cages suggest that birds would not find magnetic field information very useful in the wild. This argument can be countered by the argument that these effects are showing up in orientation cages in which the birds move only a very small distance in a highly artificial environment, and given these conditions, it is astonishing that one sees any effect at all. Clearly, we have to wait until it's possible to do more sophisticated magnetic manipulations on free-flying birds to determine whether or not they really use magnetic cues in their migrations.

These exciting experiments of the Wiltschkos and their collaborators were initially greeted with some skepticism in the United States, but in a collaborative study between the Wiltschkos and Emlen's group, conducted at Cornell University, Ithaca, New York, in 1974, Indigo Buntings were also shown to use magnetic information in their orientation (Emlen et al. 1976) About this same time, Keeton (1971, 1972) demonstrated that homing pigeons equipped with small bar magnets showed essentially normal orientation under sunny skies but were often, although not always, disoriented under overcast skies. Walcott and Green (1974), elaborating on this experiment, used a pair of coils to produce a more uniform magnetic field around the pigeon's head and found that when released under overcast skies birds exposed to one magnetic polarity flew home, whereas birds with a reversed field headed 180° away from home. This experiment was successfully repeated by Visalberghi and Alleva (1979) in Italy. These experiments, taken together, led to the idea that pigeons as well as migratory birds can use the earth's magnetic field as a compass. For the pigeons at least, the sun is used as the preferred compass reference under sunny skies, but if it were not available, the earth's magnetic field served as a backup (for a review of bird orientation and magnetic fields, see Ossenkopp and Barbeito 1978).

It is surprising enough that birds could detect the relatively weak magnetic field of the earth. However, in about 1970, William Southern (see Southern 1978 for a summary), who was working with the orientation of young Ring-billed Gulls (*Larus delawarensis*), found that their orientation was relatively variable. In examining possible sources of this variability, he looked at the fluctuations in the earth's magnetic field as measured by the *K-index*. This index is a measure of temporal changes in the earth's magnetic field. To his surprise, he found that the orientation of his gulls was good on days with relatively low K-indexes (calm magnetic fields) but became disoriented – the bearings became more scattered – on days with a high K-index (disturbed magnetic field). Furthermore, under overcast skies, any K-index greater than 1 was sufficient to cause disorientation, whereas when the sun was visible, a K-index greater than 4 was necessary to disorient the chicks. These results are particularly interesting because they suggest a sensitivity to magnetic fields vastly greater than any previously proposed. They imply that the gull chicks are responding to magnetic field variations on the order of tens of gammas, approximately one-five thousandth of the normal earth's magnetic field. Keeton et al. (1974), using homing pigeons released repeatedly from the same location, found that the direction of their average vanishing bearing varied depending upon the K-index. They reported that increases in the disturbance of the field deflected the average vanishing bearing to the left of home, even under sunny skies. Once again from the steepness of the regression line, the sensitivity of the pigeons can be estimated to be on the order of tens of gammas.

There are reports (Graue 1965; Talkington 1967, unpub.; Wagner 1976; Frei and Wagner 1976) that irregularities in the earth's magnetic field also alter pigeon orientation. With these findings in mind, Walcott (1978) released homing pigeons at magnetic anomalies, places where the earth's magnetic field is disturbed by large deposits of magnetic ore. The birds were disoriented (Walcott 1980). By releasing pigeons at a number of anomalies whose strengths varied, it was possible to correlate the accuracy of the pigeons' vanishing bearings with the degree of magnetic irregularity at the anomaly. The results of the experiment showed an association between these two variables, and once again we can estimate the sensitivity of the pigeon to magnetic fields. The results suggest that changes in the magnetic field on the order of tens of gammas are sufficient to cause a significant increase in the scatter of pigeons' vanishing bearings under *sunny* skies. These results are exceedingly interesting because they suggest that birds have a much greater sensitivity to magnetic fields than would be required for a simple compass mechanism. The earth's magnetic field has the strength of approximately 55,000 gammas, and the small variations at anomalies or due to magnetic storms would be expected to have little or no effect upon a magnetic compass. The intriguing question thus becomes, Could this high sensitivity to magnetic fields be part of a navigational system rather than part of the magnetic compass? At the moment, we simply don't know the answer to this question, but the possibility is being actively investigated (Walcott 1980, in press; B. R. Moore 1980).

It is interesting that disturbances of the earth's magnetic field as measured by the K-index deflect the average vanishing direction of experienced pigeons, whereas magnetic anomalies simply increase the scatter. One wonders whether this is related to the fact that although the earth's field varies with time during magnetic storms, the effect is relatively uniform spatially; in contrast, the irregularity of the field at anomalies is variable in space but constant in time. Perhaps it is significant that at the much weaker and more uniform magnetic anomalies in Switzerland, Wagner (1976) and Frei and Wagner (1976) found a deflection of direction but no increase in scatter of pigeon vanishing bearings. The disorientation associated with high K-values observed by Southern (1978) with young gull chicks contrast with the results obtained for pigeons. Perhaps species, age, or experience differences are important here.

If birds are using the earth's magnetic field, either as a compass or as the basis of their map, there must be some sense organ that allows them to detect it. There are several ways one could imagine such a sense organ might operate. It could be based on: (1) the induction of a current in a conductor as it cuts magnetic lines of force, (2) torque on either a permanently magnetized particle or a paramagnetic particle, or (3) by some effect such as the optical pumping mechanism described by Leask (1977).

Blakemore (1975) reported that certain bacteria contain a chain of magnetite particles. The torque exerted on these particles by the earth's magnetic field aligns the bacteria parallel to the field, and since the alignment works just as well whether the bacteria are dead or alive, it is clearly a passive process. In live bacteria, however, the flagellae rapidly propel the bacteria down to the surface of the mud where they live (Frankel et al. 1979; Denham et al. 1980).

The finding of magnetic material in bacteria led Gould and colleagues (1978) at Princeton to look for magnetic material in honey bees. They found that honey bees were indeed magnetic and contained magnetite in their abdomens. Pigeon heads were examined and also found to contain magnetic material, which several lines of evidence indicated was magnetite (Walcott et al. 1979).

Magnetic material has been found in the necks of pigeons (Presti and

Pettigrew 1980) and in dolphin heads (Zoeger and Fuller 1980; Zoeger et al. 1981). The anatomy of iron-containing tissue in pigeons has been summarized in Walcott and Walcott (in press). Although Kirschvink and Gould (1981) have summarized a number of possible mechanisms that might be used by an animal with magnetite in its tissues to detect magnetic fields, with the exception of the bacteria, there is no clear link between magnetite and the animals' orientation. Clearly, this is an exciting field, and experiments to test the possible role of magnetite in pigeon homing are currently in progress.

Olfaction

Odor is an important cue for many animals and its involvement in orientation has been demonstrated in several species. Insects, for example, frequently use odors in their orientation; the most dramatic example being that of the male silk moth who locates the female by virtue of a pheromone that she secretes and is carried by the wind. The male moth then simply has to fly upwind in order to locate the female (Fabre, cited in Schneider 1974). In a somewhat analogous way, salmon find their way back to the spawning beds where they were hatched by recognizing tiny concentrations of organic chemicals dissolved in the river water (Hasler 1960). Given these examples, it should perhaps be no surprise that birds also make use of olfactory information in finding their way. The first example of this was described by Grubb (1974; see also Billings 1968) who established that Leach's Petrels (*Oceanodroma leucorhoa*) found their way back to the island on which they nest using at least, in part, olfactory cues. Grubb noticed that birds tended to approach the island always flying upwind and then searched around for the location of their burrow when they arrived at the island. It's interesting that these examples are quite similar; animals are finding the source of an odor not by moving up a gradient but rather by orienting to either wind or water currents that bring the odor to them. Obviously, such a scheme is impractical for a homing bird, unless, of course, the home loft has a strong and distinctive odor that could be carried for hundreds of miles. Even if this were true, it would provide useful information to birds only when they were downwind of their loft. Yet in 1972 Floriano Papi and his colleagues at the University of Pisa proposed that homing pigeons did use olfactory cues in finding the loft. Papi's basic hypothesis was that pigeons learn to associate odors with the direction from which the wind is blowing. Thus, a pigeon in its loft notices that when the wind is blowing from the north, it detects odor A, when the wind is from the south, odor B, and so on, thus, building up a olfactory map of the surroundings. Slightly later, Papi added the idea that pigeons being transported to the release point paid attention to olfactory cues while in transit, allowing the birds to deduce the direction of travel and, therefore, the direction of home. This is an extremely interesting and ingenious hypothesis but one which is fundamentally much more complicated than simply flying or swimming upwind or upstream as long as one detects a characteristic odor, as moths and salmon appear to. Despite this apparent complexity, however, Papi and his group together with Hans Wallraff in Germany have now assembled an impressive array of experiments that seem to support the olfactory hypothesis (for a review, see Papi 1976; Papi et al 1980a; Wallraff 1980a). Basically three types of experiments have been used: (1) to interfere with the pigeons' olfaction, (2) to manipulate olfactory cues during the pigeon's trip to the release site, and (3) to distort the learning of appropriate wind information at home. Because the olfactory hypothesis is relatively new and because it is quite controversial, we think it is worthwhile to outline in some detail the various experiments that have been performed and their results.

Interference with olfaction. The most obvious approach here is simply to cut the olfactory nerves, release the pigeons, and compare their orientation with sham-operated controls. The first experiments of this kind were done by Papi et al. (1971) and Benvenuti et al. (1973). They found that at unfamiliar sites birds with sectioned olfactory nerves showed poorer homing success than controls, but Benvenuti et al. (1973) only found slight effects on the bird's initial orientation. Hermayer and Keeton (1979) repeated these experiments and in a series of eight releases found no consistent effects on the birds' vanishing bearings but confirmed that more experimental birds than controls were lost in releases at unfamiliar sites. In the most extensive series to date, Wallraff (1980c, 1981) released over 600 first-flight pigeons (birds that have not previously been taken away from their home loft) at various distances from their home loft. Experimental birds had both olfactory nerves sectioned, whereas controls were mainly untreated. Wallraff reports that in most releases experimentals were less well oriented than controls, and very few birds from either group returned home. A large proportion of the lost birds were recovered at various locations, and analysis of these recovery data showed that the distributions of the control recoveries were homeward oriented; the experimental birds were, in general, randomly oriented but were recovered at similar distance to the controls.

One problem with using bilateral surgery to make birds anosmic is that it seems all too possible that the surgery would have nonspecific effects that would increase the stress on the birds and lead to a reduced motivation to home. To try and equalize the trauma of surgery, an alternative technique was developed by Papi et al. (1972). Both controls and experimental birds had one olfactory nerve sectioned, and prior to release a plug was inserted in one nostril, either on the same side as the severed nerve (controls) or the opposite side (experimentals). This technique has also been used by Baldaccini et al. (1975), Fiaschi and Wagner (1976), and Papi et al. (1978b). In these four studies, few tests based on this procedure were done and variable results found regarding initial orientation. From unfamiliar sites, experimental birds generally showed poorer homing than controls. More recently, Benvenuti (1979) conducted a series of nine releases, at mainly unfamiliar sites, using this method of making birds anosmic. Both controls and experimentals were poorly oriented, and no significant difference in orientation was found between them when the results were pooled. The experimental birds, however, had a significantly poorer homing success. Papi et al. (1980b) transported birds made anosmic in this way to the release site in either aluminum or iron containers. Comparing the pooled results of the five releases of anosmic and control birds transported in the aluminum container, a significant difference in initial orientation was found. Control birds were homeward oriented, whereas anosmic birds were randomly distributed. There was no difference between the groups in terms of the numbers of birds lost, although of those that homed, control birds were slightly faster than the anosmic birds.

An alternative approach to making birds anosmic has been the use of nasal tubes to bypass the olfactory mucosa. Keeton et al. (1977) performed a series of releases at unfamiliar sites at various distances from the loft using birds fitted with either one (controls) or two tubes (experimentals). No differences were observed between the initial orientation of experimentals and controls. There was similarly no difference in homing performance between the two groups as long as the birds were released less than 25 km from the loft. At longer distances, the experimental birds homed less well than the controls. In three other series, birds made anosmic this way and released at familiar sites did as well as controls both in their vanishing directions and in homing success (Hartwick et al. 1977; Papi et al., 1978b; Benvenuti 1979). At unfamiliar sites, all investigators have reported poor homing performance on the part of the anosmic birds, but inconsistent results have been obtained for initial orientation. Although no difference between controls and experimentals were seen by Papi, et al. (1978b), in the other two studies, the experimental birds were either randomly oriented or showed a significantly different orientation from the controls.

Schmidt-Koenig and Phillips (1978) tried yet another technique. A local anesthetic was sprayed on the pigeons' olfactory mucosa prior to transport to the release site and again at the site. This treatment produced only slight and inconsistent effects on the birds' initial orientation. Homing performance was similarly little affected, but as it had been shown that the effect of the anesthetic only lasts between 1 and 2 hours, this does not seem surprising.

Taken all together, what do these anosmic tests reveal? It seems clear that the various methods of making birds anosmic often have an effect, particularly on homing performance from distant unfamiliar sites, but the effects on initial orientation have been less clear-cut. Although experiments in Italy and Germany have shown that anosmic birds are often disoriented or deflected in their orientation no similar marked effects have been seen in any of the experiments conducted in the United States at Ithaca, New York.

Keeton et al. (1977) and Papi et al. (1978b) argue that because anosmic birds in Ithaca show good initial orientation, there is no evidence that their navigation system was disturbed. Furthermore, they argue that the birds' poor homing performance from distant unfamiliar sites might not result from a disturbance of their navigation but rather be caused by stress, which reduced their motivation to home. This possibly has already been mentioned regarding the bilateral surgery technique, but it also applies to the other procedures. Although unilateral nerve section combined with nasal plugs controls for any general effects produced by the surgery, experimental birds suffer from a complete lack of respiratory feedback, whereas controls with one nostril free do not. Nasal tubes were designed to overcome this problem, but it was found (Keeton et al. 1977) that they often became blocked with mucous and consequently affected the birds' respiration.

Airplane radiotracking of anosmic Cornell University birds supports Keeton's view; of the five experimental birds for which tracks were obtained, four landed a few kilometers from the release site, whereas the fifth followed a flight path for approximately 11 km, which was similar to that flown by control birds, before it too landed (Papi et al. 1978b). In addition, several investigators have noted a tendency for anosmic pigeons to land in the vicinity of the release site (e.g., Papi 1976; Hartwick et al. 1977), which may simply reflect the birds unwillingness to fly under conditions of stress and discomfort.

Wallraff (1979, 1980c, 1981) has stressed that the recovery data obtained in these experiments show that the anosmic birds flew just as far as the controls, and, thus, anosmic birds appeared equally motivated to home. However, as Wallraff presents no data on the time course of the recoveries, his interpretation is difficult to evaluate. A different argument has been put forward by Papi et al. (1980a). They point out that although anosmic pigeons are frequently lost when they are released at unfamiliar sites, they have no difficulty homing from familiar sites. This suggests that anosmic birds are sufficiently motivated to home but at unfamiliar sites simply lack the ability to determine the direction of home. This is a compelling argument, but as Keeton in his discussion in Papi et al. (1978b) suggests, "birds with lowered motivation resulting from the discomfort of the nasal tubes might be more willing to keep going at a thoroughly familiar location than at an unfamiliar one."

Odorous substances. If pigeons are really attending to odor cues as they home, the application of a strong masking odor might be expected to cause disorientation. Benvenuti et al. (1977) plugged the pigeons' nostrils during transport so they were unable to detect any olfactory information on the outward journey and at the release site coated the beaks of experimental pigeons with α-pinene for the experimentals and Vaseline for the controls. They reported that experimental birds oriented poorly toward home and homed more slowly, whereas controls in every case oriented homeward. Fiaschi and Wagner (1976) were able to repeat these experiments in Switzerland, but Keeton and Brown (1976), Hartwick et al. (1978), and Papi et al. (1978b) found no clear effect of this treatment either on initial orientation or homing performance. Once again, there is a diversity of results in different areas and with different pigeons, all of which makes it hard to come to any clear conclusion.

In summary, then, interfering with olfaction by severing the olfactory nerves seems to have a variable effect on orientation but to definitely interfere with successful homing from distant unfamiliar sites. Nasal tubes seem to have roughly the same effect, whereas anaesthesia has no effect at all. In Italy and Switzerland, applying odorous substances to the pigeons' beak interferes with both orientation and homing, but in Germany and America, pigeons are immune to such interference.

Outward journey. The second large group of experiments involves manipulations on the journey to the release site. In the first of these, known as *detour experiments*, pigeons were taken to the same release point by quite different routes. If, for example, the release point was north, one group of pigeons was taken there via a substantial detour to the east, and another group taken there via a substantial detour to the west. Thus, for one group of birds, the actual arrival at the release site was from the southeast, and for the other group, from the southwest. If the birds had been paying attention to olfactory information on the route to the release site, one might expect the two groups to head home in quite different ways. Papi et al. (1973) performed five such releases, and, in each case, the vanishing bearings of the birds were deflected in the predicted directions. Repetitions of the experiments (Keeton 1974b; Fiaschi and Wagner 1976; Papi 1976; Papi et al. 1978b; Hartwick et al. 1978) have yielded quite variable results. Papi et al. (1978a) have summarized the Italian findings. Of 27 detour experiments, there were only 7 cases in which both detour groups were significantly oriented in different directions. In each of those 7 cases, the deflections are in the predicted directions. It is important to point out that detour effects are consistent with several models, for example, inertial (Barlow 1964, 1966) or magnetic (Wiltschko et al. 1978) navigation as well as the olfactory hypothesis. To test whether olfactory information was responsible for the effect; Papi et al. (1978a) conducted nine experiments in which birds were transported to the release site on a detour route with either one (controls) or both (experimentals) nostrils plugged. Although the pooled results of these experiments show no difference in the vanishing bearings of the two groups (both were markedly deflected to the right of home), in only one test were both groups significantly oriented. Papi et al. (1978a) interpret this result as "showing that the detour effect is due to odors sensed during the outward journey."

In other experiments, the effects of restricting the olfactory information available to the birds during their journey to the release site have been investigated. Wallraff (1980b) transported pigeons under "cueless" conditions in Germany and then in Italy (Wallraff et al. 1980). The birds were transported on a rapidly spinning turntable in an artificial magnetic field and were supplied with bottled air. No effect of this treatment was seen in Germany, and only slight effects were seen in Italy. Wallraff and Foa (1981) used charcoal filters to remove odors from the air with which pigeons were supplied during the outward journey; control birds had access to olfactory cues. Prior to release, control and experimental birds were made anosmic, using xylocaine. Because they could not obtain olfactory information at the release site, the expectation was that birds deprived of olfactory cues on the way to the release point would be less well oriented, and so they seemed to be. Unfortunately, control birds were very poorly oriented also, and the difference between the two groups, although significant, is not impressive.

Taken together then, all the results of altering the olfactory experience of pigeons on the way to the release site suggest that there is sometimes an effect but that it is neither as clear nor as repeatable as one might perhaps desire.

Deflector lofts. The third approach is that of distorting olfactory information at the home loft. This is usually accomplished by altering the direction of local winds and thus distorting the pigeons' correlation between odors and the direction from which they come (Baldaccini et al. 1975). There are two major sets of such experiments: the deflector lofts and the fan tunnels. The deflector lofts consist of a slatted loft in which large protruding panels deflect the apparent angle of the arriving wind (Baldaccini et al. 1975). They come in two forms, one that rotates the wind pattern clockwise by about 70° and the other counterclockwise by about 70°. Birds were raised in these lofts and then released with the result that their vanishing bearings were deflected in the same sense as the rotation of the winds in the experimental lofts. This experiment has been repeated in Germany by Kiepenheuer (1978) and at Cornell University by Waldvogel et al. (1978). For once, the results in all three places are in almost complete agreement; birds raised in these lofts shown the appropriate clockwise or counterclockwise deflection! Although as Papi et al. (1980a) point out, the results of the deflector loft experiments are in agreement with the olfactory hypothesis, they do not prove it unambiguously; the shields also reflect other stimuli, sounds and light, to mention two.

Waldvogel and Phillips (in press) found that deflector loft birds released under overcast skies showed a greatly reduced deflection. Under sunny skies, the deflection returned. Furthermore, Kiepenheuer (1979) showed that treating birds from the deflector lofts with xylocaine had no effect – their orientation was the same as unanaesthetized, deflector loft birds.

Waldvogel et al. (1980) showed that pigeons need not be permanent residents of the deflector lofts to show significant deflection of their initial orientation. As little as 7 days' exposure was sufficient. Using this "short term deflector loft technique," Waldvogel and Phillips (in press) have shown clearly that the cue responsible for the deflected orientation is reflected light, not olfaction. They did this by replacing the Plexiglas panels of the loft with altered panels. These panels rotated the wind in one direction but reflected light cues in the other. Short-term birds in such altered-wind lofts showed orientation that was deflected in accordance with light cues, not the wind direction. Finally, they found that there was a seasonal effect in the deflection, which disappears in late summer. This time of disappearance corresponds to the season when the sun's disk is no longer in the appropriate position at sunrise and sunset to creat reflected polarization patterns in the deflector panels. Phillips and Waldvogel (in press) propose that pigeons might use skylight polarization patterns in the calibration of their sun compass. Interestingly, Able (in press) finds that sparrows viewing the sunset sky through polarization filters show a deflection of their activity patterns. All in all then, it appears that the deflector loft effect, real though it is, probably is due to light cues rather than olfaction.

We should mention the results of a series of experiments in which birds were exposed to artificial winds generated by fans coupled with artificial odors. Birds were raised in tunnels with fans generating an artificial wind current. When the fan at one end of the loft was on, turpentine was injected into the air, and when the fan at the other end of the loft was on, olive oil was injected into the air. Thus, the birds should learn to associate characteristic odors with artificial winds from either of two directions and, being taken to a release point and treated with either olive oil or turpentine, should fly in opposite directions. This is exactly what the Italian group reports (Papi et al. 1974), but, once again, the experiment could not be repeated at Cornell University (Quine and Waldvogel, unpub.). Finally, two further fan experiments should be mentioned. Ioale' et al. (1978) and Ioale' (1980) used the same arrangement of corridors and fans, but no artificial odors were employed. Instead, when winds were blowing along the axis of the corridors, the fans were turned on, blowing either in the direction of the natural wind (control) or against it (experimentals). The rationale was that the experimental birds should learn a reversed olfactory map and show an orientation shift of 180° from the controls when released. For the 12 experiments conducted on the axis of the corridors, a dramatic shift in the orientation of the experimental birds was reported, and in seven tests, the experimentals were clearly showing the predicted 180° shift. These are impressive results, but one or two worries remain. Although Ioale' and his colleagues consider the results as supporting the olfactory theory, their case would have been more convincing if further controls had been used, for example, making the experimental birds anosmic and showing that no deflection resulted. A hint that something more then olfaction is involved comes from two releases conducted in directions roughly at right angles to the corridors. Because the birds had not received any olfactory information from these directions, olfactory theory should predict that the birds would be disoriented. However, in both experiments, controls and experimentals were well oriented toward home. Overall, the experiments provide strong circumstantial, but not conclusive, evidence in favor of olfactory theory; further investigation of the effect is needed.

The question then is where do all these experiments leave us? Papi and the Italian group consider them overwhelming evidence that pigeons do use olfactory information, and, indeed, Wallraff (1980a) has gone so far as to say that olfactory information is essential for the successful homing of pigeons and indeed forms the basis of the pigeons' map or true navigation system. Looking over the mass of evidence that Papi and his colleagues have gathered from their extensive and ingenious series of experiments, it is hard not to agree with them. Yet there are some very curious results that certainly demand further investigation.

Why do anosmic birds in some regions consistently show homeward orientation but poor homing from unfamiliar sites? Is this due to poor motivation to home or is it due to a specific navigational deficit? The results of the deflector lofts seem to be due to light reflections rather than odor, but what is really being reflected and detected? Is the detour effect really due to olfaction? The wind fan reversals certainly suggest odor cues are involved, but, if so, then why can birds orient when released at right angles to the corridor axis?

These are all open questions, and we consider that the evidence, as it stands, does not yet convincingly show that olfaction is essential to homing in pigeons. With further research, this may turn out to be the case or, as Keeton (1980) has argued, olfactory cues may be just one of the many sources of information birds may use in navigating. It may well be that birds from different regions rely on different cues to different extents; certainly pigeons from Ithaca, New York, seem to rely less on olfactory cues than do Italian birds. Whatever the answer, there are some interesting questions here that have yet to be resolved.

Other potential cues

In an exciting series of papers, Melvin Kreithen has demonstrated that the sensory world of birds, or at least homing pigeons, extends much further than we had previously anticipated. Pigeons are sensitive, for example, to tiny changes in air pressure (Kreithen and Keeton 1974b). They can detect the pressure change involved in a vertical displacement of 3 to 4 m. This should be useful to a bird in maintaining a constant altitude, but it might also be useful in detecting the pressure changes associated with weather fronts, although there is no direct evidence that pigeons do so. Secondly, pigeons are sensitive to short wavelengths of light; they can see in the near ultra-violet (Kreithen and Eisner 1978). What use birds make of this ability is totally unclear. Pigeons can also detect polarized light patterns (Kreithen and Keeton 1974a). Whether this enables them to see polarization patterns in the sky or what they use this ability for is again totally unclear. Finally, pigeons are sensitive to very low frequencies of sound, down to something of the order of 0.10 Hz. Yodlowski et al. (1977) and Kreithen and Quine (1979) have demonstrated that pigeons can detect these frequencies, and Kreithen suggests that pigeons might use very low-frequency sound sources as beacons. If they did that, the birds would have to be able to localize the source of these infrasounds, a difficult task for an animal. The distance between ears is only a tiny fraction of the wavelength of sound! Kreithen and Quine (1979), however, point out that the pigeons' frequency discrimination ability is accurate enough even at these very low frequencies for the birds to make use of the Doppler frequency shift as they flew in a circle. So far, however, there is no evidence that pigeons actually use infrasounds in their orientation.

Integration of cues

Although so far the information that birds might use in their orientation has been considered cue by cue, there is every reason to think that birds may use several cues either at the same time, sequentially, or alternately. For example, experienced homing pigeons appear to use the sun as a compass when it is available but switch to using the earth's magnetic field when it is not. In this case, there is a clear hierarchy.

For migratory birds, the situation seems more complex. Wiltschko and Wiltschko (1975) tested three species of warblers during migration. Placing them in a standard orientation cage with a clear view of the sky and a normal earth strength magnetic field resulted in normal, southerly orientation. However, if Helmholtz coils around the cage moved magnetic north to 120°, the orientation of the birds shifted to geographic northwest, the direction of magnetic south. This implies that the warblers were paying attention to the direction of the magnetic field and ignoring the stars. Nulling the horizontal component of the earth's field resulted in random orientation.

European Robins, tested in the spring, gave a somewhat different picture. The robins tested in the shifted field continued to orient as they had before. However, after 3 days, their orientation had shifted in accordance with the magnetic field. The supposition is that robins were using star patterns to maintain their orientation. Unlike the warblers, the robins maintained their orientation in the absence of the horizontal component of the earth's magnetic field. Furthermore, robins newly exposed to the shifted field and then tested with the horizontal field nulled continued to orient north. After 3 days in the shifted field, if the horizontal field were nulled, the robins' orientation was also shifted. The Wiltschkos (1975; Wiltschko 1975) interpret these results as showing that both warblers and robins use the earth's magnetic field as their primary cue. For warblers, shifting the field resulted in an immediate shift in orientation implying that the stars, if they were used, are calibrated against the direction of the magnetic field at least as often as each night. It may even be that the warblers don't use the stars at all — without magnetic field information their orientation became random. However, if this were the case, it is hard to understand how Sauer (1957) and Sauer and Sauer (1960) obtained orientation to star patterns. For the robins, the magnetic field is still of primary importance, but the stars seem to have a greater influence. Perhaps the robins cross-check the star patterns and magnetic fields less frequently; perhaps they simply put a greater reliance on stars. Whatever the case, stars are clearly more important to the robins than to the warblers.

Recently, Bingman (1981) has obtained some very interesting results in the development of orientation in Savannah Sparrows (*Passerculus sandwichensis*). He found that sparrows, reared under either the normal night sky or without the sky but with a normal magnetic field, tested in the normal earth magnetic field either with or without stars were oriented. Their orientation was bimodal with the axis of this bimodality parallel to the correct migratory direction. Why this bimodality exists is not clear; wild-caught birds were unimodal; hand-reared birds, whatever their treatment, were mostly bimodal. Despite the bimodality, clearly, the sparrows were able to use the earth's magnetic field for orientation, since shifting the field 90° also shifted the orientation a corresponding amount.

Savannah Sparrows reared indoors with no view of the night sky showed no orientation when tested outdoors in the absence of magnetic cues. Thus, experience with the night sky seems to be essential for orientation by the stars.

Bingman's most interesting result came from rearing Savannah Sparrows outdoors under the normal night sky but with a magnetic field shifted by 90°. Tested indoors in a normal earth field, such birds showed a bimodal axis of orientation 90° different from birds reared under a normal sky and magnetic field. This result means that these birds were orienting in the correct *geographic* direction in the shifted field. It implies that they were deriving correct geographic direction from some other source, most likely the stars, and then interpreting the magnetic field in this context.

These results clearly indicate that the development of the orientation system in birds may be quite a complex affair. It also emphasizes that several cues may be involved and that different species may well use them differently.

Summary

Recent years have seen a great increase in our knowledge about the cues that birds use in their migration and homing, and despite this increase in knowledge, we are still unsure about exactly what sensory cues birds use on their extraordinary journeys. Bird orientation is an exciting field with rapid progress, yet we are still at that position described by Griffin (1973) where for the basis of the pigeon's map we are unable to name even one cue as being definitely involved. Finding that first cue is a challenge for us all.

Literature cited

Able, K. P. 1974. Environmental influences on the orientation of free-flying nocturnal bird migrants. *Anim. Behav. 22*:224–238.
– 1977. The orientation of passerine nocturnal migrants following offshore drift. *Auk 94*:320–329.
– 1978. Field studies of the orientation cue hierarchy of nocturnal songbird migrants. K. Schmidt-Koenig and W. T. Keeton (eds.). *Animal migration, navigation and homing*, pp. 228–238. Berlin: Springer.
– 1980a. Mechanisms of orientation, navigation and homing. In S. A. Gauthreaux, Jr. (ed.). *Animal migration, orientation and navigation*, pp. 284–364. New York: Academic Press.
– 1980b. Evidence on migratory orientation from radar and visual observations: North America. In R. Nöhring (ed.). *Acta XVII Congressus Internationalis Ornithologici*, pp. 540–546. Berlin: Verlagder Deutschen Ornithologen-Gesellschert.
– in press. The role of polarized light in migratory orientation of White-throated Sparrow (*Zonotrichia albicollis*). In *Proceedings of the International Symposium on Avian Navigation*.

Baldaccini, N. E., S. Benvenuti, V. Fiaschi, and F. Papi. 1975. New data on the influence of olfactory deprivation on the homing behavior of pigeons. In D. Denton and J. P. Coghlan (eds.). *Olfaction and taste*, Vol. V, pp. 351–353. New York: Academic Press.

Barlow, J. S. 1964. Inertial navigation as a basis for animal navigation. *J. Theor. Biol. 6*:76–117.

– 1966. Inertial navigation in relation to animal navigation. *J. Inst. Navig. 19*:302–316.

Benvenuti, S. 1979. Impaired homing ability in anosmic pigeons. *Z. Tierpsychol. 51*:406–414.

Benvenuti, S., V. Fiaschi, and A. Foa. 1977. Homing behavior of pigeons disturbed by application of an olfactory stimulus. *J. Comp. Physiol. 120*:173–179.

Benvenuti, S., V. Fiaschi, L. Fiore, and F. Papi. 1973. Homing performances of inexperienced and directionally trained pigeons subjected to olfactory nerve section. *J. Comp. Physiol. 83*:81–92.

Billings, S. M. 1968. Homing in Leach's Petrel. *Auk 85*:36–43.

Bingman, V. P. 1981. Ontogeny of a multiple stimulus orientation system in the Savannah Sparrow (*Passerculus sandwichensis*). Ph.D. thesis, State University of New York at Albany.

Bingman, V. P., and K. Able. 1979. The sun as a cue in the orientation of the White-throated Sparrow, a nocturnal migrant. *Anim. Behav. 27*:621–625.

Bingman, V. P., K. P. Able, and P. Kerlinger. 1982. Wind drift, compensation, and the use of landmarks by nocturnal bird migrants. *Anim. Behav. 30*:49–53.

Blakemore, 1975. Magnetotactic bacteria. *Science 190*:377–378.

Bruderer, B. 1980. Radar data on the orientation of migratory birds in Europe. In R. Nöhring (ed.). *Acta XVII Congressus Internationalis Ornithologici*, pp. 547–552. Berlin: Verlag der Deutschen Ornithologen-Gesellschaft.

Cochran, W. W., G. G. Montgomery, and R. R. Graber. 1967. Migratory flights of *Hylocichla* thrushes in spring: a radio telemetry study. *Living Bird 6*:213–225.

Denham, C. R., R. P. Blakemore, and R. B. Frankel. 1980. Bulk magnetic properties of magnetotactic bacteria. *IEEE Trans. Magn. 16*:1006–1007.

Dingle, H. 1980. Ecology and evolution of migration. In S. A. Gauthreaux, Jr. (ed.). *Animal migration, orientation and navigation*, pp. 1–101. New York: Academic Press.

Eastwood, E. 1967. *Radar ornithology*. London: Methuen & Co.

Emlen, S. T. 1967a. Migratory orientation in the Indigo Bunting, *Passerina cyanea*. I. Evidence for use of celestial cues. *Auk 84*:309–342.

– 1967b. Migratory orientation in the Indigo Bunting, *Passerina cyanea*. II. Mechanisms of celestial orientation. *Auk 84*:463–489.

– 1969. Bird migration: influence of physiological state upon celestial orientation. *Science 165*:716–718.

– 1970. Celestial rotation: its importance in the development of migratory orientation. *Science 170*:1198–1201.

– 1975. Migration: orientation and navigation. In D. S. Farner and J. R. King (eds.). *Avian biology*, Vol. 5, pp. 129–220. New York; Academic Press.

– 1980. Decision making by nocturnal bird migrants: the integration of multiple cues. In R. Nöhring (ed.). *Acta XVII Congressus Internationalis Ornithologici*, pp. 553–560. Berlin: Verlag der Deutschen Ornithologen-Gesellschaft.

Emlen, S. T., W. Wiltschko, N. Demong, R. Wiltschko, and S. Bergman. 1976. Magnetic direction finding: evidence for its use in migratory Indigo Buntings. *Science 193:*505–508.

Fiaschi, V., and G. Wagner. 1976. Pigeons' homing: some experiments for testing the olfactory hypothesis. *Experientia 32:*991.

Frankel, R. B., R. P. Blakemore, and R. S. Wolfe. 1979. Magnetite in freshwater, magnetotactic bacteria. *Science 203:*1355–1356.

Frei, U., and G. Wagner. 1976. Die Anfangsorientierung von Brieftauben in erdmagnetisch gestörten Gebiet des Mont Jorat. *Rev. Suisse Zool. 83:*891–897.

Gauthreaux, S. A. 1978. Importance of the daytime flights of nocturnal migrants: redetermined migration following displacement. In K. Schmidt-Koenig and W. T. Keeton (eds.). *Animal migration, navigation and homing*, pp. 219–227. Berlin: Springer.

Gauthreaux, S. A., and K. P. Able. 1970. Wind and the direction of nocturnal songbird migration. *Nature (London) 228:*476–477.

Gould, J. L., J. L. Kirschvink, and K. S. Deffeyes. 1978. Bees have magnetic remanence. *Science 201:*1026–1028.

Graue, L. C. 1965. Initial orientation in pigeon homing related to magnetic contours. *Am. Zool. 5:*704. (abstract).

Griffin, D. R. 1952. Bird navigation. *Biol. Rev. 27:*359–400.

– 1955. Bird navigation. In A. Wolfson (ed.). *Recent studies in avian biology*, pp. 154–197. Urbana: University of Illinois Press.

– 1973. Oriented bird migration in or between opaque cloud layers. *Proc. Am. Philos. Soc. 117:*117–141.

Grubb, T. C. 1974. Olfactory navigation to the nesting burrow in Leach's Petrel. *Anim. Behav. 22:*192–202.

Gwinner, E. 1973. Circannual rhythms in birds: their interaction with circadian rhythms and environmental photoperiod. *J. Reprod. Fert. 19:*51–65.

Gwinner, E., and W. Wiltschko. 1978. Endogenously controlled changes in migratory direction of the Garden Warbler, *Sylvia borin. J. Comp. Physiol. 125:*267–273.

Hartwick, R. F., A. Foa, and F. Papi. 1977. The effect of olfactory deprivation by nasal tubes upon homing behavior in pigeons. *Behav. Ecol. Sociobiol. 2:*81–89.

Hartwick, R., J. Kiepenheuer, and K. Schmidt-Koenig. 1978. Further experiments on the olfactory hypothesis of pigeon homing. In K. Schmidt-Koenig and W. T. Keeton (eds.). *Animal migration, navigation and homing*, pp. 107–118. Berlin: Springer.

Hasler, A. D. 1960. *Underwater guideposts*. Ann Arbor: University of Michigan Press.

Hermayer, K. L., and W. T. Keeton. 1979. Homing behavior of pigeons subjected to bilateral olfactory nerve section. *Monit. Zool. Ital. 13:*303–313.

Hoffmann, K. 1954. Versuche zu der im Richtungsfinden der Vogel enthaltenen Zeitschatzung. *Z. Tierpsychol. 2:*453–475.

– 1965. Clock mechanisms in celestial orientation of animals. In J. Aschoff (ed.). *Circadian clocks*, pp. 426–441. Amsterdam: North Holland.

Ioale' P. 1980. Further investigations of the homing behavior of pigeons subjected to reverse wind direction at the loft. *Monit. Zool. Ital. 14:*77–87.

Ioale', P., F. Papi, B. Fiaschi, and N. E. Baldaccini. 1978. Pigeon navigation: effects upon homing behavior by reversing wind direction at the loft. *J. Comp. Physiol. 128:*285–295.

Keeton, W. 1969. Orientation by pigeons: is the sun necessary? *Science 165:*922–928.

– 1971. Magnets interfere with pigeon homing. *Proc. Natl. Acad. Sci. USA 68:*102–106.

– 1972. Effects of magnets on pigeon homing. In S. R. Galler, K. Schmidt-Koenig, G. J. Jacobs, and R. E. Belleville (eds.). *Animal orientation and navigation*, pp. 579–594. Washington, D.C.: Government Printing Office.

– 1974a. The orientation and navigation basis of homing in birds. *Adv. Stud. Behav. 5:*47–132.

– 1974b. Pigeon homing: no influence of outward-journey detours on initial orientation. *Monit. Zool. Ital. 8:*227–234.

– 1980. Avian orientation and navigation: new developments in an old mystery. In R. Nöhring (ed.). *Acta XVII Congressus Internationalis Ornithologici*, pp. 137–157. Berlin: Verlag der Deutschen Ornithologen-Gesellschaft.

Keeton, W. T., and A. I. Brown. 1976. Homing behavior of pigeons not disturbed by application of an olfactory stimulus. *J. Comp. Physiol. 105:*252–266.

Keeton, W. T., M. L. Kreithen, and K. L. Hermayer. 1977. Orientation of pigeons deprived of olfaction by nasal tubes. *J. Comp. Physiol. 114:*289–299.

Keeton, W. T., T. S. Larkin, and D. M. Windsor. 1974. Normal fluctuations in the earth's magnetic field influence pigeon orientation. *J. Comp. Physiol. 95:*95–103.

Kiepenheuer, J. 1978. Pigeon homing: a repetition of the deflector loft experiment. *Behav. Ecol. Sociobiol. 3:*393–395.

– 1979. Pigeon homing: deprivation of olfactory information does not affect the deflector effect. *Behav. Ecol. Sociobiol. 6:*11–22.

Kirschvink, J. L., and J. L. Gould. 1981. Biogenic magnetite as a basis for magnetic field detection in animals. *BioSystems 13:*181–201.

Kramer, G. 1951. Eine neue Methode zur Erforschung der Zugorientierung und die bisher damit erzielten Ergebnisse. In *Proceedings of the 10th International Ornithological Cogress, Uppsala*, pp. 271–280.

– 1953a. Die Sonnenorientierung der Vogel. *Verh. Dtsch. Zool. Ges. 1952:*72–84.

– 1953b. Wird die Sonnenhohe bei der Heimfindeorientierung verwertet? *J. Ornithol. 94:*201–219.

Kramer, G., and U. Von St. Paul. 1950. Ein wesentlicher Bestandteil der Orientierung der Reisetauben: die Richtungsdressur. *Z. Tierpsychol. 7:*620–631.

Kreithen, M. L., and T. Eisner. 1978. Detection of ultraviolet light by the homing pigeon. *Nature (London) 272:*347–348.

Kreithen, M. L., and W. T. Keeton. 1974a. Detection of polarized light by the homing pigeon, *Columba livia. J. Comp. Physiol. 89:*83–92.

– 1974b. Detection of changes in atmospheric pressure by the homing pigeon. *Columba livia. J. Comp. Physiol. 91:*355–362.

Kreithen, M. L., and D. Quine. 1979. Infrasound detection by the homing pigeon: a behavioral audiogram. *J. Comp. Physiol. 129:*1–4.

Lack, D. 1962. Radar evidence on migratory orientation. *Br. Birds 55:*139–158.

Leask, M. J. M. 1977. A physicochemical mechanism for magnetic field detection by migrating birds and homing pigeons. *Nature (London) 267:*144–145.

Matthews, G. V. T. 1953. Navigation in the Manx Shearwater. *J. Exp. Biol. 30:*370–396.

– 1968. *Bird navigation,* 2nd ed. Cambridge: Cambridge University Press.

Mazzeo, R. 1953. Homing of the Manx Shearwater. *Auk 70:*200–201.

Merkel, G. W., H. G. Fromme, and W. Wiltschko. 1964. Nachtvisuelles Orientierungsvermögen bei nachtlich zugunrahigen Rotkehlchen! *Vogelwarte 22:* 168–173.

Mewaldt, L. R. 1964. California sparrows return from displacement to Maryland. *Science 146:*941–942.

Michener, M., and C. Walcott. 1967. Homing of single pigeons–analysis of tracks. *J. Exp. Bio. 47:*99–131.

Miller, L. J., and C. M. Weise. 1978. Effects of altered photoperiod on migratory orientation in White-throated Sparrows–*Zonotrichia albicollis. Condor 80:*94–96.

Moore, B. R. 1980. Is the homing pigeon's map geomagnetic? *Nature (London) 285:*69–70.

Moore, F. 1978. Sunset and the orientation of a nocturnal migrant bird. *Nature (London) 274:*154–156.

– 1980. Solar cues in the migratory orientation of the Savannah Sparrow–*Passerculus sandwichensis. Anim. Behav. 28:*684–704.

Ossenkopp, K. P., and R. Barbeito. 1978. Bird orientation and the geomagnetic field. *Neurosci. Biobehav. Rev. 2:*255–270.

Papi, F. 1976. The olfactory navigation system of pigeons. *Verh. Deut. Zool. Ges. 1976:*184–205.

Papi, F., V. Fiaschi, S. Benvenuti, and N. E. Baldaccini. 1973. Pigeon homing: outward journey detours influence the initial orientation. *Monit. Zool. Ital. 7:*129–133.

Papi, F., L. Fiore, V. Fiaschi, and S. Benvenuti. 1971. The influence of olfactory nerve section on the homing capacity of carrier pigeons. *Monit. Zool. Ital. 5:*265–267.

– 1972. Olfaction and homing in pigeons. *Monit. Zool. Ital. 6:*85–95.

Papi, F., P. Ioale', V. Fiaschi, S. Benvenuti, and N. E. Baldaccini. 1974. Olfactory navigation of pigeons: the effect of treatment with odorous air currents. *J. Comp. Physiol. 94:*187–193.

– 1978a. Pigeon homing: cues detected during the outward journey influence initial orientation. In K. Schmidt-Koenig and W. T. Keeton (eds.). *Animal migration, navigation and homing,* pp. 65–77. Berlin: Springer.

Papi, F., W. T. Keeton, A. I. Brown, and S. Benvenuti. 1978b. Do American and Italian pigeons rely on different homing mechanisms? *J. Comp. Physiol. A 128:*303–317.

Papi, F., P. Ioale, V. Fiaschi, S. Benvenuti, and N. E. Baldaccini. 1980a. Olfactory and magnetic cues in pigeon navigation. In R. Nöhring (ed.). *Acta XVII Congressus Internationalis Ornithologici*, pp. 569–573. Berlin: Verlag der Deutschen Ornithologen Gesellshaft.

Papi, F., G. Mariotti, A. Foa', and D. Fiaschi. 1980b. Orientation of anosmic pigeons. *J. Comp. Physiol. 125:227–232.*

Phillips, J. B. and J. Waldvogel. in press. Reflected light cues generate the short-term deflector-loft effect. In *Proceedings of the International Symposium on Avian Navigation.*

Presti, D., and J. D. Pettigrew. 1980. Ferromagnetic coupling to muscle receptors as a basis for geomagnetic field sensitivity in animals. *Nature (London) 285:99–101.*

Richardson, W. J. 1978. Timing and amount of bird migration in relation to weather: a review. *Oikos 30:224–272.*

– 1980. Autumn landbird migration over the western Atlantic Ocean as evident from radar. In R. Nöhring (ed.). *Acta XVII Congressus Internationalis Ornithologici*, pp. 501–506. Berlin: Verlag der Deutschen Ornithologen-Gesellschaft.

Sauer, E. G. F. 1957. Die Sterneorientierung nachtlich ziehender Grasmucken (*Sylvia atricapilla, borin u. currara*). *Z. Tierpsychol. 14:29–70.*

Sauer, E. G. F., and E. M. Sauer. 1960. Star navigation of nocturnal migrating birds. The 1958 planetarium experiments. *Cold Spring Harbor Symp. Quant. Biol. 25:463–473.*

Schlichte, A. J. 1973. Untersuchungen über die Bedeutung optischer Parameter für das Heimkehrverhalten der Brieftaube. *Z. Tierpsychol. 32:257–280.*

Schmidt-Koenig, K. 1979. *Avian orientation and navigation.* New York: Academic Press.

Schmidt-Koenig, K., and J. B. Phillips. 1978. Local anesthesia of the olfactory membrane and homing pigeons. In K. Schmidt-Koenig and W. T. Keeton, (eds.). *Animal migration, navigation and homing*, pp. 119–124. Berlin: Springer.

Schmidt-Koenig, K., and H. J. Schlichte. 1972. Homing in pigeons with impaired vision. *Proc. Natl. Acad. Sci. USA 69:2446–2447.*

Schmidt-Koenig, K. and C. Walcott. 1978. Tracks of pigeons homing with frosted lenses. *Anim. Behav. 26:480–486.*

Schneider, D. 1974. The sex-attractant receptor of moths. *Sci. Am. 231(1):28–35.*

Southern, W. 1978. Orientation responses of Ring-billed Gull chicks: a reevaluation. In K. Schmidt-Koenig and W. T. Keeton (eds.). *Animal migration, navigation and homing*, pp. 311–317. Berlin: Springer.

Talkington, L. 1967. Bird navigation and geomagnetism. *Am. Zool. 7:199* (abstract).

Visalberghi, E., and E. Alleva. 1979. Magnetic influences on pigeon homing. *Biol. Bull. 156:246–256.*

Vleugel, D. A. 1954. Waarnemingen over de nachttrek van lijsters (Turdus) en hun waarschijnlijke orientering. *Limosa 27:1–9.*

Wagner, G. 1972. Topography and pigeon orientation. In S. R. Galler, K. Schmidt-Koenig, G. J. Jacobs, and R. E. Belleville (eds.). *Animal orientation and navigation*, pp. 259–273. Washington, D.C.: Government Printing Office.

– 1976. Das Orientierungsverhalten von Brieftauben im erdmagnetisch gestörten Gebiete des Chasseral. *Rev. Suisse Zool. 83:*883–890.

Walcott, C. 1978. Anomalies in the earth's magnetic field increase the scatter of pigeons' vanishing bearings. In K. Schmidt-Koenig and W. T. Keeton (eds.). *Animal migration, navigation, and homing*, pp. 141–151. New York: Springer-Verlag.

– 1980. Magnetic orientation in homing pigeons. *IEEE Trans. Magn. 16(5):* 1008–1013.

– in press. Is there evidence for a magnetic map in homing pigeons? In *Proceedings of the International Symposium on Avian Navigation.*

Walcott, C., J. Gould, and J. Kirschvink. 1979. Pigeons have magnets. *Science 205:*1027–1029.

Walcott, C., and R. Green. 1974. Orientation of homing pigeons altered by a change in the direction of an applied magnetic field. *Science 184:*180–182.

Walcott, B., and C. Walcott. in press: A search for magnetic field receptors in animals. In *Proceedings of the International Symposium on Avian Navigation.*

Waldvogel, J. A., S. Benvenuti, W. T. Keeton, and F. Papi. 1978. Homing pigeon orientation influenced by deflected winds at the home loft. *J. Comp. Physiol. 128:*297–301.

Waldvogel, J. A., and J. B. Phillips. in press. Pigeon homing: new experiments involving permanent resident deflector loft birds. In *Proceedings of the International Symposium on Avian Navigation.*

Waldvogel, J. A., J. B. Phillips, D. R. McCorkle, and W. T. Keeton. 1980. Short term residence in deflector lofts alters initial orientation of homing pigeons. *Behav. Ecol. Sociobiol. 7:*207–211.

Wallraff, H. G. 1979. Olfaction and homing in pigeons. A problem of navigation or motivation? *Naturwissenschaften 66:*269–270.

– 1980a. Olfaction as a basic component of pigeon homing. In H. Van der Starr (ed.). *Olfaction and taste*, VII, pp. 327–330. London: IRL Press.

– 1980b. Does pigeon homing depend on stimuli perceived during displacement? I. Experiments in Germany. *J. Comp. Physiol. 139:*193–201.

1980c.Olfaction and homing in pigeons: nerve section experiments, critique, hypotheses. *J. Comp. Physiol. 139:*209–224.

– 1981. The olfactory component of pigeon navigation: steps of analysis. *J. Comp. Physiol. 143:*411–412.

Wallraff, H. G., and A. Foa. 1981. Pigeon navigation: charcoal filter removes relevant information from environmental air. *Behav. Ecol. Sociobiol. 9:*67–77.

Wallraff, H. G., F. Papi, P. Ioalé, and A. Foa. 1980. Does pigeon homing depend on stimuli perceived during displacement? II. Experiments in Italy. *J. Comp. Physiol. 139:*203–208.

Williams, T., and J. Williams. 1978. Orientation of transatlantic migrants. In K. Schmidt-Koenig and W. T. Kuton (eds.). *Animal migration, navigation and homing*, pp. 239–251. Berlin: Springer.

Wiltschko, W. 1975. Interaction of stars and magnetic field in the orientation system of night migrating birds. II. Spring experiments with European Robins (*Erithacus rubecula*). *Z. Tierpsychol. 39:*265–282.

Wiltschko, W., and R. Wiltschko. 1972. Magnetic compass of European Robins. *Science 176:*62–64.

– 1975. The interaction of stars and magnetic field in the orientation system of night migrating birds. I. Autumn experiments with European warblers (Gen. *Sylvia*). *Z. Tierpsychol. 37:*337–355.

Wiltschko, R., W. Wiltschko, and W. T. Keeton. 1978. Effect of outward journey in an altered magnetic field on the orientation of young pigeons. In K. Schmidt-Koenig and W. Keeton (eds.). *Animal migration, navigation and homing,* pp. 152–161. Berlin: Springer.

Yodlowski, M. L., M. L. Kreithen, and W. T. Keeton. 1977. Detection of atmospheric infrasound by the homing pigeon. *Nature (London) 265:*725–726.

Zoeger, J., and M. D. Fuller. 1980. Magnetic material in the head of a dolphin. *Trans. Am. Geophys. Union 61:*225.

Zoeger, J., M. D. Fuller, and J. R. Dunn. 1981. Magnetic material in the head of the Common Pacific Dolphin. *Science 213:*892–894.

Commentary

KENNETH P. ABLE

It is apparent from Walcott and Lednor's summary that we have learned much about bird navigation over the past 25 years or so. Yet despite these advances, we find ourselves in the uncomfortable position of lacking a coherent, mechanistic hypothesis to explain behavior that overwhelming evidence has shown to occur. In contrast to Walcott and Lednor's balanced review of what we have learned, I will concentrate on those important things that we do not know and the assumptions that most of us make despite the fact that they have not been adequately tested. My remarks will be quite selective with no attempt at an inclusive treatment. I view my role as that of devil's advocate and my goal will be to raise questions rather than to espouse any particular point of view. The emphasis will be slanted somewhat more toward migration. Because this is a commentary, I have made no effort to be complete in citing literature; in particular, I have avoided, wherever possible, duplication of citations contained in Chapter 13.

For over three decades, students of bird navigation have used the homing pigeon as a model system. In doing so, we assume, at least implicitly, that what we learn about pigeons can be generalized to other bird species. If it cannot, there would be little reason to devote so much effort to the study of a species in which homing ability has been subjected to intense artificial selection over many generations. To date, however, the track record seems remarkably good: Both the sun and magnetic compasses appear to be similar in mechanism in pigeons and migrants. On the other hand, to what extent is the homing navigation of a pigeon analogous to any aspect of migration in wild species?

It is unequivocally documented that at least a modest number of migratory species show a high degree of year-to-year site fidelity during both the breeding and nonbreeding seasons. Further study is likely to reveal considerable ecological variability in this trait; individuals of species that occupy relatively unpredictable environments might be quite opportunistic. However, the fact that at least some kinds of migrants return year after year to specific sites suggests that they may have abilities at least qualitatively similar to those of pigeons. However, knowing that migrants return to specific places is not the same as knowing how they get there. In fact, we have no evidence for enroute navigation during migration and precious little for homing ability in typical migrant species.

Displacement followed by homing to both breeding and wintering sites has been demonstrated in a good number of species, but, in general, the proportion of individuals that homed has been low, and their homing speeds very slow (see Able 1980). Without following individual birds it is usually impossible to preclude a large search component in the homing feat. Experiments with several species of swallows have shown both initial homeward orientation and fast homing speeds, behavior roughly comparable to that of homing pigeons (e.g., Nastase 1982). Recently, I displaced breeding Wood Thrushes (*Hylocichla mustelina*) up to 25 km and followed them with radiotelemetry. Although they take several days to return, they usually make only one short flight each day, and those flights are very well oriented in the home direction. If other species behave similarly, the long homing times of previous trials may be misleading. Therefore, at least some evidence exists that indicates that typical migrants may have homing abilities similar to those of pigeons. We do not know whether or how that ability is involved in the site fidelity associated with migration in many species.

Displacement experiments with wild birds are faced with a very difficult problem: the unknown extent of the familiar area of the birds. The flight experience of homing pigeons can be controlled and one can be reasonably certain that they are being transported into completely foreign areas. With wild birds there is always some doubt. Regrettably, we have virtually no information on the movement patterns of birds in nature. It seems clear that something gained through experience with a locality is necessary for homing to that place; there is some evidence that birds must move around in the area in order to acquire whatever information they use (Löhrl 1959; Berndt and Winkel 1980). If we cannot assume that a displaced bird has been taken beyond the limit of its area of familiarity, we cannot be sure that any observed homing is a result of true navigation (Type III orientation of Griffin, 1955).

There is yet another fundamental problem with homing experiments, even those with pigeons. The very act of displacing an animal involves transporting it from home to the release point, opening the possibility that information gathered on the outward journey could be used in the homing process. In fact, there is evidence that pigeons do use such information (reviewed in Able 1980). We can, of course, attempt to prevent the pigeons from perceiving presumed relevant cues during the outward journey, and a number of such experiments have been performed. These include anesthetizing the birds, spinning them, altering or attempting to scramble magnetic field cues, and manipulating odors reaching the birds. Because we do not know what information the birds use in the homing process, however, it remains an assumption that they are unable to perceive the necessary cues. Despite the fact that effects on various aspects of homing behavior have apparently been induced by manipulating magnetic and olfactory information during the displacement journey, a respectable number of experimentals nevertheless homed in all cases. If we grant for the moment the assumption that we did preclude accurate route-based navigation, as Baker (1978) terms it, then location-based navigation must have occurred. A route-based mechanism might still be sufficient for homing but obviously not necessary. There is no reason why the two should be mutually exclusive.

As Walcott (Chapter 13) emphasizes, one of the most important things we learned during the past two decades was that a search for a unitary mechanism of orientation in birds was doomed to failure. As Keeton (1980) noted, we must apply the same reasoning to questions about navigation and "the map." Yet recently a degree of polarization has crept into the field with various groups espousing opposing navigation mechanisms as if only one can be correct. A negative result from an experiment testing the olfactory model of Papi does not negate a role for odors any more than releasing pigeons carrying magnets on a sunny day proves that a magnetic compass doesn't exist. I am not suggesting that we bury our heads in the sand of complexity and redundancy, but our chances of finding that one first explanation may well be facilitated by acknowledging at the outset that there could be several routes to the same endpoint. Wiltschko (in press) has proposed, for example, that young pigeons rely more heavily on route-specific information and switch gradually to the use of site-specific cues with increasing experience.

There is considerable circumstantial evidence that the development of homing ability requires familiarity with the goal area over a substantial period of time. In fact, two recent models of homing navigation involve the gradual formation of a familiar area map based on association between learned landmarks and the compass directions between them (Baker 1978; Wiltschko and Wiltschko 1978). This process might be extended along the migratory route and a large-scale mosaic of familiar areas accumulated. Some observed behavior of migrants (e.g., the peculiar back-and-forth movements of Canada Geese, *Brant cana-densis*, documented by Raveling, 1976) is consistent with this idea. Likewise, a familiar area map could be extended to an extensive if not worldwide grid map as described by Wiltschko and Wiltschko (1978); Baker (1978) finds a grid map unnecessary. Indeed, due to the complications discussed earlier, unequivocal evidence for the existence of a grid map is hard to come by. Displacement under cue-deprived or cue-altered conditions and release-site-specific effects on homing orientation (release site biases; Keeton 1973; magnetic anomalies: Wagner 1976; Walcott, 1978) are the best we can cite. I do not suggest abandoning the idea of a true grid map, as Baker proposes, but we should be aware that the data supporting it are slim and often equivocal.

Our attempts to unravel the compass orientation mechanisms of migrants and homing pigeons have obviously enjoyed somewhat greater success. The existence of sun, star, and magnetic compasses has been demonstrated to the satisfaction of most. Yet we have very little information on the generality of these abilities, and the interaction among the various compass cues has been studied in only a tiny sample of species. Magnetic effects on orientation by birds in cages have been replicated in a few species, but any indication of magnetic orientation by free-flying migrants has been very elusive, even under conditions in which such effects can be seen in pigeons (Able et al. 1982). This is predictable from the Wiltschko's (1975b) model, which suggests that the magnetic compass is consulted only infrequently by migrants. At the same time, the failure to find such effects remains a major inconsistency between cage and field studies.

The historical development of this field was such that stellar orientation was the first compass mechanism discovered in night migrants. As such, it came to be regarded as the primary, if not the only, orientation capability used by these birds. Appropriate orientation in tests under clear skies was presumed to be based on the stars. The documentation of other means of compass orientation, of course, rendered such results inconclusive. At the present time, what most ornithologists regard as a widespread, predominant mechanism of migratory orientation has in fact been conclusively demonstrated in only a small number of studies that employed direct manipulations of star patterns.

The early hints from Keeton's experiments with magnet-bearing pigeons that the several compasses might be interrelated were borne out by the important work done by the Wiltschkos (1975 a,b). In their scheme, the magnetic compass emerged as the innate driver, the stars serving only as reference points to help maintain directions established by magnetic cues. However, like the star compass before it (Emlen 1972), the magnetic compass has now been found to be subject to modification during early ontogenetic development, perhaps based on information from the stars themselves (Bingman 1981). The relationship between the two compasses is thus rather unclear at present, and it may even change over the life of an individual. It certainly seems clear that stellar rotation sets the bearings of the star compass, but what defines the pole point as north? Likewise, in the Wiltschkos' elegant magnetic compass model, what is the mechanism by which particular magnetic bearings are associated with geographic directions? It only obscures our understanding to say that it is innate. Developmental studies seem surely to be the route to advances in theis area.

As important as the kinds of experiments I have been discussing will be in future developments in this field, I want to end by emphasizing the continuing need for more basic descriptive information about many aspects of migration. Our hypotheses and the design of our experiments involve assumptions about the behavior of migrants in the field that are in many cases untested. We discuss the development of familiar area maps based on learned landmarks and compass bearings, yet we know virtually nothing about the movement patterns of young birds prior to departure on migration. Do they explore a sufficiently large area in a manner consistent with the hypothesis? Particularly in North America, we have remarkably little detailed information on the migratory routes and destinations of specific populations. No individual bird has ever been tracked over the course of an entire migration journey, leaving us largely in the dark concerning the nature of navigation during migration. As valuable as the various experimental systems have been, our goal is not to find out how the homing pigeon alone navigates nor to study behavior that occurs only within the confines of a paper funnel. Only with adequate comparative field data can we chart our experimental course most profitably.

Literature cited

Able, K. P. 1980. Mechanisms of orientation, navigation, and homing. In S. A. Gauthreaux, Jr. (ed.). *Animal migration, orientation, and navigation*, pp. 283–373. New York: Academic Press.

– 1982. Field studies of avian nocturnal migratory orientation. I. Interaction of sun, wind and stars as directional cues. *Anim. Behav. 30:*761–767.

Able, K. P., V. P. Bingman, P. Kerlinger, W. F. Gergits. 1982. Field studies of avian nocturnal migratory orientation. II. Experimental manipulation of orientation in White-throated Sparrows (*Zonotrichia albicollis*) released aloft. *Anim. Behav. 30:*768–773.

Baker, R. R. 1978. *The evolutionary ecology of animal migration.* New York: Holmes & Meier.

Berndt, R., and W. Winkel. 1980. Field experiments on problems of imprinting to the birthplace in the Pied Flycatcher *Ficedula hypoleuca*. In R. Nöhring (ed.). *Acta XVIII Congressus Internationalis Ornithologici*, pp. 851–854. Berlin: Verlag der Deutschen Ornithologen-Gesellschaft.

Bingman, V. P. 1981. Ontogeny of a multiple stimulus orientation system in the Savannah Sparrow (*Passerculus sandwichensis*). Ph.D. thesis, State University of New York at Albany.

Emlen, S. T. 1972. The ontogenetic development of orientation capabilities. In S. R. Galler, K. Schmidt-Koenig, G. J. Jacobs, and R. E. Belleville (eds.). *Animal orientation and navigation*, pp. 191–210. Washington, D.C.: Government Printing Office.

Griffin, D. R. 1955. Bird navigation. In A. Wolfson (ed.). *Recent studies in avian biology*, pp. 154–197. Urbana. University of Illinois Press.

Keeton, W. T. 1973. Release-site bias as a possible guide to the "map" component in pigeon homing. *J. Comp. Physiol. 86:*1–16.

Keeton, W. T. 1980. Avian orientation and navigation: new developments in an old mystery. In R. Nöhring (ed.). *Acta XVII Congressus Internationalis Ornithologici*, pp. 137–157. Berlin: Verlag der Deutschen Ornithologen-Gesellschaft.

Löhrl, H. 1959. Zur Frage des Zeitpunkts einer Prägung auf die Heimatregion beim Halsbandschnapper (*Ficedula albicollis*). *J. Ornithol. 100:*132–140.

Nastase, A. J. 1982. Orientation and homing ability of the Barn Swallow. *J. Field Ornithol. 53:*15–21.

Raveling, D. G. 1976. Migration reversal: a regular phenomenon of Canada Geese. *Science 193:*153–154.

Wagner, G. 1976. Das Orientierungsverhalten von Brieftauben im erdmagnetisch gestörten Gebiete des Chasseral. *Rev. Suisse Zool. 83:*883–890.

Walcott, C. 1978. Anomalies in the earth's magnetic field increase the scatter of pigeon's vanishing bearings. In K. Schmidt-Koenig and W. T. Keeton (eds.). *Animal migration, navigation and homing*, pp. 143–151. Berlin: Springer.

Wiltschko, R. in press. The role of outward-journey information in the orientation system of homing pigeons. In *Proceedings of the International Symposium on Avian Navigation*.

Wiltschko, W., and R. Wiltschko. 1975a. The interaction of stars and magnetic field in the orientation system of night migrating birds. I. Autumn experiments with European warblers (Gen. *Sylvia*). *Z. Tierpsychol. 37:337–355*.

– 1975b. The interaction of stars and magnetic field in the orientation system of night migrating birds. II. Spring experiments with European Robins (*Erithacus rubecula*). *Z. Tierpsychol. 39:265–282*.

– 1978. A theoretical model for migratory orientation and homing in birds. *Oikos 30:177–187*.

Index